HEAT AND MASS TRANSFER

HEAT AND MASS TRANSFER

PROF. SHYAM K. AGRAWAL

Formerly, Professor of Mechanical Engineering
Bhagalpur College of Engineering,
Bhagalpur

First Published in India by Viva Books Private Limited, 2005.

This edition published in 2005 by

Anshan Limited
6 Newlands Road
Tunbridge Wells
Kent TN4 9AT
UK.
Tel/Fax: +44 (0) 1892 557767
E-mail: info@anshan.co.uk
www.anshan.co.uk

Published by arrangement with

Viva Books Private Limited, 4262/3, Ansari Road, New Delhi - 110 002. India

British Library Cataloguing in Publication Data
A catalogue record for this book is available from the British Library

ISBN 1-904798-47-0

Printed in India.

"Let each become all that he was
created capable of being."

In memory of
Dear father, Raghunath Das Agrawal
Loving mother, Moti Devi
Adorable sister, Sita Devi
Respected brother, Ram Kumar Agrawal

Contents

the Cylinder is (i) Hollow and (ii) Solid; An Expression for Temperature Distribution in a Sphere with Internal Heat Generation when the Sphere is (i) Solid and (ii) Hollow; The Utility of Extended Surfaces; An Expression for the Temperature Distribution in a Fin of Uniform Cross-section; Significance of Fins having Insulated Tips; Fin Efficiency and Fin Effectiveness; Extended Surfaces do not always Increase the Heat Transfer Rate; An Expression for Temperature Distribution for an Annular Fin of Uniform Thickness; An Expression for Temperature Distribution for a Straight Fin of Triangular Profile

Heat Flux—Empirical Relations; Simplified Free Convection Relations for Air; Heat Transfer by Free Convection in a Limited Volume; Heat Transfer from Inclined Surfaces; Heat Transfer by Natural Convection from Rotating Cylinders and Disks

Non-Circular Cylinders; Empirical Correlation for the Evaluation of
'h' in Flow over a Sphere; Heat Transfer from Tube Bundles in
Cross-flow—Its Significance; Heat Transfer in High Speed Flow—
Aerodynamic Heating; Adiabatic Wall Temperature and Recovery
Factor—Defined; Evaluation of 'h' in High Speed Flow –
Reynolds Analogy

Essentials of Heat Transfer by Radiation; Physical Mechanism of
Thermal Radiation; Basic Terminology—Definitions;
Monochromatic Emissive Power of a Black Body—Planck's
Distribution Law; Wien's Displacement Law; Derivation of Stefan-
Boltzmann Law from Planck's Law of Radiation; Statement and
Proof of Kirchhoff's Law of Radiation; The Concept of a Black
Body; Intensity of Radiation; An Expression for Obtaining I, the
Intensity of Radiation; Wavelength Dependent Characteristics—
Solar Radiation; Greenhouse Effect; Characteristics of Real
Surfaces; Radiation Energy Exchange between Two Black Bodies
Placed in a Non-absorbing Medium—the Shape Factor; The
Reciprocity Theorem; Shape Factor Characteristics; Heat Exchange
between Gray Bodies—Radiation Network; Electrical Network
for Two Gray Bodies; Electrical Network for Three Gray Bodies;
Electrical Network for a System Consisting of Four Gray Surfaces;
Radiation—Convection System; Heat Transfer Coefficient for
Radiation; Gas Radiation; Absorptivity of Gases; Radiant Heat
Exchange between Two Infinite Parallel Planes separated by a Gray
Gas; Mean Beam Length; Heat Exchange between Gas Volume and
Black Enclosure;

Heat Exchangers: Regenerators and Recuperators; Classification
of Heat Exchangers; Expression for Log Mean Temperature
Difference—Its Characteristics; Special Operating Conditions for
Heat Exchangers; LMTD for Cross-flow Heat Exchangers; Fouling
Factors in Heat Exchangers; The Overall Heat Transfer Coefficient;
Heat Exchangers Effectiveness—Useful Parameters;
Effectiveness—NTU Relations; Heat Exchanger Design—
Important Factors; Increasing the Heat Transfer Coefficient

Condensation and Boiling; Condensation—Filmwise and
Dropwise; Filmwise Condensation Mechanism on a Vertical Plane
Surface—Assumption; An Expression for the Liquid Film
Thickness and the Heat Transfer Coefficient in Laminar Filmwise
Condensation on a Vertical Plate; An Expression for 'h' in
Turbulent Liquid Film; Condensation of Superheated Vapour; The

Preface

The present society uses the per capita consumption of energy as the yardstick for ascertaining its prosperity. Energy available in nature as thermal energy—fuels (fossil and nuclear), solar energy, geothermal energy, biomass—has to be transferred to a working fluid used in energy conversion systems. As such, heat and mass transfer processes have become an integral part of energy conversion systems. And, there will always be a need for teachable textbook in the field of heat and mass transfer, presenting the basic concepts in a logical manner and their applications to the solution of physical problems.

Chapter 1 introduces all the three modes of heat transfer. Worked out examples show how basic principles can be applied to solve problems of interest. Since a problem of practical importance involves at least two and sometimes all the three modes of heat transfer occuring simultaneously, the concept of overall heat transfer coefficient and electrical analogy have their own significance.

Chapters 2, 3 and 4 are related to heat transfer by conduction. Chapter 2 is meant for one-dimensional heat conduction through simple and composite structures of different configurations. Extended surface problems have been treated extensively.

Chapter 3 discusses transient heat conduction with emphasis on lumped parameter system and solution by Heisler charts for two- and three-dimensional systems.

Chapter 4 deals with two-dimensional heat conduction with an emphasis on shape factor. Since two- and three-dimensional problems can be easily handled by numerical technique, the basic principles involved in solving steady and unsteady problems by numerical technique are also illustrated.

Chapters 5, 6, 7 and 8 are related to heat transfer by convection. The convection heat transfer mechanism requires a solid fluid interface, and the movement of fluid molecules either by natural method or by an artificial method to transport energy from one point to another. Chapter 5 outlines the methods for evaluating the free convection heat transfer coefficient. Forced convection heat transfer coefficient in laminar and turbulent flows can be evaluated by various methods and are discussed in Chapters 6 and 7.

The flow of gases over cylinders, spheres and over a bank of cylinders have their own importance in engineering, especially in designing heat exchangers.

Moreover, in many applications especially in connection with aerospace problems, flow velocity is comparable with the local velocity of sound and therefore, additional effects like fluid compressibility and viscous dissipation should be considered. These are discussed in Chapter 8.

Chapter 9 is related to thermal radiation. After defining the terminology and different laws of radiation, some typical engineering heat transfer problems including solar radiation are solved. Later on, the energy exchange between black and gray bodies placed in transparent and absorbing medium are discussed. Electrical analogy is extensively used in solving problems. Errors in temperature measurement are also analysed.

Chapter 10 illustrates examples in which heat transfer by conduction and convection occur s imultaneously. T he a nalysis of heat e xchangers is b ased on log Mean Temperature Difference and Effectiveness-NTU method. Significant parameters in the design of heat exchangers are also discussed.

Chapter 11 deals with the process of convection in which a fluid is undergoing a change of phase. These processes are significant in condensation of a vapour or the boiling of a liquid. When a liquid changes its phase and solidifies, heat transfer by conduction is significant and dominant. Frost bites and freezing of lakes are discussed. Heat pipes are discussed because they are capable of transporting large quantities of heat energy through small cross-sectional area and with small temperature difference.

Chapter 12 deals with diffusion process in a mixture and evaporation of a liquid in an isothermal environment. Drying of solids and convective mass transfer are important in chemical industries. Simultaneous heat and mass transfer find their application in air-conditioning. Thus they are also discussed.

The text has been developed to give a better understanding of heat and mass transfer principles at the undergraduate level. SI units have been used exclusively and a variety o f standard worked out examples and m ultiple choice questions with their answers will be of great assistance to students preparing for their degree or professional examinations.

I am indebted to Prof. P. N. Maskara, Ex-Professor of Mechanical Engineering at M.N.R. Engineering College, Allahabad for his kind cooperation and valuable assistance. I am grateful to my wife Poonam, to my children Nutan, Kalpana and Dr. Ratan who have been a source of inspiration and to my grandchildren Nandini, Sarthak, Ishita, Archit and Pulkit who missed my company during their vacations.

SHYAM K. AGRAWAL

Foreword

It gives me immense pleasure to find this opportunity of writing the foreword to a valuable book written by one of my old students, Dr. Shyam Kumar Agrawal. Knowing Dr. Agrawal as I do, I could never have any doubt about his capability to bring out a publication of this nature.

While Heat and Mass Transfer is a basic subject of interest to Mechanical and Chemical Engineers, its application is also of vital importance to practising engineers. I am sure the book written by Dr. Agrawal fulfils this need to a very significant extent and therefore, would be very well received by the students, teachers and practising engineers.

This book is another feather in the glorious cap Dr. Agrawal had already earned the right to wear after publishing three books which are serving their usefulness very well indeed.

Dr. Shyam Kumar Agrawal has been utilizing his retired life in such worthwhile pursuits whereby he can rightly claim to be continuing to serve the society as he did during his active working life. I congratulate him and wish him all the best in his future endeavors.

DR. R.N. SAHAY
Director (Retd.)
Bihar Institute of Technology
Sindri, Dhanbad and
Ex-visiting Professor
BITS, Pilani (Rajasthan)

CHAPTER 1

Basic Concepts

1. Heat Energy and Heat Transfer

Heat is a form of energy in transition and it flows from one system to another, without transfer of mass, whenever there is a temperature difference between the systems. The process of heat transfer means the exchange in internal energy between the systems and in almost every phase of scientific and engineering work processes, we encounter the flow of heat energy.

2. Importance of Heat Transfer

Heat transfer processes involve the transfer and conversion of energy and therefore, it is essential to determine the specified rate of heat transfer at a specified temperature difference. The design of equipments like boilers, refrigerators and other heat exchangers require a detailed analysis of transferring a given amount of heat energy within a specified time. Components like gas/steam turbine blades, combustion chamber walls, electrical machines, electronic gadgets, transformers, bearings, etc require continuous removal of heat energy at a rapid rate in order to avoid their overheating. Thus, a thorough understanding of the physical mechanism of heat flow and the governing laws of heat transfer are a must.

3. Modes of Heat Transfer

The heat transfer processes have been categorized into three basic modes: Conduction, Convection and Radiation.

Conduction — It is the energy transfer from the more energetic to the less energetic particles of a substance due to interaction between them, a microscopic activity.

Convection — It is the energy transfer due to random molecular motion along with the macroscopic motion of the fluid particles.

Radiation — It is the energy emitted by matter which is at finite temperature. All forms of matter emit radiation attributed to changes in the electron configuration of the constituent atoms or molecules. The transfer of energy by conduction and convection requires the presence of a material medium whereas radiation does not. In fact radiation transfer is most efficient in vacuum.

All practical problems of importance encountered in our daily life involve at least two, and sometimes all the three modes occuring simultaneously. When the rate of heat flow is constant, i.e., does not vary with time, the process is called a steady state heat transfer process. When the temperature at any point in a system changes with time, the process is called unsteady or transient process. The internal energy of the system changes in such a process. When the temperature variation of an unsteady process describes a particular cycle (heating or cooling of a building wall during a 24 hour cycle), the process is called a periodic or quasi-steady heat transfer process.

Heat transfer may take place when there is a difference in the concentration of the mixture components (the diffusion thermoeffect). Many heat transfer processes are accompanied by a transfer of mass on a macroscopic scale. We know that when water evaporates, the heat transfer is accompanied by the transport of the vapour formed through an air-vapour mixture. The transport of heat energy to steam generally occurs both through molecular interaction and convection. The combined molecular and convective transport of mass is called convection mass transfer and with this mass transfer, the process of heat transfer becomes more complicated.

4. Thermodynamics and Heat Transfer—Basic Difference

Thermodynamics i s m ainly concerned w ith t he c onversion o f heat e nergy i nto other useful forms of energy and is based on (i) the concept of thermal equilibrium (Zeroth L aw), (ii) t he First L aw (the p rinciple o f c onservation of e nergy) a nd (iii) the Second Law (the direction in which a particular process can take place). Thermodynamics is silent about the heat energy exchange mechanism. The transfer of heat energy between systems can only take place whenever there is a temperature gradient and thus, heat transfer is basically a non-equilibrium phenomenon. The science of heat transfer tells us the rate at which the heat energy can be transferred when there is a thermal non-equilibrium. That is, the science of heat transfer seeks to do what thermodynamics is inherently unable to do.

However, the subjects of heat transfer and thermodynamics are highly complimentary. Many heat transfer problems can be solved by applying the priniciples of conservation of energy (the First Law).

5. Dimension and Unit

Dimensions and units are essential tools of engineering. Dimension is a set of basic entities expressing the magnitude of our observations of certain quantities. The state of a system is identified by its observable properties, such as mass, density, temperature, etc. Further, the motion of an object will be affected by the observable properties of that medium in which the object is moving. Thus a number of observable properties are to be measured to identify the state of the system.

A unit is a definite standard by which a dimension can be described. The difference between a dimension and the unit is that a dimension is a measurable property of the system and the unit is the standard element in terms of which a dimension can be explicitly described with specific numerical values.

Every major country of the world has decided to use SI units. In the study of heat transfer the dimensions are: L for length, M for mass, θ for temperature, T for time and the correesponding units are: metre for length, kilogram for mass, degree celcius (°C) or kelvin (K) for temperature and second (s) for time. The parameters important in the study of heat transfer are tabulated in Table 1.1 with their basic dimensions and units of measurement.

Table 1.1 Dimensions and units of various parameters

Parameter	Dimension	Unit
Mass	M	kilogram, kg
Length	L	metre, m
Time	T	seconds, s
Temperature	θ	kelvin, K, Celcius°C,
Velocity	L/T	metre/second, m/s,
Density	ML^{-3}	kg/m^3
Force	MLT^{-2}	Newton, N = 1 kg m/s^2
Pressure	$ML^{-1}T^{-2}$	N/m^2, Pascal, Pa
Energy, Work	$ML^2\,T^{-2}$	N-m, = Joule, J
Power	$ML^2\,T^{-3}$	J/s, Watt, W
Absolute Viscosity	$ML^{-1}\,T^{-1}$	$N\text{-}s/m^2$, Pa-s
Kinematic viscosity	L^2T^{-1}	m^2/s
Thermal Conductivity	$MLT^{-3}\,\theta^{-1}$	W/mK, W/m°C
Heat Transfer Coefficient	$MT^{-3}\,\theta^{-1}$	W/m^2K, W/m^2°C
Specific Heat	$L^2T^{-2}\,\theta^{-1}$	J/kg K, J/kg°C
Heat Flux	MT^{-3}	W/m^2

6. Mechanism of Heat Transfer by Conduction

The transfer of heat energy by conduction takes place within the boundaries of a system, or a cross the boundary of the system into another system placed in direct physical contact with the first, without any appreciable displacement of matter comprising the system, or by the exchange of kinetic energy of motion of the molecules by direct communication, or by drift of electrons in the case of heat conduction in metals. The rate equation which describes this mechanism is given by Fourier Law:

$$\dot{Q} = - kAdT/dx \tag{1.1}$$

where $\dot{Q}$ = rate of heat flow in X-direction by conduction in J/S or W,

        k = thermal conductivity of the material. It quantitatively measures the heat conducting ability and is a physical property of the material that depends upon the composition of the material, W/mK,

        A = cross-sectional area normal to the direction of heat flow, m^2,

   dT/dx = temperature gradient at the section, as shown in Fig. 1.1 The negative sign is included to make the heat transfer rate $\dot{Q}$ positive in the direction of heat flow (heat flows in the direction of decreasing temperature gradient).

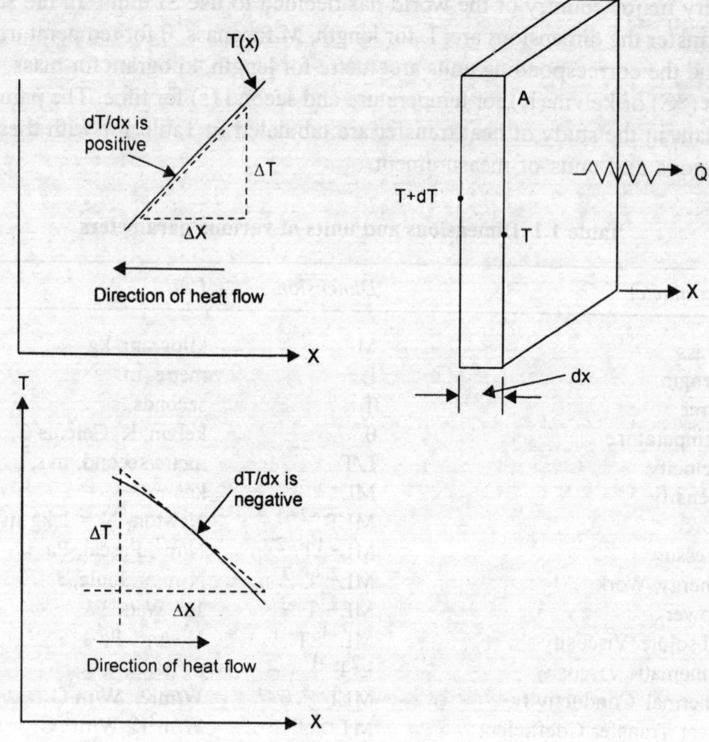

Fig. 1.1 Heat flow by conduction

7. Thermal Conductivity of Materials

Thermal conductivity is a physical property of a substance and in general, it depends upon the temperature, pressure and nature of the substance. Thermal conductivity of materials are usually determined experimentally and a number of methods for this purpose are well known.

Thermal Conductivity of Gases: According to the kinetic theory of gases, the heat transfer by conduction in gases at ordinary pressures and temperatures take place through the transport of the kinetic energy arising from the collision of the gas molecules. Thermal conductivity of gases depends on pressure when very low (<2660 Pa) or very high (>2 × 10^9 Pa). Since the specific heat of gases increases with temperature, the thermal conductivity increases with temperature and with decreasing molecular weight.

Thermal Conductivity of Liquids: The molecules of a liquid are more closely spaced and molecular force fields exert a strong influence on the energy exchange in the collision process. The mechanism of heat propagation in liquids can be conceived as transport of energy by way of unstable elastic oscillations. Since the density of liquids decreases with increasing temperature, the thermal conductivity of non-metallic liquids generally decreases with increasing temperature, except for liquids like water and alcohol because their thermal conductivity first increases with increasing temperature and then decreases.

Thermal Conductivity of Solids: (i) Metals and Alloys: The heat transfer in metals arise due to a drift of free electrons (electron gas). This motion of electrons brings about the equalization in temperature at all points of the metals. Since electrons carry both heat and electrical energy, the thermal conductivity of metals is proportional to its electrical conductivity and both the thermal and electrical conductivity decrease with increasing temperature. In contrast to pure metals, the thermal conductivity of alloys increases with increasing temperature. Heat transfer in metals is also possible through vibration of lattice structure or by elastic sound waves but this mode of heat transfer mechanism is insignificant in comparison with the transport of energy by electron gas. (ii) Nonmetals: Materials having a high volumetric density have a high thermal conductivity but that will depend upon the structure of the material, its porosity and moisture content. High volumetric density means less amount of air filling the pores of the materials. The thermal conductivity of damp material is considerably higher than the thermal conductivity of dry material because water has a higher thermal conductivity than air. The thermal conductivity of granular material increases with temperature. (Table 1.2 gives the thermal conductivities of various materials at 0°C.)

Table 1.2 Thermal conductivity of various materials at 0°C

Material	*Thermal conductivity (W/m K)*	*Material*	*Thermal conductivity (W/m K)*
Gases		*Solids : Metals*	
Hydrogen	0.175	Silver, pure	410
Helium	0.141	Copper, pure	385
Air	0.024	Aluminium, pure	202
Water vapour (saturated)	0.0206	Nickel, pure	93
Carbon dioxide	0.0146	Iron, pure	73
(thermal conductivity of helium		Carbon steel, 1%C	43
and hydrogen are much higher		Lead, pure	35
than other gases, because their		Chrome-nickel-steel	16.3
molecules have small mass and		(18% Cr, 8% Ni)	
higher mean travel velocity)			
		Non-metals	
Liquids		Quartz, parallel to axis	41.6
Mercury	8.21	Magnesite	4.15
Water*	0.556	Marble	2.08 to 2.94
Ammonia	0.54	Sandstone	1.83
Lubricating oil		Glass, window	0.78
SAE 40	0.147	Maple or Oak	0.17
Freon 12	0.073	Saw dust	0.059
		Glass wool	0.038

*water has its maximum thermal conductivity (k = 0.68 W/mK) at about 150°C

Example 1.1   The average heat transfer coefficient for air flowing through a pipe can be determined by an empirical relation

$h_c = 0.1 \, V^{0.3}/ D^{0.8}$, Btu/hr ft^2 °F,

where V is the velocity in feet/second, and D is the pipe inner diameter in feet. Calculate the value of the constant if all quantities are expressed in SI units.

Solution:
$$1 \text{ Btu} = 1055.1 \text{ J}; \ 1 \text{ hour} = 3600 \text{ s}; \ 1 \text{ feet} = 0.3048 \text{ metre}$$
$$1°F = 1/1.8 \text{ K}$$
$$\therefore \ h_c = 0.1 V^{0.3}/D^{0.8} \times (\text{Btu/hr ft}^2 \text{ °F}) \times (1055.1 \text{ J/Btu}) \times (1 \text{ hr}/3600 \text{ s})$$
$$\times (1 \text{ ft}/0.3048 \text{ m})^2 \times (1.8 \text{ °F}/1 \text{ K})$$
$$\times (1 \text{ ft}/0.3048 \text{ m})^{0.3} (1 \text{ ft}/0.3048 \text{ m})^{0.8}$$
$$= 2.095 \, V^{0.3}/D^{0.8} \text{ W/m}^2\text{K}.$$

(Since some countries are still following British thermal units (Btu), the conversion factors for commonly used quantities in heat transfer are tabulated in Table 1.3)

Table 1.3 Conversion factors for commonly used quantities in heat transfer

Quantity	SI	MKS	FPS
Area	1 m^2	1m^2	10.764 ft^2
Density	1 kg/m^3	1 kg/m^3	0.06243 Ib$_m$/ft^3
Energy	1 Joule, J	0.2388 cal	9.4787 × 10^{-4} Btu
Energy per unit	1 J/kg	0.2388 cal/kg	4.2995 × 10^{-4} Btu/lb$_m$
Force	1 Newton, N	0.1019 kg f	0.2248 lb$_f$
Heat flux	1 W/m^2	0.8598 kcal/m^{2}h	0.3171 Btu/(ft^{2}h)
Heat generation	1 W/m^3	0.8598 kcal/m^{3}h	0.09665 Btu/(ft^{3}h)
Heat transfer	1 W/(m^{2}K)	0.8598 kcal/m^{2}hK	0.1761 Btu/h.ft^{2}F
Heat transfer rate	1 J/s, 1 W	860 cal/h	3.412 Btu/h
Mass flow rate	1 kg/s	1 kg/s	2.2046 1b$_m$/s
Pressure & stress	1 N/m^2, Pa	0.1019 kg$_f$ / m^2	0.02089 1b$_f$/ft^2
Specific heat	1 J/kg. K	0.2388 cal/kgK	2.3886 × 10^{-4} Btu/1b$_m$°F
Temperature	$T(K) = T(°C) + 273$	$T(K) = T(°C) + 273$	$[T(°F) + 459.67]/1.8$
Temperature difference	1 K	1°C	1.8°F
Thermal conductivity	1 W/m.K	0.8598 kcal/h.mK	0.57782 Btu/°F.h.ft
Thermal diffusivity	1 m^2/s	1 m^2/s	10.7639 ft^2/s
Thermal resistance	1 K/W	1.1628 K.h/kcal	0.5275 °F.h/Btu
Velocity	1 m/s	1 m/s	3.2808 ft/s
Dynamic viscosity	1 Pa.s	0.1019 kg$_f$. s/m^2	5.8016 × 10^{-6} 1b$_f$h/ft^2 0.672 lb$_m$ /(ft.s)
Volume flow rate	1 m^3/s	1 m^3/s	35.3134 ft^3/s

MKS: Metric system –1 cal is energy required to raise 1 gm of water by 1°C
FPS: English system –1 Btu is energy required to raise 1 lb$_m$ of water by 1°F
1 kg$_f$ = 9.81 kg.m/s^2; 1 lb$_f$ = (1 lb$_m$ 32.2 ft/s^2)

Example 1.2   A glass panel ($k = 0.78$ W/mK) 1.5 m × 2.5 m is 16 mm thick. If its inside and outside surface temperatures are 30°C and 5°C respectively, calculate the heat loss by conduction through the panel.

Solution: The Eq. (1.1) can be integrated with the boundary conditions:

$$\text{at } x_1 = 0, \ T_1 = 30°C; \ x_2 = 16 \text{ mm}, T_2 = 5°C$$

Assuming that A and k are constants, we write

$$\dot{Q}\int_1^2 dx = kA\int_1^2 dT \ ; \qquad \therefore \ \dot{Q} = kA\,(T_1 - T_2)/(x_2 - x_1)$$

or
$$\dot{Q} = kA\,(T_1 - T_2)/L = (T_1 - T_2)/(L/kA) \tag{1.1a}$$

and
$$\dot{Q} = 0.78 \times (1.5 \times 2.5)(30 - 5) / 16 \times 10^{-3}$$

The Eq. (1.1a) is analogous to the basic equation for the flow of current in a conductor as given by Ohm's law:

$$i = \Delta V/R \tag{1.1b}$$

where i is the current (amp); ΔV is the voltage difference causing the current to flow, R is the electrical resistance.

By comparing Eqs. (1.1a) and (1.1b), we can say that the heat flow system and electrical flow system are analogous to each other. And a heat flow problem can be easily analysed experimentally by making an analogous electric circuit, measuring the electrical quantities involved, and then transferring the data back to heat quantities for obtaining the final result. Fig. 1.2 (a, b)

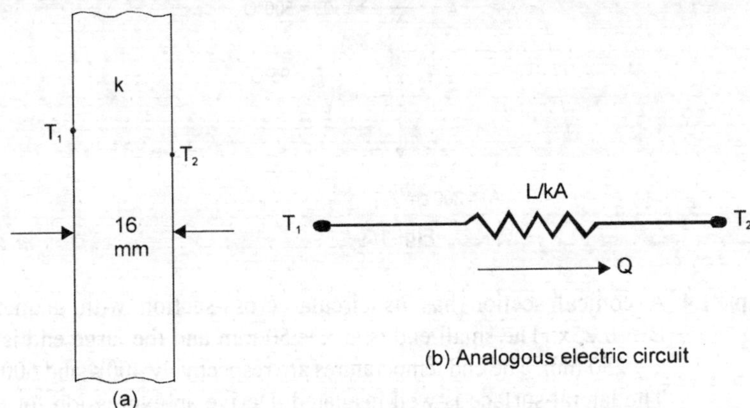

(b) Analogous electric circuit

(a)

$$T_1 - T_2 : \text{Thermal potential}$$
$$\dot{Q} : \text{Thermal current}$$
$$L/kA : \text{Thermal resistance}$$

Fig. 1.2 (a, b)

Example 1.3   A truncated cone 25 cm high is made of aluminium (k = 204 W/mK). The cross-sectional area at the top and bottom are 50 cm^2 and 200 cm^2 respectively. The lower surface is maintained at 500°C and the upper surface is at 95°C. The lateral surface is insulated. Assuming one-dimensional flow of heat, calculate the rate of heat transfer.

Solution: The truncated cone is shown in Fig. 1.3. The cross-sectional area at top is 50 cm^2 and the cross-sectional area at the bottom is 200 cm^2.

Thus, the cross-sectional area is a function of x. The area at any section is

$$A = [200 - (200 - 50)x/25] = [200 - 6x] \times 10^{-4} m^2$$

The Eq. (1.1) has to be integrated with the boundary conditions:

at $x = 0$, $T_1 = 500$°C; and at $x = 25$, $T_2 = 95$°C

$$\dot{Q} = -k(200 - 6x) \times 10^{-4}\, dT/dx$$

or $$\dot{Q} \int_0^{25} \frac{dx}{(200 - 6x)} = 10^{-4}\, k(500 - 95)$$

and $$\dot{Q} \times 0.231 = 10^{-4} \times 204 \times 405; \qquad \therefore\ \dot{Q} = 35.766\ W.$$

(A body may have various temperature and the temperature distribution within the body may be non-uniform and in that case, it is important to know the dependance of thermal conductivity on temperature.)

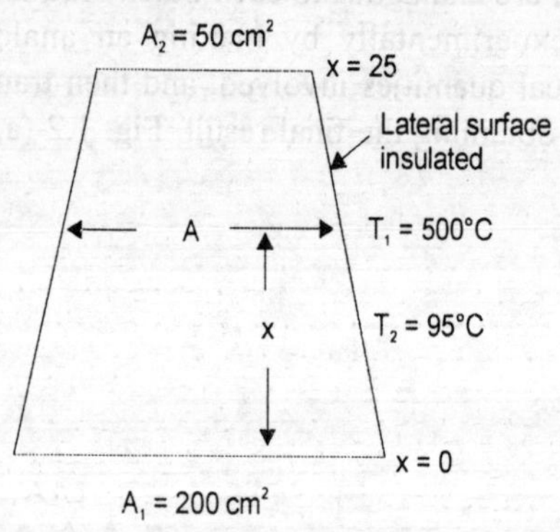

Fig. 1.3

***Example 1.4**   A conical section has its circular cross-section with diameter D = 0.25x. The small end is at x = 50 mm and the large end is at x = 250 mm. The end temperatures are respectively 400K and 600K. The lateral surface is well insulated. Derive an expression for the temperature distribution assuming one-dimensional conditions and calculate the rate of heat flow through the cone.

(CS 1987)

Solution: $\therefore\ \dot{Q} = -kA\,(dT/dx) = -\dfrac{k\pi D^2}{4}\dfrac{dT}{dx} = -k(0.049)x^2 \cdot \dfrac{dT}{dx}$

____

* variable cross-sectional area

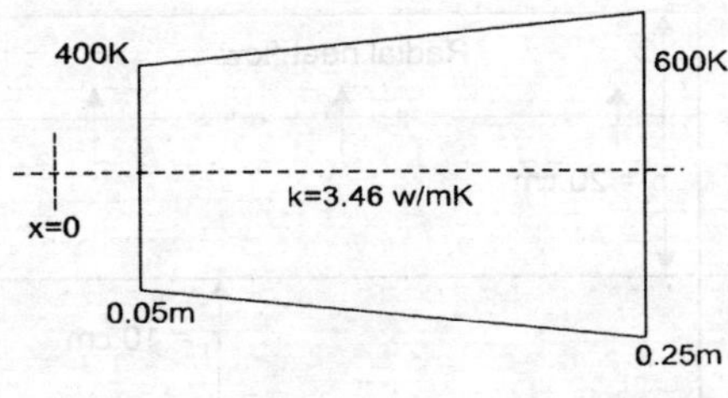
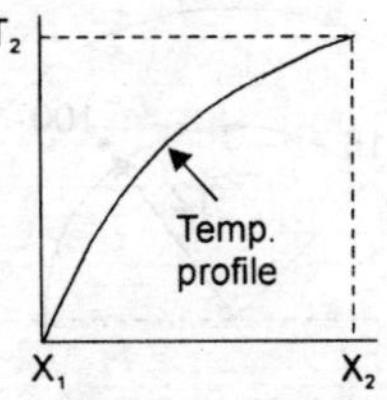

Fig. 1.4

Separating the variables,

$$dT = \frac{\dot{Q}}{0.049\,k} \cdot dx / x^2$$

Upon integration, $(T - T_1) = [\{\dot{Q}/(0.049\,k)\}\ \{(1/x) - (1/x_1)\}]$

Also, $\qquad\qquad T_2 = T_1 - Q/(0.049\,k) \cdot [(1/x_1) - (1/x_2)]$

and $\qquad\qquad \dot{Q} = 0.049\,k\,(T_1 - T_2) / [(1/x_1) - (1/x_2)]$

Therefore, $\qquad\qquad T = T_1 + (T_1 - T_2)\,[(1/x - 1/x_1) / (1/x_1 - 1/x_2)]$

$$\dot{Q} = 0.049 \times 3.46 \times (400 - 600) / (1/0.05 - 1/0.25)$$

$$= 2.12\ W$$

$$\frac{dT}{dx}\ (\text{at } x = 0.05\ m) = 5001.8°C/m; \qquad (\text{at } x = 0.25m) = 692.25°C/m$$

Example 1.5 During an experiment to determine the thermal conductivity of a material used in a thick cylindrical shell (inner radius 10 cm, outer radius 20 cm), two thermocouples were inserted, one at radius equal to 12 cm and the other at radius equal to 18 cm. The temperatures recorded were 100°C and 50°C respectively. Calculate the thermal conductivity of the material and the temperature at the inner and outer surface of the shell if the heat transfer rate per metre length of the shell was 600 W.

Solution: The cylindrical shell shown in Fig 1.5 has inner radius 10 cm and the outer radius $r_2 = 20$ cm. The length is L metre. The cross-sectional area normal to the direction of heat flow is $2\pi\,rL$ and the rate equation can be written as:

$$\dot{Q} = -kA\,dT/dx = -k\,2\pi rL\,dT/dr$$

Separating the variables, $\dfrac{Q}{2\pi\,Lk} \displaystyle\int_1^2 dr/r = \int_1^2 dT$

The boundary conditions are: at $r = r_1$, $T = T_1$ and at $r = r_2$, $T = T_2$.

$$\frac{\dot{Q}}{2\pi\,Lk}\ln(r_2/r_1) = (T_1 - T_2) \quad \text{and} \quad \dot{Q} = (T_1 - T_2)/[\ln(r_2/r_1)/2\pi Lk]$$

We can draw an analogous electric circuit, Fig. 1.6

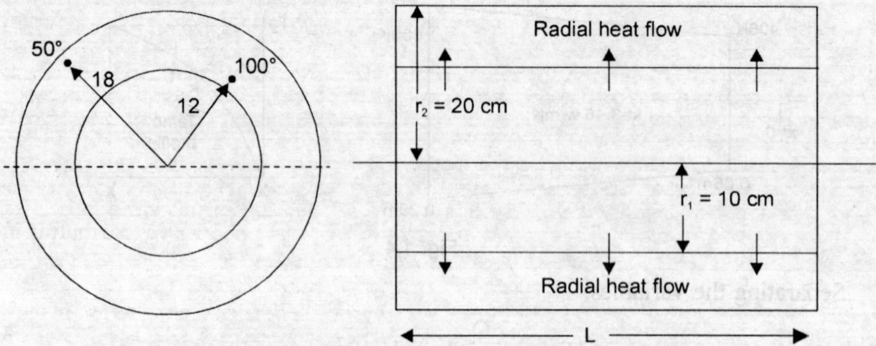

Fig. 1.5 Radial heat flow in a cylindrical shell

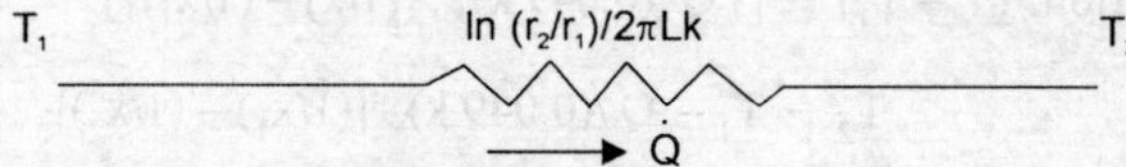

Fig. 1.6 Analogous electric circuit for radial heat transfer through a cylindrical shell

Therefore, $k = \dot{Q} \ln (r_2/r_1) / [2\pi L (T_1 - T_2)]$

After substituting the values,

$$k = [(600 \ln (18/12)] / [2\pi (100 - 50)] = 0.774 \text{ W/mK}$$

The temperature at the inner radius: $= 100 + 600 \ln (1.2)/2\pi \times 0.774$
$$= 122.49°C$$

The temperature at the outer radius: $= 50 - 600 \ln (20/18)/2\pi \times 0.774$
$$= 37°C.$$

Example 1.6   The temperature at the inner radius ($r_i = 5$ cm) is 125°C and at the outer radius ($r_0 = 10$ cm) is 60°C. Calculate the rate of heat flow through the spherical shell when the thermal conductivity of the shell material is 2.0 W/mK. What would be the temperature of the shell material at radius equal to 7.5 cm.

Solution: The spherical shell is shown in Fig. 1.7. The cross-sectional area normal to the direction of heat flow would be $4\pi r^2$ and the rate equation can be written as:

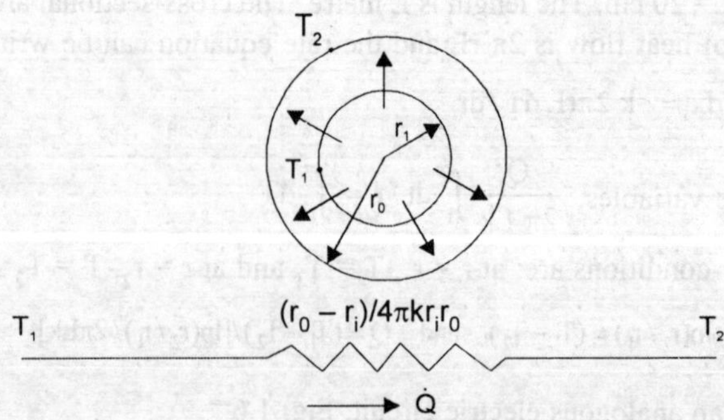

Fig. 1.7 Radial heat flow in a spherical shell with its equivalent electric circuit

$$\dot{Q} = -k\,4\pi r^2\,dT/dr$$

Separating the variables, we get: $\dot{Q}\int_1^2 dr/r^2 = -4\pi\,k\int_1^2 dT$

The boundary conditions are : at $\quad r = r_i,\ T_1 = 125°C$
$$r = r_0,\ T_2 = 60°C$$

$$\dot{Q}\,(r_0 - r_i)\,r_i\,r_0 = 4\pi\,k\,(T_1 - T_2)$$

and $\quad \dot{Q} = (T_1 - T_2)/[(r_0 - r_i)/4\pi k r_0 r_i]$

$$= \frac{125 - 60}{\dfrac{(10-5)\times 10^{-2}}{4\times 3.142 \times 2\times 0.05\times 0.10}} = 163.4\,\text{W}$$

at $\quad r = 7.5$ cm, $T = T_1 - \dot{Q}\,(r - r_i)/4\pi k\,r\,r_i$

$$= 125 - \frac{163.4\times 2.5\times 10^{-2}}{4\times 3.142\times 2\times 5\times 7.5\times 10^{-4}}$$

$$= 81.88°C.$$

Example 1.7 A 100 W lamp is buried in a soil. The bulb is lighted and a thermocouple embedded at a distance of 30 cm from the centre of the bulb measures a temperature of 60°C. If the temperature at the free surface of the soil is 20°C, estimate the thermal conductivity of the soil.

Solution: Assuming that the bulb can be treated as a spherical body and steady state conditions prevail, we can write

$$\dot{Q} = (T_1 - T_2)\,4\pi k/(1/r_i - 1/r_0)$$

when $r_i = 0.3$m, $T_1 = 60°C$ and $\quad$ at $r_0 = \infty$, $T_2 = 20°C$

or, $\quad 100 = \dfrac{(60 - 20)\,4\times 3.142\times k}{1/0.3}$

$$k = \frac{3.33\times 100}{4\times 3.142\times 40} = 0.662\ \text{W/mK}\ .$$

8. Convective Heat Transfer Coefficient—Newton's Law of Cooling

We are surrounded by fluids and fluids are always in direct physical contact of solids. The exchange of heat energy from a heated solid surface to a cold fluid or from a heated fluid to a cold solid surface is a very important process in engineering heat transfer. Convection is the name given to the process by which heat energy is transferred between a solid and a fluid flowing past it. The heat transfer rate equation used to describe this mechanism was first suggested by Newton and is known as Newton's law of cooling. Or,

$$\dot{Q}_c = hA(T_W - T_\infty) = hA(\Delta T) \qquad (1.2)$$

where $\dot{Q}_c$ is the rate of heat transfer by convection, W and
h is the average convective heat transfer coefficient which depends on the rate

of mixing of one portion of fluid particles with the other portion of fluid particles due to the movement of the mass of the fluid. It also depends upon the physical properties of fluid, geometry of the surface and the temperature difference. A is the surface area normal to the direction of heat flow, (m^2).

$(T_W - T_\infty)$ is the temperature difference between the temperature of the solid surface and the bulk temperature of the fluid, Fig. 1.8.

(The approximate values of h in W/m^2 °C is given in Table 1.4 for different conditions.)

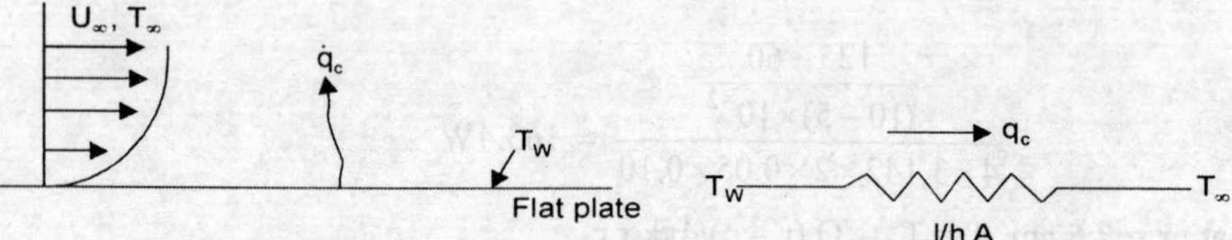

Fig. 1.8 Heat transfer from a wall/flat plate by convection to a fluid at temperature T∞ and its analogous electric circuit

Table 1.4 Approximate values of convective heat transfer coefficient*

Process	h (W/m² K)
Natural Convection	
Still gases	4.0 to 15
Still liquids	100 to 900
Forced Convection	
Air	12 to 200
Liquids	100 to 3500
Boiling liquids	2000 to 25000
Condensing vapours	4000 to 25000

Example 1.8   During a hot summer day, the ambient temperature is 40°C and the temperature at the outside surface of a 20 cm thick wall (k = 0.7 W/mK) is 36°C. What would be the convective heat transfer coefficient at the outer surface of the wall if the rate of heat transfer through the wall is 80 W/m^2. Calculate the temperature at the inside surface of the wall and the convective heat transfer coefficient at the inner surface if the room temperature is 10°C.

Solution: Since the temperature of the surroundings is more than the inside temperature of the room, heat energy will flow from outside to inside. The variation of temperature in the direction of heat flow is shown in Fig. 1.9. At the outside surface of the wall, the temperature of the fluid (air) will decrease from 40°C to 36°C in a thin region near the wall and similarly, the temperature at the inside surface of the wall will decrease from T_2 to 10°C.

In most practical cases, the fluid temperature is constant throughout its bulk, apart from a thin film near the solid surface bounding the fluid.

---

* When the values of convective heat transfer coefficient are very large, heat transfer by convection dominates over heat transfer by conduction, such as in steam generators.

Under steady state conditions, the rate of heat transfer is 80 W/m². From Newton's Law:

$$\dot{Q} = hA \, (T_\infty - T_W)$$

$$\dot{Q}/A = 80 = h \, (T_\infty - T_W) = h \, (40 - 36)$$

or $h = 80 / 4 = 20 \ W/m^2 \, °C$

The heat energy flowing through the wall by conduction will also be 80 W/m²

$$\dot{Q}/A = (T_W - T_2)/L/k$$

or, $80 = (36 - T_2) / 0.2 / 0.7;$ $\therefore \ T_2 = 13.14°C$

The convective heat transfer coefficient at the inside surface of the wall

$$h = \frac{\dot{Q}/A}{\Delta T} = \frac{80}{(13.14 - 10)} = 25.48 \ \ W/m^2°C$$

The overall heat transfer coefficient U is written as

$$\dot{Q}/A = U(T_\infty - T_i) \Rightarrow U = 80/(40 - 10) = 2.667 \ W/m^2K.$$

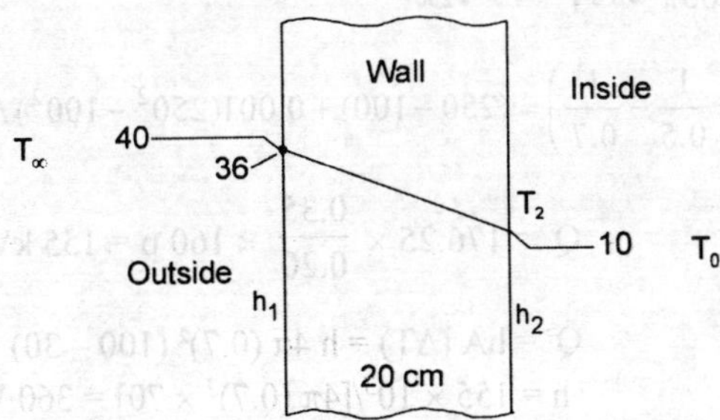

Fig. 1.9 Temperature variation for heat transfer from one fluid to another through a dividing wall

***Example 1.9** A plane wall has thickness L and its two surfaces are maintained at temperatures T_1 and T_2. If the thermal conductivity of the material varies with the temperature and is given by $k = k_0 \, (1 + \alpha T)$, derive an expression for steady state heat transfer rate.

Solution: The Eq. (1.1) can be written as

$$\dot{Q} = -kA \, dT/dx = -k_0(1 + \alpha T)A \, dT/dx$$

Separating the variables, we get

$$(\dot{Q}/A) \int_1^2 dx = -k_0 \int_1^2 (1 + \alpha T) \, dT$$

$$(\dot{Q}/A) = k_0/L[(T_1 - T_2) + \frac{1}{2}\alpha(T_1^2 - T_2^2)]$$

$$\dot{q} = (\dot{Q}/A) = k_0[(T_1 - T_2)/L][1 + \alpha(T_1 + T_2)/2]$$

$$= k_m (T_1 - T_2)/L$$

where k_m is the conductivity at the mean temperature, $T_m = (T_1 + T_2)/2$

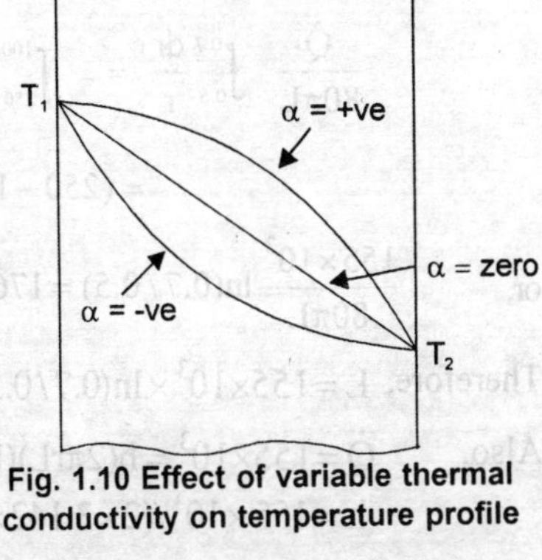

Fig. 1.10 Effect of variable thermal conductivity on temperature profile

---

*Variable thermal conductivity

The temperature profile within the wall is shown in Fig 1.10. The heat transfer rate can be obtained by using the relation developed for constant thermal conductivity evaluated at arithmetic mean temperature.

Example 1.10 A hollow spherical shell (inner radius 0.5 m, outer radius 0.7 m, thermal conductivity $k = 40 (1 + 0.001T)$ is used to store a liquid at 250°C. The temperature at the outer surface of the sphere is 100°C. Calculate the heat flow rate through the spherical shell and the convective heat transfer coefficient at the outer surface if the ambient temperature is 30°C.

Solution: By Fourier Law: $\dot{Q} = - kA \, dT/dr$

$$= - 40 (1 + 0.001 \, T) \, 4\pi r^2 . \, dT/dr$$

Separating the variables,

$$\frac{\dot{Q}}{160\pi} \int_{0.5}^{0.7} \frac{dr}{r^2} = - \int_{250}^{100} (1 + 0.001 \, T) \, dT$$

or

$$\frac{\dot{Q}}{160 \, \pi} \left(\frac{1}{0.5} - \frac{1}{0.7} \right) = (250 - 100) + 0.001(250^2 - 100^2)/2 = 176.25$$

and,

$$\dot{Q} = 176.25 \times \frac{0.35}{0.20} \times 160 \, p = 155 \text{ kW}$$

again,

$$\dot{Q} = hA \, (\Delta T) = h \, 4\pi \, (0.7)^2 \, (100 - 30)$$

Therefore,

$$h = 155 \times 10^3/[4\pi \, (0.7)^2 \times 70] = 360 \text{ W/m}^2\text{°C}.$$

Example 1.11 The temperature at the inside surface of a cylindrical shell (inner radius 0.5m, outer radius 0.7 m, thermal conductivity $k = 40 (1 + 0.001T)$ is 250°C. The temperature at the outer surface of shell is 100°C. Estimate the length of the cylinder if the rate of heat loss is the same as in Ex. 1.10, and the heat transfer coefficient at the outer surface when the ambient temperature is 30°C.

Solution: From Fourier Law: $\dot{Q} = - kA \, dT / dr = - 40 (1 + 0.001 \, T) \, 2\pi rL \, dT / dr$
Separating the variables,

$$\frac{\dot{Q}}{80\pi L} \int_{0.5}^{0.7} \frac{dr}{r} = - \int_{250}^{100} (1 + 0.001 \, T) \, dT$$

$$= (250 - 100) + (250^2 - 100^2) \times 0.001/2 = 176.25$$

or,

$$\frac{155 \times 10^3}{80\pi L} \ln(0.7/0.5) = 176.25$$

Therefore, $L = 155 \times 10^3 \times \ln(0.7/0.5)/(80 \times 3.142 \times 176.25) = 1.177 \text{ m}$

Also, $\dot{Q} = 155 \times 10^3 = h(2\pi rL)(100 - 30)$

$h = 155 \times 10^3 /(2 \times 3.142 \times 0.7 \times 1.77 \times 70) = 427.68 \text{ W/m}^2\text{K}.$

9. The Nature of Thermal Radiation

Thermal radiation is electromagnetic radiation emitted in a wavelength band between 0.1 μ to 100 μ (1μ = 10^{-6} m) solely as a result of the temperature of the surface and therefore, the heat transfer by radiation does not require a matter-filled intervening space, as in conduction and convection. The rate of heat transfer by conduction and convection is proportional to the difference in the temperature between the hot source and the cold sink, and, the amount of heat transferred does not depend on the absolute magnitude of the temperature as long as the difference in the temperature is the same. This is not the case with thermal radiation. The quantity of heat exchanged by radiation is proportional to the difference of the fourth power of the absolute temperature of the radiating bodies. Thus for a given temperature difference, the heat transferred is much greater at higher temperature than at lower temperature.

The exact character of radiant heat emission is not completely agreed upon, but it is known that an ideal emitter, called a black body, emits thermal radiation according to Stefan-Boltzmann Law:

$$E_b = \sigma T^4 \tag{1.3}$$

where E_b is the emissive power of a black body and is the energy emitted per unit surface area per unit time, W/m².

σ is the Stefan-Boltzmann constant, 5.668 × 10^{-8} W/m² K⁴.

T is the temperature of the body on the absolute scale, K.

Non-black surfaces radiate energy according to the equation $E = \epsilon E_b$ where ϵ is the emissivity, a property of the surface and its value ranges from 0.0 (an ideal reflector) to 1.0 (an ideal emitter). The exchange of energy by radiation from a solid to its infinite surrounding is given by the rate equation, Fig. 1.10.

$$\dot{Q}_r = \epsilon \sigma A(T_S^4 - T_\infty^4) \tag{1.4}$$

where $\dot{Q}_r$ = rate of heat flow by radiation, T_S is the absolute surface temperature and T_∞ is the absolute ambient temperature.

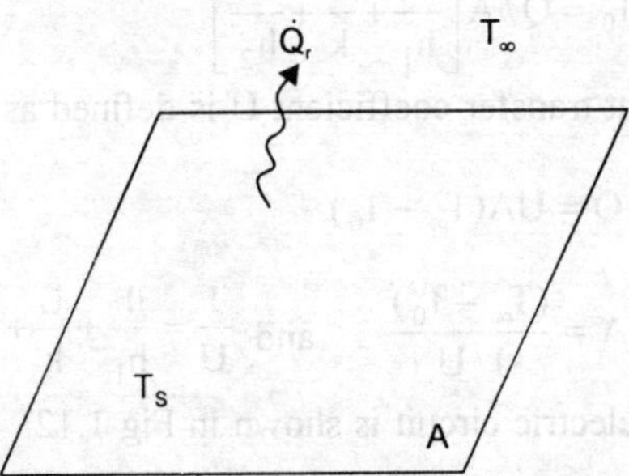

Fig. 1.11 Radiation from a surface area A to surroundings at T_∞

Example 1.12 A black surface placed on the moon absorbs incident solar radiation at the rate of 950 W/m². Calculate the equilibrium temperature if the temperature of the surroundings is (i) 0°C, (ii) 25°C.

Solution: Applying the principles of energy conservation, the energy absorbed by the body must be equal to the energy exchanged with the surroundings. Let T_S K be the temperature of the surface. From Eq (1.4).

Or, (i) $\dot{Q}_r = 1 \times 5.668 \times 10^{-8} (T_S^4 - 273^4) = 950$ ($\because \in = 1$ for black surface)

Therefore, $T_S = 386.5$ K $= 113.5°$C

(ii) $\dot{Q}_r = 950 = 1 \times 5.668 \times 10^{-8} (T_S^4 - 298^4)$

and $T_S = 396.2$ K, $\equiv 123.2°$C

We can define a radiation heat transfer coefficient as:

$$h_r = \dot{Q}_r / \Delta T = 950 / (396.2 - 298)$$
$$= 9.67 \ W/m^2 K.$$

10. The Overall Heat Transfer Coefficient

In most of the engineering applications, heat transfer involves at least two or perhaps all the three modes of heat transfer. In Ex. 1.8, the temperature of the outside air is higher than the temperature of the air inside the room and as such heat energy flows from outside to inside. Since all real fluids are viscous, there is always a thin layer of fluid that adheres to the wall on both sides and the heat energy transferred through this thin layer is by convection only. And the heat transfer involves two modes—conduction and convection. The rate equations are written as:

$$\dot{Q} = h_1 A(T_\infty - T_W) = kA(T_W - T_2)/L = h_2 A(T_2 - T_0)$$

or, $(T_\infty - T_W) = \dfrac{\dot{Q}/A}{h_1}$ . . . (a); $(T_W - T_2) = \dfrac{\dot{Q}/A}{L/k}$ . . . (b); and

$(T_2 - T_0) = \dfrac{\dot{Q}/A}{h_2}$ . . . (c)   (h_1 and h_2 are convective heat transfer coefficients at outside and inside surface)

Upon addition of (a) + (b) + (c)

$$T_\infty - T_0 = \dot{Q}/A \left[\frac{1}{h_1} + \frac{L}{k} + \frac{1}{h_2} \right]$$

The term overall heat transfer coefficient U is defined as

$$\dot{Q} = UA(T_\infty - T_0)$$

or $\dot{Q}/A = \dfrac{(T_\infty - T_0)}{1/U}$ and $\dfrac{1}{U} = \dfrac{1}{h_1} + \dfrac{L}{k} + \dfrac{1}{h_2}$

and the equivalent electric circuit is shown in Fig 1.12.

$$T_\infty \qquad\qquad T_W \qquad\qquad T_2 \qquad\qquad T_0$$
$$1/h_1 \qquad\qquad L/k \qquad\qquad 1/h_2$$

Fig. 1.12 The equivalent electric circuit for combined heat transfer process

$$\left[\frac{1}{U} = \frac{1}{h_1} + \frac{L}{k} + \frac{1}{h_2} \right]$$

Example 1.13 A 0.625 cm thick plank of wood ($k = 0.166$ W/mK) floats in a large pool of water. The convective heat transfer coefficients between the top surface of the wood and the air, and between the bottom surface of the wood and the water are 10 W/m²°C and 450 W/m²°C respectively. The wood absorbs 470 W/m² of solar radiation. The temperature of air is 25°C and that of water is 10°C. Ignoring radiation from wood, determine the rate of heat transfer from wood to air.

Solution: Let the temperature of the top and bottom surface of the wood be T_1 and T_2 respectively. The equivalent electric circuit is shown in Fig. 1.13. The heat energy flowing from the top surface of wood to air is $\dot{Q}_1$ W/m² and $\dot{Q}_2$ W/m² flows to the water through the wood.

The thermal resistances are: $R_1 = 1/h_1 = 1/10 = 0.1$

$$R_2 = L/k = 37.65 \times 10^{-3}$$

$$R_3 = 1/h_2 = 1/450 = 2.22 \times 10^{-3}$$

(The unit of thermal resistances is m²K/W)

$$\dot{Q}_1 = (T_1 - 25)/R_1 = (T_1 - 25)/0.1 = 10(T_1 - 25)$$

$$\dot{Q}_2 = (T_1 - 10)/(R_2 + R_3) = (T_1 - 10)/(37.65 \times 10^{-3} + 2.22 \times 10^{-3})$$

$$= 25.08(T_1 - 10)$$

Since $\quad \dot{Q}_1 + \dot{Q}_2 = 470$ W/m²

$$10(T_1 - 25) + 25.08(T_1 - 10) = 470$$

Therefore, $\quad T_1 = 27.68$°C

$$\dot{Q}_1 = 10(T_1 - 25) = 26.8 \text{ W/m}^2; \quad \dot{Q}_2 = 470 - 26.8 = 443.2 \text{ W/m}^2$$

and $\quad T_2 = T_1 - \dot{Q}_2 R_2 = 27.68 - 443.2 \times 37.65 \times 10^{-3} = 10.99$ °C

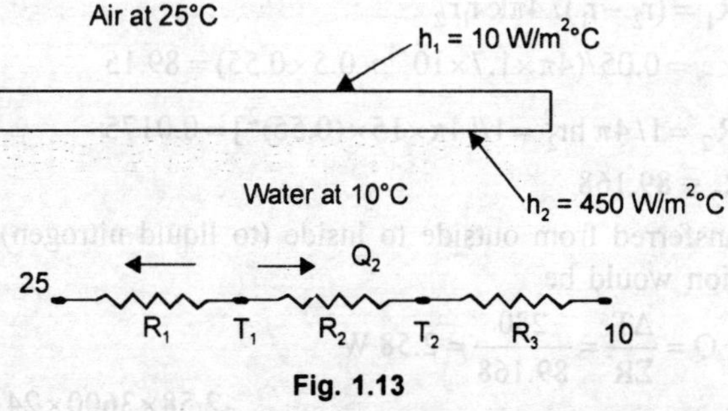

Fig. 1.13

Example 1.14 The top surface of a 2.5 cm thick hot plate ($\epsilon = 0.8$, $k = 40$ W/mK, surface area 0.35 m²) loses 350 W by radiation and 650 W by convection to the surroundings at 20°C. Calculate the temperature at the top and bottom surface of the hot plate and the convective heat transfer coefficient.

Solution: Let T_1 and T_2 are the temperatures at the bottom and top surface of the hot plate. The energy lost by radiation from the top surface is given by

$$\dot{Q}_r = \epsilon \, A\sigma(T_2^4 - T_\infty^4)$$

$$= 0.80 \times 0.35 \times 5.668 \times 10^{-8}(T_2^4 - 293^4)$$

or $\qquad T_2^4/10^8 = \dfrac{350}{0.80 \times 0.35 \times 5.668} + (2.93^4) = 294.27;$ which gives

$$T_2 \cong 141.17°C$$

Energy lost by convection, $\dot{Q}_c = hA(\Delta T)$

or, $\qquad 650 = h \times 0.35 \times (141.17 - 20);$ and $h = 15.33$ W/m^2°C

Energy lost by conduction from the bottom face to top face of the hot plate,

$$\dot{Q} = \dot{Q}_c + \dot{Q}_r = 1000 \text{ W};$$

and $\qquad 1000 = (T_1 - T_2)/L/kA = (T_1 - 141.17)/(0.025/40 \times 0.35)$

or, $\qquad T_1 = 141.17 + 1.79 = 142.96°C.$

Example 1.15 A spherical vessel 1 m in diameter is used to store liquid nitrogen at –200°C. Calculate the amount of nitrogen vapourized per day if the container is provided with a 5 cm thick layer of superinsulation k = 0.00017 W/mK, when the outside temperature is 30°C, the convective heat transfer coefficient at the outer surface is 15 W/m^2 °C and 200 kJ of energy is required to vapourize 1 kg of nitrogen at that temperature.   *(Similar to CS 87)*

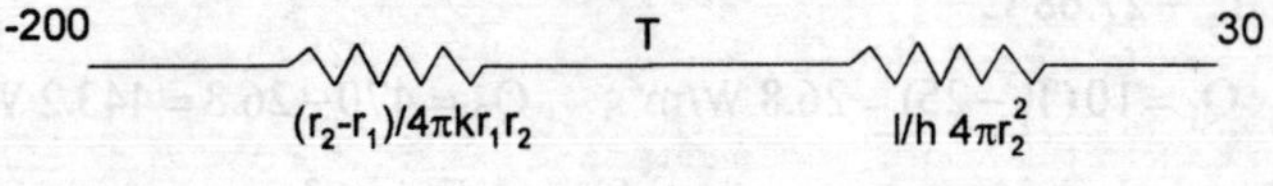

$$-200 \quad\text{~~~~~~}\quad T \quad\text{~~~~~~}\quad 30$$
$$(r_2\text{-}r_1)/4\pi kr_1 r_2 \qquad\qquad 1/h\,4\pi r_2^2$$

Fig. 1.14

Solution: The equivalent electric circuit is shown in Fig. 1.14

$$R_1 = (r_2 - r_1)/4\pi k \, r_1 r_2$$

$$= 0.05/(4\pi \times 1.7 \times 10^{-4} \times 0.5 \times 0.55) = 89.15$$

$$R_2 = 1/4\pi \, hr_2^2 = 1/[4\pi \times 15 \times (0.55)^2] = 0.0175$$

and $\quad R_1 + R_2 = 89.168$

The heat transferred from outside to inside (to liquid nitrogen) by conduction and convection would be

$$\dot{Q} = \frac{\Delta T}{\Sigma R} = \frac{230}{89.168} = 2.58 \text{ W}$$

and the mass of nitrogen vapourized per day $= \dfrac{2.58 \times 3600 \times 24}{200 \times 1000} = 1.14 \text{ kg}$

The temperature at the outside surface of the superinsulation would be

$$T = 30 - 2.58 \times 0.0175 = 29.95°C.$$

(The superinsulating materials consist of multiple layers of highly reflective materials separated by insulating spacers. The entire system is evacuated to minimise air conduction.)

Example 1.16 A communication satellite can be considered as a 100 cm diameter sphere. It is continuously exposed to solar radiation having an intensity of 1000 W/m^2. The satellite surface can be covered with two materials having $\alpha_2 = \epsilon_1 = 0.5$ and $\alpha_1 = \epsilon_2 = 0.05$, such that the surface temperature is maintained at 35°C. Estimate how much its surface should be covered with each insulating material. Take the outer space as a black body at 0K.

Solution: Let x part of the sphere is covered with insulation having $\epsilon_1 = 0.5$, $\alpha_1 = 0.05$ and $(1 - x)$ part with $\epsilon_2 = 0.05$, $\alpha_2 = 0.5$. The sphere receives solar radiation at the rate of 1000 W/m^2 . The energy absorbed by the surface would be:

$$1000A[x \times 0.05 + (1-x) \times 0.5] = 1000A[0.5 - 0.45x]$$

Energy emitted by the surface is $A\sigma \in T^4$

$$= A[x \times 0.5 + (1-x) \times 0.05] \times 5.667 \times 10^{-8}(273 + 25)^4$$

By making an energy balance

$$(0.45 x + 0.05)\, 510.25 = 500 - 450x, \text{ therefore } x = 0.698$$

or, 69.8% of the surface should be covered with insulation having emissivity 0.5 (absorptivity $\alpha = 0.05$) and 30.2% of the surface should be covered with insulation $\epsilon = 0.05$ and absorptivity $\alpha = 0.5$.

Example 1.17 A steel plate, inserted in an adiabatic wall, generates heat at a rate of 2kW/m^3. Estimate the maximum plate temperature, Fig. 1.15

Solution: Let the surface area of the steel plate inserted in the adiabatic wall is A m^2. Under steady state condition, the amount of heat generated within the plate has to be convected a way to the surroundings.

Or, $\quad$ h A $(T_2 - 25)$ = amount of heat generated = 2000 × A × 0.05 = 100A

∴ $\quad$ $T_2 = 25 + 100/6 = 41.67°C$

The heat energy will flow from the bottom surface to the top surface by conduction. A s such,

$$\dot{Q} = (T_1 - T_2)\,/L/kA = 100\ A$$
$$T_1 - T_2 = 100\ L/k = 1000 \times 0.05/40 = 0.125, \text{ and}$$
$$T_1 = 41.67 + 0.125 = 41.8°C$$

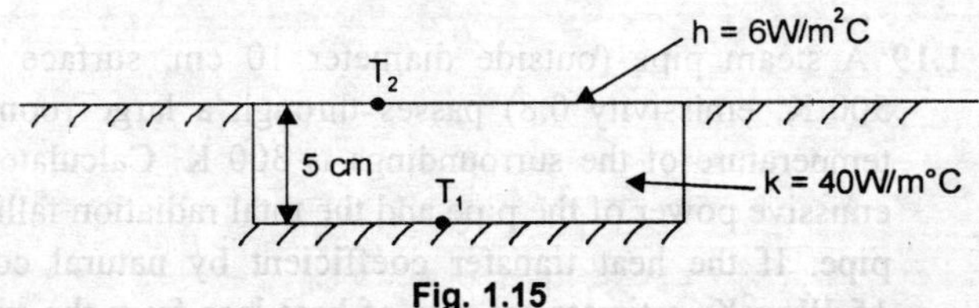

Fig. 1.15

α stands for the absorptivity of the material, defined as the fraction of the incident radiant energy absorbed by the material.

Example 1.18 The temperature of hot gases inside the combustion chamber of a gas turbine is 1050 K. The furnace walls receive heat by radiation and convection from hot gases. The furnace wall is 25 cm thick, k = 0.9 W/mK, convective heat transfer coefficient at the inside and outside surface of the wall is 6 W/m^{2}K and 100 W/m^{2}K respectively. The ambient temperature is 310 K. Estimate the temperature at the two faces of the wall and the absorptivity at the inner surface for a heat flow rate of 2500 W/m^2.

Solution: The physical configuration of the wall and its analogous electric circuit is shown in Fig. 1.16.

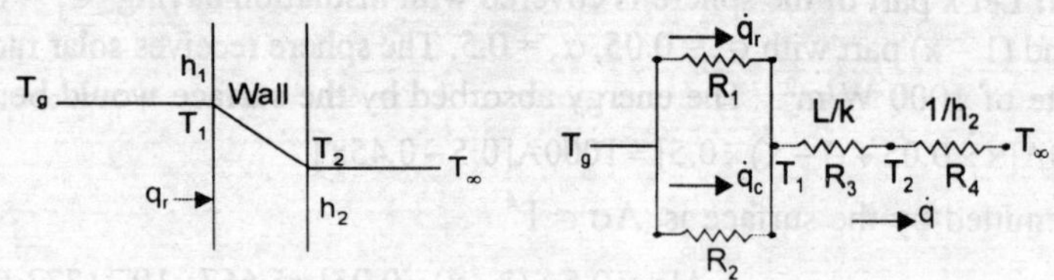

Fig. 1.16

Since the inside surface of the wall receives heat energy by radiation and convection, the thermal resistances R_1 and R_2 are in parallel.

The resistance R_4 = 1/h_2 = 1/100

$$T_2 = T_\infty + \dot{q} \times R_4 = 310 + 2500/100 = 335 \text{ K}$$

The resistance R_3 = L/k = 0.25/0.9; $\quad T_1 = T_2 + \dot{q} \times R_3$

$$= 335 + 2500 \times 0.25/0.9$$
$$= 1029.4 \text{K}$$

Heat energy received by convection, $\quad \dot{q}_c = h(\Delta T) = 6 \times (1050 - 1029.4)$
$$= 123.33 \text{ W/m}^2$$

Heat energy received by radiation, $\quad \dot{q}_r = 2500 - 123.33 = 2376.67$
$$= \alpha \times 5.667 \times 10^{-8} (1050^4 - 1029.4^4)$$

Therefore, absorptivity, $\quad\quad\quad \alpha = 0.453$.

The thermal resistance due to radiation, $R_1 = (\Delta T)/q_r$

$$= (1050 - 1029.4)/2376.67$$
$$= 8.67 \times 10^{-3} \text{ m}^2 \text{ K/W}.$$

Example 1.19 A steam pipe (outside diameter 10 cm, surface temperature 500 K, emissivity 0.8) passes through a large room where the temperature of the surroundings is 300 K. Calculate the surface emissive power of the pipe and the total radiation falling upon the pipe. If the heat transfer coefficient by natural convection is 15 W/m^{2}K, estimate the rate of heat loss from the surface of the pipe.

Solution: The surface emissive power can be obtained by using Eq. (1.3)

or, $\quad E = \varepsilon \sigma T^4 = 0.8 \times 5.668 \times 10^{-8} \times 500^4 = 2834 \text{ W/m}^2$

The surface area of the pipe is very small compared with the size of the enclosure, the large enclosure can always be treated as a black body. Therefore, the total radiation falling upon the pipe surface is given by

$$G = \sigma T^4 = 5.668 \times 10^{-8} \times 300^4 = 459.1 \text{ W/m}^2$$

Heat loss by the pipe due to radiation is given by Eq. (1.4)

$$\dot{Q}_r = \varepsilon \sigma A (T_S^4 - T_\infty^4)$$
$$= 0.8 \times 5.66 \times 10^{-8} (3.142 \times 0.1 \times 1)(500^4 - 300^4)$$
$$\dot{Q}_r / L = 775.04 \text{ W/metre length of the pipe}$$

Heat loss by pipe due to natural convection is

$$\dot{Q}_c / L = hA(\Delta T) = 15 \times (3.142 \times 0.1)(500 - 300)$$
$$= 942.6 \text{ W/ metre length of the pipe}$$

Total loss $= 775.04 + 942.6 = 1717.64 \text{ W/m}$.

Example 1.20 A circular conducting rod, diameter D and length L and electrical resistance per unit length R_e, is in thermal equilibrium with its surroundings. Obtain an expression to compute the variation in temperature of the rod with time when an electric current I is passed through the rod.

Solution: Initially the rod is in thermal equilibrium with its surroundings. This equilibrium will get disturbed when the electric current flows through the rod. Applying the principles of conservation of energy, we write

Energy generated within the rod = Energy convected away to the
$\qquad\qquad\qquad\qquad\qquad$ surroundings + The energy stored
$\qquad\qquad\qquad\qquad\qquad$ within the element.

Energy generated within the rod due to flow of electric current is : $\Rightarrow I^2 R_e L$

Energy radiated away to the surroundings + Energy convected away

$$= \varepsilon \sigma (\pi DL)(T^4 - T_\infty^4) + h(\pi DL)(T - T_\infty)$$

Rate of energy storage within the element

$$= \frac{d}{dt}(\rho V c \, T), \text{ where V is the volume.}$$

Therefore, $I^2 R_e L = \varepsilon \sigma (\pi DL)(T^4 - T_\infty^4) + h(\pi DL)(T - T_\infty) + \rho \cdot c \left(\dfrac{\pi D^2}{4}\right) L \dfrac{dT}{dt}$

and $\qquad \dfrac{dT}{dt} = \dfrac{I^2 R_e - \pi D \left[\varepsilon \sigma \left(T^4 - T_\infty^4\right) + h\left(T - T_\infty\right) \right]}{\rho \cdot C \left(\pi D^2 / 4\right)}$

Example 1.21 A cubical container contains ice at 0°C. The container is placed in a room at 25°C. Assuming steady state, one-dimensional conduction through each wall (thickness L, thermal conductivity k), estimate the time required to completely melt the ice.

Solution: During melting of ice, the temperature of ice will remain at 0°C, therefore, the temperature difference across the wall of the container will remain constant and equal to 25°C.

The rate of heat flow by conduction can be written as

$$\dot{Q} = k\,A(25 - 0.0)/L$$

where the area $A = 6B^2$, B being the length of each side of container.

The amount of energy inflow during the time t is, then

$$\dot{Q}xt = k(6B^2)\,25t/L$$

The amount of heat energy required to melt M kg of ice during that time t
$$= \text{Mass of ice} \times h_{sf}$$

where h_{sf} = latent energy associated with conversion from solid to liquid state. Applying the principle of energy conservation, we have

$$M \times h_{sf} = k(6B^2)\,25t/L$$

or, $\qquad\qquad t = Mh_{sf}L/(150\,k\,B^2).$

Example 1.22 Hot water flows with a velocity of 0.1 m/s in a 100 m long 0.1m diameter pipe. Heat loss from the pipe outer wall is uniform and is equal to 420 W/m² If the inlet temperature of water is 80°C, calculate the water temperature at exit. Neglect pipe wall thickness.

Solution: Mass flow rate of water $\rho AV = 10^3 \times 3.142 \times (0.05)^2 \times 0.1 = 0.7855\,\text{kg/s}$

By making an energy balance, $\dot{m} \times c \times (80 - T) = \dot{Q} = 420 \times \text{area}$
$$= 0.7855 \times 4200 \times (80 - T)$$
$$= 420 \times 3.142 \times 0.1 \times 100$$
or, $\qquad\qquad\qquad\qquad\qquad\qquad\qquad T = 76°C.$

Example 1.23 The left and right hand face of a 1 m thick plane wall is maintained at 500 K and 300 K respectively. If the thermal conductivity of the wall varies as $k = 0.25\,(1 + 0.02T)$ where k is in W/mK and T is in K, calculate (a) the heat transfer rate (b) the temperature at a section 0.4 m from the left face (c) the distance from the left face where the temperature is 400 K and (d) the average thermal conductivity of the wall material during this temperature range.

Solution: (a) From Fourier Law: $\quad \dot{Q}/A = -k\,dT/dx$
$$= -0.25\,(1 + 0.02\,T)\,dT/dx$$

By separating the variables, $\dot{Q}/A \displaystyle\int_0^1 dx = -0.25 \int_{500}^{300} (1 + 0.02T)dT$

$$= 0.25\,(T + 0.01T^2)\,\Big|_{300}^{500}$$

which gives, $\qquad\qquad\qquad\qquad \dot{Q}/A = 450\ \text{W/m}^2$

(b) When x = 0.4m; $\qquad\qquad \dot{Q}/A\,(0.4) = 0.25\,[500 - T + 0.01\,(500^2 - T^2)]$

or, $\quad 0.01\ T^2 + 0.25\ T = 125 + 2500 - 450 \times 0.4$
$$= 2445.$$

$$T = [-0.25 \pm (0.25^2 + 4 \times 0.01 \times 2445)^{1/2}]/(2 \times 0.01)$$

Since the temperature cannot be negative, we take the positive value and,
$\quad$ T = 482 K.

(c) Let the distance from the left face be x metres where the temperature is 400K.

or, $\quad \dot{Q}/A\ (x) = 0.25\ [500 - 400 + 0.01\ (500^2 - 400^2)] = 250$
$$x = 250/450 = 0.555\text{m from the left face}$$

(d) Let the average conductivity is k_m, then by Fourier Law

$$\dot{Q}/A = k_m\ (T_1 - T_2)/L = k_m\ (500 - 300)/1$$
$$k_m = 450/200 = 2.25\ \text{W/mK}.$$

Example 1.24 A 4 mm thick aluminium plate is initially at 25°C. Its top surface is suddenly exposed to ambient air at 20°C and to solar radiation having an incident flux of 900 W/m². The plate absorbs 8% of the incident radiation and has an emissivity of 0.25. The convection heat transfer coefficient between the surface and the air is h = 20 W/m² K. Estimate the initial rate of change of the plate temperature. What would be the equilibrium temperature of the plate when steady- state conditions are reached? Take ρ = 2700 kg/m³, and C = 900 J/kgK.

Solution: Let the surface area of the plate be 1 m². The quantity of heat energy lost by the plate by convection $= hA(\Delta T)$
or, $\quad \dot{Q}_c = 20 \times 1 \times (25 - 20) = 100\ \text{W}$
Enery lost by radiation, $\quad \dot{Q}_r = \varepsilon\sigma(T_1^4 - T_\infty^4) = 0.25 \times 5.67 \times 10^{-8}(298^4 - 293^4)$
$$= 7.315\ \text{W}$$
$\quad$ Total energy loss $= \dot{Q} = 107.315\ \text{W}$

Energy received by radiation = 0.8 × incident flux = 720 W

Rate of energy storage, $\quad \dot{Q} = 720 - 107.315 = 612.685\ \text{W}$

$$= \text{mass} \times \text{specific heat} \times \text{time rate of change in temperature}$$
$$= m \times C \times dT/dt$$

$\therefore \quad dT/dt = 612.685/(4 \times 10^{-3} \times 1 \times 2700 \times 900) = 0.063\ \text{K/s}$

The temperature will increase.

Let the equilibrium temperature of the plate be T K.

Energy lost by connection, $\dot{Q}_c = hA(\Delta T) = 20 \times 1 \times (T - 293)$

Energy lost by radiation, $\quad \dot{Q}_r = \varepsilon\sigma A(T^4 - 293^4)$
$$= 0.25 \times 1 \times 5.67 \times 10^{-8}(T^4 - 293^4)$$

Energy received by radiation = 720 W.

By making an energy balance:

$$20\,(T - 293) + 0.25 \times 5.67 \times 10^{-8}\,(T^4 - 293^4) = 720$$

Solving by trial and error, $T = 579\ K \equiv 306°C$.

11. Evaporation is a Cooling Process

When the relative humidity, ϕ, is less than 1, the wet bulb temperature is lower than the dry bulb temperature. This is because of evaporation at the surface of the wet bulb thermometer. The rate of cooling by rapid evaporation is very high because each evaporating gram of water draws at least 2258 J from water left behind. This is an enormous amount of energy compared to 4.182 J per degree C that is drawn from each gram of water that cools by thermal conduction. Thus, evaporation is really a cooling process. If we place two bowls of water each containing the same amount, and one at about 100°C and the other at about 65°C, inside a refrigerator at the same time, the hot water will freeze before the warm water because of more loss of energy due to evaporation.

12. Insulating Materials—Types and Characteristics

Insulating materials are usually applied to reduce the rate of heat flow and therefore, these materials should have a low thermal conductivity. This objective is achieved by trapping air or a gas either inside small cavities in a solid or by filling that space, across which the heat flow rate is to be reduced, with small solid particles and t rapping a ir b etween t he p articles. T hus, t he t hermal c onductivity o f s uch materials is not really the property of the material and it depends upon the mechanism of heat transmission.

The insulating materials can be divided into three categories:

(a)   *Fibrous*—These materials contain small diameter particles of filaments of low density and are poured into a gap to fill it loosely. Therefore, they would have very high porosity (90%). The common fibrous materials are fiberglass (used for temperatures below 200°C), mineral wool (used for temperatures below 700°C), refractory fibres such as Al_2O_3 (alumina) and SiO_2 (silica) for use between 700°C and 1700°C.

(b)   *Cellular*—These a re c losed—or o pen c ell m aterials a nd a re a vailable a s extended flexible or rigid boards. This insulating material has low heat capacity, low density, and relatively good compressive strength.

(c)   *Granular*—These are inorganic material particles bonded into desired shapes or to be used as powders. Examples are: perlite powder, diatomaceous silica and vermiculite.

When the trapped gases in cellular materials are condensed or frozen to create a partial vacuum, the effectiveness of the insulation improves and they become suitable for use at cryogenic temperatures. Fibrous and granular insulating materials can also be evacuated to eliminate conduction and convection and their effective thermal conductivity appreciably decreases.

While selecting an insulating material, the relevant properties to be considered are: effective thermal conductivity, density, the upper limit of temperature, structural rigidity, degradation, chemical stability and cost.

13. Semiconductors and Superconductors

The classification of a substance as a conductor or as an insulator is based on how tightly the atoms of the substance can hold their electrons. Some materials, such as germanium and silicon, are neither good conductors nor good insulators. These materials are fair insulators in their pure crystalline form but increase tremendously in conductivity when even one atom in 10 million is replaced with an impurity which adds or removes an electron from the crystal structure. Such materials sometimes behave as insulators and sometimes as conductors, and are called semiconductors.

At temperatures near absolute zero, certain metals acquire infinite conductivity. These are called superconductors. Superconductivity at high temperature (above 100K) was discovered in a non-metallic compound also. Electric current passes through a superconductor without losing energy and there is no heat loss when charges flow.

SUMMARY

1. Heat transfer is energy in transit due to a temperature difference.
2. The transfer of heat energy can take place by Conduction, Convection and Radiation. The process of heat transfer can be steady, unsteady or cyclic.
3. In transient heat transfer process, the internal energy of the system changes.
4. Basic laws of heat transfer are:

Conduction	Diffusion of energy due to random molecular motion	Fourier law	$\dot{Q} = -kAdT/dx$
Convection	Diffusion of energy due to bulk motion of fluid particles along with diffusion of energy due to random molecular motion	Newton's law	$\dot{Q} = hA(\Delta T)$
Radiation	Energy transfer by electromagnetic waves	Stefan-Boltzmann law	$\dot{Q} = \varepsilon\sigma AT^4$

5. Thermal conductivity of a material is its physical property, depends upon temperature, pressure, structure and porosity of the material. The thermal conductivity of a gas increases with increasing temperature and decreasing molecular weight because the mean molecular speed increases with temperature. The thermal conductivity of non-mettalic liquids generally decreases with increasing temperature except for water and glycerine. In solids, the transport of thermal energy is due to two effects: the migration of free electrons and lattice vibration waves, and these effects are additive.

In pure metals, the electrical resistivity is low and therefore, the conductivity depends upon migration of free electrons. In alloys, the transport of thermal energy depends on migration of free electrons and on the lattice vibrations. For non-metallic solids, the thermal conductivity depends upon lattice vibration, Fig. 1.17 (a, b, c).

6.   Convective heat transfer coefficient is a property of flow and also depends upon the physical properties of fluid and geometry of surface.

7.   All surfaces emit thermal energy by radiation within a wavelength band. The amount of energy emitted is proportional to the fourth power of the absolute temperature and does not require a matter-filled intervening space, as in conduction and convection.

8.   Heat transfer problems can be easily solved by drawing analogous electric circuits.

9.   The subjects of heat transfer and thermodynamics are highly complimentary. Many heat transfer problems require direct application of the law of the conservation of energy (the first law of thermodynamics).

10. Evaporation is a cooling process.

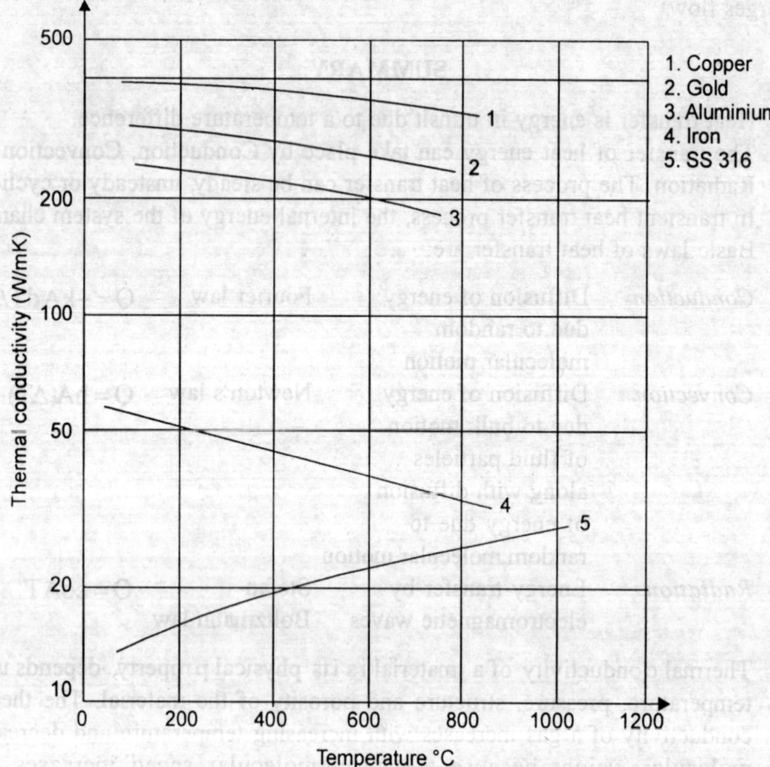

Fig. 1.17 (a) Variation of thermal conductivity with temperature

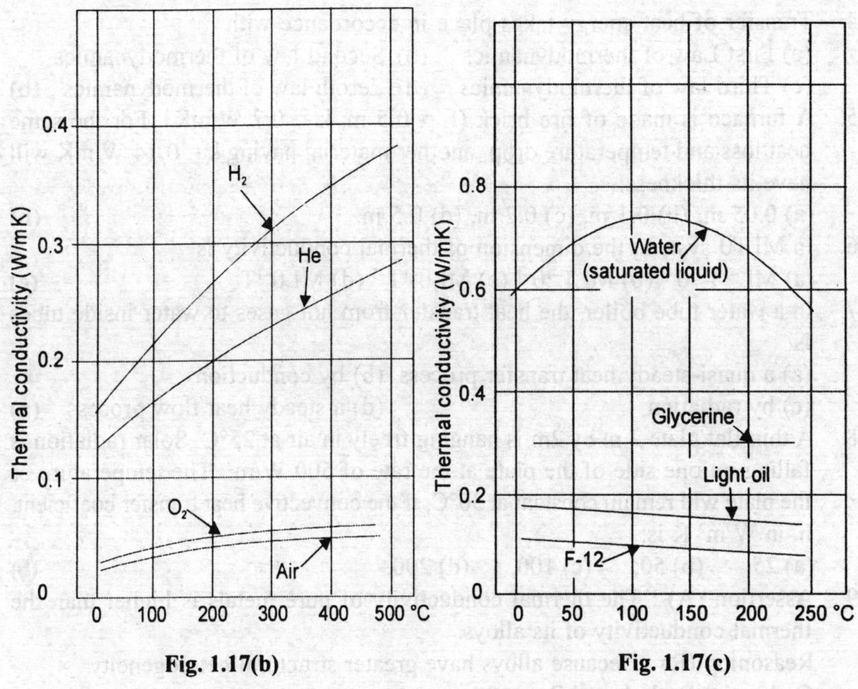

Fig. 1.17(b)   **Fig. 1.17(c)**

MULTIPLE CHOICE QUESTIONS

1. Match List I with List II and choose the answer from the code:

   List I   List II

   A. Gases   1. Transport of energy by electron gas

   B. Liquids   2. Volumetric density

   C. Porus solid   3. Unstable elastic collision

   D. Metals   4. Random collision of molecules

   CODE

	A	B	C	D
(a)	4	2	3	1
(b)	2	3	4	1
(c)	3	4	4	1
(d)	4	3	2	1

   (d)

2. Assertion (A): The thermal conductivity of liquids decreases with temperature except for water and glycerin.

   Reasoning (R): Because water and glycerin are heavily associated liquids.

   Code: (a) Both A and R are false   (b) Both A and R are true

         (c) A is true and R is false   (d) A is false, R is true   **(b)**

3. Electric bulbs transfer heat energy by

   (a) conduction only   (b) convection only

   (c) radiation only   (d) both (a) and (c)   **(d)**

4. Transfer of heat energy takes place in accordance with
    (a) First Law of thermodynamics   (b) Second law of thermodynamics
    (c) Third law of thermodynamics   (d) Zeroth law of thermodynamics **(b)**
5. A furnace is made of fire brick (L = 0.5 m, k = 0.7 W/mK). For the same heat loss and temperature drop, another material having k = 0.14 W/mK will have its thickness:
    (a) 0.05 m, (b) 0.1 m, (c) 0.2 m, (d) 0.5 m.   **(b)**
6. In MLTθ system, the dimension of thermal conductivity is
    (a) $ML^{-1}T^{-1}\theta^{-3}$ (b) $MLT^{-1}\theta^{-1}$ (c) $ML\theta^{-1}T^{-3}$ (d) $ML\theta^{-1}T^{-2}$ **(c)**
7. In a water tube boiler, the heat transfer from hot gases to water inside tubes is
    (a) a quasi-steady heat transfer process (b) by conduction
    (c) by radiation   (d) a steady heat flow process   **(a)**
8. A thin flat plate 2 m by 2m is hanging freely in air at 25°C. Solar radiation is falling on one side of the plate at the rate of 500 W/m². The temperature of the plate will remain constant at 30°C, if the convective heat transfer coeficient, h, in W/m² K is:
    (a) 25,   (b) 50,   (c) 100,   (d) 200.   **(b)**
9. Assertion (A) : The thermal conductivity of pure metals is higher than the thermal conductivity of its alloys.
    Reasoning (R) : Because alloys have greater structural heterogeneity.
    Code: (a) Both A and R are false
           (b) Both A and R are true but R is not the correct explanation of A
    (c) Both A and R are true and R is the correct explanation of A
    (d) A is true and R is false.   **(c)**
10. Arrange the thermal conductivity of the following materials in ascending order:
    copper, mercury, silver, water.
    Code:
    (a) copper, silver, mercury, water   (b) mercury, water, copper, silver
    (c) water, mercury, copper, silver   (d) silver, copper, mercury, water   **(c)**
11. The Fourier Law Q = – kA dT/dx assumes
    (a) constant value of thermal conductivity
    (b) constant and uniform temperature at the surface of a wall
    (c) steady-state one-dimensional flow
    (d) all of the above
    (e) only (b) and (c)   **(e)**
12. The thermal conductivity of a damp brick is higher than the thermal conductivity of dry brick, because
    (a) the thermal conductivity of air is less than that of water
    (b) In damp brick, the heat is transferred by convection due to the capillary movement of water within the porus material
    (c) both (a) and (b)
    (d) none of the above   **(c)**
13. The thermal conductivity of powdery and porus material
    (a) decreases with increasing volumetric density

(b) increases with increasing volumetric density
(c) is independent of its volumetric density
(d) none of the above   **(b)**

14. The thermal conductivity of a structure like concrete, stone, etc, may vary from sample to sample because of variations in
    (a) structure, (b) density, (c) composition, (d) porosity,
    (e) all of the above.   **(e)**

15. Insulating materials are used in cases in which one wishes to obstruct the flow of heat between an enclosure and its surroundings. Low temperature insulations are:
    (a) asbestos, (b) glass wool, (c) magnesia, (d) diatomaceous earth.   **(b)**

16. The difference in temperature between the two sides of a plane wall can be increased by
    (a) decreasing the thermal conductivity of the material
    (b) increasing the heat flow rate
    (c) either (a) or (b)   (d) both (a) and (b)   **(d)**

17. A 0.5 m thick plane wall has its two surfaces kept at 300 and 200 °C. Thermal conductivity of the wall varises linearly with temperature and its values at 300°C and 200°C are 25 W/mK and 15 W/mK respectively. Then the steady state heat flux through the wall is
    (a) 8 kW/m², (b) 5 kW/m², (c) 4 kW/m², (d) 3 kW/m².   **(c)**

18. A masonary wall (k = 0.75 W/mK) transmits 80% of the heat rate through another wall (k = 0.25 W/mK, thickness 100 mm). If the temperature difference across both the walls are the same, the thickness of the masonary wall would be
    (a) 150 mm, (b) 240 mm, (c) 300 mm, (d) 375 mm.   **(c)**

19. Let M represents the molecular weight and T represents the temperature of the gas. the thermal conductivity of the gas will decrease when
    (a) both M and T increase   (c) M decreases and T increases
    (b) both M and T decrease   (d) M increases and T decreases   **(d)**

20. Match List I with List II and select the correct answer using the code given below

List I		List II
A.	Granular insulation	1. Vermiculite
B.	Semiconductors	2. Germanium
C.	Low electrical resistivity	3. Liquid metals
D.	Nuclear power plants	4. Pure metals

    CODE

	A	B	C	D
(a)	2	1	4	3
(b)	1	2	4	3
(c)	1	2	3	4
(d)	4	3	2	1

                                                                          (b)

NUMERICALS

1. The inner surface of a 25 cm brick wall (k = 0.865 W/mK) of a furnace is maintained at temperature 820°C and the outer surface temperature is 170°C. Calculate the heat loss per m² of the wall area. If the ambient temperature is 25°C, what would be the convective heat transfer coefficient at the outer surface of the wall. **(2.25 kW/m², 15.5 W/m²K)**

2. The thermal conductivity of a wall of thickness L is given by the equation $k = k_0 + 2bT + 3CT^2$ where k_0 is the thermal conductivity at 0°C. Show that the steady state heat flow through the wall with surface temperatures T_1 and T_2 is given by $\dot{Q}/A = [k_0 + b_1 (T_1 + T_2) + C (T_1^2 + T_1T_2 + T_2^2)] (T_1 - T_2)/L$

3. A glass window 60 cm by 30 cm is 16 mm thick. If the inside and outside air temperatures are 15°C and 40°C and respectively, Calculate the gain in heat energy through the window. The convective heat transfer coefficient on the inside and outside surfaces are 10 W/m² K and 100 W/m² K.
    k = 0.78 W/mK **(34.48 W)**

4. A pipe, internal diameter 20 cm, outer diameter 30 cm, k = 30 W/mK carries hot gases at 200°C. The convective heat transfer coefficient at the inside surface of the pipe and at the outer surface is 100 W/m² K and 15 W/m² K. Calculate the energy lost to the surroundings at 25°C per metre length of the pipe, the temperature at the two pipe surfaces and the overall heat transfer coefficient based on the pipe outer diameter.
    (1.97 kW/m, 168.63°C, 164.4°C, 11.95 W/m²K)

5. A 10 cm diameter sphere (ρ = 7800 kg/m³, C = 0.5 kJ/kgK), surface emissivity 0.6, is heated in a furnace to a temperature of 350°C. It is placed in a large room at 25°C. Calculate the loss of heat energy by radiation. Neglecting radiation, calculate the initial rate of cooling if the convective heat tranfer coefficient is 15 W/m² K. **(152.5W, 0.075W)**

6. A metal plate absorbs 80% of the incident solar radiation having an intensity of 950 W/m². Assuming that the lower surface of the plate is insulated and the emissivity of the plate surface is 0.2, calculate the temperature of the plate if the convective heat transfer coefficient at the surface is 15 W/m²K and the temperature of the surrounding is 25°C. **(344 K)**

7. A vertical square plate 50 cm on a side, is maintained at 80°C and is exposed to room air at 25°C. The surface emissivity is 0.8 and the convective heat transfer coefficient is 15 W/m²K. Calculate the amount of energy required to maintain the temperature of the plate. **(292.86 W)**

8. The temperature recorded by thermocouples embedded at radii 10 cm and 20 cm of a spherical shell are 200°C and 150°C respectively. Calculate the thermal conductivity of the material if 150 W of energy is supplied to the system.
    (2.387 W/mK)

9. A long conducting rod of diameter D and electrical resistance per unit lenth R is initially in thermal equilibrlum with the surroundings. When an electrical current I is passed through the rod, the equilibrium is disturbed. Develop an expression to compute the rate of change of temperatune with time. List the assumptions made.

CHAPTER 2

Steady State Conduction— One Dimension

1. The General Heat Conduction Equation for an Isotropic Solid with Constant Thermal Conductivity

Any physical phenomenon is generally accompanied by a change in space and time of its physical properties. The heat transfer by conduction in solids can only take place when there is a variation of temperature, in both space and time. Let us consider a small volume of a solid element as shown in Fig. 2.1(a). The dimensions are: Δx, Δy, Δz along the X-, Y-, and Z- coordinates.

First we consider heat conduction in the X-direction. Let T denote the

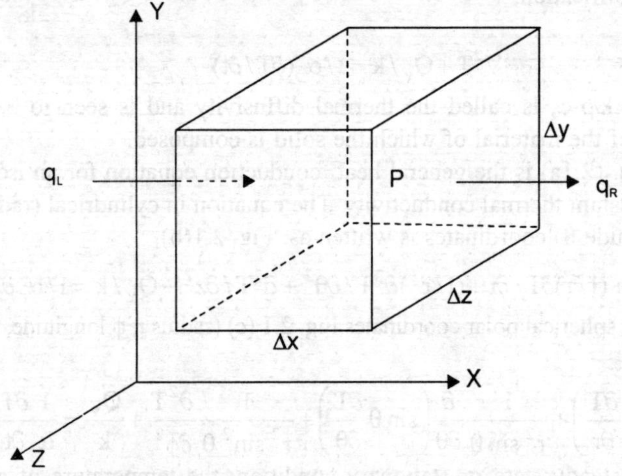

Fig. 2.1(a) Elemental volume in cartesian coordinate

temperature at the point P (x, y, z) located at the geometric centre of the element. The temperature gradient at the left hand face $(x - \Delta x/2)$ and at the right hand face $(x + \Delta x/2)$, using the Taylor's series, can be written as:

$$\partial T / \partial x \mid_L = \partial T / \partial x - \partial^2 T / \partial x^2 . \Delta x / 2 + \text{higher order terms}$$

$$\partial T / \partial x \mid_R = \partial T / \partial x + \partial^2 T / \partial x^2 . \Delta x / 2 + \text{higher order terms}.$$

The net rate at which heat is conducted out of the element in X-direction assuming k as constant and neglecting the higher order terms,

we get
$$- k \Delta y \Delta z \left[\frac{\partial T}{\partial x} + \frac{\partial^2 T}{\partial x^2} \frac{\Delta x}{2} - \frac{\partial T}{\partial x} + \frac{\partial^2 T}{\partial x^2} \frac{\Delta x}{2} \right]$$

$$= -k \Delta y \Delta z \Delta x \left(\frac{\partial^2 T}{\partial x^2} \right)$$

Similarly for Y- and Z-direction,

we have $-k \Delta x \Delta y \Delta z \, \partial^2 T / \partial y^2$ and $-k \Delta x \Delta y \Delta z \, \partial^2 T / \partial z^2$.

If there is heat generation within the element as $\dot{Q}_v$ per unit volume and the internal energy of the element changes with time, by making an energy balance, we write

$$\begin{array}{ccc} \text{Heat generated within} & \text{Heat conducted away} & \text{Rate of change of internal} \\ \text{the element} & - \quad \text{from the element} & = \quad \text{energy within the element} \end{array}$$

or,
$$\dot{Q}_v (\Delta x \Delta y \Delta z) + k(\Delta x \Delta y \Delta z)(\partial^2 T / \partial x^2 + \partial^2 T / \partial y^2 + \partial^2 T / \partial z^2)$$
$$= \rho \, c \, (\Delta x \Delta y \Delta z) \, \partial T / \partial t$$

Upon simplification, $\partial^2 T / \partial x^2 + \partial^2 T / \partial y^2 + \partial^2 T / \partial z^2 + \dot{Q}_v / k = \dfrac{\rho c}{k} \partial T / \partial t$

or,
$$\nabla^2 T + \dot{Q}_v / k = 1/\alpha \; (\partial T / \partial t) \tag{2.1a}$$

where $\alpha = k/\rho \cdot c$, is called the thermal diffusivity and is seen to be a physical property of the material of which the solid is composed.

The Eq. (2.1a) is the general heat conduction equation for an isotropic solid with a constant thermal conductivity. The equation in cylindrical (radius r, axis Z and longitude θ) coordinates is written as: Fig. 2.1(b),

$$\partial^2 T / \partial r^2 + (1/r)\partial T / \partial r + (1/r^2)\partial^2 T / \partial \theta^2 + \partial^2 T / \partial z^2 + \dot{Q}_v / k = 1/\alpha \; \partial T / \partial t \tag{2.1b}$$

And, in spherical polar coordinates Fig. 2.1 (c) (radius r, ϕ longitude, θ colatitude) is

$$\frac{1}{r^2} \frac{\partial}{\partial r}\left(r^2 \frac{\partial T}{\partial r} \right) + \frac{1}{r^2 \sin \theta} \frac{\partial}{\partial \theta}\left(\sin \theta \frac{\partial T}{\partial \theta} \right) + \frac{1}{r^2 \sin^2 \theta} \frac{\partial^2 T}{\partial \phi^2} + \frac{\dot{Q}_v}{k} = \frac{1}{\alpha} \frac{\partial T}{\partial t} \tag{2.1c}$$

Under steady state or stationary condition, the temperature of a body does not vary with time, i.e. $\partial T / \partial t = 0$. And, with no internal generation, the equation (2.1) reduces to

$$\nabla^2 T = 0 \tag{2.2}$$

'It should be noted that Fourier law can always be used to compute the rate of heat transfer by conduction from the knowledge of temperature distribution even

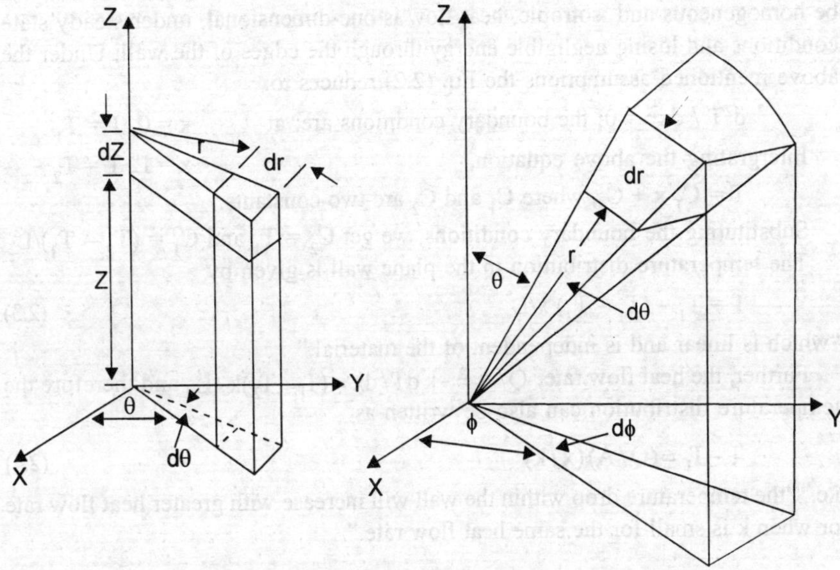

Fig. 2.1(b) Elemental volume in cylindrical coordinate (c) Spherical coordinate

for unsteady condition and with internal heat generation.'

2. One-dimensional Heat Flow

The term 'one-dimensional' is applied to heat conduction problem when:
- (i)   Only one space coordinate is required to describe the temperature distribution within a heat conducting body;
- (ii) Edge effects are neglected;
- (iii) The flow of heat energy takes place along the coordinate measured normal to the surface.

3. Thermal Diffusivity and its Significance

Thermal diffusivity is a physical property of the material, and is the ratio of the material's ability to transport energy to its capacity to store energy. It is an essential parameter for transient processes of heat flow and defines the rate of change in temperature. In general, metallic solids have higher α, while non-metallics, like paraffin, have a lower value of α. Materials having large α respond quickly to changes in their thermal environment, while materials having lower α respond very slowly, take a longer time to reach a new equilibrium condition.

TEMPERATURE DISTRIBUTION IN 1-D SYSTEMS

4. A Plane Wall

A plane wall is considered to be made out of a constant thermal conductivity material and extends to infinity in the Y- and Z-direction. The wall is assumed to be homogeneous and isotropic, heat flow is one-dimensional, under steady state conditions and losing negligible energy through the edges of the wall. Under the above mentioned assumptions the Eq. (2.2) reduces to

$d^2T / dx^2 = 0$; the boundary conditions are: at $\quad\quad$ $x = 0, T = T_1$

Intergrating the above equation, $\quad\quad\quad\quad\quad\quad\quad\quad$ $x = L, T = T_2$

$T = C_1 x + C_2$, where C_1 and C_2 are two constants.

Substituting the boundary conditions, we get $C_2 = T_1$ and $C_1 = (T_2 - T_1)/L$
The temperature distribution in the plane wall is given by

$$T = T_1 - (T_1 - T_2) \, x/L \tag{2.3}$$

"which is linear and is independent of the material."

Further, the heat flow rate, $\dot{Q}/A = -k \, dT/dx = (T_1 - T_2)k/L$, and therefore the temperature distribution can also be written as

$$T - T_1 = (\dot{Q}/A)(x/k) \tag{2.4}$$

i.e., "the temperature drop within the wall will increase with greater heat flow rate or when k is small for the same heat flow rate."

5. A Cylindrical Shell—Expression for Temperature Distribution

In the cylindrical system, when the temperature is a function of radial distance only and is independent of azimuth angle or axial distance, the differential equation (2.2) would be, (Fig. 2.2)

$$d^2T/dr^2 + (1/r)\, dT/dr = 0$$

with boundary conditions: at $r = r_1$, $T = T_1$ and at $r = r_2$, $T = T_2$.
The differential equation can be written as:

$$\frac{1}{r}\frac{d}{dr}(r \, dT/dr) = 0. \text{ Or, } \frac{d}{dr}(r \, dT/dr) = 0.$$

upon integration, $T = C_1 \ln(r) + C_2$, where C_1 and C_2 are the arbitrary constants.

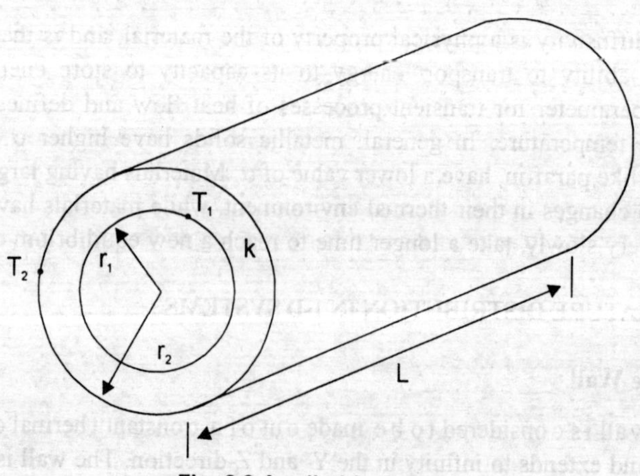

Fig. 2.2 A cylindrical shell

By applying the boundary conditions,
$$C_1 = (T_2 - T_1)/\ln(r_2/r_1)$$
and $\quad C_2 = T_1 - \ln(r_1).\ (T_2 - T_1)/\ln(r_2/r_1)$

The temperature distribution is given by
$$T = T_1 + (T_2 - T_1).\ \ln(r/r_1)/\ln(r_2/r_1) \text{ and}$$

$$\dot{Q}/L = -\ kA\ dT/dr$$

$$= 2\pi k(T_1 - T_2)/\ln(r_2/r_1) \tag{2.5}$$

From Eq. (2.5) it can be seen that the temperature varies logarithmically through the cylinder wall in contrast with the linear variation in the plane wall.

If we write Eq. (2.5) as $\dot{Q} = kA_m(T_1 - T_2)/(r_2 - r_1)$, where

$$A_m = 2\pi(r_2 - r_1)L/\ln(r_2/r_1)$$

$$= (A_2 - A_1)/\ln(A_2/A_1),$$

where A_2 and A_1 are the outside and inside surface areas respectively. The term A_m is called 'Logarithmic Mean Area' and the expression for the heat flow through a cylindrical wall has the same form as that for a plane wall.

6. Spherical and Parallelopiped Shells—Expression for Temperature Distribution

Conduction through a spherical shell is also a one-dimensional steady state problem if the interior and exterior surface temperatures are uniform and constant. The Eq. (2.2) in one-dimensional spherical corrdinates can be written as

$$(1/r^2)\frac{d}{dr}(r^2 dT/dr) = 0 \text{ , with boundary conditions,}$$

at $\qquad r = r_1, T = T_1; \quad$ at $r = r_2, T = T_2$

or, $\qquad \dfrac{d}{dr}(r^2 dT/dr) = 0$

and upon intergration, $T = -\ C_1/r + C_2$, where C_1 and C_2 are constants. Substituting the boundary conditions,

$$C_1 = (T_1 - T_2)\ r_1 r_2/(r_1 - r_2), \text{ and } C_2 = T_1 + (T_1 - T_2)\ r_1 r_2/r_1\ (r_1 - r_2)$$

The temperature distribution in the spherical shell is given by

$$T = T_1 - \left\{ \frac{(T_1 - T_2)\ r_1 r_2}{(r_2 - r_1)} \right\} \times \left\{ \frac{(r - r_1)}{r\ r_1} \right\} \tag{2.6}$$

and the temperature distribution associated with radial conduction through a sphere is represented by a hyperbola. The rate of heat conduction is given by

$$\dot{Q} = 4\pi k(T_1 - T_2)r_1 r_2/(r_2 - r_1) = k(A_1 A_2)^{1/2}(T_1 - T_2)/(r_2 - r_1) \tag{2.7}$$

where $\quad A_1 = 4\pi r_1^2$ and $A_2 = 4\pi r_2^2$

If A_1 is approximately equal to A_2 i.e., when the shell is very thin,

$$\dot{Q} = kA(T_1 - T_2)/(r_2 - r_1); \text{ and } \dot{Q}/A = (T_1 - T_2)/\Delta r/k$$

which is an expression for a flat slab.

[The above equation (2.7) can also be used as an approximation for parallelopiped shells which have a smaller inner cavity surrounded by a thick wall, such as a small furnace surrounded by a large thickness of insulating material, although the heat flow especially in the corners, cannot be strictly considered one-dimensional. It has been suggested that for $(A_2/A_1) > 2$, the rate of heat flow can be approximated by the above equation by multiplying the geometric mean area $A_m = (A_1 A_2)^{\frac{1}{2}}$ by a correction factor 0.725.]

7. Composite Surfaces

There are many practical situations where different materials are placed in layers to form composite surfaces, such as the wall of a building, cylindrical pipes or spherical shells having different layers of insulation. Composite surfaces may involve any number of series and parallel thermal circuits.

8. Heat Transfer Rate through a Composite Wall

Let us consider a general case of a composite wall as shown in Fig. 2.3. There are 'n' layers of different materials of thicknesses L_1, L_2, etc and having thermal conductivities k_1, k_2, etc. On one side of the composite wall, there is a fluid A at temperature T_A and on the other side of the wall there is a fluid B at temperature T_B. The convective heat transfer coefficients on the two sides of the wall are h_A and h_B respectively. The system is analogous to a series of resistances as shown in the figure.

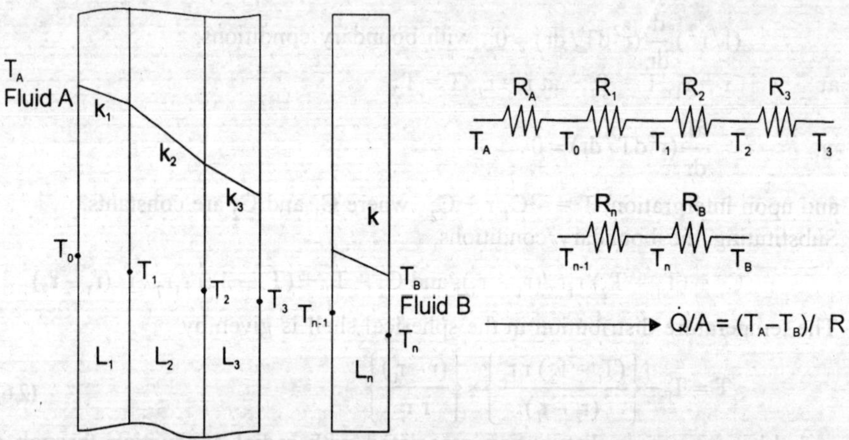

Fig. 2.3 Heat transfer through a composite wall

9. The Equivalent Thermal Conductivity

The process of heat transfer through composite and plane walls can be more conveniently compared by introducing the concept of 'equivalent thermal conductivity', k_{eq}. It is defined as:

$$k_{eq} = \left(\sum_{i=1}^{n} L_i \right) \Big/ \sum_{i=1}^{n} (L_i / k_i) \qquad\qquad (2.8)$$

$$= \frac{\text{Total thickness of the composite wall}}{\text{Total thermal resistance of the composite wall}}$$

And, its value depends on the thermal and physical properties and the thickness of each constituent of the composite structure.

Example 2.1   A furnace wall consists of 150 mm thick refractory brick (k = 1.6 W/mK) and 150 mm thick insulating fire brick (k = 0.3 W/mK) separated by an air gap (resistance 0.16 K/W). The outside wall is covered with a 10 mm thick plaster (k = 0.14 W/mK). The temperature of hot gases is 1250°C and the room temperature is 25°C. The convective heat transfer coefficient for gas side and air side is 45 W/m²K and 20 W/m²K. Calculate (i) the rate of heat flow per unit area of the wall surface (ii) the temperature at the outside and inside surface of the wall and (iii) the rate of heat flow when the air gap is not there.

Solution: Using the nomenclature of Fig. 2.3, we have per m² of the area, $h_A = 45$, and $R_A = 1/h_A = 1/45 = 0.0222$; $h_B = 20$,   and $R_B = 1/20 = 0.05$
Resistance of the refractory brick, $R_1 = L_1/k_1 = 0.15/1.6 = 0.0937$
Resistance of the insulating brick, $R_3 = L_3/k_3 = 0.15/0.30 = 0.50$
The resistance of the air gap, $R_2 = 0.16$
Resistance of the plaster, $R_4 = 0.01/0.14 = 0.0714$
Total resistance = 0.8973, m²K/W

Heat flow rate $= \Delta T / \Sigma R = (1250 - 25)/0.8973 = 1366.2$ W/m²

Temperature at the inner surface of the wall $= T_A - 1366.2 \times 0.0222 = 1222.25$
Temperature at the outer surface of the wall $= T_B + 1366.2 \times 0.05 = 93.31$°C
When the air gap is not there, the total resistance would be
$$0.8973 - 0.16 = 0.7373$$
and the heat flow rate = (1250 – 25) /0/7373 = 1661.46 W/m²
The temperature at the inner surface of the wall = 1250 – 1660.46 × 0.0222
$$= 1213.12°C$$

i.e., when the air gap is not there, the heat flow rate increases but the temperature at the inner surface of the wall decreases.
The overall heat transfer coefficient U with and without the air gap is

$$U = (\dot{Q}/A)/\Delta T$$

$$= 1366.2 /(1250 - 25) = 1.115 \; Wm^2\,°C$$

and $1661.46/1225 = 1.356$ W/m²°C

The equivalent thermal conductivity of the system without the air gap
$$k_{eq} = (0.15 + 0.15 + 0.01)/(0.0937 + 0.50 + 0.0714)$$
$$= 0.466 \; W/mK.$$

Example 2.2   A brick wall (10 cm thick, k = 0.7 W/m°C) has plaster on one side of the wall (thickness 4 cm, k = 0.48 W/m°C). What thickness of an insulating material (k = 0.065 W/m°C) should be added on the other side of the wall such that the heat loss through the wall is reduced by 80 percent.

Solution: When the insulating material is not there, the resistances are:

$$R_1 = L_1/k_1 = 0.1/0.7 = 0.143$$

and $R_2 = 0.04/0.48 = 0.0833$

Total resistance = 0.2263

Let the thickness of the insulating material is L_3. The resistance would then be

$$L_3/0.065 = 15.385\ L_3$$

Since the heat loss is reduced by 80% after the insulation is added.

$$\frac{\dot{Q}\ \text{with insulation}}{\dot{Q}\ \text{without insulation}} = 0.2 = \frac{R\ \text{without insulation}}{R\ \text{with insulation}}$$

or, the resistance with insulation = 0.2263/0.2 = 01.1315

and, $15.385\ L_3 = 1.1315 - 0.2263 = 0.9052$

$$L_3 = 0.0588\ m = 58.8\ mm.$$

Example 2.3   An ice chest is constructed of styrofoam (k = 0.033 W/mK) having inside dimensions 25 by 40 by 100 cm. The wall thickness is 4 cm. The outside surface of the chest is exposed to air at 25°C with h = 10 W/m²K. If the chest is completely filled with ice, calculate the time for ice to melt completely. The heat of fusion for water is 330 kJ/kg.

Solution: If the heat loss through the corners and edges are ignored, we have three pairs of walls through which conduction heat transfer will occur.

(a) 2 walls each having dimensions 25 cm × 40 cm × 4 cm
(b) 2 walls each having dimensions 25 cm × 100 cm × 4 cm
(c) 2 walls each having dimensions 40 cm × 100 cm × 4 cm

The surface area for convection heat transfer (based on outside dimensions)

$$2\ (33 \times 48 + 33 \times 108 + 48 \times 108) \times 10^{-4} = 2.0664\ m^2$$

Resistance due to conduction and convection can be written as

$$2\left(\frac{0.04}{0.033 \times 0.25 \times 0.4} + \frac{0.04}{0.033 \times 0.25 \times 1} + \frac{0.04}{0.033 \times 0.4 \times 1}\right) + \frac{1}{10 \times 2.0664}$$

$$= 40 + 0.0484 = 40.0484\ K/W$$

$$\dot{Q} = \Delta T/\Sigma R$$

$$= (25 - 0.0)/40.0484 = 0.624\ W$$

Inside volume of the container = 0.25 × 0.4 × 1 = 0.1 m³
Mass of ice stored = 800 × 0.1 = 80 kg; taking the density of ice as 800 kg/m³. The time required to melt 80 kg of ice is

$$t = \frac{80 \times 330 \times 1000}{0.624 \times 3600 \times 24} = 490\ \text{days.}$$

Example 2.4   A composite furnace wall is to be constructed with two layers of materials (k_1= 2.5 W/m°C and k_2 = 0.25 W/m°C), The convective heat transfer coefficient at the inside and outside surfaces are expected to be 250 W/m²°C and 50 W/m²°C respectively. The temperature of gases and air are 1000 K and 300 K. If the interface temperature is 650 K, Calculate (i) the thicknesses of the two materials when the total thickness does not exceed 65 cm and (ii) the rate of heat flow. Neglect radiation.

Solution: Let the thickness of one material (k = 2.5 W/mK) is x m, then the thickness of the other material (k = 0.25 W/mK) will be (0.65 – x)m.

For steady state condition, we can write

$$\frac{\dot{Q}}{A} = \frac{1000-650}{\dfrac{1}{250}+\dfrac{x}{2.5}} = \frac{1000-300}{\dfrac{1}{250}+\dfrac{x}{2.5}+\dfrac{(0.65-x)}{0.25}+\dfrac{1}{50}}$$

$\therefore$ 700 (0.004 + 0.4x) = 350 {0.004 + 0.4x + 4 (0.65 – x) + 0.02}

(i)   6x = 3.29 and x = 0.548 m

and the thickness of the other material = 0.102 m.

(ii) $\dot{Q}$ /A = (350)/ (0.004 + 0.4 × 0.548) = 1.568 kW/m².

Example 2.5   A composite wall consists of three layers of thicknesses 300 mm, 200 mm and 100 mm with thermal conductivities 1.5, 3.5 and k_3 W/mK respectively. The inside surface is exposed to gases at 1200°C with convection heat transfer coefficient as 30W/m²K. The temperature of air on the other side of the wall is 30°C with convective heat transfer coefficient 10 Wm²K. If the temperature at the outside surface of the wall is 180°C, calculate the temperature at other surface of the wall, the rate of heat transfer and the overall heat transfer coefficient.

Solution: The composite wall and its equivalent thermal circuits is shown in the figure.

The heat energy will flow from hot gases to the cold air through the wall. From the electric circuit, we have

$$\dot{Q}/A = h_2(T_4 - T_0) = 10\times(180-30)$$

$$= 1500 \text{ W/m}^2$$

also, $\dot{Q}/A = h_1(1200-T_1)$

$$T_1 = 1200 - 1500/30 = 1150°C$$

$$\dot{Q}/A = (T_1 - T_2)/L_1/k_1$$

$$T_2 = T_1 - 1500\times0.3/1.5 = 850$$

Similarly, $\dot{Q}/A = (T_2 - T_3)/(L_2/k_2)$

$$T_3 = T_2 - 1500\times0.2/3.5 = 764.3°C$$

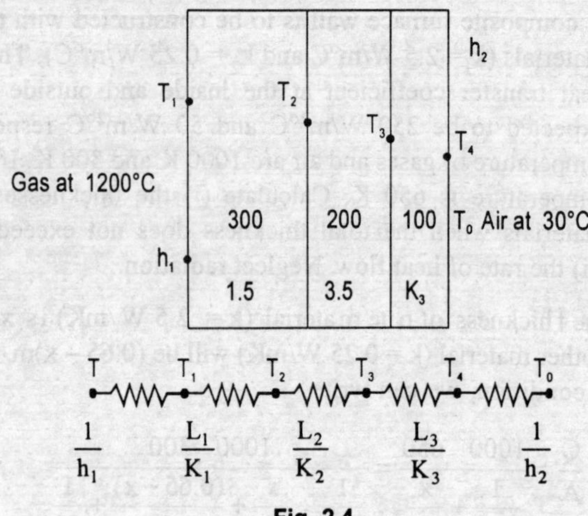

Fig. 2.4

and $\quad \dot{Q}/A = (T_3 - T_4)/(L_3/k_3)$

$\quad L_3/k_3 = (764.3 - 180)/1500$ and $k_3 = 0.256$ W/mK

Check:

$\dot{Q}/A = (1200 - 30)/\sum R$; where $\sum R = 1/h_1 + L_1/k_1 + L_2/k_2 + L_3/k_3 + 1/h_2$

$\quad \Sigma R = 1/30 + 0.3/1.5 + 0.2/3.5 + 0.1/0.256 + 1/10 = 0.75$

and $\quad \dot{Q}/A = 1170/0.78 = 1500$ W/m²

The overall heat transfer coefficient, $U = 1/\sum R = 1/0.78 = 1.282$ W/m²K

Since the gas temperature is very high, we should consider the effects of r adiation a lso. Assuming t he h eat t ransfer c oefficient d ue t o r adiation = 3 0 W/m²K the electric circuit would be:

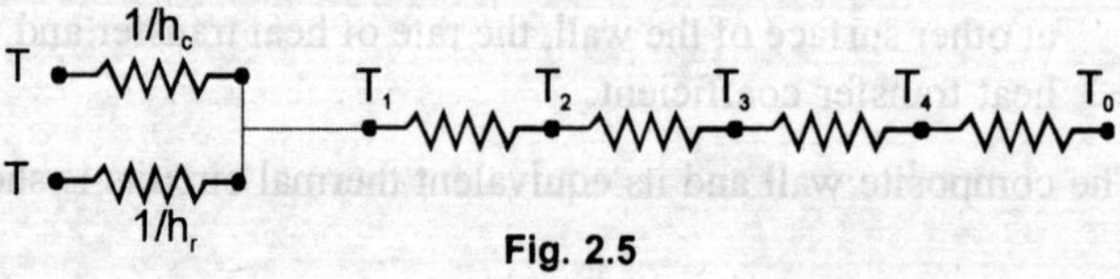

Fig. 2.5

The combined resistance due to convection and radiation would be

$$\frac{1}{R} = \frac{1}{R_1} + \frac{1}{R_2} = \frac{1}{\frac{1}{h_c}} + \frac{1}{\frac{1}{h_r}} = h_c + h_r = 60 \text{ W/m}^2{}^\circ C$$

$\therefore \quad \dot{Q}/A = 1500 = 60(T - T_1) = 60(1200 - T_1)$

$\therefore \quad T_1 = 1200 - \dfrac{1500}{60} = 1175°C$

again, $\therefore \quad \dot{Q}/A = (T_1 - T_2)/L_1/k_1 \Rightarrow T_2 = T_1 - 1500 \times 0.3/1.5 = 875°C$

and $\quad T_3 = T_2 - 1500 \times 0.2/3.5 = 789.3°C$

$\quad L_3/k_3 = (789.3 - 180)/1500; \quad \therefore k_3 = 0.246$ W/mK

$$\Sigma R = \frac{1}{60} + \frac{0.3}{1.5} + \frac{0.2}{3.5} + \frac{0.1}{0.246} + \frac{1}{10} = 0.78$$

and
$$U = 1 / \Sigma R = 1.282 \ \text{W/m}^2\text{K} \cdot$$

Example 2.6   A flat roof (12 m × 20 m) of a building has a composite structure. It consists of a 15 cm lime-khoa plaster covering ($k = 0.17$ W/m°C) over a 10 cm cement concrete ($k = 0.92$ W/m°C). The ambient temperature is 42 °C. The outside and inside heat transfer coefficients are 30 W/m² °C and 10 W/m² °C. The top surface of the roof absorbs 750 W/m² of solar radiant energy. The temperature of the space may be assumed to be 260 K. Calculate the temperature of the top surface of the roof and the amount of water to be sprinkled uniformly over the roof surface such that the inside temperature is maintained at 18°C.

Solution: The physical system is shown in Fig. 2.6 and it is assumed we have one-dimensional flow, properties are constant and steady state conditions prevail.

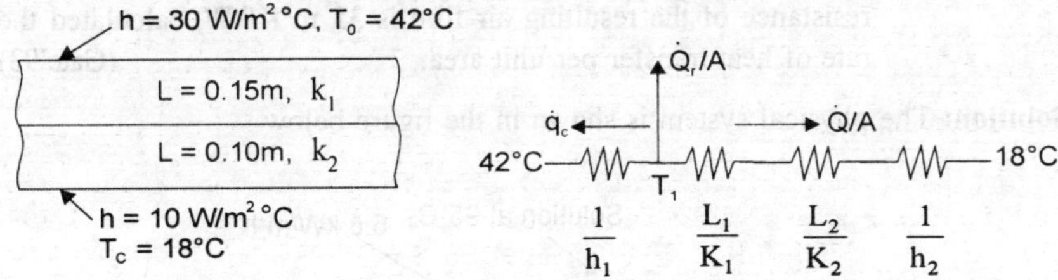

Fig. 2.6

Let the temperature of the top surface be T_1 °C.
Heat lost by the top surface by convection to the surroundings is

$$\dot{Q}_c / A = h(\Delta T) = 30 \times (T_1 - 42) = (30T_1 - 1260)$$

Heat energy conducted inside through the roof = $(\Delta T / \Sigma R)$

or,
$$\frac{\dot{Q}}{A} = \frac{T_1 - 18}{\dfrac{L_1}{k_1} + \dfrac{L_2}{k_2} + \dfrac{1}{h_2}} = (T_1 - 18) \Big/ \left(\frac{0.15}{0.17} + \frac{0.1}{0.92} + \frac{1}{10} \right)$$

$$= 0.918 \, (T_1 - 18)$$

Assuming that the top surface of the roof behaves like a black body, energy lost by radiation

$$\dot{Q}_r / A = \sigma [(T_1 + 273)^4 - 260^4]$$

$$= 5.67 \times 10^{-8} (T_1 + 273)^4 - 259.1$$

By making an energy balance on the top surface of the roof,
Energy coming in = Energy going out

$$750 = (30T_1 - 1260) + 0.918 \, (T_1 - 18) + 5.67 \times 10^{-8} (T_1 + 273)^4 - 259.1$$

or, $$2285.624 = 30.918 \, T_1 + 5.67 \times 10^{-8} (T_1 + 273)^4$$

Solving by trial and error, $T_1 = 53.4°C$, and the total energy conducted through the roof per hour is

$$0.918 (53.4 - 18) \times (12 \times 20) \times 3600 = 28077.58 \text{ kJ/hr}$$

Assuming the latent heat of vaporization of water as 2430 kJ/kg, the quantity of water to be sprinkled over the surface such that it evaporates and consumes 28077.58 kJ/hr, is

$$\dot{M}_W = 28077.58/2430 = 11.55 \text{ kg/hr}.$$

Example 2.7 An electric hot plate is maintained at a temperature of 350°C and is used to keep a solution boiling at 95°C. The solution is contained in a cast iron vessel (wall thickness 25 mm, k = 50 W/mK) which is enamelled inside (thickness 0.8 mm, k = 1.05 W/mK). The heat transfer coefficient for the boiling solution is 5.5 kW/m²K. Calculate (i) the overall heat transfer coefficient and (ii) heat transfer rate.

If the base of the cast iron vessel is not perfectly flat and the resistance of the resulting air film is 35 m²K/kW, calculated the rate of heat transfer per unit area. (Gate'93)

Solution: The physical system is shown in the figure below.

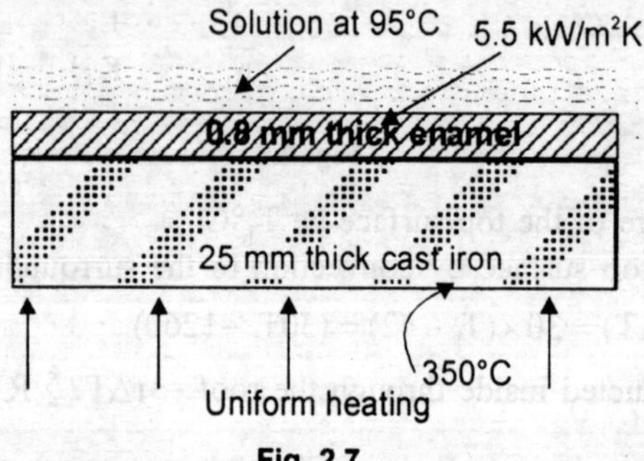

Fig. 2.7

Under steady state conditions,

$$\dot{Q}/A = U(\Delta T) = \frac{(\Delta T)}{1/U}, \text{ where U is the overall heat transfer coefficient.}$$

$$= \frac{(\Delta T)}{R} = \frac{(\Delta T)}{\dfrac{L_1}{k_1} + \dfrac{L_2}{k_2} + \dfrac{1}{h}}$$

Therefore,

$$1/U = \frac{L_1}{k_1} + \frac{L_2}{k_2} + \frac{1}{h}$$

$$= \left(\frac{0.025}{50} + \frac{0.0008}{1.05} + \frac{1}{5500} \right) = 0.00144$$

$$U = 692.65 \ \text{W/m}^2\text{K}$$

$$\dot{Q}/A = U(\Delta T) = 692.65 \times (350 - 95) = 176.65 \ \text{kW/m}^2$$

With the presence of air film at the base, the total resistance to heat flow would be:

$$0.00144 + 0.035 = 0.03644 \ \text{m}^2\text{K/W}$$

and the rate of heat transfer, $\dot{Q}/A = 255/0.03644 = 7 \ \text{kW/m}^2$.

(Fig. 2.8 shows a combination of thermal resistance placed in series and parallel for a composite wall having one-dimensional steady state heat transfer. By drawing analogous electric circuits, we can solve such complex problems having both parallel and series thermal resistances.)

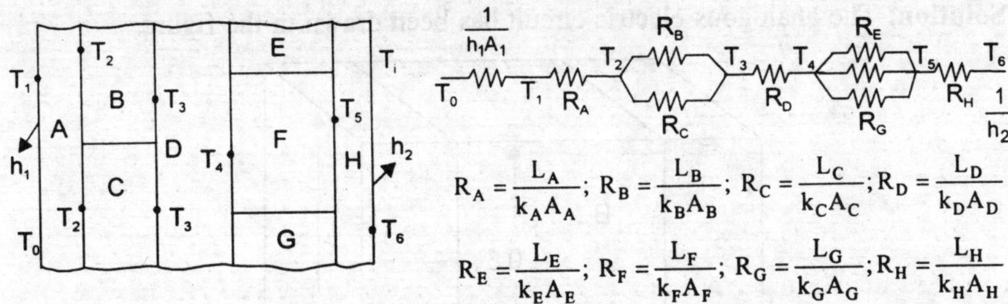

$$R_A = \frac{L_A}{k_A A_A}; \ R_B = \frac{L_B}{k_B A_B}; \ R_C = \frac{L_C}{k_C A_C}; \ R_D = \frac{L_D}{k_D A_D}$$

$$R_E = \frac{L_E}{k_E A_E}; \ R_F = \frac{L_F}{k_F A_F}; \ R_G = \frac{L_G}{k_G A_G}; \ R_H = \frac{L_H}{k_H A_H}$$

Fig. 2.8 Series and parallel one-dimensional heat transfer through a composite wall with convective heat transfer and its electrical analogous circuit

Example 2.8 A door (2 m × 1 m) is to be fabricated with 4 cm thick card board (k = 0.2 W/mK) placed between two sheets of fibre glass board (each having a thickness of 40 mm and k = 0.04 W/mK). The fibre glass boards are fastened with 50 steel studs (25 mm diameter, k = 40 W/mK). Estimate the percentage of heat transfer flow rate through the studs.

Solution: The thermal circuit with steel studs can be drawn as in Fig. 2.9.

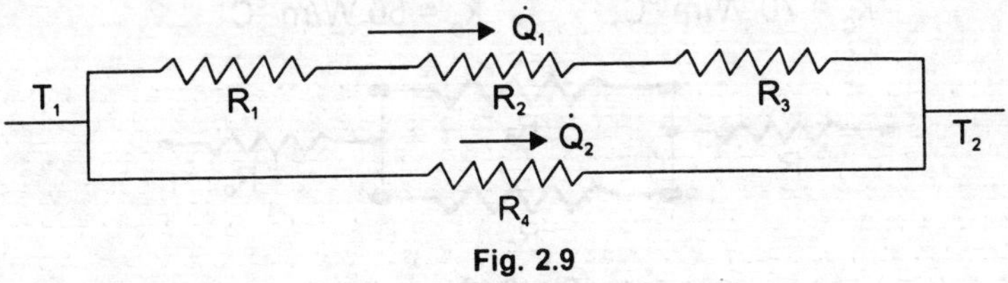

Fig. 2.9

The cross-sectional area or the surface area of the door for the heat transfer is $2 \ \text{m}^2$. The cross-sectional area of the steel studs is

$$50 \times \pi/4 \ (0.025)^2 = 0.02455 \ \text{m}^2$$

and the area of the door − area of the steel studs = 2.0 − 0.02455

$$= 1.97545$$

R_1, the resistance due to fibre glass board on the outside = L/kA

$$= 0.04/(0.04 \times 1.97545)$$

$$= 0.506$$

R_2, the resistance due to card board = 0.101

R_3, the resistance due to fibre glass board on the inside = 0.506

R_4, the resistance due to steel studs = L/kA = 0.12/ (40 × 0.2455) = 0.1222

With reference to Fig 2.9, $\dot{Q}_1 = (T_1 - T_2)/\Sigma R = (T_1 - T_2)/1.113$

and, $\qquad\qquad\qquad \dot{Q}_2 = (T_1 - T_2)/0.1222$

Therefore, $\qquad \dot{Q}_2/(\dot{Q}_1 + \dot{Q}_2) = 8.1833/9.0818 = 0.9$

i.e., 90 percent of the heat transfer will take place through the studs.

Example 2.9 Find the heat transfer rate per unit depth through the composite wall sketched. Assume one dimensional heat flow.

Solution: The analogous electric circuit has been drawn in the figure.

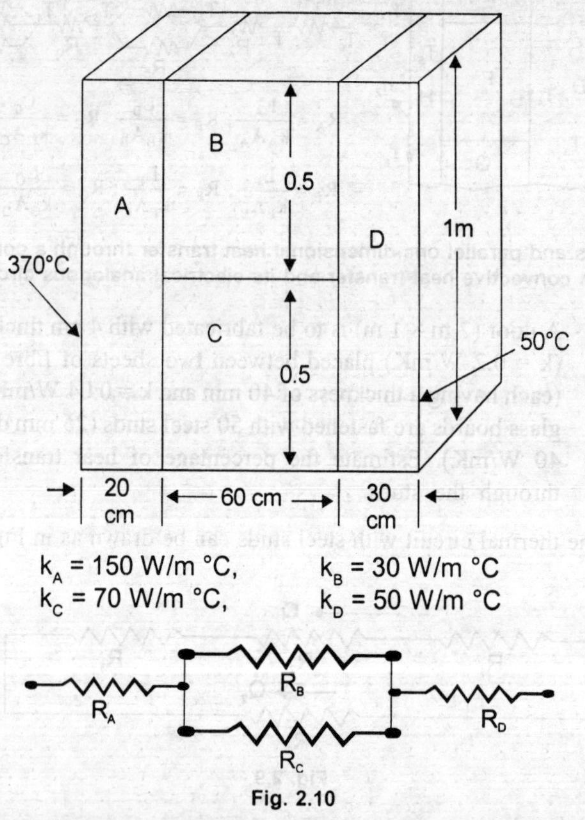

Fig. 2.10

$R_A = 0.2/150 = 0.00133$

$R_B = 0.6/(30 × 0.5) = 0.04$

$R_C = 0.6/(70 × 0.5) = 0.017$

$R_D = 0.3/50 = 0.006$

$1/R_B + 1/R_C = 1/R_{BC} = 83.82$

Therefore, $R_{BC} = 1/83.82 = 0.0119$

Total resistance to heat flow = 0.00133 + 0.0119 + 0.006 = 0.01923

Rate of heat transfer per unit depth = (370-50)/ 0.01923 = 16.64 kW/m.

10. The Significance of Biot Number

Let us consider steady state conduction through a slab of thickness L and thermal conductivity k. The left hand face of the wall is maintained at constant temperature T_1 and the right hand face is exposed to ambient air at T_0, with convective heat transfer coefficient h. The analogous electric circuit will have two thermal resistances: R_1 = L/k and R_2 = l/h. The drop in temperature across the wall and the air film will be proportional to their resistances, that is, (L/k)/(1/h) = hL/k.

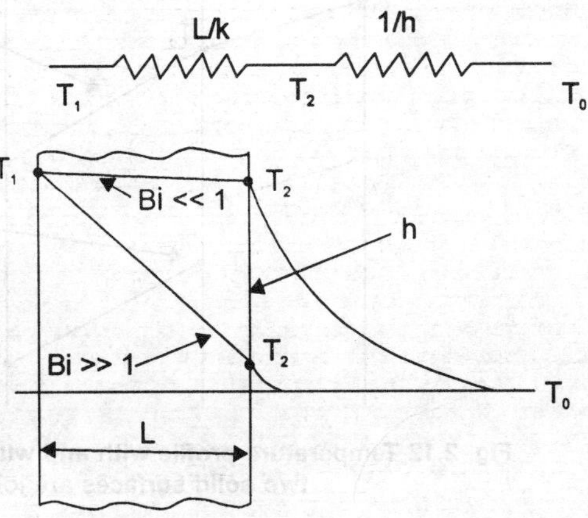

Fig. 2.11 Effect of Biot number on temperature profile

This dimensionless number is called 'Biot Number'. or,

$$Bi = \frac{Conduction\ resistance}{Convection\ resistance}$$

When Bi >> 1, the temperature drop across the air film would be negligible and the temperature at the right hand face of the wall will be approximately equal to the ambient temperature. Similarly, when Bi <<1, the temperature drop across the wall is negligible and the transfer of heat will be controlled by the air film resistance.

11. The Concept of Thermal Contact Resistance

Heat flow rate through composite walls are usually analysed on the assumptions that – (i) there is a perfect contact between adjacent layers, and (ii) the temperature at the interface of the two plane surfaces is the same. However, in real situations, this is not true. No surface, even a so-called 'mirror-finish surface', is perfectly smooth in a microscopic sense. As such, when two surfaces are placed together, there is not a single plane of contact. The surfaces touch only at limited number of spots, the aggregate of which is only a small fraction of the area of the surface or 'contact area'. The remainder of the space between the surfaces may be filled with air or other fluid. In effect, this introduces a resistance to heat flow at the interface. This resistance is called 'thermal contact resistance' and causes a temperature drop between the materials at the interfaces as shown in Fig. 2.12. (That is why, Eskimos make their houses having double ice walls separated by a thin layer of air, and in winter, two thin woolen blankets are more comfortable than one woolen blanket having double thickness.)

Example 2.10 A furnace wall consists of an inner layer of fire brick 25 cm thick k = 0.4 W/mK and a layer of ceramic blanket insulation, 10 cm thick, k = 0.2 W/mK. The thermal contact resistance between the

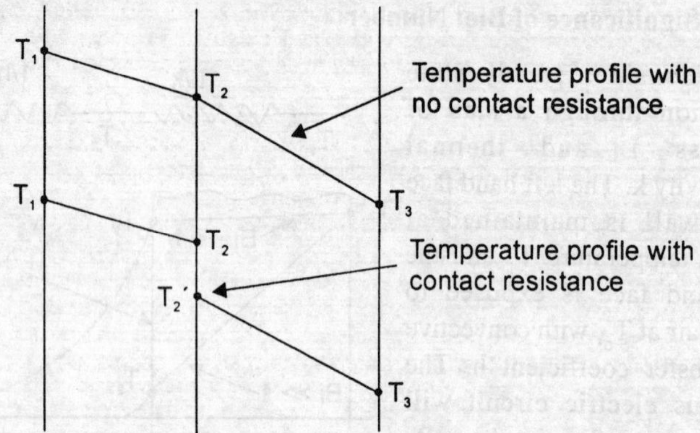

Fig. 2.12 Temperature profile with and without contact resistance when two solid surfaces are joined together

two walls at the interface is 0.01 m²K/W. Calculate the temperature drop at the interface if the temperature difference across the wall is 1200K.

Solution: The resistance due to inner fire brick = L/k = 0.25/0.4 = 0.625
The resistance of the ceramic insulation = 0.1/0.2 = 0.5
Total thermal resistance = 0.625 + 0.01 + 0.5 = 1.135

Rate of heat flow, $\dot{Q}/A = \Delta T/\Sigma R = 1200/1.135 = 1057.27 \text{ W/m}^2$

Temperature drop at the interface, $\Delta T = (\dot{Q}/A) \times R = 1057.27 \times 0.01 = 10.57 \text{ K}$.

Example 2.11 A 20 cm thick slab of aluminium (k = 230 W/mK) is placed in contact with a 15 cm thick stainless steel plate (k = 15 W/mK). Due to roughness, 40 percent of the area is in direct contact and the gap (0.0002 m) is filled with air (k = 0.032 W/mK). The difference in temperature between the two outside surfaces of the plate is 200°C. Estimate (i) the heat flow rate, (ii) the contact resistance, and (iii) the drop in temperature at the interface.

Solution: Let us assume that out of 40% area in direct contact, half the surface area is occupied by steel and half is occupied by aluminium.
The physical system and its analogous electric circuits is shown in Fig. 2.13.

$$R_1 = \frac{0.2}{230 \times 1} = 0.00087, \qquad R_2 = \frac{0.0002}{230 \times 0.2} = 4.348 \times 10^{-6}$$

$$R_3 = \frac{0.0002}{0.032 \times 0.6} = 1.04 \times 10^{-2}, \qquad R_4 = \frac{0.0002}{15 \times 0.2} = 6.667 \times 10^{-5}$$

and $\quad R_5 = \dfrac{0.15}{(15 \times 1)} = 0.01$

Again, $\quad 1/R_{2,3,4} = 1/R_2 + 1/R_3 + 1/R_4$

$$= 2.3 \times 10^5 + 96.15 + 1.5 \times 10^4 = 24.5 \times 10^4$$

Therefore, $\quad R_{2,3,4} = 4.08 \times 10^{-6}$

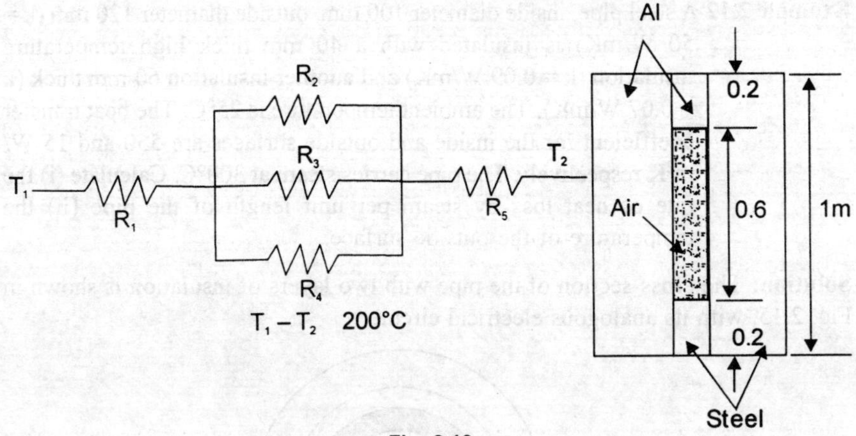

Fig. 2.13

Total resistance, $\Sigma R = R_1 + R_{2,3,4} + R_5$

$$= 870 \times 10^{-6} + 4.08 \times 10^{-6} + 10000 \times 10^{-6} = 1.0874 \times 10^{-2}$$

Heat flow rate, $\dot{Q} = 200/1.087 \times 10^{-2} = 18.392$ kW per unit depth of the plate

Contact resistance, $R_{2,3,4} = 4.08 \times 10^{-6}$ mK/W

Drop in temperature at the interface, $\Delta T = 4.08 \times 10^{-6} \times 18392 = 0.075°C$.

12. An Expression for the Heat Transfer Rate through a Composite Cylindrical System

Let us consider a composite cylindrical system consisting of two coaxial cylinders, radii r_1, r_2, and r_2 and r_3, thermal conductivities k_1 and k_2, the convective heat transfer coefficients at the inside and outside surfaces h_1 and h_2 as shown in the figure. Assuming radial conduction under steady state conditions we have:

$R_1 = 1/h_1 A_1 = 1/2\pi r_1 L h_1$

$R_2 = \ln(r_2/r_1)2\pi L k_1$

$R_3 = \ln(r_3/r_2)2\pi L k_2$

$R_4 = 1/h_2 A_2 = 1/2\pi r_3 h_2 L$

And $\dot{Q}/2\pi L = (T_1 - T_0)/\Sigma R$

$$= (T_1 - T_0)/\left[(1/h_1 r_1 + \ln(r_2/r_1)/k_1 + \ln(r_3/r_2)/k_2 + 1/h_2 r_3)\right]$$

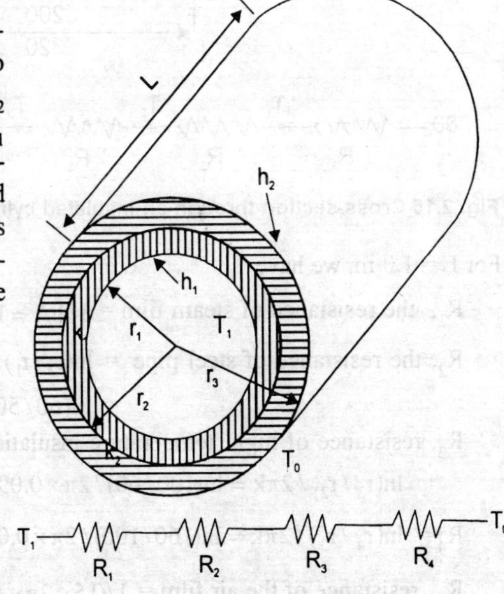

Fig. 2.14

Example 2.12 A steel pipe, inside diameter 100 mm, outside diameter 120 mm (k = 50 W/mK) is insulated with a 40 mm thick high temperature insulation (k = 0.09 W/mK) and another insulation 60 mm thick (k = 0.07 W/mK). The ambient temperature is 25°C. The heat transfer coefficient for the inside and outside surfaces are 550 and 15 W/m²K respectively. The pipe carries steam at 300°C. Calculate (i) the rate of heat loss by steam per unit length of the pipe (ii) the temperature of the outside surface.

Solution: The cross-section of the pipe with two layers of insulation is shown in Fig. 2.15, with its analogous electrical circuit.

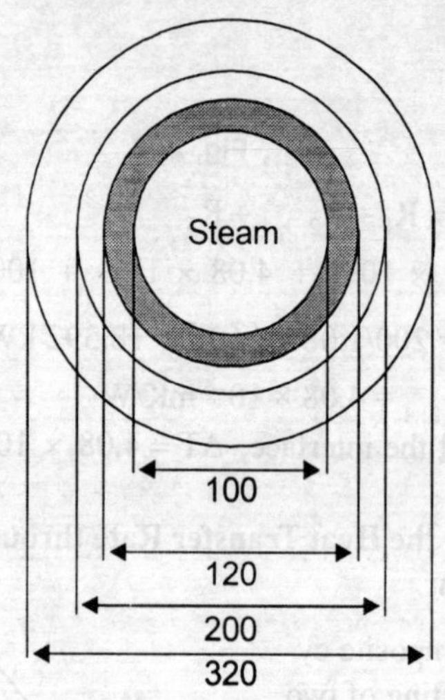

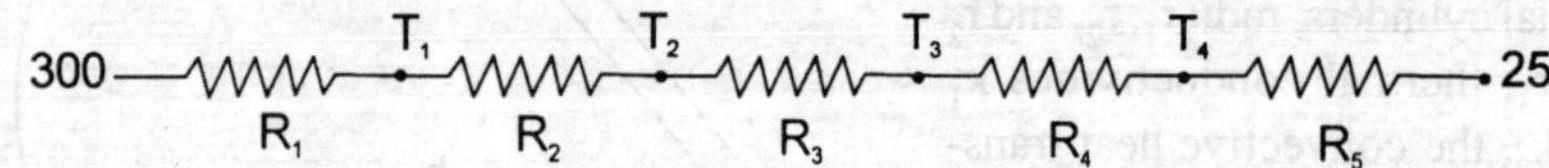

Fig. 2.15 Cross-section through an insulated cylinder, thermal resistances in series

For L = 1.0 m, we have

R_1, the resistance of steam film = $1/hA = 1/(500 \times 2\pi \times 50 \times 10^{-3}) = 0.00579$

R_2, the resistance of steel pipe = $\ln(r_2/r_1)/2\pi k$

$$= \ln(60/50)/2\pi \times 50 = 0.00058$$

R_3, resistance of high temperature insulation

$$\ln(r_3/r_2)/2\pi k = \ln(100/60)/2\pi \times 0.09 = 0.903$$

$R_4 = \ln(r_4/r_3)/2\pi k = \ln(160/100)/2\pi \times 0.07 = 1.068$

R_5, resistance of the air film = $1/(15 \times 2\pi \times 160 \times 10^{-3}) = 0.0663$

The total resistance = 2.04367

and $\dot{Q} = \Delta T/\Sigma R = (300 - 25)/2.04367 = 134.56$ W per metre length of pipe.

Temperature at the outside surface, $T_4 = 25 + R_5\dot{Q} = 25 + 134.56 \times 0.0663$
$$= 33.92°C$$

When the better insulating material (k = 0.07, thickness 60 mm) is placed first on the steel pipe, the new value of R_3 would be

$R_3 = \ln(120/60)/2\pi \times 0.07 = 1.576$; and the new value of R_4 will be

$R_4 = \ln(160/120)/2\pi \times 0.09 = 0.5087$

The total resistance = 2.15737 and $\dot{Q} = 275/2.15737 = 127.47$ W per m length

(Thus the better insulating material be applied first to reduce the heat loss.)

The overall heat transfer coefficient, U, is obtained as $U = \dot{Q}/A\Delta T$

The outer surface area = $\pi \times 320 \times 10^{-3} \times 1 = 1.0054$

and $\qquad\qquad\qquad$ U = 134.56/(275 × 1.0054) = 0.487 W/m^{2}K.

Example 2.13 A steam pipe 120 mm outside diameter and 10 m long carries steam at a pressure of 30 bar and 0.99 dry. Calculate the thickness of a lagging material (k = 0.99 W/mK) provided on the steam pipe such that the temperature at the outside surface of the insulated pipe does not exceed 32°C when the steam flow rate is 1 kg/s and the dryness fraction of steam at the exit is 0.975 and there is no pressure drop.

Solution: The latent heat of vaporization of steam at 30 bar = 1794 kJ/kg.

The loss of heat energy due to condensation of steam $\qquad$ = 1794(0.99 – 0.975)
$$= 26.91 \text{ kJ/kg.}$$

Since the steam flow rate is 1 kg/s, the loss of energy $\qquad$ = 26.91 kW.

The saturation temperature of steam at 30 bar is 233.84°C and assuming that the pipe material offers negligible resistance to heat flow, the termperature at the outside surface of the uninsulated steam pipe or at the inner surface of the lagging material is 233.84°C. Assuming one-dimensional radial heat flow through the lagging material, we have

$$\dot{Q} = (T_1 - T_2)/[\ln(r_2/r_1)] \, 2\pi L k$$

or, 26.91 × 1000 (W) = (233.84 – 32) × 2π × 10 × 0.99/ln(r_2/60)

$\qquad$ ln (r_2/60) = 0.4666

$\qquad\qquad$ r_2/60 = exp (0.4666) = 1.5946

$\therefore$ $\qquad\qquad$ r_2 = 95.68 mm and the thickness = 35.68 mm.

Example 2.14 A wire, diameter 0.5 mm length 30 cm, is laid coaxially in a tube (inside diameter 1 cm, outside diameter 1.5 cm, k = 20 W/mK). The space between the wire and the inside wall of the tube behaves like a hollow tube and is filled with a gas. Calculate the thermal conductivity of the gas if the current flowing through the wire is 5 amps and voltage across the two ends is 4.5 V, temperature of the wire is 160°C, convective heat transfer coefficient at the outer surface of the tube is 12 W/m^{2}K and the ambient temperature is 300 K.

Solution: Assuming steady state and one-dimensional radial heat flow, we can draw the thermal circuit as shown in Fig, 2.16.

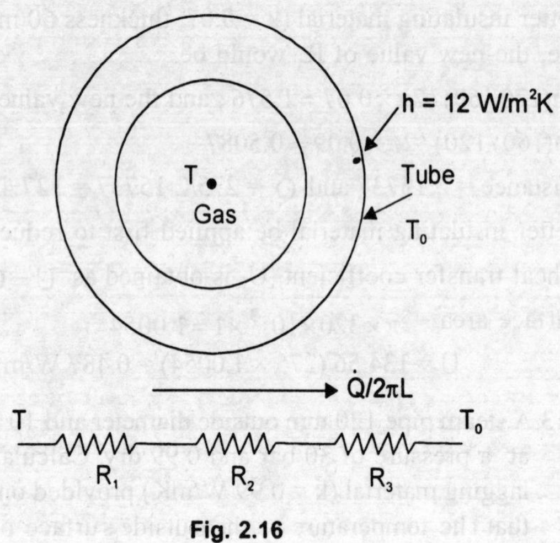

Fig. 2.16

The rate of heat transfer through the system,

$$\dot{Q}/2\pi L = VI/2\pi L = (4.5\times5)/(2\times3.142\times0.3) = 11.935 \ (W/m)$$

R_1, the resistance due to gas $= \ln(r_2/r_1)/k$

$$= \ln(0.01/0.0005)/k = 2.996/k$$

R_2, resistance offered by the matallic tube $= \ln(r_3/r_2)/k$

$$= \ln(1.5/1.0)/20 = 0.02$$

R_3, resistance due to fluid film at the outer surface

$$1/hr_3 = 1/(12\times1.5\times10^{-2}) = 5.556$$

and $\dot{Q}/2\pi L = \Delta T/\Sigma R = [(273 + 160) - 300]/\Sigma R$

Therefore, $\Sigma R = 133/11.935 = 11.1437$, and

$$R_1 = 2.9996/k = 11.1437 - 0.02 - 5.556 = 5.568$$

or, $\qquad k = 2.996/5.568 = 0.538 \ W/mK.$

Example 2.15 A steam pipe (inner diameter 16 cm, outer diameter 20 cm, k = 50 W/mK) is covered with a 4 cm thick insulating material (k = 0.09 W/mK). In order to reduce the heat loss, the thickness of the insulation is increased to 8mm. Calculate the percentage reduction in heat transfer assuming that the convective heat transfer coefficient at the inside and outside surfaces are 1150 and 10 W/m²K and their values remain the same.

Solution: Assuming one-dimensional radial conduction under steady state,

$$\dot{Q}/2\pi L = \Delta T/\Sigma R$$

R_1, resistance due to steam film $= 1/hr = 1/(1150 \times 0.08) = 0.011$

R_2, resistance due to pipe material $= \ln (r_2/r_1)/k$
$$= 1n\ (10/8)/50\ = 0.00446$$

R_3, resistance due to 4 cm thick insulation $= \ln (r_3/r_2)/k$
$$= \ln (14/10)/0.09 = 3.738$$

R_4, resistance due to air film $= 1/hr = 1/(10 \times 0.14) = 0.714$

Therefore, $\dot{Q}/2\pi L = \Delta T/(0.011 + 0.00446 + 3.738 + 0.714) = 0.22386\Delta T$

When the thickness of the insulation is increased to 8 cm, the values of R_3 and R_4 will change.
$$R_3 = \ln (r_3/r_2)/k = \ln (18/10)/0.09 = 6.53;\ \text{and}$$
$$R_4 = 1/hr = 1/(10 \times 0.18) = 0.556$$

Therefore, $\dot{Q}/2\pi L = \Delta T/(0.011 + 0.00446 + 6.53 + 0.556)$
$$= \Delta T/7.1 = 0.14084\Delta T$$

Percentage reduction in heat transfer $= \dfrac{(0.22386 - 0.14084)}{0.22386} = 0.37 \equiv 37\%$.

Example 2.16 A small hemispherical oven is built of an inner layer of insulating fire brick 125 mm thick (k = 0.31 W/mK) and an outer covering of 85% magnesia 40 mm thick (k = 0.05 W/mK). The inner surface of the oven is at 1073 K and the heat transfer coefficient for the outer surface is 10 W/m²K, the room temperature is 20°C. Calculate the rate of heat loss through the hemisphere if the inside radius is 0.6m.

Solution: The resistance of the fire brick $= (r_2 - r_1)/2\pi k r_1 r_2$
$$= \frac{0.725 - 0.6}{2\pi \times 0.31 \times 0.6 \times 0.725} = 0.1478$$

The resistance of 85% magnesia $= (r_3 - r_2)/2\pi k r_2 r_3$
$$= \frac{0.765 - 0.725}{2\pi \times 0.05 \times 0.725 \times 0.765} = 0.2295$$

The resistance due to fluid film at the outer surface $= 1/hA$
$$= \frac{1}{10 \times 2\pi \times (0.765 \times 0.765)} = 0.0272$$

Rate of heat flow, $\dot{Q} = \Delta T/\Sigma R = \dfrac{800 - 20}{0.1478 + 0.2295 + 0.272} = 1930$ W.

Example 2.17 A cylindrical tank with hemispherical ends is used to store liquid oxygen at $- 183°C$. The diamater of the tank is 1.5 m and the total length is 8 m. The tank is covered with a 10 cm thick layer of insulation. Determine the thermal conductivity of the insulating material so that the boil off rate does not exceed 10 kg/hr. The latent heat of vapourization of liquid oxygen is 214 kJ/kg. Assume that the outer surface of insulation is at 27°C and the thermal resistance of the wall of the tank is negligible.   (ES-94)

Solution: The maximum amount of heat energy that flows by conduction from outside to inside = Mass of liquid oxygen × Latent heat of vaporization

$$= 10 \times 214 = 2140 \text{ kJ/hr} = 2140 \times 1000/3600 = 594.44 \text{ W}$$

Length of the cylindrical part of the tank $= 8 - 2r = 8 - 1.5 = 6.5\text{m}$

since the thermal resistance of the wall does not offer any resistance to heat flow, the temperature at the inside surface of the insulation can be assumed as $-183°C$ whereas the temperature at the outside surface of the insulation is $27°C$.

Heat energy coming in through the cylindrical part, $\dot{Q}_1 = \dfrac{\Delta T}{\dfrac{\ln(r_2 / r_1)}{2\pi Lk}}$

or, $\qquad \dot{Q}_1 = \dfrac{(27+183)\times 2\pi \times 6.5k}{\ln(8.5/7.5)} = 68531.84 \text{ k}$

Heat energy coming in through the two hemispherical ends,

$$\dot{Q}_2 = 2\times(\Delta T \times 2\pi k \; r_2 r_1 /(r_2 - r_1)) = \frac{2\times 210 \times 2\pi k \times 0.85 \times 0.75}{0.10}$$

$$= 16825.4 \text{ k}$$

Therefore, $594.44 = (68531.84 + 16825.4)$ k; or, $k = 6.96 \times 10^{-3}$ W/mK.

Example 2.18 A spherical vessel, made out of 2.5 cm thick steel plate is used to store 10 m³ of a liquid at 200°C for a thermal storage system. To reduce the heat loss to the surroundings, a 10 cm thick layer of insulation (k = 0.07 W/mK) is used. If the convective heat transfer coefficient at the outer surface is 10 W/m²K and the ambient temperature is 25°C, calculate the rate of heat loss neglecting the thermal resistance of the steel plate.

If the spherical vessel is replaced by a 2 m diameter cylindrical vessel with flat ends, calculate the thickness of insulation required for the same heat loss.

Solution: Volume of the spherical vessel $= 10 \text{ m}^3 = \dfrac{4\pi r^3}{3}$ $\qquad \therefore r = 1.336 \text{ m}$

Outer radius of the spherical vessel, $r_2 = 1.3363 + 0.025 = 1.361\text{m}$
Outermost radius of the spherical vessel after the insulation $= 1.461 \text{ m}$
Since the thermal resistance of the steel plate is negligible, the temperature at the inside surface of the insulation is 200°C.
Thermal resistance of the insulating material $= (r_3 - r_2)/4\pi k \; r_3 r_2$

$$= \frac{0.1}{4\pi \times 0.07 \times 1.461 \times 1.361} = 0.057$$

Thermal resistance of the fluid film at the outermost surface $= 1/hA$

$$= 1/[10 \times 4\pi \times (1.461)^2] = 0.00373$$

Rate of heat flow $= \Delta T / \Sigma R = (200 - 25)/(0.057 + 0.00373)$
$$= 2873.8 \text{ W}$$

Volume of the insulating material used $= = (4/3)\pi(r_3^3 - r_2^3) = 2.5m^3$

Volume of the cylindrical vessel $= 10m^3 = \dfrac{\pi}{4}(d)^2 L$; $\therefore L = 10/\pi = 3.183$ m

Outer radius of cylinder without insulation $= 1.0 + 0.025 = 1.025$ m
Outermost radius of the cylinder (with insulation) $= r_3$
Therefore, the thickness of insulation $= r_3 - 1.025 = \delta$
Resistance, the heat flow by the cylnderical element

$$= \frac{\ln(r_3/1.025)}{2\pi Lk} + 1/hA = \frac{\ln(r_3/1.025)}{2\pi \times 3.183 \times 0.07} + \frac{1}{10 \times 2\pi \times r_3 \times 3.183}$$

$$= 0.714\ \ln\,(r_3/1.025) + 0.005/r_3$$

Resistance to heat flow through sides of the cylinder

$$= 2\delta/kA + 1/hA = \frac{2(r_3 - 1.025)}{0.07 \times \pi \times 1} + \frac{1}{10 \times 2 \times \pi}$$

$$= 9.09\ (r_3 - 1.025) + 0.0159$$

For the same heat loss, $\Delta T/\Sigma R$ would be equal in both cases, therefore,

$$\frac{1}{0.06073} = \frac{1}{0.714\ \ln(r_3/1.025) + 0.005/r_3} + \frac{1}{9.09(r_3 - 1.025) + 0.0159}$$

Solving by trial and error, $(r_3 - 1.025) = \delta = 9.2$ cm
and the volume of the insulating material rquired $= 2.692\ m^3$.

13. Concept of Critical Thickness of Insulation

The addition of insulation at the outside surface of small pipes may not reduce the rate of heat transfer. When an insulation is added on the outer surface of a bare pipe, its outer radius, r_0, increases and this increases the thermal resistance due to conduction logarithmically whereas the thermal resistance to heat flow due to fluid film on the outer surface decreases linearly with increasing radius, r_0. Since the total thermal resistance is proportional to the sum of these two resistances, the rate of heat flow may not decrease as insulation is added to the bare pipe.

Fig. 2.17 shows a plot of heat loss against the insulation radius for two different cases. For small pipes or wires, the radius r_1 may be less than r_c and in that case, addition of insulation to the bare pipe will increase the heat loss until the critical radius is reached. Further additon of insulation will decrease the heat loss rate from this peak value. The insulation thickness $(r^* - r_1)$ must be added to reduce the heat loss below the uninsulated rate. If the outer pipe radius r_1 is greater than the critical radius r_c any insulation added will decrease the heat loss.

14. Expression for Critical Thickness of Insulation for a Cylindrical Pipe

Let us consider a pipe, outer radius r_1, as shown in Fig. 2.18. An insulation is added such that the outermost radius is r, a variable, and the insulation thickness is $(r - r_1)$. We assume that the thermal conductivity, k, for the insulating material

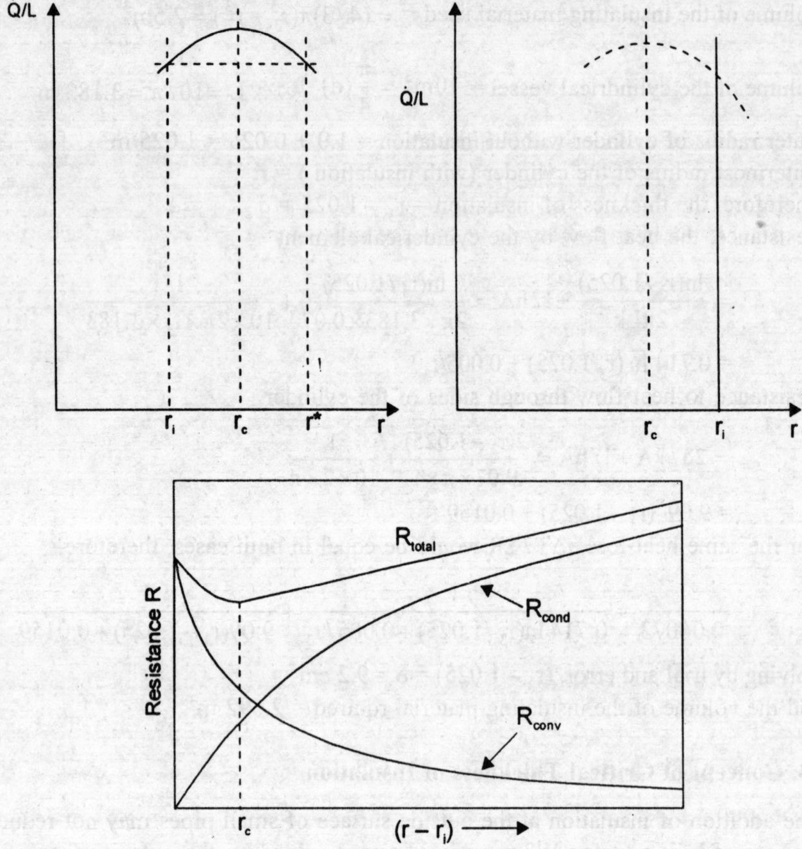

Fig. 2.17 Critical thickness for pipe insulation

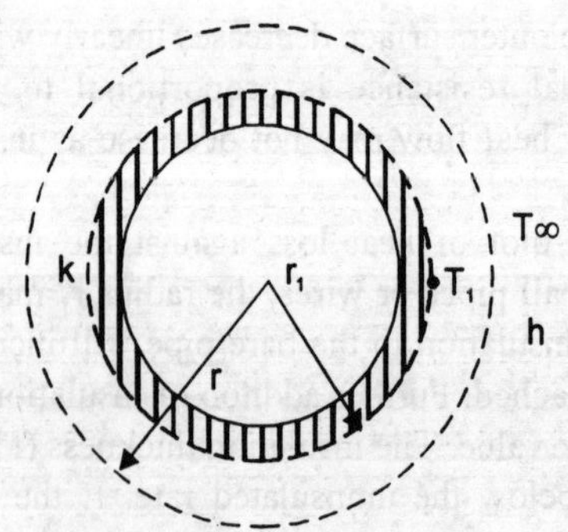

Fig. 2.18 Critical thickness of insulation for a pipe

is very small in comparison with the thermal conductivity of the pipe material and as such the temperature T_1, at the inside surface of the insulation is constant. It is further assumed that both h and k are constant. The rate of heat flow, per unit length of pipe, through the insulation is then,

$\dot{Q}/L = 2\pi(T_1 - T_\infty)/(\ln(r/r_1)/k + 1/hr)$, where T_∞ is the ambient temperature.

An optimum value of the heat loss is found by setting $\dfrac{d(\dot{Q}/L)}{dr} = 0$.

or,
$$\frac{d(\dot{Q}/L)}{dr} = 0 = -\frac{2\pi(T_1 - T_\infty)(1/kr - 1/hr^2)}{(\ln(r/r_1)/k + 1/hr^2}$$

or,
$$(1/kr) - (1/hr^2) = 0, \text{ and } r = r_c = k/h \tag{2.10}$$

where r_c denote the 'critical radius' and depends only on thermal quantities k and h.

If we evaluate the second derivative of (Q/L) at $r = r_c$, we get

$$\left.\frac{d^2(\dot{Q}/L)}{dr^2}\right|_{r=r_c} = -2\pi(T_1 - T_\infty)\left[\frac{\dfrac{k}{hr} + \ln\left(\dfrac{r}{r_1}\right)\left(\dfrac{2k}{hr} - 1\right) - 2\left(1 - \dfrac{k}{hr}\right)^2}{\dfrac{1}{kr}\left(\dfrac{k}{h} + r\ln\left(\dfrac{r}{r_1}\right)\right)}\right]_{r=r_c}$$

$$= -[2\pi(T_1 - T_\infty)h^2/k]/[1 + \ln r_c/r_1]^2$$

which is always a negative quantity. Thus, the optimum radius, $r_c = k/h$ will always give a maximum heat loss and not a minimum.

15. An Expression for the Critical Thickness of Insulation for a Spherical Shell

Let us consider a spherical shell having an outer radius r_1 and the termperature at that surface T_1. An insulation is added such that the outermost radius of the shell is r, a variable. The thermal conductivity of the insulating material, k, the convective heat transfer coefficient at the outer surface, h, and the ambient temperature T_∞ is constant. The rate of heat transfer through the insulation on the spherical shell is given by

$$\dot{Q} = \frac{(T_1 - T_\infty)}{(r - r_1)/4\pi k\, r\, r_1 + 1/h\, 4\pi r^2}$$

$$\frac{d\dot{Q}}{dr} = 0 = \frac{4\pi(T_1 - T_\infty)(1/kr^2 - 2/hr^3)}{[(r - r_1)/k\, r\, r_1 + 1/hr^2]^2}$$

which gives, $1/kr^2 - 2/hr^3 = 0$;

or,
$$r = r_c = 2\,k/h \tag{2.11}$$

---

*The existence of a critical radius requires a change in the heat transfer area in the direction of heat flow. For a plane wall the area perpendicular to the direction of heat flow is constant and there is no critical insulation thickness.

Example 2.19 Hot gases at 175°C flow through a metal pipe (outer diameter 8 cm). The convective heat transfer coefficient at the outside surface of the insulation (k = 0.18 W/mK) is 2.6 W/m²K and the ambient temperature is 25°C. Calculate the insulation thickness such that the heat loss is less than the uninsulated case.

Solution: *(a) Pipe without Insulation*

Neglecting the thermal resistance of the pipe wall and due to the inside convective heat transfer coefficient, the temperature of the pipe surface would be 175°C.

$$\dot{Q}/L = h \times 2\pi r(T_1 - T_\infty) = 2.6 \times 2\pi \times 0.04(175 - 25) = 98 \text{ W/m}$$

*(b) Pipe Insulated, Outermost Radius, r**

$$\dot{Q}/L = 98 = (T_1 - T_\infty)/\left(\frac{\ln\ (r^*/4)}{2\pi \times 0.18} + \frac{100}{2.6 \times 2\pi \times r^*} \right)$$

or $\dfrac{150}{98} = 0.884 \ln\ (r^*/4) + 6.12/r^*$; which gives $r^* = 13.5$ cm.

Therefore, the insulation thickness must be more than 9.5 cm.

(Since the critical thickness of insulation is r_c = k/h = 0.18/2.6 = 6.92 cm, and is greater than the radius of the bare pipe, the required insulation thickness must give a radius greater than the critical radius.)

If the outer radius of the pipe was more than the critical radius, any addition of insulating material will reduce the rate of heat transfer. Let us assume that the outer radius of the pipe is 7 cm $(r > r_c)$.

$\dot{Q}/L$, without insulation = $hA(\Delta T) = 2.6 \times 2 \times 3.142 \times 0.07 \times (175 - 25)$

$$= 171.55 \text{ W/m}$$

By adding 4 cm thick insulation, outermost radius = 7.0 + 4.0 = 11.0 cm.

and $\quad \dot{Q}/L = (175 - 25)/\left[\dfrac{\ln(11/7)}{2\pi \times 0.18} + \dfrac{1}{2.6\pi \times 2 \times 0.11} \right] = 133.58$ W/m.

Reduction in heat loss = $\dfrac{171.55 - 133.58}{171.55} = 0.22$ or 22%.

Example 2.20 An electric conductor 1.5 mm in diameter at a surface temperature of 80°C is being cooled in air at 25°C. The convective heat transfer coefficient from the conductor surface is 16 W/m²K. Calculate the surface temperature of the conductor when it is covered with a layer of rubber insulation (2 mm thick, k = 0.15 W/mK) assuming that the conductor carries the same current and the convective heat transfer coefficient is also the same. Also calculate the increase in the current carrying capacity of the conductor when the surface temperature of the conductor remains at 80°C.

Solution: When there is no insulation,

$$\dot{Q}/L = hA(\Delta T) = 16 \times 2 \times 3.142 \times 0.75 \times 10^{-3} = 4.147 \text{ W/m}$$

When the insulation is provided, the outermost radius $= 0.75 + 2 = 2.75$ mm

$$\dot{Q}/L = 4.147 = (T_1 - 25)/\left(\frac{\ln 2.75/0.75}{2\pi \times 0.15} + \frac{1000}{16 \times 2\pi \times 2.75}\right)$$

or $\qquad T_1 = 45.71°C$

i.e., the temperature at the outer surface of the wire decreases because the insulation a dds a r esistance.

The critical radius of insulation, $r_c = k/h = 0.15/16 = 9.375$ mm

i.e., when an insulation of thickness $(9.375 - 0.75) = 8.625$ mm is added, the heat tansfer rate would be the maximum and the conductor can carry more current. The heat transfer rate with outermost radius equal to $r_c = 9.375$ mm

$$\dot{Q}/L = (80 - 25)/\left(\frac{\ln 9.375/0.75}{2\pi \times 0.15} + \frac{1000}{16 \times 2\pi \times 9.375}\right) = 14.7 \text{ W/m}$$

The rate of heat transfer is proportional to (current)^2, the new current I_2 would be:

$$I_2/I_1 = (14.7/4.147)^{1/2} = 1.883$$

or, the current carrying capacity can be increased 1.883 times. But the maximum current capacity of wire would be limited by the permissible temperature at the centre of the wire.

The surface temperature of the conductor when the outermost radius with insulation is equal to the critical radius, is given by

$$\dot{Q}/L = 4.147 = (T - 25)\Big/\left(\frac{\ln(9.375/0.75)}{2\times 3.142\times 0.15} + \frac{1000}{16\times 2\times 3.142\times 9.375}\right)$$

or $\qquad T = 40.83°C.$

Example 2.21 Develop an expression for the outer radius of a current carrying wire with and without insulation such that the rate of heat ransfer and the temperature at the outer surface of the wire remains the same.

Solution: Let us assume : $r_0 =$ outer radius of the bare wire,

$\qquad\qquad\qquad\quad r_2 =$ outer radius after the insulation is added

$\qquad\qquad\qquad\quad T_0 =$ temperature of the outer surface of wire

$\qquad\qquad\qquad\quad T_\infty =$ temperature of the surroundings,

$\qquad\qquad\qquad\quad h =$ convective heat transfer coefficient

When the wire is uninsulated,

$$\dot{Q}/L = h \times 2\pi \times r_0(T_0 - T_\infty) = (T_0 - T_\infty)/\frac{(1)}{2\pi r_0 h}$$

When the wire is insulated,

$$\dot{Q}/L = (T_0 - T_\infty)/(1/2\pi r_2 h) + [\ln(r_2/r_0)/2\pi k)]$$

Thus, $\dfrac{1}{2\pi\, r_0 h} = \dfrac{1}{2\pi\, r_2 h} + \dfrac{\ln(r_2/r_0)}{2\pi k}$

and, $\quad 1/(r_0 h) = 1/(r_2 h) + \ln(r_2/r_0)/k$

or $\quad \ln(r_2/r_0) = \dfrac{k}{hr_0}(1 - r_0/r_2)$

Example 2.22 A current carrying conductor, radius 1 mm can be covered with either of the two insulating materials ($k_1 = 0.15$ W/mK and $k_2 = 0.1$ W/mK). The convective heat transfer coefficient at the outer surface is 15 Wm^2K. Suggest the insulating material for the wire to carry more current.

Solution: Critical radii for the two insulating materials would be:

$$r_{c_1} = k_1/h = 0.15/15 = 10\,mm; \quad r_{c_2} = k_2/h = 0.1/15 = 6.67\,mm$$

Assuming that the outer surface temperature and the ambient temperature reamains the same, the minimum thermal resistance would permit the maximum current to flow.

(i) Resistance with no insulation $= 1/hr = 1/(15 \times 1 \times 10^{-3}) = 66.67$

(ii) Resistance with better insulation ($k = 0.1$ W/mK)

$$= [\ln(r_2/r_1)/k] + (1/hr) = \frac{\ln(6.67/1)}{0.1} + \frac{1000}{15 \times 6.67} = 28.97$$

(iii) Resistance with inferior insulation ($k = 0.15$ W/mK)

$$= [\ln(r_2/r_1)/k] + (1/hr) = \frac{\ln(10/1)}{0.15} + \frac{1000}{15 \times 10} = 22$$

Thus an inferior insulating material would permit more current to flow.

Example 2.23 An electric conductor 10 mm in diameter is insulated to increase its current carrying capacity by 25 percent. If the ambient temperature and the temperature at conductor surface is 25°C and 80°C respectively and remains constant, calculate the thermal conductivity of the insulating material. Take h = 15 W/m^{2}K.

Solution: (i) Conductor uninsulated:
$$\dot{Q}_1 = \dot{Q}/2\pi L = hr(T_1 - T_\infty) = 15 \times 5 \times 10^{-3}(80 - 25) = 4.125\,W/m$$

(ii) Conductor insulated, outermost radius r* = k/h

$$Q_2 = Q/2\pi L = (T_1 - T_\infty)/\left(\frac{\ln(r*/r)}{k} + \frac{1}{hr*}\right) = 55/\left(\frac{\ln(r*/5)}{k} + \frac{1000}{15r*}\right)$$

since the current carrrying capacity increases by 25%,

$$(1.25)^2 = \dot{Q}_2/\dot{Q}_1 = 55k/(\ln(13.33k) + 1) \times 4.125, \quad \text{after putting r* = k/h,}$$

or, $8.533k = \ln(13.33k) + 1$

Solving by trial and error, k = 0.0385 W/mK.

16. Systems with Internal Heat Generation

Systems with heat sources (or sinks) are encountered in many processes, such as, liberation of heat energy by an electric current flowing through a conductor, volumetric liberation of heat from the heat liberating elements of atomic reactors as a consequence of nuclear reactions taking place, liberation or absorption of heat during many chemical reactions, etc. While investigating heat transfer processes in such cases, it is essential to know the intensity of volumetric liberation (or absorption) of heat energy. For steady-state processes $\partial T/\partial t = 0$, the differential equation of conduction in the presence of internal heat generation takes the form

$$\nabla^2 T + \dot{Q}_v/k = 0$$

17. An Expression for the Temperature Distribution in a Slab with Heat Generation under Steady State Condition

For one-dimensional steady state heat conduction through a plane wall, the equation is written as

$$d^2T/dx^2 + \dot{Q}_v/k = 0$$

Upon integration, $\qquad dT/dx + (\dot{Q}_v/k)x = C_1$

and $\qquad\qquad T = -(\dot{Q}_v/k)x^2/2 + C_1 x + C_2 \qquad\qquad\qquad (2.12)$

where the constants C_1 and C_2 are obtained by applying the prescribed boundary conditions.

Example 2.24 A 3.0 cm thick plate has heat generated uniformly at the rate of 5×10^5 W/m^3. Calculate (i) the temperature at the centre of the plate for k = 20W/mK and the left face and right face is maintained at 200°C and 50°C respectively, (ii) the maximum temperature in the plate when the left face is insulated and the right face is at 200°C.

Solution: The temperature distribution in a slab or flat plate is given by Eq. (2.12):

$$T = -(\dot{Q}_v/k)\,x^2/2 + C_1 x + C_2$$

Let the boundary conditions be: at x = 0, T = T_1, at x = L, T = T_2.
Substituting the two boundary conditions, $C_2 = T_1$, and

$$T_2 = -(\dot{Q}_v/k)\frac{L^2}{2} + C_1 L + T_1$$

$$\therefore \qquad C_1 = (T_2 - T_1)/L + (\dot{Q}_v/k)\frac{L}{2}$$

Therefore, $T - T_1 = (\dot{Q}_v/2k)(Lx - x^2) + (T_2 - T_1)x/L$

(i) For $\qquad$ L = 0.03m, x = 0.015 and T_1 = 200°C, T_2 = 50°C
$\qquad\qquad$ T = 127.81°C, the temperature at the centre of the plate.

(ii) When the left face is insulated, the temperature distribution can be obtained by applying the boundary conditions:
at $\qquad$ x = 0, $\quad$ dT/dx = 0
and at $\quad$ x = L, $\qquad$ T = T_2. Thus the temperature distribution is:

$$T = T_2 - (\dot{Q}_v/2k)(x^2 - L^2)$$

and this system is equivalent to a system where the plane wall has a thickness equal to 2L and both faces are being maintained at the same temperature, Fig. 2.19.

The equation for temperature distribution tells us that the maximum temperature will occur at x = 0 and will be equal to

$$T = T_2 + (\dot{Q}_v/2k)L^2 = 200 + \frac{5 \times 10^5}{2 \times 20}(0.03)^2 = 211.25°C.$$

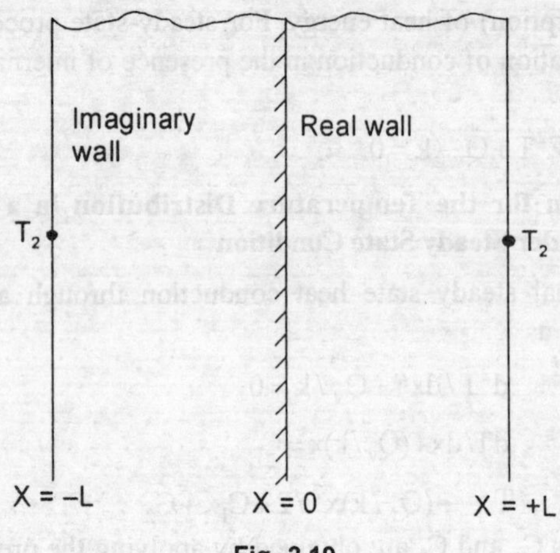

Fig. 2.19

Example 2.25 Electric heater wires are installed in a solid wall having a thickness of 8 cm (k = 2.5 W/mK). The right face is maintained at 300°C and is exposed to an environment with h = 50 W/m²K and T_∞ = 30°C while the left face is exposed to h = 75 W/m²K and T_∞ = 50°C. Calculate the rate of heat generation.

Solution: The right face will lose heat to the surrounding by convection

$$\dot{Q}_R = h(\Delta T) = 50 \times (300 - 30) = 13500 \text{ W/m}^2$$

The heat transferred by conduction from the right face to the left face

$$= (300 - T) / (L/k), \text{ where T is the temperature at the left face}$$
$$= h (T - 50)$$

Solving for T, we get T = 123.53°C

The heat energy lost by left face, $\dot{Q}_L = h(\Delta T) = 75 (123.53 - 50) = 5514.7$ W/m²

Total loss = 13500 + 5514.7 = 19014.7 W/m²

∴ Rate of heat generation, $\dot{Q}_v = 19014.7 \times 0.08 = 1521.17$ W/m³.

Example 2.26 A plane wall 6.0 cm thick generates heat internally at the rate of 0.3 MW/m³. The left side of the wall is insulated and the right side is exposed to an environment at 95°C. The convective heat transfer coefficient between the wall and the environment is 570 W/m²K. The thermal conductivity of the wall is 21W/mK. Calculate the temperature of the wall at the insulated face..

Solution: The governing equation for heat transfer is $d^2T/dx^2 + \dot{Q}_v /k = 0$.

Upon intergration, $dT/dx + \dot{Q}_v /k \cdot x = C_1$

The boundary condition : $dT/dx = 0$ at x = 0, (left face insulated); ∴ $C_1 = 0$

∴ $\quad dT/dx = -(\dot{Q}_v /k)x$; $\quad$ at x = L, $-k(dT/dx) = h(T_2 - T_\infty)$,

where T_2 is the temperature of the wall at the right face.

$$\therefore \qquad \dot{Q}_v L = 570(T_2 - 95) = 0.3 \times 10^6 \times 0.06 = 18000$$
$$\therefore \qquad T_2 = 95 + 18000/570 = 126.58\ °C.$$

Again, $dT/dx = -\dot{Q}_v\, x/k$

$$T = (-\dot{Q}_v/k)(x^2/2) + C_2, \qquad \text{at } x = L,\ T = 126.58$$
$$\therefore \qquad C_2 = 126.58 + \dot{Q}_v/L^2/2k = 126.58 + 25.71 = 152.79$$

and $\qquad T = 152.29 - \dot{Q}_v\, x^2/2k.$

This tells us that the maximum temperature at $x = 0$ will be 152.29°C.

Example 2.27 A shielding wall of a nuclear reactor receives a gamma-ray flux such that heat is generated within the wall according to the relation $\dot{Q}_v = \dot{Q}_0 \exp(-\alpha x)$ where $\dot{Q}_0$ is the heat generation at the inner face of the wall exposed to gamma ray flux and α is a constant. Using this relation for heat generation, derive an expression for the temperature distribution in a wall of thickness L, where the inside and outside temperatures are maintained at T_1 and T_2.

Solution: The governing equation is : $d^2T/dx^2 + \dfrac{\dot{Q}_0}{k}\exp(-\alpha x) = 0$

The boundary conditions are: $x = 0,\ T = T_1$, and $x = L,\ T = T_2$.

Upon integration: $T = -(\dot{Q}_0/k)\exp(-\alpha x)/\alpha^2 + C_1 x + C_2$

Substituting the first boundary condition, $C_2 = T_1 + \dot{Q}_0/(k\alpha^2)$

and $\qquad T_2 = -\dot{Q}_0 \exp(-\alpha L)/(k\alpha^2) + C_1 L + T_1 + \dot{Q}_0/(k\alpha^2)$

$$= -\dot{Q}_0\big[1 - \exp(-\alpha L)\big]/(k\alpha^2) + T_1 + C_1 L$$

Therefore, $C_1 = (T_2 - T_1)/L + \dot{Q}_0\Big[1 - \exp(-\alpha L)/(\alpha^2 kL)\Big]$

and the temperature distribution is given by

$$T - T_1 = \dot{Q}_0\Big[1 - \exp(-\alpha x)/(\alpha^2 k)\Big] + \frac{\dot{Q}_0\big[1 - \exp(-\alpha L)\big]}{\alpha^2 k} \cdot \frac{x}{L} + (T_2 - T_1)\ x/L$$

We can also obtain an expression for the maximum temperature in the wall

$$\frac{dT}{dx} = \frac{\dot{Q}_0}{k}\exp(-\alpha x) + C_1 = 0\ ;$$

or, $\qquad C_1 = \dfrac{\dot{Q}_0 \exp(-\alpha x)}{\alpha k} = (T_2 - T_1)\ /L + \dfrac{\dot{Q}_0\big[1 - \exp(-\alpha L)\big]}{\alpha^2 kL}$

$$\therefore \qquad \exp(-\alpha x) = \frac{(T_2 - T_1)\alpha k}{L\dot{Q}_0} + \frac{\big[\exp(-\alpha L) - 1\big]}{\alpha L}$$

The value of x has to be determined from the equation

$$x = -\frac{1}{\alpha}\ln\left(\frac{(T_2 - T_1)\,\alpha k}{L\dot{Q}_0} + \frac{\{\exp(-\alpha L) - 1\}}{\alpha L}\right).$$

Example 2.28 A plane wall, 10 cm thick and having a thermal conductivity $k = 30$ W/mK, has internal heat generation at the rate of 5×10^6 W/m³. The temperature of the surrounding air is 30°C. Estimate the convective heat transfer coefficient at the surfaces of the wall, the temperature at the surface of the wall if the maximum temperature in the wall is limited to 300°C.

Solution: When a plane wall having a thickness 2L has its both faces at the same temperature, the temperature distribution is given by

$$T = T_2 - (\dot{Q}_v / 2k)(x^2 - L^2)$$

where T_2 is the temperature at the wall surface (see Ex. 2.24, Fig. 2.19) and the maximum temperature occurs at the mid plane of the wall.

Therefore, $\quad 300 = T_2 + \dot{Q}_v L^2 / 2k$, and $T_2 = 300 - \dfrac{5 \times 10^6 \times (0.05)^2}{2 \times 30}$

$$= 91.67 \ °C$$

The heat generated within the wall would be transferred to the surroundings by convection,

or, $\quad\quad 0.1 \times 5 \times 10^6 = 2h\,(91.67 - 30)$

and $\quad\quad\quad\quad h = 4053.8$ W/m²K $= 4.05$ kW/m²K.

Example 2.29 A mild steel plate ($k = 25$ W/mK, 8mm thick) has a heat generation rate of 10 MW/m³ and is placed between two brass plates (4 mm thick, $k = 150$ W/mK). The convective heat transfer coefficient at the outside surfaces of the brass plates is 225 W/m²K and the ambient temperature is 35 °C. Calculate the maximum temperature in the steel plate and the temperature at the steel and brass plates interface.

Solution: The rate of heat generation within the steel plate $= 10 \times 10^6 \times 0.008$
$$= 80000 \text{ W/m}^2$$

By symmetry Fig. 2.20, the heat transfer to the surrounding from each side of the plate $= 40000$ W/m².

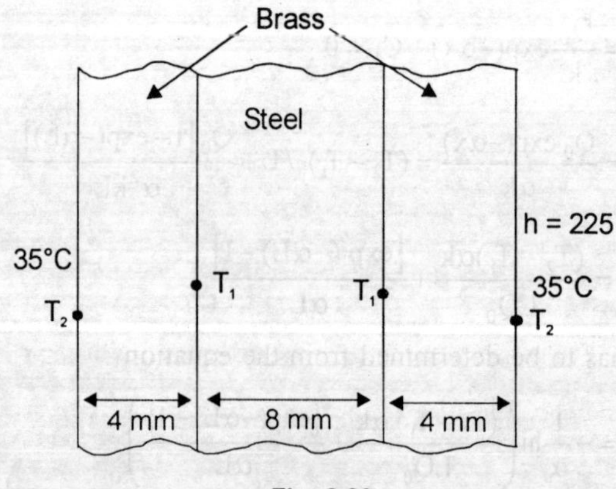

Fig. 2.20

If the temperature at the outer surface of the brass plates is T_2, then

$$\dot{Q} = 40000 = h\,(T_2 - T_0) = 225 \times (T_2 - 35)$$
$$T_2 = 212.78°C$$

If the temperature at the steel-brass interface is T_1, the heat flow by conduction through the brass p late is

$$\dot{Q} = (T_1 - T_2)/\,L/k = (T_1 - 212.78)\,/\,(0.004/150) = 40000$$

or, $T_1 = 212.78 + 40000 \times 0.004\,/150 = 213.85°C$

Again, for symmetrical conditions, the maximum temperature will occur at the mid plane of the steel plate and will be equal to

$$T = T_1 + \dot{Q}_v\,L^2/8k, \text{ where } L \text{ is the width of the steel plate}$$
$$213.85 + 10 \times 106 \times (0.008)^2\,/\,8 \times 25 = 217.05°C.$$

Example 2.30 A composite wall consists of two materials : A (thickness 0.05 m, k = 75 W/mK), and B (thickness 0.02 m, k = 150 W/mK). Heat energy is being generated in material A at the rate of 1.5×10^6 W/m³ and the inner surface of the wall A is insulated while the outer surface of the wall B is cooled by water, h = 1000 W/m²K. The ambient temperature is 30°C. Determine the temperature at the insulated surface, the interface and the outer surface.

Solution: By making an energy balance on the control volume about material B, we have heat flux into material at $x = L_A$ and that must be equal to the heat flux from the material B due to convection at $x = L_A + L_B$ or, $Q' = h(T_2 - T_0)$, as shown in the figure. Since the left face of the wall is insulated, the energy generated within the wall m ust leave through the r ight face of t he wall, $x = L_A$.

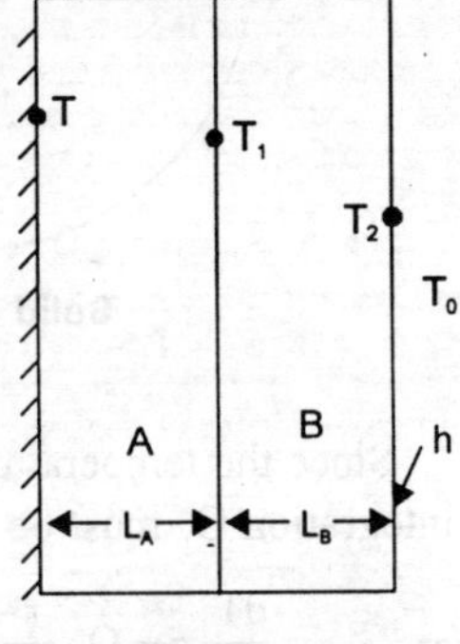

Fig. 2.21

$$\dot{Q}_v L_A = Q'$$

and, $T_2 = T_0 + \dot{Q}_v L_A\,/h$

$$= 30 + \frac{1.5 \times 10^6 \times 0.05}{1000} = 105°C.$$

The temperature at the insulated surface, $T = T_1 + \dot{Q}_v L_A^{\,2}\,/\,2k_A$

Since the material A is having heat generation, it cannot be represented by a thermal circuit element. The equivalent thermal circuit is shown in Fig. 2.22. Thus,

$$T_1 = T_0 + Q'(L_B\,/k_B + 1/h)$$

$$= 30 + 1.5 \times 10^6 \times 0.05 \times \left(\frac{0.02}{150} + \frac{1}{1000}\right) = 115°C$$

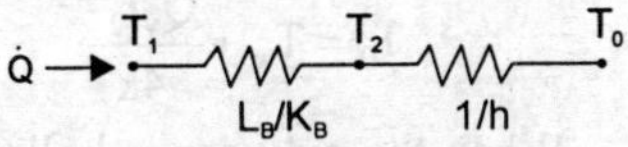

and T, the temperature at the insulated face is

Fig. 2.22

$$= 115 + \frac{1.5 \times 10^6 \times (0.05)^2}{2 \times 75} = 140°C.$$

18. Expression for Temperature Distribution in a Cylinder with Heat Generation when the Cylinder is (i) Hollow and (ii) Solid

Let us consider a cylinder, (Fig. 2.23) outer radius r_2 having uniformly distributed heat sources and constant thermal conductivity. Assuming one-dimensional, steady state radial flow, the governing equation can be written as:

$$\frac{1}{r}\frac{d}{dr}\frac{(rdT)}{dr}+\frac{\dot{Q}_v}{k}=0$$

or, $\quad \dfrac{d}{dr}\dfrac{(rdT)}{dr}+\dfrac{\dot{Q}_v r}{k}=0$, $\qquad$ Upon integration, $\quad \dfrac{rdT}{dr}+\dfrac{\dot{Q}_v r^2}{2k}=C_1$

and, $\quad \dfrac{dT}{dr}+\dot{Q}_v\cdot\dfrac{r}{2k}=\dfrac{C_1}{r}$

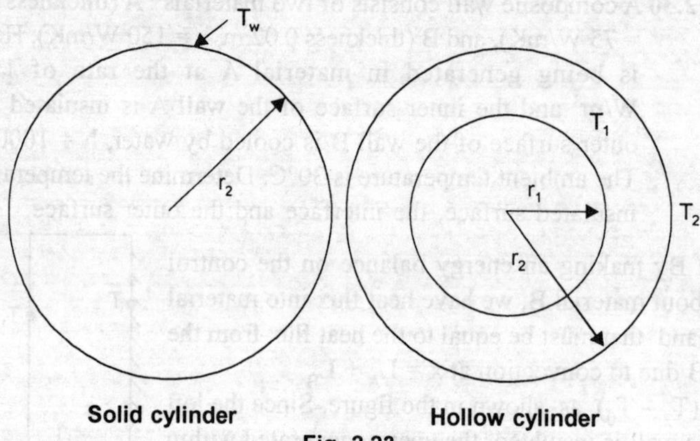

Fig. 2.23

Since the temperature gradient dT/dr has to be finite as $r \to o$, the constant of integration C_1 must be equal to zero for a solid cylinder.

or, $\quad \dfrac{dT}{dr}=-\dot{Q}_v\dfrac{r}{2k}$ $\qquad$ The boundary conditions are:

$\qquad\qquad\qquad\qquad\qquad\qquad$ at $r = r_2$, $T = T_w$, which gives

and, $\quad T=-\dot{Q}_v\dfrac{r^2}{4k}+C_2$ $\qquad \therefore \ C_2=T_w+\dot{Q}_v\dfrac{r_2^2}{4k}$

The temperature distribution is given by: $T=T_w+\dfrac{\dot{Q}_v(r_2^2-r^2)}{4k}$ $\qquad$ (2.13)

and the temperature at the centre of a solid cylinder ($r = 0$) will be

$$T_0=T_w+\frac{\dot{Q}_v r_2^2}{4k} \qquad\qquad (2.14)$$

When the cylinder is hollow, the boundary conditions are :

$\qquad$ at $r = r_1$, $T = T_1$; and at $r = r_2$, $T = T_2$, and $\quad T = -\dfrac{\dot{Q}_v r^2}{4k} + C_1 \ln(r) + C_2$; which reduces to

$$T = T_2 + \dot{Q}_v \frac{(r_2^{\;2} - r^2)}{4k} + C_1 \ln(r/r_2),$$

where $\quad C_1 = \dfrac{(T_1 - T_2) + \dot{Q}_v(r_1^{\;2} - r_2^{\;2})/4k}{\ln(r_2/r_1)}$ $\qquad\qquad\qquad$ (2.15)

Example 2.31 A hollow cylindrical copper conductor (30 mm outside diameter, 15 mm inside diameter, k = 380 W/mK) has a current density of 40 A/mm^2. The external surface is covered with a uniform layer of insulation (k = 0.3 W/mK, thickness 10 mm) and the ambient temperature is 15°C. Assuming radial conduction, calculate the heat energy to be removed per unit time by forced cooling on the inside surface of the conductor.

Assume that the temperature of the insulation does not exceed 130°C at any point, convective heat transfer coefficient at the outside surface is 40 W/m^{2}K and the electrical resistivity of copper 2×10^{-5} Ω mm.

Solution: The heat generated per unit volume due to the current flowing is

$\dot{Q}_v = \dfrac{I^2 R}{AL}$, where I is the current, R is the resistance, A is the cross-sectional area and L is the length.

The resistance, R = δL/A, where δ = Electrical resistivity

$$\dot{Q}_v = \frac{I^2}{A}\frac{\delta}{L}\frac{L}{A} = (I/A)^2\delta = 40^2 \times 2 \times 10^{-5} \text{ W/mm}^3 = 32 \times 10^6 \text{ W/m}^3$$

The maximum temperature of the insulation (= 130°C) will occur at the interface between the insulation and the copper tube. Hence for the insulation,

$$Q' = \Delta T / R = (130 - T)/[\ln(25/15)/2 \times 0.3 \times \pi]$$

or, 0.271 Q' = 130 – T, where T is the temperature at the outside surface of the insulation.

Also, $\quad Q' = h \times 2\pi r(T - 15) = 40 \times 2\pi \times 25 \times 10^{-3}(T - 15)$

$$= 6.284 \ (T - 15)$$

Thus, 130 – T = 0.271Q' and T – 15 = 0.159Q'. $\qquad$ Or, $\quad$ Q' = 267.44 W.

Total heat generated = $Q_v \times$ volume = $32 \times 10^6 \times \dfrac{\pi}{4}(0.03^2 - 0.015^2) \times 1$

$$= 16967 \text{ W/m length}$$

And, energy to be removed from the inside = 16967 – 267.44 = 16.7 kW.

Example 2.32 A 5 mm diameter wire 30 cm long has a voltage of 5 Volts impressed on it. The outer surface temperature of the wire is maintained at 95°C. Calculate the temperature at the centre of the wire. The resistivity of the wire is $70 \times 10^{-6}\,\Omega m$, thermal conductivity is 22.5 W/mK. A 10 mm thick insulation is provided on the wire and the convective heat transfer coefficient at the surface is 100 W/m²K, the ambient temperature is 25 °C. Calculate the temperature at the outer surface when the rate of heat generation remains the same. What should be the thermal conductivity of the insulation material?

Solution: The resistance of the wire, $R = \delta L / A = \dfrac{70 \times 10^{-6} \times 0.3 \times 4}{\pi (5)^2 \times 10^{-6}} = 1.07\,\Omega$

Current flowing through the system, $I = V/R = 5/1.07 = 4.673$ Amp

$$\dot{Q}_v = VI / \text{volume} = \frac{5 \times 4.673 \times 4}{\pi (5)^2 \times 10^{-6} \times 0.3} = 3.97 \times 10^6 \text{ W/m}^3$$

The temperature at the centre of the wire is given by, Eq. (2.14)

$$T_0 = T_w + \dot{Q}_v (r_2)^2 / 4k = 95 + \frac{3.97 \times 10^{-6}}{4 \times 22.5} \times (2.5)^2 \times 10^{-6}$$
$$= 95.276\text{°C}$$

Thermal resistance due to convective heat transfer coefficient is

$$1 / hA = 1/(100 \times 2\pi \times 12.5 \times 10^{-3} \times 0.3) = 0.424$$

Heat energy transferred by convection = Total heat energy generated = VI

$\therefore \qquad VI = 5 \times 4.673 = (T_s - 25)/0.424$; which gives, $T_s = 25 + 9.9 = 34.9\text{°C}$

When the temperature at the surface of the wire is maintained at 95°C and the temperature at the surface of the insulation is 34.9°C, this would require the thermal conductivity of the insulating material, k, as given by

$$\frac{\ln(r_2 / r_1)}{2\pi Lk} \times VI = \Delta T = (95 - 34.9) = 60.1$$

or, $\qquad k = 1n(5)/(2 \times 3.14 \times 0.3 \times 2.57) = 0.332$ W/mK.

Example 2.33 In a cylindrical fuel rod of a nuclear reactor, the internal heat generation is given by the equation $\dot{q} = \dot{q}_0 (1 - (r/R)^2)$, where
R = outer radius of the solid fuel rod,
r = radial distance from the axis of rod,
$\dot{q}_0$ = heat generation per unit volume at the centre
Calculate the temperature drop from the centre-line to the surface of a 3 cm rod having k = 20 W/mK if the rate of heat removal from the surface is 2.9×10^6 W/m².

Solution: When the heat generation due to nuclear reactions is not uniform, the internal heat generation is usually written as:

$$\dot{q}_v = \dot{q}_0 (1 - (r/R)^2)$$

and the governing equation for radial heat transfer reduces to

$$\frac{1}{r}\frac{d}{dr}\left(r\frac{dT}{dr}\right) = -\frac{\dot{q}_0}{k}(1 - r^2/R^2)$$

$$\frac{d}{dr}\left(r\frac{dT}{dr}\right) = -\dot{q}_0\frac{(r - r^3/R^2)}{k}$$

Upon integration, $\left(r\dfrac{dT}{dr}\right) = -\dfrac{\dot{q}_0}{k}\left(\dfrac{r^2}{2} - \dfrac{r^4}{4R^2}\right) + C_1$

Since at $r = 0$, dT/dr is finite, therefore, C_1 must be zero.

and $T = C_2 - \dfrac{\dot{q}_0}{k}\left(\dfrac{r^2}{4} - \dfrac{r^4}{16R^2}\right)$, if $T = T_0$ at $r = 0$, $C_2 = T_0$; and

$$T - T_0 = \frac{\dot{q}_0}{k}\left(\frac{r^4}{16R^2} - \frac{r^2}{4}\right), \text{ gives the temperature distribution.}$$

The temperature at the surface of the rod ($r = R$) is given by

$$T_w = T_0 - \frac{3R^2}{16k}\dot{q}_0; \text{ and the temperature drop, } T_0 - T_w = \frac{3R^2}{16k}\dot{q}_0$$

The rate of heat removal, $\dot{Q}/A = 2.9 \times 10^6 \text{ W/m}^2 = -k\left(\dfrac{dT}{dr}\right)_{r=R} = \dot{q}_0 R/4$

Thus, $\dot{q}_0 = \left(\dfrac{\dot{Q}}{A}\right)\dfrac{4}{R} = \dfrac{2.9\times10^6\times4}{1.5\times10^{-2}} = 7.73\times10^8 \text{ W/m}^3$

And the temperature drop, $T_0 - T_w = 7.73\times10^8 \times \dfrac{3\times(1.5\times10^{-2})^2}{16\times20} = 1630°C.$

Example 2.34 A chemical reaction is being carried out at constant pressure in a packed bed between two co-axial cylinders with radius 1.25 cm and 1.5 cm. The entire inner wall is at a uniform temperature of 500°C and the surface is insulated. The heat energy released during reaction is uniform and the rate is 2×10^6 W/m^3. The thermal conductivity of the fuel bed is 0.5 W/mK. Calculate the temperature of the outer surface.

Solution: The governing equation can be written as:

$$\frac{d}{dr}\left(r\frac{dT}{dr}\right) = -\dot{Q}_v\frac{r}{k}$$

and the boundary conditions are: at $r = r_1$, $T = T_1 = 500°C$, $dT/dr = 0$.

Upon integration, $\dfrac{dT}{dr} = \dfrac{-\dot{Q}_v r}{2k} + C_1/r$, $\therefore C_1 = \dfrac{\dot{Q}_v r_1^2}{2k}$ ($\because$ at $r = r_1$, $dT/dr = 0$)

By integrating again, $T = \dfrac{\dot{Q}_v}{2k}\left(r_1^2\ln(r) - \dfrac{r^2}{2}\right) + C_2$

By applying the second boundary condition, at $r = r_1$, $T = 500$

$$500 = \frac{\dot{Q}_v}{2k}\left(r_1^2 \ln(r_1) - \frac{r_1^2}{2}\right) + C_2$$

$$C_2 = 500 - \frac{\dot{Q}_v\left(r_1^2 \ln(r_1) - r_1^2/2\right)}{2k}$$

and

$$T = 500 + \dot{Q}_v \frac{r_1^2 \ln(r/r_1)}{2k} - \dot{Q}_v \frac{\dot{Q}_v r_1^2(r^2/r_1^2 - 1)}{4k}$$

at $r = r_2 = 1.5$ cm; $T = 500 + \dfrac{2\times10^6 \times(1.25\times10^{-2})^2 \ln(1.5/1.25)}{2\times0.5}$

$$= -2\times10^6 \times(1.25\times10^{-2})^2 \times \frac{1}{2}[(1.5/1.25)^2 - 1]$$

$$= 500 + 56.97 - 68.75 = 488.22°C.$$

Example 2.35 An oil acts as lubricant ($\mu = 10$N–s/m^2, $k = 230$ W/mK, $\rho = 1.22 \times 10^3$ kg/m^3) between two co-axial cylindrical surfaces (outer diameter 10 cm, inner diameter 9.95 cm). The outer cylinder rotates at 10,000 rpm. Calculate the maximum temperature in the oil if both wall temperatures are maintained at 75°C. The rate of heat generation per unit volume $\dot{Q}_v$ due to viscous dissipation may be assumed to be $\mu(du/dy)^2$.

Solution: Since the clearance between the two cylinders is very small, the variation in velocity may be assumed linear (Fig. 2.23a & 2.23b), i.e.,

$$du/dy = V/c, \text{ where } V = 2\pi r_2 \, N/60 \text{ and } c = 0.025 \text{ cm}$$

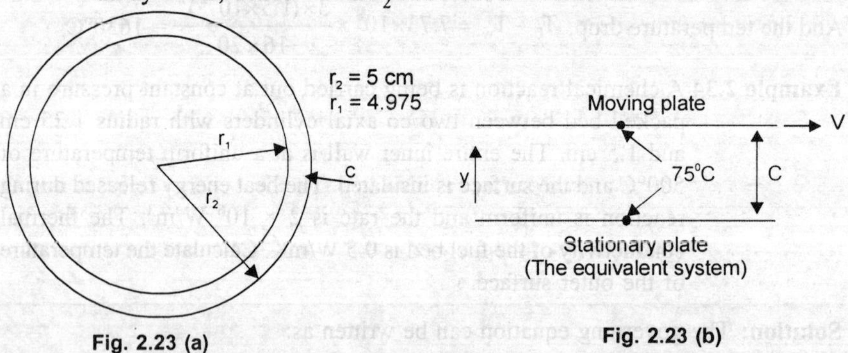

Fig. 2.23 (a) Fig. 2.23 (b)

The energy generation is $\dot{Q}_v = \mu(V/c)^2$ and the governing equation is

$$\frac{d^2T}{dy^2} = -\frac{\mu}{k}(V/c)^2 \quad \text{[for the equivalent system of Fig. 2.23(b)]}$$

The boundary conditions are: at $y = 0$, $T = 75°C$; at $y = c$, $T = 75°C$

Upon integration, $\qquad T = -(\mu/k)(V^2/c^2)(y^2/2) + C_1 y + C_2$

Applying the boundary conditions, $C_2 = 75$, and $C_1 = \mu V^2/(2k \cdot c)$ and

$$T = 75 + [\mu V^2/(2k \cdot c)] (y - y^2/c);$$

for maximum temperature, $dT/dy = 0$, gives $y = c/2$, and $T_{max} = 75 + \mu V^2/8k$

$$\therefore \quad T_{max} = 75 + \frac{10}{8 \times 230}(2\pi \times \frac{0.05 \times 10000}{60})^2 = 75 + 14.9 = 89.9°C.$$

19. An Expression for Temperature Distribution in a Sphere with Internal Heat Generation when the Sphere is (i) Solid and (ii) Hollow

Assuming that (i) steady state conditions prevail,

(ii) one-dimensional radial heat conduction holds,

(iii) thermal conductivity is uniform, and

(iv) the system has uniform heat generation,

the governing equation in spherical coordinates reduces to

$$\frac{1}{r^2}\frac{d}{dr}\left(r^2\frac{dT}{dr}\right) + \frac{\dot{Q}_v}{k} = 0$$

Upon integration, $\quad r^2\dfrac{dT}{dr} = \dfrac{-\dot{Q}_v r^3}{3k} + C_1$

or, $\quad \dfrac{dT}{dr} = \dfrac{-\dot{Q}_v r}{3k} + \dfrac{C_1}{r^2}$

and $\quad T = -(\dot{Q}_v r^2)/6k - C_1/r + C_2 \hfill (2.16)$

Case I : When the sphere is solid, the temperature gradient at the centre of the sphere ($r = 0$), will be finite and in that case, $C_1 = 0$.

If T_w is the temperature at the outer surface, $r = R$,

$$T_w = \frac{-\dot{Q}_v R^2}{6k} + C_2; \text{ which gives, } C_2 = T_w + \frac{\dot{Q}_v R^2}{6k}$$

and the temperature distribution would be: $\quad T - T_w = \dfrac{\dot{Q}_v(R^2 - r^2)}{6k} \hfill (2.17)$

The temperature will be maximum at $r = 0$, and is given by

$$T_0 = T_w + \frac{\dot{Q}_v R^2}{6k} \hfill (2.18)$$

Case II : When the sphere is hollow (inner radius r_1 and outer radius r_2) the boundary conditions would be:

at $r = r_1$, $dT/dr = 0$, because the sphere is a closed surface; $r = r_2$, $T = T_w$

Therefore, $\quad 0 = \dfrac{-\dot{Q}_v}{3k}r_1 + C_1/r_1^2$ which gives $C_1 = \dfrac{\dot{Q}_v r_1^3}{3k}$

$$T = -\frac{\dot{Q}_v r^2}{6k} - \frac{\dot{Q}_v r_1^3}{3kr} + C_2$$

Applying the second boundary condition,

$$T_w = -\frac{\dot{Q}_v r_2^{\,2}}{6k} - \frac{\dot{Q}_v r_1^{\,3}}{2kr_2} + C_2$$

Therefore $C_2 = T_w + \dfrac{\dot{Q}_v}{3k}\left(\dfrac{r_2^{\,2}}{2} + \dfrac{r_1^{\,3}}{3r_2}\right)$

and the temperature distribution is given by

$$T = T_w + (\dot{Q}_v / 6k)(r_2^{\,2} - r^2) + (\dot{Q}_v r_1^{\,3} / 3k)(1/r_2 - 1/r) \tag{2.19}$$

·**Example 2.36** A solid sphere (radius R = 0.5 m) has an internal energy generation rate 2×10^6 W/m³. If the thermal conductivity of the material is 40 W/mK and the convective heat transfer coefficient at the surface of the sphere is 10kW/m² K, calculate the temperature at the centre and at the outer surface. The ambient temperature is 30°C.

Solution: For a solid sphere, the temperature at the centre is given by Eq. (2.18)

$$T_0 = T_w + \dot{Q}_v R^2 / 6k \,;$$

The rate of heat generation = $\dot{Q}_v$ × Volume of the sphere

$$= 2 \times 10^6 \times \frac{4}{3}\pi(0.5)^3 = 1.047 \times 10^6 \text{ W}$$

The heat energy convected away from the surface to the surroundings

$$hA(\Delta T) = 1.047 \times 10^6$$

$$\therefore \quad \Delta T = 1.047 \times 10^6 /[10 \times 10^3 \times 4\pi(0.5)^2] = 33.32$$

Therefore, $T_w = T_\infty + 33.32 = 63.32$

and the temperature at the centre of the sphere,

$$T_0 = 63.32 + \dot{Q}_v R^2 / 6k = 63.32 + 2 \times 10^6 \times (0.5)^2 /(6 \times 40)$$
$$= 2146.65°C.$$

Example 2.37 A hollow sphere (inner diameter 16 cm, oouter diameter 24 cm, k = 40 W/mK) has internal heat generation rate of 2×10^7 W/m³. The inside surface of the sphere may be treated as insulated and the temperature at the outside surface is 375°C. Calculate the maximum temperature in the solid and the convective heat transfer coefficient if the ambient temperature is 40°C.

Solution: The temperature distribution in a hollow sphere with internal heat generation is given by, Eq. (2.19)

$$T = T_w + \dot{Q}_v (r_2^{\,2} - r^2)/6k - \dot{Q}_v r_1^{\,3}(1/r - 1/r_2)/3k$$

The maximum temperature will occur at the inner radius and will be

$$T_1 = 375 + \frac{2 \times 10^7}{6 \times 40}\left[(0.12)^2 - (0.08)^2\right] - \frac{2 \times 10^7}{3 \times 40}\left[\frac{1}{0.08} - \frac{1}{0.12}\right](0.08)^3$$
$$= 375 + 666.67 - 355.56 = 686.11°C$$

Volume of the hollow sphere $= \dfrac{4\pi}{3} \times \left[(0.12)^3 - (0.08)^3 \right] = 5.09 \times 10^{-3}$

Total energy gnerated, $\quad Q' = 2 \times 10^7 \times 5.09 \times 10^{-3} = 101800$ W

By making an energy balance, $101800 =$ heat energy convected away

$$= hA\,(375 - 40)$$
$$h = 101800/[4\pi(0.12)^2 \times 335]$$
$$= 1730 \text{ W/m}^2.$$

20. The Utility of Extended Surfaces

Extended surfaces are attached to the primary surface to increase the rate of heat transfer between a structure and its surrounding ambient fluid. That is, extended surfaces increase artificially the surface area for heat transmission and if the surface is proportioned properly, the net result will be an increase in the heat transmission rate between the structure and the ambient fluid.

21. An Expression for the Temperature Distribution in a Fin of Uniform Cross-section

Let us consider an extended surface, classed as straight fin or spine as shown in Fig. 2.24. Since the cross-sectional area for heat flow (normal to the length

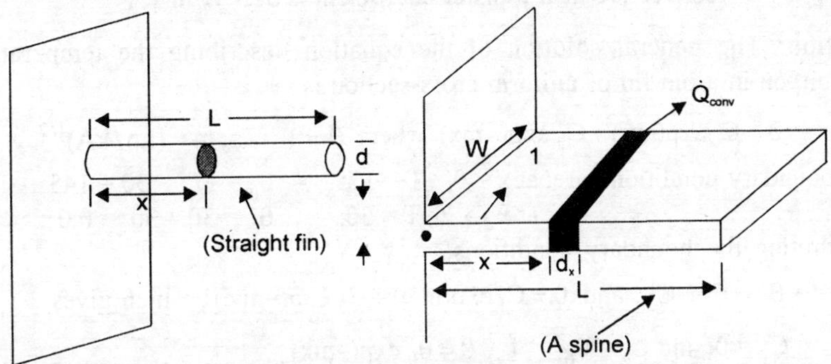

Fig. 2.24 A straight fin of uniform thickness or spine of constant cross-section

coordinate) for both profiles, is constant, they can be treated identically.

The assumptions made in the analysis are:

    (i) one-dimensional heat flow: thickness of the fin is small in comparison with the length and width of the fin,

    (ii) thermal conductivity is constant,

    (iii) no heat generation within the fin,

    (iv) steady state,

    (v) convective heat transfer coefficient is constant over the entire length.

By making an energy balance for the control volume (an element of the fin at some distance x, thickness dx), we have

$\dot{Q}_x$ = heat conducted into the element at $x = -kA\,(dT/dx)$.

$\dot{Q}_x + dx$ = heat conducted out of the element at $x + dx = -kA\,(dt/dx)_{x + dx}$

$\dot{Q}_c$ = heat convected out of the element between x and $x + dx$.

Thus, $\quad \dot{Q}_x = \dot{Q}_{x+dx} + \dot{Q}_c$; Using Taylor's Series expansion,

or, $\quad -kA\,(dT/dx) = -kA\,(dT/dx) - kA\,(d^2T/dx^2)\,dx + hp\,dx\,(T - T_\infty)$

where higher order terms have been neglected and T_∞ is the average temperature of the surrounding fluid. Upon simplification

$$d^2T/dx^2 - (hp/kA)\,(T - T_\infty) = 0.$$

This equation must be satisfied at every point in the fin and if the convective heat transfer coefficient and thermal conductivity are constant, the equation can be easily solved.

Let $\quad\quad \theta = T - T_\infty$, and $m = (hp/kA)^{1/2}$, then we have

$$d^2\theta/dx^2 - m^2\theta = 0 \text{ and the general solution is,}$$

$$\theta = C_1 \exp(mx) + C_2 \exp(-mx) \quad\quad\quad (2.20)$$

and the constants C_1 and C_2 are to be determined by the prescribed boundary conditions.

Example 2.38 One end of a very long aluminium rod is connected to a wall at 175°C. The other end is exposed to ambient temperature at 30°C. Calculate the total heat energy dissipated by the rod if the rod is 4 mm in diameter, thermal conductivity 155 W/mK and the convective heat transfer coefficient is 325 W/m²K.

Solution: The general solution of the equation describing the temperature distribution in a pin fin of uniform cross-section is

$$\theta = C_1 \exp(mx) + C_2 \exp(-mx) \text{ where } \theta = T - T_\infty, m = (hp/kA)^{1/2}$$

The boundary conditions are: at x = 0, $\quad$ T = 175, $\quad\quad \theta_0 = 175 - 30 = 145$

$$\text{at } x \to \infty, T = 30, \quad\quad \theta_\infty = 30 - 30 = 0.0$$

Substituting the boundary conditions,

$$\theta_0 = C_1 + C_2, \text{ and } 0 = C_1 \exp(m\infty) + C_2 \exp(-m\infty) ; \text{ which gives,}$$

$$C_1 = 0, \text{ and } C_2 = \theta_0; \quad\quad \theta = \theta_0 \exp(-mx)$$

$$m = (hp/kA)^{1/2} = \sqrt{\left[(325 \times \pi \times 4 \times 10^{-3})/155 \times \frac{\pi(4 \times 10^{-3})^2}{4}\right]}$$

$$= 45.79$$

and the temperature distribution in the rod is (Fig. 2.25)

(T − 30) /145 = exp (−45.79 x); or, T = 30 + 145 exp (−45.79x)

The heat flowing by conduction across the root of the fin has to be transmitted by convection from the surface of the rod to the fluid,

As such, $\quad \dot{Q}_{fin} = -kA(dT/dx)_{x=0} = \int_0^\infty hp(T - T_\infty)\,dx$

Since $\quad\quad \theta = \theta_0 \exp(-mx), (d\theta/dx)_{x=0} = (dT/dx)_{x=0} = -m\theta_0$

$\therefore \quad\quad -kA(dT/dx)_{x=0} = kA(hp/kA)^{1/2}\theta_0 = (hpkA)^{1/2}\theta_0$

Rate of heat loss = $(325 \times \pi \times (4 \times 10^{-3}) \times 155 \times \pi(4 \times 10^{-3})^2/4)^{1/2}(145)$

$$= 12.93 \text{ W.}$$

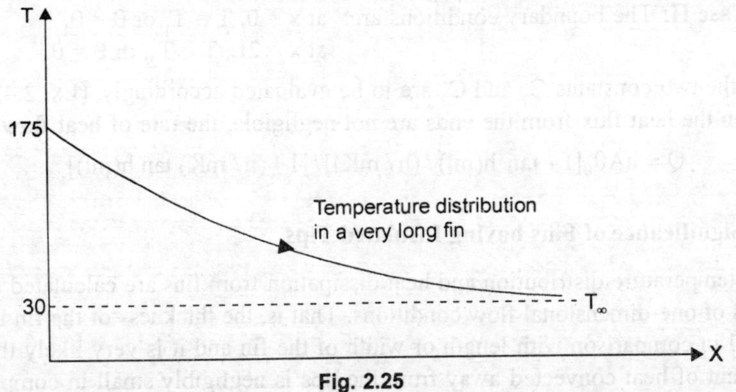

Fig. 2.25

Example 2.39 A thin circular fin of length L has its two ends at temperature T_1 and T_2. The ambient temperature is T_∞. Obtain an expression for one-dimensional temperature distribution when (i) $T_1 = T_2$ and (ii) $T_1 \neq T_2$.

Solution: The general solution for temperature distribution is given by
$$\theta = C_1 \exp(mx) + C_2 \exp(-mx)$$

Case I : When $T_1 = T_2$, the system can be replaced by an equivalent fin of length, L, (Fig. 2.26) such that dT/dx = 0, at x = L and in that case the boundary conditions are : at x = 0, $\theta = \theta_0 = T_1 - T_\infty$ and, at x = L, $d\theta/dx = 0$

Substituting the boundary conditions, $\theta_0 = C_1 + C_2$
$$d\theta / dx = 0 = mC_1 \exp(mL) - mC_2 \exp(-mL)$$
$$C_1 = C_2 \exp(-2mL)$$

or, $\theta_0 = C_2 \exp(-2mL) + C_2$ which gives,

$C_2 = [\theta_0 / 1 + \exp(-2mL)]$ and $C_1 = \theta_0 \exp(-2mL)/[1 + \exp(-2mL)]$

Since $\exp(mL) \times \exp(-mL) = 1$, the temperature distribution would be
$$\theta = [\theta_0 / \{\exp(mL) - \exp(-mL)\}] \times [\exp\{-m(L-x)\} + \exp\{m(L-x)\}]$$
and $\theta = \theta_0 \cos hm(L-x)/\cos h(mL)$ (2.21)

and the heat flow rate from the rod would be
$$\dot{Q} = -kA \ d\theta/dx \big|_{x=0} = \sqrt{(hpkA)} \ \theta_0 \tan h \ (mL) \qquad (2.22)$$

where tan h is the hyperbolic tangent.

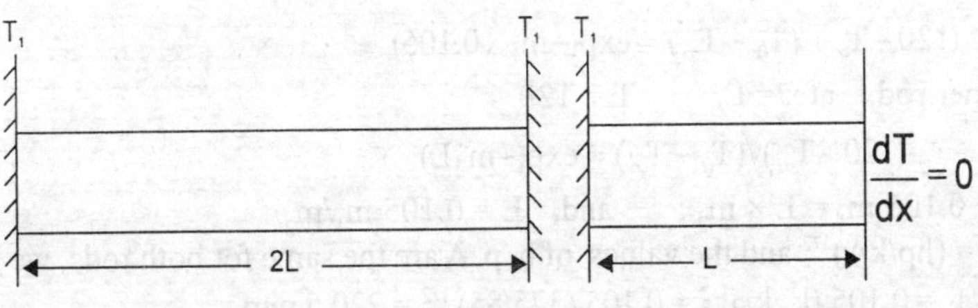

Fig. 2.26

Case II: The boundary conditions are: at $x = 0$, $T = T_1$ or $\theta = \theta_1$
$$\text{at } x = 2L,\ T = T_2,\ \text{or } \theta = \theta_2$$

and the two constants C_1 and C_2 are to be evaluated accordingly. (Ex. 2.42)
When the heat flux from the ends are not negligible, the rate of heat flow

$$Q = hA\theta_0[1 + \tan h(ml)/(h/mK)]/[1 + (h/mK)\tan h(ml)] \qquad (2.23)$$

22. Significance of Fins having Insulated Tips

The temperature distribution and heat dissipation from fins are calculated on the basis of one-dimensional flow conditions. That is, the thickness of the fin is very small in comparison with length or width of the fin and it is very likely that the amount of heat convected away from the tips is negligibly small in comparison with the heat convected away form the other surfaces. Thus, the expression describing the temperature distribution is considerably simplified.

Example 2.40 The temperature at two points 15 cm apart of a thin long aluminium rod is 175 and 125°C. Calculate the convective heat transfer coefficient when the ambient temperature is 30°C. ($k = 155$ W/mK)

Solution: The temperature distribution in a long rod is given by

$$\theta = \theta_0 \exp(-mx) \quad \text{when } x = 0,\ \ \theta = \theta_0 = 175 - 30 = 145$$

$$x = 0.15\text{m},\ \ \theta = \theta_1 = 125 - 30 = 95$$

Therefore, $95 = 145 \exp(-m \times 0.15)$;   and, $0.15\,m = \ln(145/95) = 0.4228$
$hp/kA = m^2 = 7.947$, $\therefore\ \ m = 2.819$

$$h = 7.947 \times 155 \times (\pi/4)(5 \times 10^{-3})^2 /[\pi(5 \times 10^{-3})] = 1.54 \text{ W/m}^2\,°\text{C}.$$

Example 2.41 Two long rods of the same diameter, one made of brass ($k = 85$ W/mK) and the other made of copper ($k = 375$ W/mK) have one of their ends inserted into the furnace. Both rods are exposed to the same environment. At a distance of 105 mm away from the furnace end the temperature of the brass rod is 120°C. At what distance from the furnace end, the same temperature would be attained by the copper rod? *(ES 93)*

Solution: The temperature distribution in a very long thin rod is given by

$$\theta = \theta_0 \exp(-mx)\ ;\ \text{or},\ (T - T_\infty)/(T_0 - T_\infty) = \exp(-mx)$$

For brass rod, at $x = 0.105$ m, $T = 120$, therefore,

$$(120 - T_\infty)/(T_0 - T_\infty) = \exp(-m_1 \times 0.105)$$

For copper rod,   at $x = L$, $T = 120$

$$\therefore \qquad (120 - T_\infty)/(T_0 - T_\infty) = \exp(-m_2 L)$$

or, $0.105\,m_1 = L \times m_2$;   and, $L = 0.105\,m_1/m_2$
Since $m = (hp/kA)^{1/2}$ and the values of h, p, A are the same for both rods, we have:
$$L = 0.105\,(k_2/k_1)^{1/2} = 0.105\,(375/85)^{1/2} = 220.5 \text{ mm}.$$

Example 2.42 A steel rod (10 mm square, k = 40 W/mK, length 20 cm) has its two ends maintained at 225°C and 100°C. The convective heat transfer coefficient is 25 W/m²K. Calculate the temperature at the mid location when $T_\infty = 30°C$.

Solution: The temperature distribution along a thin rod is given by
$$\theta = C_1 \exp(m,x) + C_2 \exp(-mx)$$
The boundary conditions are : at $x = 0, \theta = \theta_1 = 225 - 30 = 195$
$$x = 0.2m, \theta = \theta_2 = 100 - 30 = 70$$

$$\theta_1 = 195 = C_1 + C_2$$
$$\theta_2 = 70 = C_1 \exp(0.2m) + C_2 \exp(-0.2m)$$

Solving for C_1 and C_2, we have, $C_1 = \left[\theta_2 \exp(0.2m) - \theta_1\right] / \left[\exp(-0.4m) - 1\right]$

and $\qquad C_2 = \theta_1 - C_1$

$$m = (hp/kA)^{1/2} = \left[(25 \times 4 \times 10 \times 10^{-3})/(40 \times 100 \times 10^{-6})\right]^{1/2} = 15.81$$

$$C_1 = \frac{\left[70 \exp(0.2 \times 15.81) - 195\right]}{\left[\exp(0.4 \times 15.81) - 1\right]} = 2.62 \; ; \qquad C_2 = 195 - 2.62 = 192.38$$

$$\theta = 2.62 \exp(15.81 \cdot x) + 192.38 \exp(-15.81 \cdot x)$$

at the mid location, $x = 0.1$ m, therefore,
$$\theta = 2.62 \exp(15.81 \times 0.1) + 192.38 \exp(-15.81 \times 0.1) = 52.31$$
$$T = 30 + 52.31 = 82.31°C.$$

Example 2.43 A thin rod (perimeter p, thermal conductivity k, cross-setional area A) connects two walls maintained at temperature T_1 and T_2. Heat energy is generated at a uniform rate in the rod and the convective heat transfer coefficient at the surface is h. Obtain an expression for the temperature distribution.

Solution: We take a small element of the rod at a distance x from the left wall ($x = 0$) as shown in the figure.

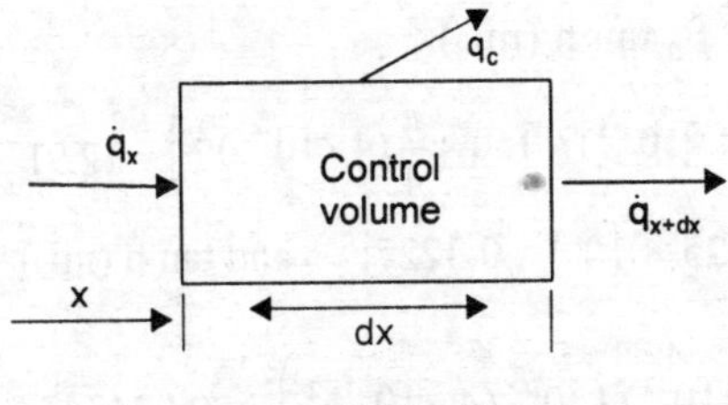

Fig. 2.27

$$\dot{q}_x = -kA(dT/dx); \quad \dot{q}_{x+dx} = -kA(dT/dx) - kA(d^2T/dx^2)dx$$

$$\dot{q}_c = hpdx(T - T_\infty) \text{ heat energy generated in the element } = \dot{q}_v A \cdot dx$$

By making an energy balance:

$$-kA(dT/dx) + \dot{q}_v A dx = -kA(dT/dx) - kA(d^2T/dx^2)dx + hpdx(T - T_\infty)$$

or, $\quad (d^2T/dx^2) + \dot{q}_v/k - (hp/kA)(T - T_\infty) = 0$

Let $\quad \theta = (T - T_\infty)$ and $m^2 = hp/kA$, we have

$$d^2\theta/dx^2 + \dot{q}_v/k - m^2\theta = 0$$

We choose another variable such that $\theta' = \theta - \dot{q}_v/km^2$, and we get :

$$d^2\theta'/dx^2 + \dot{q}_v/k - m^2(\theta' + \dot{q}_v/km^2) = 0 \; ;$$

which gives, $d^2\theta'/dx^2 - m^2\theta' = 0$

The general solution is: $\theta' = C_1 \exp(mx) + C_2 \exp(-mx)$

The boundary conditions are:

at $\qquad x = 0, \ \theta_1 = (T_1 - T_\infty)$, which gives $\theta'_1 = (T_1 - T_\infty) - \dot{q}_v/km^2$

at $\qquad x = L, \ \theta_2 = (T_2 - T_\infty)$, which gives $\theta'_2 = (T_2 - T_\infty) - \dot{q}_v/km^2$

Substituting the boundary conditions,

$$\theta'_1 = C_1 + C_2 \; ; \qquad \theta'_2 = C_1 \exp(mL) + C_2 \exp(-mL),$$

Solving for C_1 and C_2, we get:

$$C_1 = [\theta'_2 \ \exp(mL) - \theta'_1]/[\exp(2mL) - 1]$$

$$C_2 = [\theta'_1 \ \exp(2mL) - \theta'_2 \exp(mL)]/[\exp(2mL) - 1] \; .$$

Example 2.44 Thin circular pin fins (diameter 4 mm, length 25 mm, k = 150 W/ mK) are attached to wall at temperature 200°C. The convective heat transfer coefficient is 25 W/m²K, the ambient temperature being 25°C. Calculate the number of fins required per unit area of the wall such that the rate of heat transfer increases by 75 percent.

Solution: The rate of heat transfer from the wall without the fin is

$$\dot{q} = h(\Delta T) = 25 \times (200 - 25) = 4375 \ \text{W/m}^2$$

Rate of heat transfer from the wall with fins = $1.75 \times 4375 = 7565.23$ W/m²

Assuming that there is no transfer of heat energy from the tip of the fins, i.e., the end is insulated, the rate of heat transfer is given by :

$$\dot{q} = (hpkA)^{1/2} \theta_0 \tan h \, (mL) \; ;$$

$$m = (25 \times \pi \times 4 \times 10^{-3}) / \left(150 \times \frac{\pi}{4}(4 \times 10^{-3})^2 \right) = 12.91$$

$$mL = 12.91 \times 25 \times 10^{-3} = 0.3227; \quad \text{and } \tan h \, (mL) = 0.312;$$

$$(hpkA)^{\frac{1}{2}} = \left(25\pi(4 \times 10^{-3}) \, 150 \frac{\pi}{4}(4 \times 10^{-3})^2 \right)^{\frac{1}{2}} = 0.02434$$

$$\theta_0 = 200 - 25 = 175$$

$\dot{q}$, per fin = $0.02434 \times 175 \times 0.312 = 1.33$ W

If the number of fins is N, $\dot{q} = 1.33$ N

The area occupied by N fin = $N \times \dfrac{\pi}{4}(d)^2 = N \times \dfrac{\pi}{4}(4 \times 10^{-3})^2 = 12.57 \times 10^{-6}\,N$

The area of the wall where the fins are not there = $(1 - 12.57 \times 10^{-6}N)$

Total heat transfer = Heat transfer from the fin + Heat transfer from wall

$$7656.25 = 1.33\,N + 25\,(1 - 12.57 \times 10^{-6}N)\,(175)$$

$$= 1.33\,N + 4375 - 0.055\,N$$

or, $1.275\,N = 3250.25$; therefore, $N = 2550$ per m^2 of the wall area.

Example 2.45 A steel rod (5 cm in diameter, 90 cm long, k = 40 W /mK) is attached to a wall. The rod dissipates heat energy (45W) to the surroundings at 30°C. The convective heat transfer coefficient is 15 W/m^{2}K. Calculate the temperature of the wall. Assume that the rod loses heat by convection from the tip also.

Solution: The boundary condition for the problem would be at $x = 0$, $\theta = \theta_0$ and at $x = L$, $-kA\,(dT/dx)_{x=L} = hA\,(T - T_\infty)$, where T is the temperature at $x = L$

The solution to the governing equation:

$$d^2\theta / dx^2 - m^2\theta = 0; \quad \therefore \quad \theta = C_1 \exp(mx) + C_2 \exp(-mx)$$

Substituting the boundary conditions, $\theta = \theta_0 = C_1 + C_2$

$$(d\theta / dx)_{x=L} = mC_1 \exp(mL) - mC_2 \exp(-mL) = -h\theta / k$$

$$= (h / k)[C_1 \exp(mL) + C_2 \exp(-mL)]$$

Solving for C_1 and C_2 we get,

$$\theta / \theta_0 = \frac{[\cos h\,m(L - x) + (h / mk)\sin h\,(m(L - x)]}{\cos h\,(mL) + (h / mk)\sin h\,(mL)}$$

The heat loss from the fin, $\dot{q} = -kA(d\theta / dx)_{x=0}$

$$= \sqrt{(hpkA)}\,\theta_0 \, \frac{\sin h\,(mL) + (h / mk)\cos h\,(mL)}{\cos h\,(mL) + (h / mk)\sin h\,(mL)}$$

$$= \sqrt{(hpkA)}\,\theta_0 \, \frac{\tan h\,(mL) + (h / mk)}{1 + (h / mk)\tan h\,(mL)}$$

$$m = (hp / kA)^{1/2} = [(15 \times \pi \times 5 \times 10^{-2}) / (40 \times \pi \times (5 \times 10^{-2})^2 / 4)]^{1/2} = 5.48$$

$$mL = 4.9932 \text{ and } \tan h\,(mL) = 1.0$$

$$(hpkA)^{1/2} = \left[15 \times \pi \times 5 \times 10^{-2} \times 40 \times \pi \times (5 \times 10^{-2})^2 / 4\right]^{1/2} = 0.43$$

$$45 = 0.43\,\theta_0; \quad \theta_0 = 45/0.43 = 104.65 \quad \text{and } T_0 = 104.65 + 30 = 134.65.$$

Example 2.46 Ten straight fins of uniform cross-sectional area (width 25 mm, thickness 0.8 mm, k = 50 W/mK) are attached longitudinally to a cylinder (5 cm in diameter, 1 m long). Calculate the rate of heat transfer if the convective heat transfer coefficient is 15 W/m^{2}K and the ambient temperature is 25°C. The surface temperature of the cylinder is 175°C, Fig. 2.28.

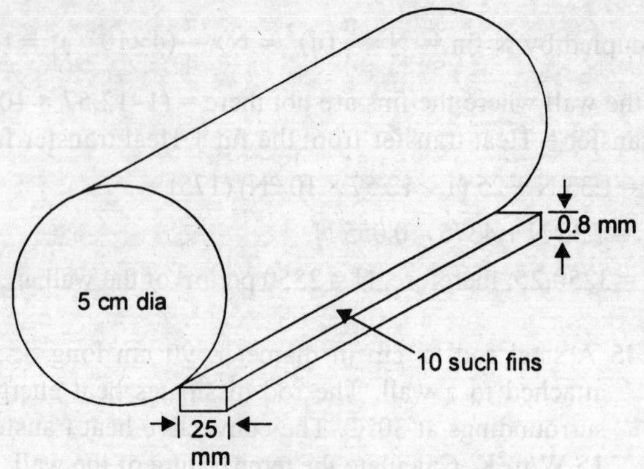

Fig. 2.28

Solution: The cross-sectional area normal to the direction of heat flow
$$= 1.0 \times 0.8 \times 10^{-3} = 0.8 \times 10^{-3} = m^2;$$
Perimeter, $p = 2 (1 + 0.0008) = 2.016 \text{ m}$
Since straight fins like, a thin plate, are attached along the length of a wall, such fins are rectangular fins and heat transfer from such a fin is analysed like pin fins having their ends insulated.

$\dot{q}$ for each fin/pin $= (hpkA)^{1/2} \; \theta_0 \tan h \; (mL)$

$$(hpkA)^{1/2} = (15 \times 2.016 \times 50 \times 0.8 \times 10^{-3})^{1/2} = 1.1$$

$$\theta_0 = 175 - 25 = 150$$

$$m = (hp/kA)^{1/2} = [(15 \times 2.016)/(50 \times 0.8 \times 10^{-3})]^{1/2} = 27.49$$

$$mL = 27.49 \times 25 \; 10^{-3} = 0.6874$$

$\tan h \; (mL) = 0.596$

Total loss from 10 fins $= 10 \times 1.1 \times 150 \times 0.596 = 983.4 \text{ W}.$

Example 2.47   A mercury in glass thermometer is placed in a well (150 mm deep) protruding from the walls of a pipe. The well is made of a material 1.0 mm thick, $k = 140 \text{ W/mK}$. The temperature of the gas flowing through the pipe is measured by the thermometer as 100°C. The pipe wall temperature is 50°C. The convective heat transfer coefficient over the well is 40 $\text{W/m}^2\text{K}$. Estimate the true temperature of the gas.

Solution: The thermometer will measure the temperature of the liquid/fluid flowing through the pipe and near the tip of the well, Fig. 2.29 If the temperature of the fluid flowing through the pipe is greater than the temperature of the pipe wall, the heat energy will flow from the gas to the pipe through the well and the temperature recorded by the thermometer will always be lower than the true temperature of the gas.

We can consider the well as a pin fin protruding from a wall and can assume that there is no heat transfer from the bottom of the well, i.e., the tip of the fin is insulated, the temperature distribution is given by (see Ex. 2.39).

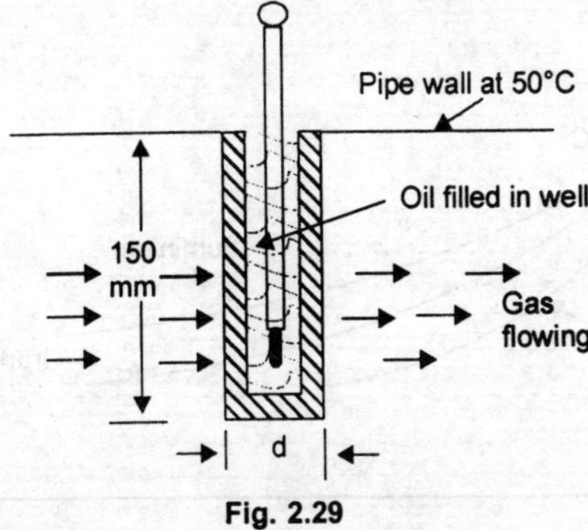

Fig. 2.29

$$\theta = \theta_0 \cos h\; m(L-x)/\cos h\;(mL) \quad \text{and when}\quad x = L,\quad \theta = \theta_0/\cos h\;(mL)$$

$$m = (hp/kA)^{1/2} = [(40 \times \pi \times (2r)/(140 \times 2\pi r(dr)]^{1/2} = 16.9$$

where dr is the thickness of the well material

$$mL = 16.9 \times 15 \times 10^{-3} = 2.535 \text{ and } \cos h\;(mL) = 6.35$$

$$\theta/\theta_0 = 1/6.35 = 0.1675.$$

or,
$$\frac{100 - T_\infty}{50 - T_\infty} = 0.1575, \qquad \therefore \quad T_\infty = 109.34\,°C$$

(In order to reduce the error in temperature measurement, we must have lower thermal conductivity of the well material, a very thin and long well and the convective heat transfer coefficient, h, should be as high as possible.)

Example 2.48 Three rods, one made of glass (k = 1.2 W/mK), one made of aluminium (k = 200 W/mK) and one made of wrought iron (k = 52 W/mK), all having the same diameter 10 mm and length 300 mm, are heated to a temperature of 200°C at one end. If the rods protrude into air at 30°C and if the heat transfer coefficient at the surface is 10 W/m²K, calculate the temperature distribution and the heat flow rate when the tip is insulated.

Solution: For cylindrical rods, the parameter, $m = (hp/kA)^{1/2} = (4h/kd)^{1/2}$

For glass, $m = (4 \times 10/1.2 \times 0.01)^{1/2} = 53.73$
For iron, $m = (4 \times 10 /52 \times 0.01)^{1/2} = 8.77$
For aluminium, $m = (4 \times 10/200 \times 0.01)^{1/2} = 4.472$

Since L = 0.3 m, the temperature distribution, (Fig. 2.30) for fins having insulated tips is given by:

Glass : $T = 30 + 170\,[\cos h\;\{57.73\,(0.3 - x)\}/\cos h\;(17.319)]$
Iron : $T = 30 + 170\,[\cos h\;\{8.77\,(0.3 - x)\}/\cos h\;(2.63)]$
Aluminium : $T = 30 + 170\,[\cos h\;(4.472\,(0.3 - x))/\cos h\;(1.342)$

Heat flow rates are given by : $\dot{q} = (hpkA)^{1/2}\,\theta_0 \tan h\;(mL)$

For glass : $(10 \times \pi \times 0.01 \times 1.2 \times \dfrac{\pi}{4}(0.01)^2) \times 170 \times \tan h\;(17.318) = 0.925\ W.$

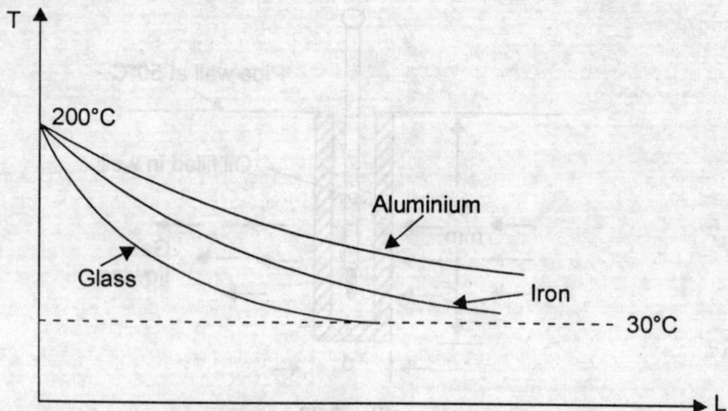

Fig. 2.30 Temperature distribution (glass is a very poor conductor and therefore heat flow rate is minimum with maximum temperature gradient at the root

For iron : $\left[10 \times \pi \times 0.01 \times 52 \times \dfrac{\pi}{4}(0.01)^2\right] \times 170 \times \tan h\,(2.63) = 6.027$ W

For aluminium, similarly we get, $\dot{q} = 11.82$ W

(Since there is no heat loss from the tip of an infinitely long rod/fin, an estimate of the validity of this approximation may be made by comparing the two equations:

$$\dot{q} = (hpkA)^{1/2}\,\theta_0 \tan h\,(mL) \quad \text{and} \quad \dot{q} = (hpkA)^{1/2}\,\theta_0$$

To get a satisfactory approximation, the above two equations would provide equivalent results if $\tan h\,(mL) \geq 0.99$ or $(mL) \geq 2.65$. Hence a rod/fin may be assumed to be infinitely long if $L \geq L_e = 2.65/m$.

If we desire that the temperature at the end of the fin is almost equal to the ambient temperature, we should have

$$\dfrac{\theta_x = L}{\theta_0} = \exp(-mL) \leq 0.01, \text{ and } mL > 4.6\;;$$

and L for the three cases would then be greater than 4.6/m.

Example 2.49 Calculate the length of the well (Example 2.47) such that the error in measuring the temperature by the thermometer is 2 percent.

Solution: Let the true temperature of the gas be 110°C.

The thermometer reading should then be = $0.98 \times 110 = 107.8°C$

or, θ at $x = L = 110 - 107.8 = 3.2$ and θ_0, at $x = 0, = 110 - 50 = 60$

   $\theta/\theta_0 = 1/\cos h\,(mL)$

or, $\cos h\,(mL) = 60/3.2 = 18.75$; $mL = 3.6236$

and $L = 3.6236/16.9 = 0.214$ m

   $= 214$ mm.

(If the required length of the well is more than the pipe diameter, it is necessary that the temperature measuring well is attached to the pipe obliquely.)

23. Fin Efficiency and Fin Effectiveness

Fins or extended surfaces increase the heat transfer area and consequently, the amount of heat transfer is increased. The temperature at the root or base of the fin is the highest and the temperature along the length of the fin goes on decreasing. Thus, the fin would dissipate the maximum amount of heat energy if the temperature all along the length remains equal to the temperature at the root. Thus, the fin efficiency is defined as:

$$\eta_{fin} = \text{(actual heat transferred)} / \text{(heat which would be transferred if the entire fin area were at the root temperature)}$$

In some cases, the performance of the extended surfaces is evaluated by comparing the heat transferred with the fin to the heat transferred without the fin. This ratio is called 'fin effectiveness' E and it should be greater than 1, if the rate of heat transfer has to be increased with the use of fins.

For a very long fin, effectiveness $E = \dot{Q}_{with\ fin} / \dot{Q}_{without\ fin}$

$$= (hpkA)^{1/2}\theta_0 / hA\theta_0 = (kp/hA)^{1/2}$$

And $\qquad \eta_{fin} = (hpkA)^{1/2}\theta_0 /(hpL\theta_0) = (hpkA)^{1/2}/(hpL)$

$$\therefore \quad \frac{E}{\eta_{fin}} = \frac{(kp/hA)^{1/2}}{(hpkA)^{1/2}} \times hpL = \frac{pL}{A} = \frac{\text{Surface area of fin}}{\text{Cross-sectional area of the fin}}$$

i.e., effectiveness increases by increasing the length of the fin but it will decrease the fin efficiency.

Expressions for Fin Efficiency for Fins of Uniform Cross-section

1. Very long fins: $(hpkA)^{1/2}(T_0 - T_\infty)/[hpL(T_0 - T_\infty)] = 1/mL$
2. For fins having insulated tips:

$$\frac{(hpkA)^{1/2}(T_0 - T_\infty)\tan h\ (mL)}{hpL(T_0 - T_\infty)} = \frac{\tan h\ (mL)}{mL}$$

For rectangular fins, mL can be expressed as

$$mL = \sqrt{(hp/kA)}.L = \sqrt{[h(2z + 2t)/(kzt)]}L$$

where z is the depth and t is the thickness.

If $\qquad z \gg t, \quad mL = (2.z.h/kt)^{1/2}L$;

and for circular fins, $mL = (hp/kA)^{1/2} L = (4\ h/kd)^{1/2}L$

3. Fins losing heat energy from tips also:

$$(A/pL)\ [1 + \tan h\ (mL)/(h/mk)] / [1 + (h/mk)\tan h\ (mL)]$$

Example 2.50 The total efficiency for a finned surface may be d efined as the ratio of the total heat transfer of the combined area of the surface and fins to the heat which would be transferred if this total area were maintained at the root temperature T_0. Show that this efficiency can be calculated from

$$\eta_t = 1 - A_f / A(1 - \eta_f) \text{ where } \eta_t = \text{ total efficiency, } A_f = \text{ surface}$$

area of all fins, A = total heat transfer area, η_f = fin efficiency.

Solution: Fin efficiency,

$$\eta_f = \frac{\text{Actual heat transferred}}{\text{Heat that would be transferred if the entire fin were at the root temperature}}$$

or, $\eta_f = \dfrac{\text{Actual heat transfer}}{hA_f(T_0 - T_\infty)}$

$\therefore$ Actual heat transfer from finned surface $= \eta_f hA_f(T_0 - T_\infty)$

Actual heat transfer from unfinned surface which are at the root temperature:

$$h(A - A_f)(T_0 - T_\infty)$$

Actual total heat transfer $= h(A - A_f)(T_0 - T_\infty) + \eta_f hA_f(T_0 - T_\infty)$

By the definition of total efficiency,

$$\eta_t = \left[h(A - A_f)(T_0 - T_\infty) + \eta_f hA_f(T_0 - T_\infty) \right] / \left[hA(T_0 - T_\infty) \right]$$

$$= \frac{(A - A_f) + \eta_f hA_f}{A} = 1 - A_f/A + \eta_f A_f/A$$

$$= 1 - (A_f/A)(1 - \eta_f).$$

24. Extended Surfaces do not always Increase the Heat Transfer Rate

The installation of fins on a heat transferring surface increases the heat transfer area but it is not necessary that the rate of heat transfer would increase. For long fins, the rate of heat loss from the fin is given by $(hpkA)^{1/2}\theta_0 = kA(hp/kA)^{1/2}\theta_0 = kAm\theta_0$. When $h/mk = 1$, $Q = hA\theta_0$ which is equal to the heat loss from the primary surface with no extended surface. Thus, when $h = mk$, an extended surface will not increase the heat transfer rate from the primary surface whatever be the length of the extended surface.

For $h/mk > 1$, $\dot{Q} < hA\theta_0$ and hence adding a secondary surface reduces the heat transfer, and the added surface will act as an insulation. For $h/mk < 1$, $\dot{Q} > hA\theta_0$, and the extended surface will increase the heat transfer, Fig. 2.31. Further, $h/mk = (h^2.kA/k^2hp)^{1/2} = (hA/kP)^{1/2}$, i.e. when $h/mk < 1$, the heat transfer would be more effective when h/k is low for a given geometry.

25. An Expression for Temperature Distribution for an Annular Fin of Uniform Thickness

In order to increase the rate of heat transfer from cylinders of air-cooled engines and in certain type of heat exchangers, annular fins of uniform cross-section are employed. Fig. 2.32 shows such a fin with its nomenclature.

In the analysis of such fins, it is assumed that:

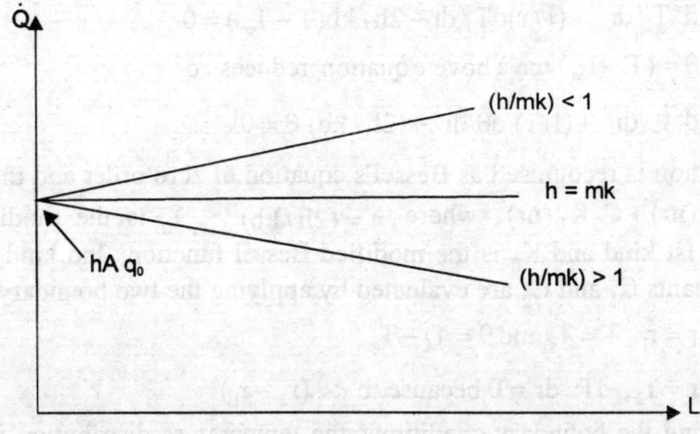

Fig. 2.31

(For increasing the heat transfer rate by fins, we should have (i) higher value of thermal conductivity, (ii) a lower value of h, fins are therefore generally placed on the gas side, (iii) perimeter/cross-sectional area should be high and this requires thin fins.)

(i)   the thickness 'b' is much smaller than the radial length $(r_2 - r_1)$ so that one-dimensional radial conduction of heat is valid;

(ii) steady state condition prevails.

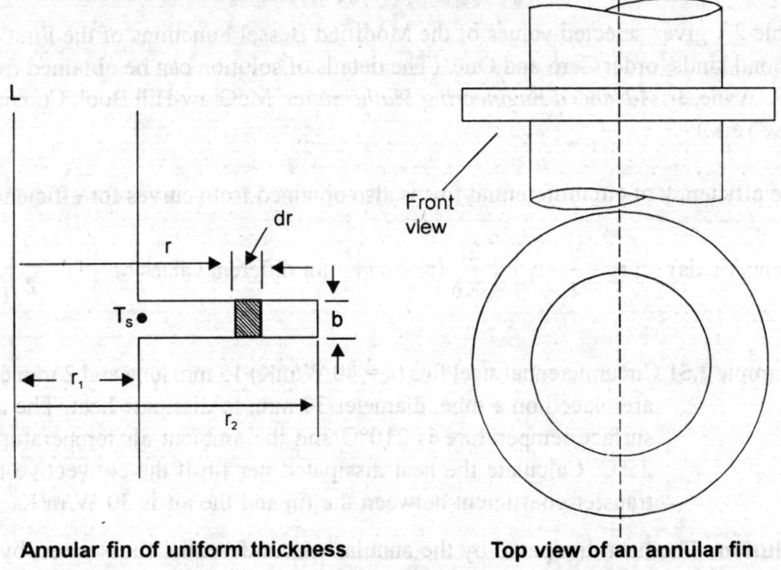

Annular fin of uniform thickness   **Top view of an annular fin**

Fig. 2.32

We choose an annular element of radius r and radial thickness dr. The cross-sectional area for radial heat conduction at radius r is $2\pi rb$ and at radius $r + dr$ is $2\pi (r + dr)b$. The surface area for convective heat transfer for the annulus is $2(2\pi\, r.dr)$. Thus, by making an energy balance,

$$-k2\pi rb\frac{dT}{dr} = -k2\pi(r+dr)b\left(\frac{dT}{dr} + \frac{d^2T}{dr^2}dr\right) + h \times 4\pi r.dr(T - T_\infty)$$

or, $\qquad d^2T/dr^2 + (1/r)dT/dr - 2h/kb(T - T_\infty) = 0$

Let, $\qquad \theta = (T - T_\infty)$ the above equation reduces to

$$d^2\theta/dr^2 + (1/r)\,d\theta/dr - (2h/kb)\,\theta = 0.$$

The equation is recognised as Bessel's equation of zero order and the solution is $\theta = C_1 I_0(nr) + C_2 K_0(nr)$, where $n = (2h/kb)^{1/2}$, I_0 is the modified Bessel function, Ist kind and K_0 is the modified Bessel function, 2nd kind, zero order. The constants C_1 and C_2 are evaluated by applying the two boundary conditions:

at $\qquad r = r_1,\; T = T_s$ and $\theta = T_s - T_\infty$

at $\qquad r = r_2,\; dT/dr = 0$ because $b << (r_2 - r_1)$

By applying the boundary conditions, the temperature distribution is given by

$$\frac{\theta}{\theta_0} = \frac{I_0(nr)K_1(nr_2) + K_0(nr)I_1(nr_2)}{I_0(nr_1)K_1(nr_2) + K_0(nr_1)I_1(nr_2)} \tag{2.24}$$

$I_1(nr)$ and $K_1(nr)$ are Bessel functions of order one.
And the rate of heat transfer is given by:

$$\dot{Q} = 2\pi\, knb\, \theta_0 r_1 \frac{K_1(nr_1)I_1(nr_2) - I_1(nr_1)K_1(nr_2)}{K_0(nr_1)I_1(nr_2) + I_0(nr_1)K_1(nr_2)} \tag{2.25}$$

Table 2.1 gives selected values of the Modified Bessel Functions of the First and Second kinds, order Zero and One. (The details of solution can be obtained from: C.R. Wylie, Jr: *Advanced Engineering Mathematics*, McGraw-Hill Book Company, New York.)

The efficiency of circumferential fins is also obtained from curves for efficiencies

(along Y-axis) $\sim \left(r_2 + \dfrac{b}{2} - r_1 \right)^{\frac{3}{2}} \left(\dfrac{2h}{Kb}(r_2 - r_1) \right)^{\frac{1}{2}}$ for different values of $\left(r_2 + \dfrac{b}{2} \right)\Big/ r_1$.

Example 2.51 Circumferential steel fins (k = 40 W/mK) 15 mm long and 2 mm thick are placed on a tube, diameter 30 mm, to dissipate heat. The tube surface temperature is 210°C and the ambient air temperature is 25°C. Calculate the heat dissipated per fin if the convective heat transfer coefficient between the fin and the air is 30 W/m²K.

Solution: The heat dissipated by the annular/ circumferential fin is given by

$$\dot{Q} = 2\pi b\, kn\, \theta_0 r_1 \frac{K_1(nr_1)I_1(nr_2) - I_1(nr_1)K_1(nr_2)}{K_0(nr_1)I_1(nr_2) + I_0(nr_1)K_1(nr_2)}$$

$$n = (2h/kb)^{1/2}$$

$$= (2 \times 30/40 \times 2 \times 10^{-3})^{1/2} = 27.386$$

$$nr_1 = 27.386 \times 15 \times 10^{-3} = 0.41; \qquad nr_2 = 27.386 \times 30 \times 10^{-3} = 0.82$$

Table 2.1 Modified Bessel Functions of the First and Second Kinds

x	$e^{-x}I_0(x)$	$e^{-x}I_1(x)$	$e^xK_0(x)$	$e^xK_1(x)$
0.0	1.0000	0.0000	∞	∞
0.2	0.8269	0.0823	2.1407	5.8334
0.4	0.6974	0.1368	1.6627	3.2587
0.6	0.5993	0.1722	1.4167	2.3739
0.8	0.5241	0.1945	1.2582	1.9179
1.0	0.4657	0.2079	1.1445	1.6361
1.2	0.4198	0.2152	1.0575	1.4429
1.4	0.3831	0.2185	0.9881	1.3010
1.6	0.3533	0.2190	0.9309	1.1919
1.8	0.3289	0.2177	0.8828	1.1048
2.0	0.3085	0.2153	0.8416	1.0335
2.2	0.2913	0.2121	0.8056	0.9738
2.4	0.2766	0.2085	0.7740	0.9229
2.6	0.2639	0.2046	0.7459	0.8790
2.8	0.2528	0.2007	0.2706	0.8405
3.0	0.2430	0.1968	0.6978	0.8066
3.2	0.2343	0.1930	0.6770	0.7763
3.4	0.2264	0.1892	0.6579	0.7491
3.6	0.2193	0.1856	0.6404	0.7245
3.8	0.2129	0.1821	0.6243	0.7021
4.0	0.2070	0.1787	0.6093	0.6816
4.2	0.2016	0.1755	0.5953	0.6627
4.4	0.1966	0.1724	0.5823	0.6453
4.6	0.1919	0.1695	0.5701	0.6292
4.8	0.1876	0.1667	0.5586	0.6142
5.0	0.1835	0.1640	0.5478	0.6003
5.2	0.1797	0.1614	0.5376	0.5872
5.4	0.1762	0.1589	0.5279	0.5749
5.6	0.1728	0.1565	0.5188	0.5633
5.8	0.1696	0.1542	0.5101	0.5525
6.0	0.1666	0.1520	0.5019	0.5422
6.4	0.1611	0.1479	0.4865	0.5232
6.8	0.1561	0.1441	0.4724	0.5060
7.2	0.1515	0.1405	0.4595	0.4905
7.6	0.1473	0.1372	0.4476	0.4762
8.0	0.1434	0.1341	0.4366	0.4631
8.4	0.1398	0.1312	0.4264	0.4511
8.8	0.1365	0.1285	0.4168	0.4399
9.2	0.1334	0.1260	0.4079	0.4295
9.6	0.1305	01235	.3995	0.4198
10.0	0.1278	0.1213	0.3916	0.4108

From Tables and by interpolation:

$I_1(nr_2) = 0.447$; $I_1(nr_1) = 0.22$; $I_0(nr_1) = 1.043$
$K_1(nr_2) = 0.836$; $K_1(nr_1) = 2.14$; $K_0(nr_1) = 1.1$

$$\dot{Q} = 2\pi \times 40 \times 27.386 \times 2 \times 10^{-3}(210 - 25) \times 15 \times 10^{-3}$$

$$\times \bullet \frac{2.14 \times 0.447 - 0.22 \times 0.836}{1.1 \times 0.447 + 1.043 \times 0.836}$$

$$= 21.65 \text{W}.$$

The selection of a suitable fin profile depends on the weight, cost, available space and the heat transfer characteristics. For minimum weight, every part of the fin must be utilized equally, i.e. the specific rate of heat flow is constant throughout the fin. This requires heat flow lines to be equally spaced and paralell to the axis of the fin.

26. An Expression for Temperature Distribution for a Straight Fin of Triangular Profile

Let us consider a triangular fin as shown in Fig. 2.33. In the analysis, the coordinate is selected in such a way that the origin coincides with the tip of the triangular fin. The other assumptions are:

(i)   the temperature at the root of the fin is T_S, the ambient temperature T_∞,
(ii) the width 'b' at the root is very small in comparison with the length L for one-dimensional flow,
(iii) the cross-sectional area varies with x.

Thus, for unit width of the fin, the cross-sectional area, $A = (b \cdot x / L) \times 1$

and the surface area for convective heat transfer, $S = 2\left[x^2 + (bx/2L)^2\right]^{1/2}$

$$= 2x(1 + b/2L)^{1/2} \equiv 2x$$
$$(\because b \ll L)$$

The governing equation for heat transfer can be written as:

$$d^2\theta/dx^2 + 1/x\, d\theta/dx - p^2\theta/x = 0, \qquad \text{where } p^2 = 2hL/kb, \ \theta = T - T_\infty$$

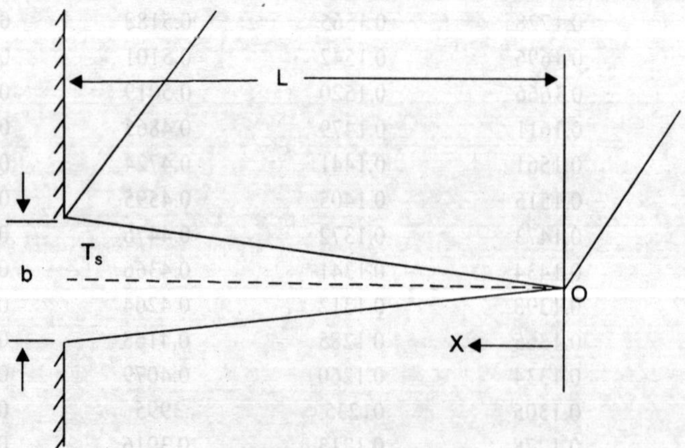

Fig. 2.33 A triangular fin

The general solution of the Bessel's equation is written as

$$\theta = C_1 I_0(2px^{1/2}) + C_2 K_0(2px^{1/2})$$

and the two constantss C_1 and C_2 are to be evaluated from the two boundary conditions at $x = 0$, and $x = L$.

The modified Bessel function of second kind, K_0 tends to infinity whereas the temperature has to be finite. Therefore, at $x = 0$, $C_2 = 0$;

and at $x = L$, $\theta = \theta_0 = T_s - T_\infty$, and $C_1 = \theta_0 / I_0(2pL^{1/2})$; and

$$\theta = \theta_0 I_0(2px^{1/2}) / I_0(2pL^{1/2})$$

$$\therefore \quad \dot{Q} = (-kA \ d\theta/dx) \text{ at } x = L = \frac{-kb\theta_0 p}{L^{1/2}} \frac{I_1(2pL^{1/2})}{I_0(2pL^{1/2})} \tag{2.26}$$

(The minus sign in the above equation is due to the direction of the coordinate x.)

Example 2.52 The engine cylinder of a motor cycle is made of aluminium alloy ($k = 186$ W/mK) and is 0.15 m long and outside diameter 50 mm. The temperature of the outer surface is 500K and the ambient air temperature is 300K. To increase the rate of heat transfer, five annular fins have been provided. The length and the thickness of the fins is 20 mm and 6 mm respectively. The fins are equally spaced. Estimate the increase in the heat transfer rate due to fins if the convective heat transfer coefficient is 50 W/m²K.   *(CS 1985)*

Solution: Using Fig. 2.32, the data given are:

$k = 186$ W/mK; $r_1 = 25 \times 10^{-3}$m; $r_2 = 45 \times 10^{-3}$m; $b = 6 \times 10^{-3}$m;

$h = 50$ W/m²K; $n = (2h/kb)^{1/2} = (2 \times 50/186 \times 6 \times 10^{-3})^{1/2} = 9.466$;

$nr_1 = 0.236$; $nr_2 = 0.426$

From Bessel Functions Tables and by interpolation:

$I_1(nr_2) = 0.2183$; $I_1(nr_1) = 0.1191$; $I_0(nr_1) = 1.0155$;

$K_1(nr_2) = 2.07$; $K_1(nr_1) = 4.31$; $K_0(nr_1) = 1.638$;

The rate of heat transfer from each fin would be [Eq. (2.25)]

$$\dot{Q} = (2 \times 3.142 \times 186 \times 9.466 \times 6 \times 10^{-3} \times 200 \times 25 \times 10^{-3})$$

$$(4.31 \times 0.2183 - 0.1191 \times 2.07)/(1.638 \times 0.2183 + 1.0155 \times 2.07)$$

$$= 93.7\text{W; and } \dot{Q} \text{ for 5 such fins} = 5 \times 93.7 = 468.5 \text{ W}$$

Heat transfer from those surfaces where fins are not there $= hA (\Delta T)$

$$= 50 \times 3.142 \times 50 \times 10^{-3} \ (0.15 - 0.03) \times 200 = 185 \text{ W}$$

Total heat transfer $= 468.5 + 185 = 653.5$ W

Heat transfer without fins $= hA (\Delta T) = 50 \times 3.142 \times 50 \times 10^{-3} \times 0.15 \times 200$

$$= 235.65$$

$\therefore$ Increase $= 653.5 - 235.65$

$$= 417.85\text{W}.$$

SUMMARY

1. The general heat conduction equation for an isotropic solid with constant thermal conductivity and internal heat generation is given by

$$\nabla^2 T + \dot{Q}_v / k = 1/\alpha \, \partial T / \partial t$$

2. Thermal diffusivity is a physical property of the material and defines the rate of change in temperature in the material.

3. The temperature distribution in a plane wall depends upon the heat flow rate through the wall, the thermal conductivity of the material and the wall thickness.

4. The temperature distribution for radial flow in a cylindrical shell is logarithmic, whereas in a spherical shell it is hyperbolic. Therefore the thermal stresses in cylindrical shells are less severe.

5. In composite cylindrical systems, when an insulation is provided to reduce the heat loss, it is always better that a superior insulating material is applied first on the surface of the cylinder.

6. Since air is a poor conductor of heat energy, two thin blankets would provide more comfort than a single blanket of double thickness.

7. Extended surfaces or fins have wide industrial applications and are provided to increase the rate of heating or cooling. Fin materials should have high values of thermal conductivity.

8. Fin efficiency is defined as
   η = actual heat transferred /heat which would be transferred if the entire fin area were at the root temperature.

9. For (h/mk) > 1, the extended surfaces will reduce the heat transfer rate. When (h/mk) = 1, an extended surface will not increase the rate of heat transfer.

10. When (h/mk) <1, the heat transfer from an extended surface would be more effective for a lower h/k for a given geometry.

Equations for temperature distribution and rate of heat transfer for fins of uniform cross-section—

Case	Tip condition	Temperature distribution, θ/θ_0	Fin heat transfer rate
1.	Infinite fin	exp (−mx)	$\sqrt{(hpkA)}\,\theta_0$
2.	Insulated tip	cos h m(L − x)/cos h mL	$\sqrt{(hpkA)}\,\theta_0$ tan h mL
3.	Convection heat transfer	$\dfrac{\cos h\, m(L-x)+(h/mk)\sin h\, m(L-x)}{\cos h\, mL+(h/mk)\sin h\, mL}$	$\dfrac{\sqrt{hpkA}\ \theta_0[\sin h\, mL + h/mK \cos h\, mL]}{[\cos h\, mL + h/mK \sin h\, mL]}$
4.	Fixed temperature $\theta(L) = \theta_L$	$(\theta_L/\theta_0)\dfrac{\sin h\, mx + \sin h\, m(L-x)}{\sin h\, mL}$	$\sqrt{(hpkA)}\theta_0\dfrac{\cos h\, mL - \theta_L/\theta_0}{\sin h\, mL}$

MULTIPLE CHOICE QUESTIONS

1. Upto the critical radius of insulation
   (a) Convection heat loss will be less than conduction heat loss
   (b) heat flux will decrease
   (c) added insulation will increase heat loss
   (d) added insulation will decrease heat loss   **(c)**

2. Given the following data:
   Inside and outside heat transfer coefficients 25 W/m^2K, thermal conductivity of 15 cm thick brick = 0.15 W/mK. The overall heat transfer coefficient in W/m^2K, will be closer to
   (a) inverse of heat transfer coefficient
   (b) heat transfer coefficient
   (c) thermal conductivity of brick
   (d) heat transfer coefficient based on thermal conductivity of brick alone   **(c)**

3. A composite slab has two layers of different materials having thermal conductivyt k_1 and k_2. If each layer has the same thickness, the equivalent thermal conductivity of the slab would be:
   (a) $k_1 + k_2$
   (b) $(k_1 + k_2)/k_1 k_2$
   (c) $2k_1 k_2 / (k_1 + k_2)$
   (d) $k_1 \, k_2$ **(c)**

4. The rate of heat transfer from a long fin (k = 0.17 W/cm K) of constant cross-sectional area 2 cm^2 perimeter 5 cm exposed to an ambient temperature 40°C (h = 0.0025 W/cm^2K), base temperature 100°C, is
   (a) 6.8 W   (b) 3.9 W   (c) 1.7 W   (d) 0.17 W   **(b)**

5. The wall of a building absorbs solar energy as shown in the figure. Assuming that the ambient temperature inside and outside are equal and considering steady state, the equivalent electrical circuit will be

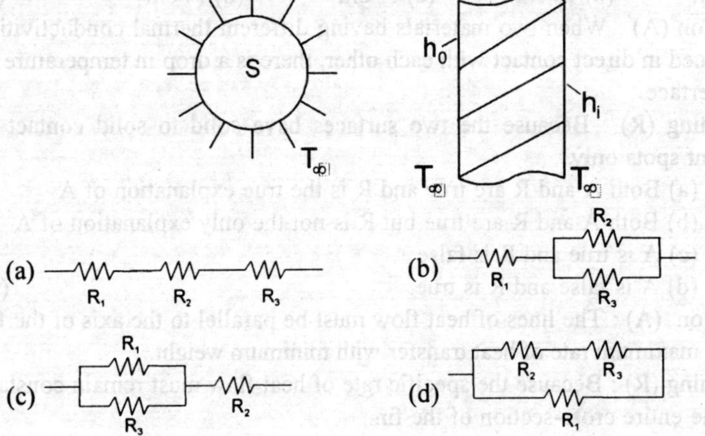

Fig. 2.34

where $R_1 = \dfrac{1}{h_0 A}$; $R_2 = \dfrac{L}{kA}$; $R_3 = \dfrac{1}{h_i A}$ **(d)**

6. If the radius of any current carrying wire is less than the critical radius, then addition of electrical insulation will enable the wire to carry a higher current because
   (a) the thermal resistance of the insulation is reduced
   (b) the thermal resistance of the insulation is increased
   (c) the heat loss from the wire would increase
   (d) the heat loss from the wire would decrease. **(c)**

7. Extended surfaces are used to increase the rate of heat transfer. When the convective heat transfer coefficient, h = mk, an addition of extended surface will
   (a) increase the heat transfer
   (b) decrease the rate of heat transfer
   (c) not increase the rate of heat transfer **(c)**
   (d) increase the rate of heat transfer when the length of the fin is very large.

8. If r_i and r_0 are the inner and outer radius of a hollow cylinder, the heat transfer by conduction through the cylindrical wall is proportional to
   (a) $r_0 - r_i$ (b) $r_0 + r_i$
   (c) $1/\ln(r_0/r_i)$ (d) $(r_0 - r_i)/\ln(r_0/r_i)$ **(c)**

9. The critical radius of insulation for a spherical shell is
   (a) h/k   (b) 2h/k   (c) k/h   (d) 2k/h **(d)**

10. Mean radius of a hollow spherical shell of inner radius r_i and outer radius r_0, for the same rate of heat transfer, is given by
   (a) $(r_0 + r_i)/2$ (b) $r_0 r_i$ (c) $(r_0 r_i)/2$ (d) $(r_0 r_i)^{1/2}$ **(c)**

11. A long pipe 4 cm O.D. and covered with 2 cm thick insulation carries a hot fluid. In order to reduce the rate of heat transfer by one third of the present rate, the thickness of the same insulation would then be
   (a) 8 cm   (b) 10 cm   (c) 12 cm   (d) 14 cm **(d)**

12. Assertion (A) : When two materials having different thermal conductivities are placed in direct contact with each other, there is a drop in temperature at the interface.
   Reasoning (R) : Because the two surfaces have solid to solid contact at different spots only.
   Code : (a) Both A and R are true and R is the true explanation of A
         (b) Both A and R are true but R is not the only explanation of A
         (c) A is true and R is false
         (d) A is false and R is true. **(b)**

13. Assertion (A) : The lines of heat flow must be parallel to the axis of the fin having maximum rate of heat transfer with minimum weight.
   Reasoning (R) : Because the specific rate of heat flow must remain constant over the entire cross-section of the fin.
   Code : (a) Both A and R are true
         (b) A is true and R is false
         (c) A is false and R is true
         (d) Both A and R are false. **(a)**

14. The thermal conductivity of a material is given by the relation
   $k = k_0(1 + \alpha T)$. Then the temperature at the mid-plane of the wall would be:

(a) equal to the average of the temperature at the two end faces

(b) more than the average of the temperature at the two end faces

(c) less than the average of the temperature at the two end faces   **(b)**

(d) dependent upon the difference in temperature between the two end faces.

15. Assertion (A) : It would be incorrect to use the conduction resistance concept with internal heat generation effects.

Reasoning (R) : Because the heat transfer rate is not independent of the spatial coordinate.

Code: (a) Both A and R are false   (b) A is true, R is false

        (c) Both A and R are true   (d) A is false, R is true   **(c)**

16. A long cylindrical rod, radius R, loses energy from the surface as q' W/m^2. The uniform internal heat generation rate would be

(a) q'/R   (b) q'/R^2 (c) 2q'/R   (d) q'/2R   **(c)**

17. Consider the following statements:

(1) The existence of a critical radius requires that heat transfer area should change in the direction of heat transfer

(2) We need not be concerned with the effects of a critical radius for a spherical system when the convection heat transfer coefficient is very small

(3) For electrical conductors carrying current the critical radius is a function of the thermal conductivity of the conductor and the convective heat transfer coefficient at the surface.

Of these statements:

(a) only 1 is correct   (b) only 2 is correct

(c) 1 and 2 are correct   (d) 1, 2 and 3 are correct   **(a)**

18. A plane wall, thickness L, has uniform internal heat generation. If the left hand face is insulated, the maximum temperature in the wall will occur at x =

(a) L   (b) L/2   (c) L/4   (d) 0   **(d)**

19. Assertion (A) : In a hollow spherical shell, the temperature gradient at the inner radius will always be zero.

Reasoning (R): Because the inner sphere is a closed surface.

Code : (a) A is false, R is true   (b) Both A and R and false

        (c) Both A and R are true   (d) A is true, R is false   **(c)**

20. If A_1 and A_2 are the internal and external surface area of a hollow cylindrical pipe, the logarithmic area is given by

(a) $(A_2 + A_1) / \ln(A_2/A_1)$ (b) $(A_2 - A_1) / \ln(A_2/A_1)$

(c) $\ln(A_2/A_1) / (A_1 + A_2)$ (d) $(A_1 + A_2) / 2$ **(b)**

21. The insulated tip temperature of a rectangular longitudinal fin will be equal to

(a) $\theta_0 \tan h\,(mL)$ (b) $\theta_0 / \sin h\,(mL)$

(c) $\theta_0 \tan h\,(mL)/(mL)$ (d) $\theta_0 / \cos h\,(mL)$ **(d)**

where $\theta_0 = T_s - T_\infty$

22. The effectiveness, E, for a very long fin

(i) increases with increasing length of the fin

(ii) is equal to the efficiency when the surface area of the fin is equal to its cross-sectional area.

(iii) should always be greater than 1.

Of these statements :

(a) only (i) is correct $\qquad$ (b) only (i) and (ii) are correct

(c) only (i) and (iii) are correct $\qquad$ (d) all are correct $\qquad$ **(d)**

NUMERICALS

1. The inner surface of a 23 cm brick wall (k = 0.865 W/mK) is kept at 820°C and the outer surface temperature is 170°C. Calculate the heat loss per unit area of the wall.

   An insulating brick wall (23 cm thick, k = 0.26 W/mK) is added to the outside of the furnance wall. Calculate the reduction in heat loss and the interface temperature if the ambient temperature is 20°C and the convective heat transfer coefficient at the outer surface is 11.9 W/m^{2}K.

   (2.45 kW/m^2, 73.6%, 648°C, 74°C)

2. An insulating wall of a cold room is composed of a 15 cm thick material (k = 1 W/mK) and an unknown thickness of cork (k = 0.045 W/mK). If the inside and outside wall temperatures are -10°C and 25°C, determine the thickness of the cork to keep the heat loss to 8 W/m^2.

   (19 cm)

3. A double-glazed window, 1.4 m high and 2.3 m wide, consists of two layer of glass each 3 mm thick, separated by an air gap 20 mm thick. The window is installed in the wall of a room in which the air temperature is maintained at 23°C and the outer surface of the window is exposed to air at 2°C. The inside and outside heat transfer coefficient is 5.7 W/m^{2}K and 9.1 W/m^{2}K. The thermal conductivity of glass is 0.76 W/mK.

   Assuming that the air in the gap between the two laycrs of glass is a stationary layer of air (k = 0.026 W/mK), calculate the heat transfer rate through the window and the temperature at the room side surface of the glass.

   (63 W, 19.6°C)

4. The inner surface of a 23 cm brick wall of a furnace is kept at 820°C, and it is found that the outcr surface temperature is 170°C. Calculate the heat loss per m^2 of wall area given that the conductivity is 0.865 W/mK.

   An insulating brick wall 23 cm thick, conductivity 0.26 W/mK is added to the outside of the furnance wall. Calculate the reduction in heat loss and the brick interface and outer surface temperatures. Assume that the inner surface temperature remains unchanged, that the surroundings are at 20°C, and the combined convection and radiation heat transfer coefficient for the inner surface is 11.9 W/m^{2}K.

   (2450 W/m^2, 73.6 percent, 648°C, 74°C)

5. Hot water flows through a 5 cm I.D. and 6.5 cm O.D. steel pipe (k = 50 W/mK) The average temperature of the hot water is 95°C and the ambient air temperature is 15°C. The convective heat transfer coefficient at the inside and outside surface of the pipe are 1kW/m^{2}K and 25 W/m^{2}K. Determine the minimum thickness of insulation (k = 0.085 W/mK) required so that the outer surface of the insulation is at 40°C and the rate of heat loss does not exceed 150 W/m length of the pipe.

6. A steam main of 150 mm O.D. containing wet steam at 28 bar is insulated with an insulating material (thickness 40 mm, k = 0.09 W/mK and an outer layer of another insulating material (25 mm thick, k = 0.06 W/mk). The convective heat transfer coefficient at the outer surface is 17 W/m²K Neglecting the thermal resistance of the pipe material, calculate the rate of heat loss per metre length of the pipe (if the ambient temperature is 20°C), and the temperature at the outer surface.

   (0.156 kW/m, 30.5°C)

7. A 5 cm O.D. pipe is to be covered with two layers of insulation each 5 cm thick. If the thermal conductivity of one material is 4 times that of the other, and the inner and outer surface temperatures of the composite insulation remains fixed, calculate the ratio of the heat transfer when the better insulating material is applied first to the pipe to the heat transfer when the better insulating material forms the outer layer.

8. A spherical pressure vessel, 1 m inside diameter is made of 20 mm steel plate (k = 48 W/mK). The vessel has two layers of insulation: the first one is 25 mm thick, k = 0.047 W/mK and the next one is 10 mm thick, k = 0.21 W/mK. The heat transfer coefficient at the outside surface is 20 W/m²K. Neglecting radiation, calculate the rate of heat loss from the sphere when the inside surface is at 500°C and the room temperature is 20°C.

   (2.744 kW)

9. A spherical tank 1 m O.D. is maintained at 120°C. The heat transfer coefficent at the surface is 25 W/m²K and the ambient temperature is 15°C. What thickness of insulation, k = 0.036 W/mK, should be added so that the outer temperature of insulation is 35°C? Estimate the percentage reduction in heat loss after the insulation is applied.

10. A hollow sphere, (30 cm I.D, 60 cm O.D. and k = 50 W/mK) contains a liquid chemical mixture releasing heat energy as 30 kW. Calculate the inside surface temperature and the heat transfer coefficient if the temperature at the outside surface is 100°C and the ambient temperature is 25°C.

    (259°C, 353.6 W/m²°C)

11. A 5 cm thick plane wall (k = 10 W/mK) generates heat energy at the rate of 1 MW/m³.The left hand face of the wall is insulated and the right hand face is exposed to air at 30°C with heat transfer coefficient as 1 kW/m²K. Estimate the maximum temperature in the wall and the temperature at the right face.

12. Heat is generated in a 2.5 cm-square copper rod at the rate of 35 MW/m³. The rod is exposed to the surroundings at 25°C and the convective heat transfer coefficient is 4000 W/m² °C. Calculate the surface temperature of the rod.   **(80°C)**

13. A stainless steel wire, 3 mm in diameter and k = 20 W/mK, is 1 metre long and carries a current of 200 amperes, the resistivity being 70 μ-ohm-cm. The wire is submerged in a liquid at 120°C and the surface heat transfer coefficient is 400 W/m²K. Calculate the temperature at the centre of the wire.

    (240.75°C)

14. An electric cable (k = 150 W/mK, 25 mm diameter) carries a current of 200 amps and has a resistance of 0.4×10^{-4} ohm per cm length. The ambient and surface temperatures are 20°C and 210°C. Calculate the surface heat transfer coefficient and the maximum temperature in the wire.

    (10.7 W/m²K, 210.17°C)

15. A steel fin (k = 54 W/mK) having a cross-section of an equilateral traiangle (each side 5 mm) is 8 cm long. The temperature at the root of the fin is 400°C and the ambient temperature is 50°C. Calculate the heat dissipated by the fin if the surface heat transfer coefficient is 90 W/m²K.

    (9.8 W)

16. An aluminium fin (k = 200 W/mK) 18 mm thick and 20 cm long has a root temperature of 300°C. The ambient temperature is 25°C and the surface heat transfer coefficient is 30 W/m²K. Determine the fin efficiency and the heat dissipated from the fin per unit depth.

17. An aluminium fin, k = 200 W/mK, 1.6 mm thick is placed radially on a circular tube 2.5 cm O.D. The fin is 6.4 mm long. The tube wall temperature is at 150°C whereas the surroundings are at 25°C. Calculate the loss by the fin if the convective heat transfer coefficient is 25 W/m²K.

18. Two plates are kept at 190°C and 50°C in a room where the air temperature is 20°C. The plates are connected by 4 metallic bolts (k = 100 W/m°C) each 1.25 cm in diameter and 30 cm long. The convective heat transfer coefficient is 10 W/m²K. Calculate the quantity of heat energy conducted away from the hotter plate and the quantity of heat energy reaching the colder plate.

19. A thermometer well is made of a steel tube 1 mm thick 150 mm in length (k = 60 W/mK). Estimate the true temperature of the fluid stream if the pipe wall temperature is 50°C, the temperature recorded by the thermometer is 100°C and the surface heat transfer coefficient is 30 W/m²C.

20. One end of a long rod is inserted into a furnace and the other end is in air at 30°C. The temperature measured at two different points on the rod, separated by 10 cm, are 127°C and 93°C. The diameter of the rod is 2 cm, surface heat transfer coefficient 1 7.4 W /m²C. Estimate the thermal conductivity o f the rod.

    (187.5 W/mK)

CHAPTER 3

Unsteady State Conduction Heat Transfer

1. Transient State Systems—Defined

The process of heat transfer by conduction where the temperature varies with time and with space coordinates, is called 'unsteady or transient'. All transient state systems may be broadly classified into two categories:

(a) Non-periodic Heat Flow System – the temperature at any point within the system changes as a non-linear function of time.

(b) Periodic Heat Flow System – the temperature within the system undergoes periodic changes which may be regular or irregular but definitely cyclic.

There are numerous problems where changes in conditions result in transient temperature distributions and they are quite significant. Such conditions are encountered in – manufacture of ceramics, bricks, glass and heat flow to boiler tubes, metal forming, heat treatment, etc.

2. Biot and Fourier Modulus—Definition and Significance

Let us consider an initially heated long cylinder (L >> R) placed in a moving stream of fluid at $T_\infty < T_s$, as shown in Fig. 3.1(a). The convective heat transfer coefficient at the surface is h, where,

$$Q = hA(T_s - T_\infty)$$

This energy must be conducted to the surface, and therefore,

$$\dot{Q} = -kA(dT/dr)_{r=R}$$

or, $\qquad h(T_s - T_\infty) = -k(dT/dr)_{r=R} \approx -k(T_c - T_s)/R$

where T_c is the temperature at the axis of the cylinder.

By rearranging, $(T_s - T_c)/(T_s - T_\infty) = h\,R/k$. $\hspace{2cm}$ (3.1)

The term, hR/k, is called the 'BIOT MODULUS'. It is a dimensionless number and is the ratio of internal heat flow resistance to external heat flow resistance and plays a fundamental role in transient conduction problems involving surface convection effects. It provides a measure of the temperature drop in the solid relative to the temperature difference between the surface and the fluid.

For Bi << 1, it is reasonable to assume a uniform temperature distribution across a solid at any time during a transient process.

Fourier Modulus – It is also a dimensionless number and is defined as

$$Fo = \alpha t / L^2 \qquad (3.2)$$

where L is the characteristic length of the body, α is the thermal diffusivity, and t is the time.

The Fourier modulus measures the magnitude of the rate of conduction relative to the change in temperature, i.e., the unsteady effect. If Fo <<1, the change in temperature will be experienced by a region very close to the surface.

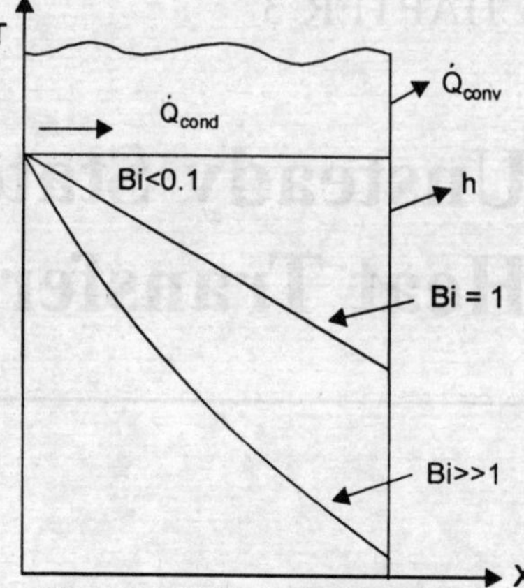

Fig. 3.1 Effect of Biot Modulus on steady state temperature distribution in a plane wall with surface convection.

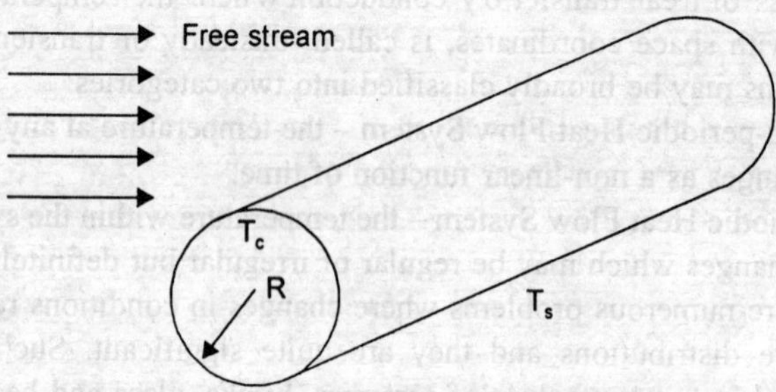

Fig. 3.1(a) Nomenclature for Biot Modulus

3. Lumped Capacity System—Necessary Physical Assumptions

We know that a temperature gradient must exist in a material if heat energy is to be conducted into or out of the body. When Bi < 0.1, it is assumed that the internal thermal resistance of the body is very small in comparison with the external resistance and the transfer of heat energy is primarily controlled by the convective heat transfer at the surface. That is, the temperature within the body is approximately uniform. This idealised assumption is possible, if

(a) the physical size of the body is very small,

(b) the thermal conductivity of the material is very large, and

(c) the convective heat transfer coefficient at the surface is very small and there is a large temperature difference across the fluid layer at the interface.

4. An Expression for Evaluating the Temperature Variation in a Solid Using Lumped Capacity Analysis

Let us consider a small metallic object which has been suddenly immersed in a fluid during a heat treatment operation. By applying the first law of thermodynamics, we have

Heat flowing out of the body = Decrease in the internal thermal energy of
during a time dt   the body during that time dt

or, $$hA_s(T-T_\infty)dt = -\rho CVdT$$

where A_s is the surface area of the body, V is the volume of the body and C is the specific heat capacity.

or, $$(hA/\rho CV)dt = -dT/(T-T_\infty)$$

with the initial condition being: at $t = 0$, $T = T_s$.

The solution is : $(T-T_\infty)/(T_s-T_\infty) = \exp(-hA/\rho CV)t$ (3.3)

Fig. 3.2 depicts the cooling of a body (temperature distribution ~ time) using lumped thermal capacity system. The temperature history is seen to be an exponential decay.

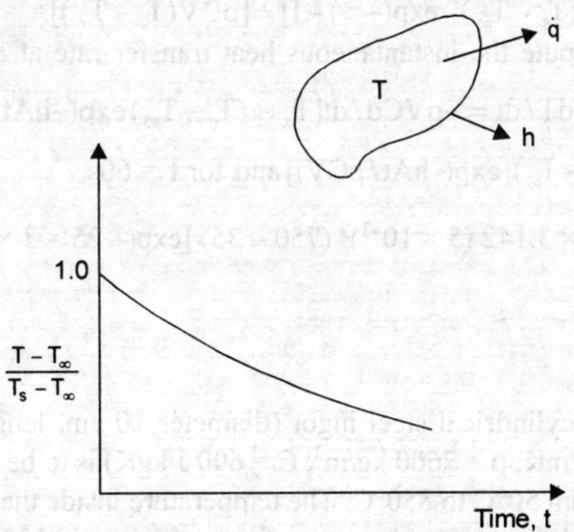

Fig. 3.2 Cooling of a lumped capacity system

We can express

$$Bi \times Fo = (hL/k)\times(\alpha t/L^2) = (hL/k)(k/\rho C)(t/L^2) = (hA/\rho CV)t,$$

where V/A is the characteristic length L.
And, the solution describing the temperature variation of the object with respect to time is given by

$$(T-T_\infty)/(T_s-T_\infty) = \exp(-Bi\cdot Fo)$$ (3.4)

Example 3.1   Steel balls 10 mm in diameter ($k = 48$ W/mK), ($C = 600$ J/kgK) are cooled in air at temperature 35°C from an initial temperature of 750°C. Calculate the time required for the temperature to drop to 150°C when $h = 25$ W/m²K and density $\rho = 7800$ kg/m3.

Solution: Characteristic length, $L = V/A = 4/3\pi r^3/4\pi r^2 = r/3 = 5\times10^{-3}/3$ m
                  $Bi = hL/k = 25\times5\times10^{-3}/(3\times48) = 8.68\times10^{-4} \ll 0.1$,

Since the internal resistance is negligible, we make use of lumped capacity analysis: Eq. (3.4),

$$(T - T_\infty) / (T_s - T_\infty) = \exp(-Bi\ Fo); (150 - 35) / (750 - 35) = 0.16084$$

$$\therefore \qquad Bi \times Fo = 1.827; Fo = 1.827/(8.68 \times 10^{-4}) = 2.1 \times 10^3$$

or, $\qquad \alpha t/L^2 = k/(\rho CL^2)t = 2100$ and $t = 568 = 0.158$ hour

We can also compute the change in the internal energy of the object as:

$$U_0 - U_t = -\int_0^t \rho CV dT = \int_0^t -\rho CV(T_s - T_\infty)(-hA/\rho CV)\exp(-hAt/\rho CV)dt$$

$$= -\rho CV(T_s - T_\infty)[\exp(-hAt/\rho CV) - 1] \qquad (3.5)$$

$$= -7800 \times 600 \times (4/3)\pi (5 \times 10^{-3})^3 (750 - 35)(0.16084 - 1)$$

$$= 1.47 \times 10^3 \text{ J} = 1.47 \text{ kJ}.$$

If we allow the time 't' to go to infinity, we would have a situation that corresponds to steady state in the new environment. The change in internal energy will be

$$U_0 - U_\infty = [\rho CV(T_s - T_\infty)\ \exp(-\infty) - 1] = [\rho CV(T_s - T_\infty)].$$

We can also compute the instantaneous heat transfer rate at any time.

or, $\quad \dot{Q} = -\rho VCdT/dt = -\rho VCd/dt[T_\infty + (T_s - T_\infty)\exp(-hAt/\rho CV)]$

$$= hA(T_s - T_\infty)[\exp(-hAt/\rho CV)] \text{ and for } t = 60s,$$

$$\dot{Q} = 25 \times 4 \times 3.142 (5 \times 10^{-3})^2 (750 - 35) [\exp(-25 \times 3 \times 60/5 \times 10^{-3} \times$$

$$7800 \times 600)]$$

$$= 4.63 \text{ W}.$$

Example 3.2 A cylindrical steel ingot (diameter 10 cm. length 30 cm, k = 40 W/mK. $\rho = 7600$ kg/m^3, C = 600 J/kgK) is to be heated in a furnace from 50°C to 850°C. The temperature inside the furnace is 1300°C and the surface heat transfer coefficient is 100W/m^{2}K. Calculate the time required.

Solution: Characteristic length. $L = V/A = \pi r^2 L/2\pi r(r + L) = rL/2(r + L)$

$$= 5 \times 10^{-2} \times 30 \times 10^{-2}/ (2 (5 + 30) \times 10^{-2})$$

$$= 2.143 \times 10^{-2} \text{ m}.$$

$$Bi = hL/k = 100 \times 2.143 \times 10^{-2}/40 = 0.0536 \ll 0.1$$

$$Fo = \alpha t/L^2 = (k/\rho C) \times (t/L^2)$$

$$= 40 \times t/(7600 \times 600 \times [2.143 \times 10^{-2}]^2) = 1.91 \times 10^{-2}t$$

and $\qquad (T - T_\infty)/(T_s - T_\infty) = \exp(-Bi\ Fo)$

or, $\qquad (850 - 1300) / (50 - 1300) = 0.36 = \exp(-Bi\ Fo)$

$\therefore \qquad Bi\ Fo = 1.02$

and $\qquad Fo = 19.06$ and $t = 19.06/(1.91 \times 10^{-2}) = 16.63$ min.

(The length of the ingot is 30 cm and it must be removed from the furnace after a period of 16.63 min, therefore, the speed of the ingot would be $0.3/16.63 = 1.8 \times 10^{-2}$ m/min.)

Example 3.3 A block of aluminium (2cm × 3cm × 4cm, k = 180 W/mK, $\alpha = 10^{-4} m^2/s$) initially at 300°C is cooled in air at 30°C. Calculate the temperature of the block after 3 min. Take h = 50W/m^2K.

Solution: Characteristic length, L = $[2 \times 3 \times 4/(2(2 \times 3 + 2 \times 4 + 3 \times 4))] \times 10^{-2}$
$$= 4.6 \times 10^{-3} \text{ m.}$$
$$Bi = hL/k = 50 \times 4.6 \times 10^{-3}/180 = 1.278 \times 10^{-3} << 0.1$$

$$Fo = \alpha t / L^2 = 10^{-4} \times 180/(4.6 \times 10^{-3})^2 = 850$$

$$\exp(-Bi\ Fo) = \exp(-1.278 \times 10^{-3} \times 850) = 0.337$$

$$(T - T_\infty)/(T_s - T_\infty) = (T - 30)/(300 - 30) = 0.337$$

$$\therefore \qquad\qquad T = 121.1°C.$$

Example 3.4 A copper wire 1 mm in diameter initially at 150°C is suddenly dipped into water at 35°C. Calculate the time required to cool to a temperature of 90°C if h = 100 W/m^2K. What would be the time required if h = 40 W/m^2K. (for copper : k = 370 W/mK, ρ = 8800 kg/m^3, C = 381 J/kgK.

Solution: The characteristic length for a long cylindrical object can be approximated as r/2. As such,
$$Bi = hL/k = 100 \times 0.5 \times 10^{-3}/(2 \times 370) = 6.76 \times 10^{-5} << 0.1$$

$$Fo = \alpha t / L^2 = (k/\rho C) \times (t/L^2)$$
$$= [370t/(8800 \times 381 \times (0.25 \times 10^{-3})^2] = 1760t$$
$$\exp(-Bi\ Fo) = (T - T_\infty)/(T_s - T_\infty)$$
$$= (90 - 35)/(150 - 35) = 0.478$$
$$Bi\ Fo = 0.738 = 6.76 \times 10^{-5} \times 1760\ t; \qquad\qquad \therefore \quad t = 6.2s$$
when $\qquad$ h = 40 W/m^2K, Bi = 2.7 × 10^{-5} and 2.7 × 10^{-5} × 1760 t = 0.738;
or, $\qquad\qquad$ t = 15.53s.

Example 3.5 A metallic rod (mass 0.1 kg, C = 350 J/kgK, diameter 12.5 mm, surface area 40cm^2) is initially at 100°C. It is cooled in air at 25°C. If the temperature drops to 40°C in 100 seconds, estimate the surface heat transfer coefficient.

Solution: $hA/\rho CV = hA/mC = h \times 40 \times 10^{-4}/(0.1 \times 350) = 1.143 \times 10^{-4} h$

and, $\qquad$ $hAt/\rho CV = 1.143 \times 10^{-4} h \times 100 = 1.143 \times 10^{-2} h$

$$(T - T_\infty)/(T_s - T_\infty) = (40 - 25)/(100 - 25) = 0.2$$
$$\therefore \qquad \exp(-1.143 \times 10^{-2}h) = 0.2$$
or, $\qquad\qquad$ 1.143 × 10^{-2}h = 1.6094, and h = 140W/m^2K.

5. Response Time of a Temperature Measuring Instrument

A temperature measuring instrument should attain the temperature of the source, to which it is exposed, as quickly as possible. The response time is the time taken by the temperature measuring instrument to attain the source temperature. In lumped heat capacity analysis, the index $(hA/\rho CV)t$ should be very large so that the exponential term reaches zero very fast.

The term $(\rho CV/hA)$ has the units of time and it is often referred to as the time constant of the system. At time, t, equal to one time constant

$$(T - T_\infty)/(T_s - T_\infty) = \exp(-1.0) = 0.368$$

i.e., at the end of one time constant, the difference in the temperature of the body and that of the ambient fluid would be 36.8 percent of the initial temperature difference. Therefore, for all practical purposes, a system is said to attain steady state after a time of 4 time constants has elapsed.

or, $(T - T_\infty)/(T_s - T_\infty) = \exp(-4) = 0.018.$

(The response time or the time constant for different sizes of thermocouples usually varies between 0.04 seconds to 2.5 seconds depending upon the mass flow rates of the ambient fluid which controls the surface heat transfer coefficient.)

Example 3.6   Determine the time constant for a copper constantan thermocouple with configuration as shown in Fig. 3.3 The thermocouple is exposed to an air stream with h = 30 W/m²K. Calculate the time required for the thermocouple to record 150°C from an initial temperature of 25°C when the air temperature is 155°C. (k = 175 W/mK, C = 418.7 J/kgK, ρ = 8960 kg/m³)

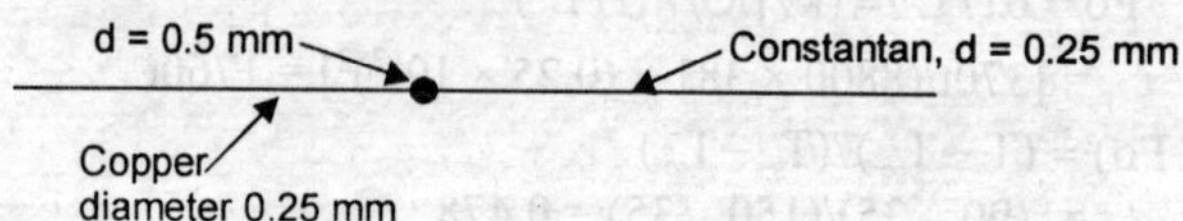

Fig. 3.3 A thermocouple

Solution: The thermal response is dictated by the beed temperature. Neglecting conduction through the wires, the beed temperature is controlled by the convection heat transfer at the surface.

The characteristic length, L = V/A = R/3 for a sphere = 0.25/3 mm

Time constant = $\rho CV/hA = \rho CL/h = 8960 \times 418.7 \times 0.25 \times 10^{-3}/(3\times30) = 1.785$s

   Bi = hL/k = $30 \times 0.0833 \times 10^{-3}/175 = 1.428 \times 10^{-5} \ll 0.1$

Using lumped system analysis,

$$(T - T_\infty)/(T_s - T_\infty) = (150 - 155)/(25 - 155) = \exp(-t/1.785)$$

∴ $t = 5.815$s.

Example 3.7   A mercury in glass thermometer 2.55 mm diameter is placed in an air stream where the surface heat transfer coefficient is 60 W/m²K. Considering cylindrical thermometer bulbs, estimate the time required to reach three-fourth its initial value. k = 10 W/mK, α = 0.017 m²/hr for mercury.

Solution: $\rho CV/hA = (k/\alpha)(V/hA) = (k/\alpha)(R/2h)$; for a cylinder, L = R/2

$$= (10/0.017)(0.125 \times 10^{-2}/2 \times 60) \times 3600 = 22.06s$$

$(T - T_\infty)/(T_s - T_\infty) = 0.75 = \exp(-t/22.06)$; ∴ $t = 22.06\ln(4/3) = 6.34$s.

(Mercury in glass thermometers can record slower unsteady temperature changes.)

Example 3.8   A 2 mm diameter copper wire is used as a thermocouple. The initial temperature is 150°C. Calculate the time required for the thermocouple to attain thermal equilibrium when it is suddenly immersed in water (temperature 40°C, h = 80 W/m^{2}K) and in air (temperature 40°C, h = 10 W/m^{2}K). Take for copper: k = 390 W/mK, C = 384 J/ kgK, ρ = 8940 kg/m^3.

Solution: For a cylindrical wire, the characteristic length, L = R/2 = 1 mm. Biot modulus, when immersed in water, = hL/k = 80 × 1 × 10^{-3}/390 << 0.1. The Biot modulus will be much smaller when the wire is immersed in air, therefore we can safely neglect the internal resistance.

(a) Immersed in water: time constant = $\rho CV / hA = \rho CL / h$

$$= \frac{8940 \times 384 \times 10^{-3}}{80} = 42.912 \text{s}$$

For   T = 50°C,

Using lumped system analysis, $(T - T_\infty)/(T_s - T_\infty) = (50 - 40)/(150 - 40)$
$$= 0.091$$

or   exp (–t/42.912) = 0.091; or, –t/42.912 = –2.398;   and t = 1.715 min.

(b) Immersed in air: time constant = 8940 × 1 × 10^{-3} × 384/10 = 427.2s

The time required to record a temperature of 50°C, is then

exp (–t/427.2) = 0.091;   or, –t/427.2 = –2.398   and t = 17.07 min.

(A 2 mm diameter themocouple wire is not suitable to measure rapid changes in the temperature of the medium. However, a better result can be obtained by using wire of much smaller diameter.)

Also, to record a temperature within 2% of the ambient fluid, the time would be 4 × 42.912 = 2.86 min for water and 4 × 427.2 = 28.48 min for air.

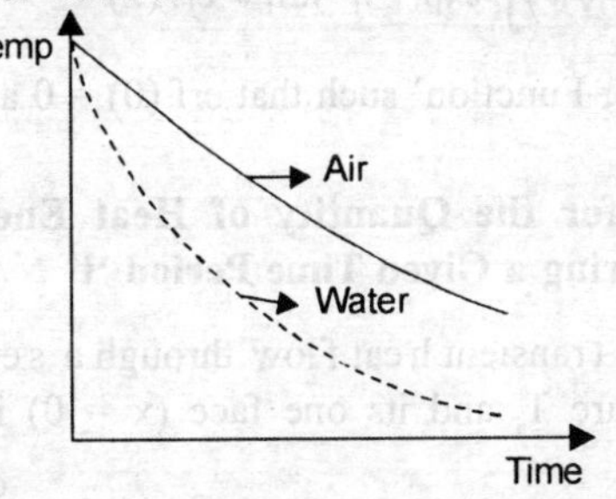

**Fig. 3.4 Temperature response of thermocouple in Ex. 3.8
after immersion in air and water**

Example 3.9   Calculate the junction diameter of a copper themocouple, initially at 25°C and is placed in a gas stream at 200°C and when it measures a temperature of 198°C within 5 seconds. For copper, ρ = 8940 kg/m^3, C = 384 J/kgK, k = 390 W/mK and the convective heat transfer coefficient = 400 W/m^{2}K.

Solution: The dimensionless temperature after 5 seconds
$$= (198 - 200) / (25 - 200) = 0.0114285$$
Assuming one lump approximation, exp (–t/t*) = 0.0114285

For $\quad$ t = 5 seconds, the time constant, t* = (−5)/1n(0.0114285) = 1.11816

$\quad\quad$ t* = ρ C L/h and therefore, the characteristic length, L would be

$\quad\quad$ L = 1.11816 × 400/8940 × 384 = 1.30285 × 10⁻⁴ m

For a spherical junction, D = 6L = 6 × 1.30285 × 10⁻⁴m = 0.782 mm

Check : Biot modulus = hL/k = 400 × 1.30285 × 10⁻⁴/390 << 0.1.

6. A Semi-infinite Solid—Definition

A semi-infinite body is shown in Fig. 3.5. The body is assumed to be infinite in the Y- and Z-directions and extends to + ∞ only in the X-direction. The term semi-infinite arises because of the finite termination of the body at x = 0, Further, it is assumed that :

$\quad$ (a) the temperature distribution in the body initially (at t = 0) is uniform.

$\quad$ (b) the temperature of the body at x = 0 suddenly changes to T_0 for all time, t > 0.

$\quad$ (c) there is always a section of the body where the temperature remains unchanged.

$\quad$ (d) the heat transfer coefficient, h, at the face x = 0 is constant and uniform.

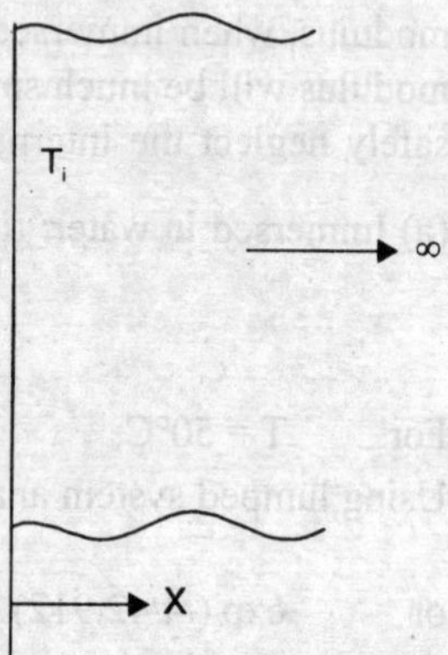

Fig. 3.5 A semi-infinite body

7. Error Function—Definition

The integral $\int_0^z \exp(-\eta^2)d\eta$ where η is a dummy variable can be solved analytically and the expression

$$\left(2/\sqrt{\pi}\right)\int_0^z \exp(-\eta^2)d\eta = \text{erf}(z) \tag{3.6}$$

is called 'Gauss Error Function' such that erf (0) = 0 and as $z \rightarrow \infty$, erf(z) $\rightarrow 1$.

8. An Expression for the Quantity of Heat Energy Added to a Semi-infinite Solid During a Given Time Period 't'

For one-dimensional transient heat flow through a semi-infinite solid which is initially at temperature T_i and its one face (x = 0) is suddenly changed to a temperature T_0, the governing equation is given by $\dfrac{\partial^2 T}{\partial x^2} = \dfrac{1}{\alpha}\dfrac{\partial T}{\partial t}$ with relevent boundary condition as:

$\quad\quad$ for t > 0: T (0, t) = T_0 and T (∞, t) = T_i

and the initial condition: at t = 0, T (x, 0) = T_i

$\quad\quad$ Let θ = T − T_0 and then the governing equation reduces to:

$$\frac{\partial^2 \theta}{\partial x^2} = \frac{1}{\alpha}\frac{\partial \theta}{\partial t} \quad\quad \text{with conditions : } t = 0, \ \theta = T_i - T_0 = \theta_o$$

and $\quad\quad\quad\quad\quad\quad\quad\quad\quad\quad$ t > 0, $\theta(0, t) = 0$

$\quad\quad\quad\quad\quad\quad\quad\quad\quad\quad\quad\quad$ $\theta(\infty, t) = \theta_o$

We choose a variable $\eta = \dfrac{x}{2\sqrt{\alpha t}}$ and substitute it in the governing equation to get $\dfrac{\partial^2 \theta}{\partial \eta^2} + 2\eta \dfrac{d\theta}{d\eta} = 0$

Upon intergration, $\ln\left(\dfrac{d\theta}{d\eta}\right) = -\eta^2 + \text{const.};$ or $\dfrac{d\theta}{d\eta} = C_2 \exp(-\eta^2)$

and $\theta = C_1 + C_2 \displaystyle\int \exp(-\eta^2)d\eta = C_1 + (C_2\sqrt{\pi}/2)\,\text{erf}(\eta)$

In order to evaluate the two constants, we apply the initial and boundary conditions:

when $\theta = 0$ at $\eta = 0$, $C_1 = 0$

$\qquad \theta = \theta_0$ at $\eta \to \infty$, $C_2 = 2\theta_0/\sqrt{\pi}$

$\therefore \quad \theta = \theta_0\,\text{erf}(\eta)$

and,
$$\frac{(T-T_0)}{(T_i - T_0)} = \frac{2}{\sqrt{\pi}} \int_0^{x/2\sqrt{\alpha t}} \exp(-\eta^2)d\eta \qquad (3.7)$$

The surface heat flow Q_0 at any instant of time is calculated from Fourier law:

$$\dot{Q}_0(t) = -kA\left.\frac{\partial T}{\partial x}\right|_{x=0} = -kA(T_i - T_0)/\sqrt{\pi\alpha t}$$

$$= +kA(T_0 - T_i)/\sqrt{\pi\alpha t} \qquad (3.8)$$

and $\dfrac{\partial T}{\partial x}$ is maximum when $t = x^2/2\alpha$

The total heat added is equal to the change in the internal energy of the semi-infinite solid during the time 't'. Therefore upon integration,

$$\dot{Q} = U_T - U_0 = \int_0^\tau \dot{Q}_0 dt = \int_0^\tau \frac{kA(T_0 - T_i)}{\sqrt{(\pi\alpha t)}}\,dt = \frac{2kA(T_0 - T_i)}{\sqrt{(\pi\alpha/t)}} \qquad (3.9)$$

Example 3.10 The initial temperature of the ground soil ($\alpha = 9.29 \times 10^{-3}\,\text{m}^2/\text{h}$) is 30°C. During winter, the surface temperature suddenly drops to -15°C. Estimate the depth to which the freezing temperature penetrates during a 10 hour period.

Solution: The dimensionless temperature distribution is

$$\frac{T(x,t) - T_0}{T_i - T_0} = \frac{0 - (-15)}{30 - (-15)} \doteq \frac{15}{45} = 0.333 = \text{erf}(\eta) = \text{erf}\left(\frac{x}{2\sqrt{\alpha t}}\right)$$

$$\therefore \qquad 0.333 = \text{erf}\left(x/2\sqrt{(9.29 \times 10^{-3} \times 10)}\right)$$

From Table 3.1 : $\text{erf}(\eta) = 0.333$ gives $\eta = 0.30458$, $x/2(\alpha t) = 0.30458$ and $x = 18.6$ cm. If $t = 10$ days, $x = 0.909$ m.

(Therefore the water pipes should be laid below a depth of 1.0 m to avoid freezing of water in pipes.)

Example 3.11 A large steel slab ($k = 40$ W/mK, $\alpha = 4.18 \times 10^{-2}$ m^2/h) initially at a temperature of 325°C, has its surface temperature lowered to 50°C. Calculate the time required for the temperature to reach 120°C at a depth of 2.5 cm. Also determine the total heat removed from the slab per unit area during this period.

Solution: The slab can be considered as a semi-infinite solid. The dimensionless temperature ratio is:

$$[T(x,t) - T_0]/(T_i - T_0) = (120 - 50)/(325 - 50)$$

$$= 0.2545 = \mathrm{erf}(x/\sqrt{(4\alpha t)})$$

From Table 3.1: $x/\sqrt{(4\alpha t)} = 0.23$ and therefore,

$$t = \left(\frac{0.025}{0.23}\right)^2 \Big/ (4 \times 4.18 \times 10^{-2}) = 4.23 \ \mathrm{min}$$

The quantity of heat removed per unit area is, Eq (3.9)

$$\dot{Q}/A = 2k(T_0 - T_i)/\sqrt{\pi\alpha/t} = 2 \times 40 \times 275/\sqrt{\pi 4.18 \times 10^{-2} \times 60/4.23} = 58\mathrm{MJ/m}^2.$$

Example 3.12 A large aluminium slab ($k = 180$ W/mK, $\alpha = 8.4 \times 10^{-5}$ m^2/s) is initially at a uniform temperature of 40°C. The surface temperature is suddenly raised to 300°C. Calculate the temperature at a depth of 2.5 cm after 60 seconds. What would be the temperature at that depth if a constant heat flux of 0.2 MJ/m^2 is added.

Solution: $x/\sqrt{(4\alpha t)} = 2.5 \times 10^{-2}/\sqrt{4 \times 8.4 \times 10^{-5} \times 60} = 0.176$

From Table 3.1 $\mathrm{erf}[x/\sqrt{(4\alpha t)}] = 0.1965$

Dimensionless temperature ratio, $[T(x, t) - 300]/(40 - 300) = 0.1965$
or, $T = 248.9$°C

When a constant heat flux $\dot{Q}_0$ is applied on a semi–infinite slab, the governing equation is $\dfrac{\partial^2 T}{\partial x^2} = \dfrac{1}{\alpha}\dfrac{\partial T}{\partial t}$ and is solved with boundary conditions

$$\dot{Q}_0/A = -k\partial T/\partial x\big|_{x=0}; \text{ and initial condition } T(x, 0) = T_i$$

The solution is given by

$$T - T_i = \frac{2Q_0}{kA}\sqrt{\frac{\alpha t}{\pi}}\exp(-x^2/4\alpha t) - \frac{Q_0 x}{kA}\left[1 - \mathrm{erf}(x/\sqrt{4\alpha t})\right]$$

For $x = 0.025$ m   and $t = 60$ s,

$$T = 40 + [(2 \times 0.2 \times 10^6)/180] + \sqrt{(8.4 \times 10^{-5} \times 60/3.142)} \ \bullet$$

$$[\exp(-0.176^2) - (0.2 \times 10^6 \times 0.025/180)(1 - 0.1965)]$$

$$= 40 + 86.28 - 22.32 = 103.96°\mathrm{C}$$

The temperature at the surface of the slab can be evaluated as

$$T = T_i + (2\dot{Q}_0/kA)\sqrt{(\alpha t/\pi)}$$

$$= 40 + 89 = 129°\mathrm{C} \cdot$$

Table 3.1 The Error Function

$\dfrac{x}{2\sqrt{\alpha\tau}}$	$\text{erf}\left(\dfrac{x}{2\sqrt{\alpha\tau}}\right)$	$\dfrac{x}{2\sqrt{\alpha\tau}}$	$\text{erf}\left(\dfrac{x}{2\sqrt{\alpha\tau}}\right)$	$\dfrac{x}{2\sqrt{\alpha\tau}}$	$\text{erf}\left(\dfrac{x}{2\sqrt{\alpha\tau}}\right)$
0.00	0.0000	0.76	0.71754	1.52	0.96841
0.02	0.02256	0.78	0.73001	1.54	0.97059
0.04	0.04511	0.80	0.74210	1.56	0.97263
0.06	0.06762	0.82	0.75381	1.58	0.97455
0.08	0.09008	0.84	0.76514	1.60	0.97635
0.10	0.11246	0.86	0.77610	1.62	0.97804
0.12	0.13476	0.88	0.78669	1.64	0.97962
0.14	0.15695	0.90	0.79691	1.66	0.98110
016	0.17901	0.92	0.80677	1.68	0.98249
0.18	0.20094	0.94	0.81627	1.70	0.98379
0.20	0.22270	0.96	0.82542	1.72	0.98500
0.22	0.24430	0.98	0.83423	1.74	0.98613
0.24	0.26570	1.00	0.84270	1.76	0.98719
0.26	0.28690	1.02	0.85084	1.78	0.98817
0.28	0.30788	1.04	0.85865	1.80	0.98909
0.30	0.32863	1.06	0.86614	1.82	0.98994
0.32	0.34913	1.08	0.87333	1.84	0.99074
0.34	0.36936	1.10	0.88020	1.86	0.99147
0.36	0.38933	1.12	0.88079	1.88	0.99216
0.38	0.40901	1.14	0.89308	1.90	0.99279
0.40	0.42839	1.16	0.89910	1.92	0.99338
0.42	0.44749	1.18	0.90484	1.94	0.99392
0.44	0.46622	1.20	0.91031	1.96	0.99443
0.46	0.48466	1.22	0.91553	1.98	0.99489
0.48	0.50275	1.24	0.92050	2.00	0.995322
0.50	0.52050	1.26	0.92524	2.10	0.997020
0.52	0.53790	1.28	0.92973	2.20	0.998137
0.54	0.55494	1.30	0.93401	2.30	0.998857
0.56	0.57162	1.32	0.93806	2.40	0.999311
0.58	0.58792	1.34	0.94191	2.50	0.999593
0.60	0.60386	1.36	0.94556	2.60	0.999764
0.62	0.61941	1.38	0.94902	2.70	0.999866
0.64	0.63459	1.40	0.95228	2.80	0.999925
0.66	0.64938	1.42	0.95538	2.90	0.999959
0.68	0.66278	1.44	0.95830	3.00	0.999978
0.70	0.67780	1.46	0.96105	3.20	0.999994
0.72	0.69143	1.48	0.96365	3.40	0.999998
0.74	0.70468	1.50	0.96610	3.60	1.000000

Example 3.13 A car weighing 14 kN is travelling at a speed of 60 kmph. It is stopped in 6 seconds with brake bands of 275 cm^2 area each pressing against steel drums ($k = 60$ W/mK, $\alpha = 0.045$ m^2/h). Assuming that the brake lining and the drum surfaces are at the same temperature and the heat energy is dissipated by flowing across the surface of the drums (assumed to be very thick). Calculate the maximum temperature rise.

Solution: The kinetic energy of the car $= \frac{1}{2}\, mV^2$
$$= \frac{1}{2}\,(14000/9.81)\,(60000/3600)^2$$
$$= 1.982 \times 10^5 \text{ J}$$

The energy will flow to the four steel drums in 6 seconds and it will increase the internal energy of the drum. Assuming that drums act like a semi-infinite body, the total change in internal energy is given by

$$\dot{Q} = U_T - U_0$$
$$= 2kA(T_0 - T_i)/\sqrt{(\alpha t/\pi)} \quad \text{where } \alpha = 0.045/3600 \text{ m}^2/\text{s}$$
$$\therefore \quad (T_0 - T_i) = \dot{Q}\sqrt{(\pi\alpha/t)}/(2kA)$$
$$= \frac{1.982\times10^5\,(3.142\times0.045/6\times3600)^{1/2}}{2\times60\times4\times275\times10^{-4}} = 38.4°C.$$

Example 3.14 A semi-infinite slab of steel ($k = 50$ W/mK, $\alpha = 0.045$ m^2/h) has its plane face into perfect thermal contact with another infinite slab of different material ($k = 1$ W/mK, $\alpha = 0.0005 m^2$/h). Their initial temperatures were 125°C and 30°C respectively. Calculate the equilibrium temperature of the surface of contact.

Solution: Let the equilibrium temperature at the surface of contact is T. Therefore, the steel slab which was initially at 125°C has its face at temperature T now. The instantaneous heat flow would be

$$\dot{Q}_0 = kA(T_i - T)/\sqrt{(\pi\alpha t)}$$

The same amount of heat would flow to the other slab whose temperature has changed from 30°C to T. Therefore,

$$k_1 A(T_i - T)/\sqrt{(\pi\alpha_1 t)} = k_2 A(T - T_i)/\sqrt{(\pi\alpha_2 t)}$$
or, $\quad 50\,(125 - T)\,/\,(0.045)^{1/2} = 1\,(T - 30)\,/\,(0.0005)^{1/2}$
which gives $\qquad\qquad\qquad T = 101°C.$

(Transient heating or cooling of a body where there is a sudden change in temperature of the fluid surrounding the body has great practical significance, such as, in quenching Trof metals during heat treatment, in drying operation or in baking process, there is a constant convective heat transfer from the surface of a solid. The walls of an uncooled rocket motor are suddenly exposed to hot gases after the propellants are ignited and the operating time is limited by the resulting thermal stresses and the temperature rise at the inner wall.)

9. An Expression for the Temperature-Time History for Transient Heat Flow in a Large Plane Wall of Thickness '2L'

Let us consider a very large plate of finite thickness 2L in the x-direction with an initially uniform temperature T_i, as shown in Fig. 3.6. Since the physical configuration, boundary conditions and initial temperature distribution are symmetrical about the vertical line, the problem can be analysed by considering the body as a semi-infinite solid of thickness L with insulated boundary at $x = 0$.

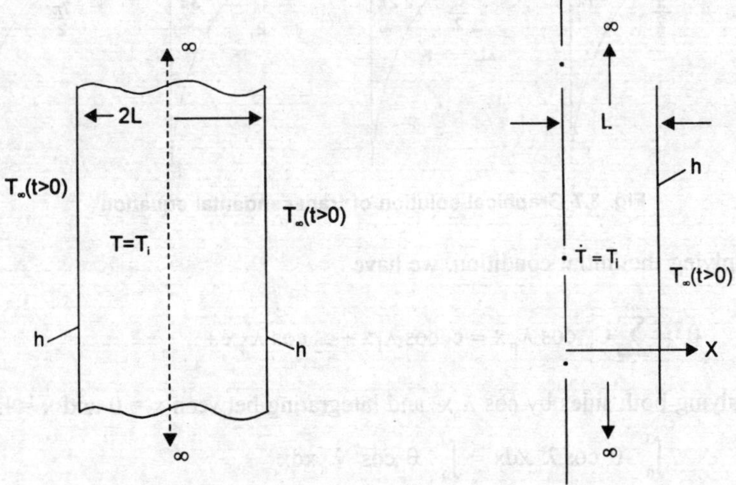

Fig. 3.6 One-dimensional system with convective boundary condition

The governing equation is,

$$\frac{\partial^2 \theta}{\partial x^2} = \frac{1}{\alpha} \frac{\partial \theta}{\partial t}$$

where $\theta = T - T_\infty$ and the conditions to be satisfied are:

$$\partial\theta/\partial x = 0 \quad \text{at } x = 0; \qquad \partial\theta/\partial x = -h\theta/k \text{ at } x = L$$

and at $t = 0$, $\theta = \theta_i = T_i - T_\infty$

Using the classical separation of variables technique, the general solution of the above equation is

$$\theta = \exp(-\lambda^2 \alpha t)(C_1 \sin \lambda x + C_2 \cos \lambda x)$$

where $-\lambda^2$ is the separation parameter and a constant.

Since $\partial\theta/\partial x = 0$ at $x = 0$, we get $\exp(-\lambda^2 \alpha t)\,\lambda\,(C_1 \cos \lambda x - C_2 \sin \lambda x) = 0$

at $x = 0$ and $C_1 = 0$,

or, $\theta = C_2 \exp(-\lambda^2 \alpha t) \cos \lambda x$.

By applying the second boundary condition, we get

$$C_2 \exp(-\lambda^2 \alpha t)\,\lambda\,(-\sin \lambda x)\big|_{x=L} = -(h/k)\,(C_2 \exp(-\lambda^2 \alpha t) \cos \lambda x)\big|_{x=L}$$

or, $\lambda \sin \lambda L = h/k \cos \lambda L$ and $\cos(\lambda L) = L/Bi$

The above equation is a transcendental equation and there are infinite number of values of λ which will satisfy it. The graphical solution is shown in Fig. 3.7. The temperature distribution, $\theta/\theta_i = f(Fo, Bi, x/L)$.

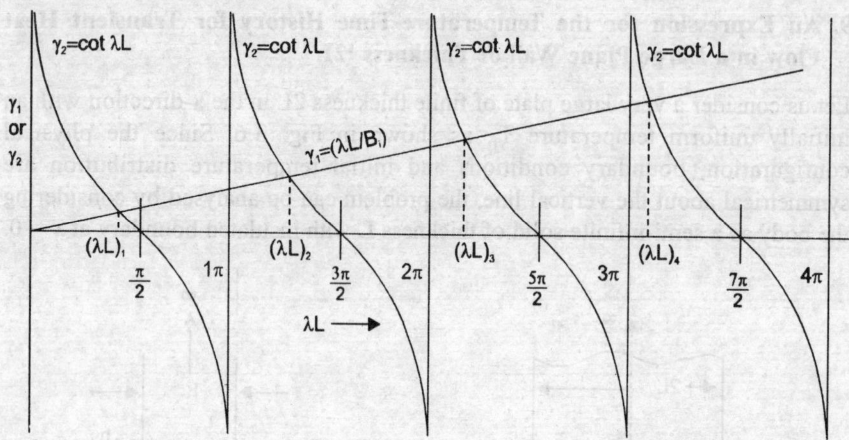

Fig. 3.7 Graphical solution of transcendental equation

By applying the initial condition, we have

$$\theta_i = \sum_1^\infty C_n \cos \lambda_n x = c_1 \cos \lambda_1 x + c_2 \cos \lambda_2 x + \dots$$

Multiplying both sides by $\cos \lambda_n x$ and integrating between $x = 0$ and $x = L$

$$\int_0^L \theta_i \cos \lambda_n x \, dx = \int_0^L \theta_i \cos^2 \lambda_n x \, dx$$

which gives $C_n = 2 \, \theta_i \sin \lambda_n L / (\lambda_n L + \sin \lambda_n L \cos \lambda_n L)$
The final solution for temperature distribution is

$$\theta = 2\theta_i \sum_{n=1}^\infty \exp\left[-(\lambda_n L)^2 (\alpha t / L^2)\right] \frac{\sin \lambda_n L \cos \lambda_n x}{\lambda_n L + \sin \lambda_n L \cos \lambda_n L} \qquad (3.10)$$

Example 3.15 A rocket nozzle is constructed of a special alloy steel 6.35 mm thick. The thermal conductivity $k = 31.1$ W/mK, thermal diffusivity $\alpha = 6.4 \times 10^{-6}$ m²/s and surface heat transfer coefficient at the flame side, $h = 8525$ W/m²K. The uniform flame temperature is 2000°C. The initial temperature of the nozzle material is 40°C. Calculate the time required for the gas side surface temperature not to exceed 1000°C.

Solution: Assuming that the outer surface of the nozzle as insulated, the dimensionless temperature ratio is
$$\theta/\theta_i = (1000 - 2000) / (40 - 2000) = 0.51$$
$$Bi = h \, L/k = 8525 \times 6.35 \times 10^{-3}/31.1 = 1.74$$
From mathematical tables: $\lambda_1 L = 1.06$; $\lambda_2 L = 3.56$
∴ $\lambda_1 = 1.06 \times 1000/6.35 = 167$; and $\lambda_2 = 3.56 \times 1000/6.35 = 562$
And the constants: $C_1 = 2 \, \theta_i \sin \lambda_1 L \, (\lambda_1 L + \sin \lambda_1 L \cos \lambda_1 L) = 0.576 \, \theta_i$
and $C_2 = 2 \, \theta_i \sin \lambda_2 L \, (\lambda_2 L + \sin \lambda_2 L \cos \lambda_2 L) = 0.188 \, \theta_i$

$$\theta/\theta_i = 0.576 \exp(-(167)^2 (6.4 \times 10^{-6}t) + 0.188 \exp(-562)^2 \times 6.4 \times 10^{-6}\,t)$$
$$= 0.51$$

Solving by trial and error, $t = 1$ second.

10. Heisler Charts — Their Significance in Solving Transient Heat Conduction Problems

Heisler charts are analytical solutions in terms of graphs and are extremely helpful in obtaining temperature distribution within a body. The solution of governing equations for transient heat conduction for heat flow and temperature distribution in plane walls, long cylinders and spheres having finite internal and surface resistances have been evaluated numerically by many investigators but the graphical results obtained by Heisler are among the most widely used for temperature distribution and heat flow. The temperature distribution and heat flow are dependent on Fourier and Biot modulus and the position inside the body.

Example 3.16 A 10 cm thick steel slab ($k = 40$ W/mK) very wide and long initially at 200°C, is suddenly exposed to a convective-fluid environment at 40°C. the average heat transfer coefficient, $h = 275$ W/m²K and thermal diffusivity $\alpha = 0.0455$ m²/h. Determine the temperature at the centreline and the temperature at 4 cm depth after 10 min of exposure and the total heat removed during that 10 min.

Solution: Fig. 3.8(a) shows the centre-line transient temperature in a symmetrical plate of half thikcness L, the symmetry being both geometrical and thermal. Fig. 3.8(b) is a plot of θ/θ_c, that is, the ratio of $(T - T_\infty)$ at any position x/L to that at the centre of the plate at a given demensionless time (Fo). Fig. 3.8(c) represents the integrated heat transfer from the plate in terms of Biot and Fourier number where $Q_i = \rho C V \theta_i$, is the initial energy difference.

$$\theta_i = T_i - T_\infty = 200 - 40 = 160: L = 5 \text{ cm} = 0.05 \text{ m}; t = 10 \text{ min} = 0.167 \text{ hr}$$
$$\text{Fo} = \alpha t/L^2 = 0.0455 \times 0.167 / (0.05)^2 = 3.0$$
$$1/\text{Bi} = k/hL = 40/(275 \times 0.05) = 2.91$$

From Fig. 3.8(a): $(T_c - T_\infty)/(T_i - T_\infty) = 0.45$

$$T_c = T_\infty + 0.45 \,(T_i - T_\infty)$$

or, $\qquad\qquad\qquad\qquad\qquad T_c = 40 + 0.45 \times 160 = 112°C$

From Fig. 3.8(b) : at x/L = 0.01/0.05 = 0.2   and 1/Bi = 2.91,

$\qquad\theta/\theta_c = 0.98;\quad \theta = 0.98 \times 72 = 70.56,\quad$ and $\quad T = 110.56°C$

From Fig. 3.8(c); for Fo = 3.0, and Bi = 1/2.91 = 0.3436

$\qquad Q/Q_i = 0.6;\quad Q_i = \rho C V \theta_i = 7800 \times 473 \times A \times 2L \times 160$

$$= 5.9 \times 10^4 \text{ kJ/m}^2$$

$\therefore \qquad\qquad\qquad\qquad Q = 0.6 \times 5.9 \times 10^4$

$$= 3.54 \times 10^4 \text{ kJ/m}^2.$$

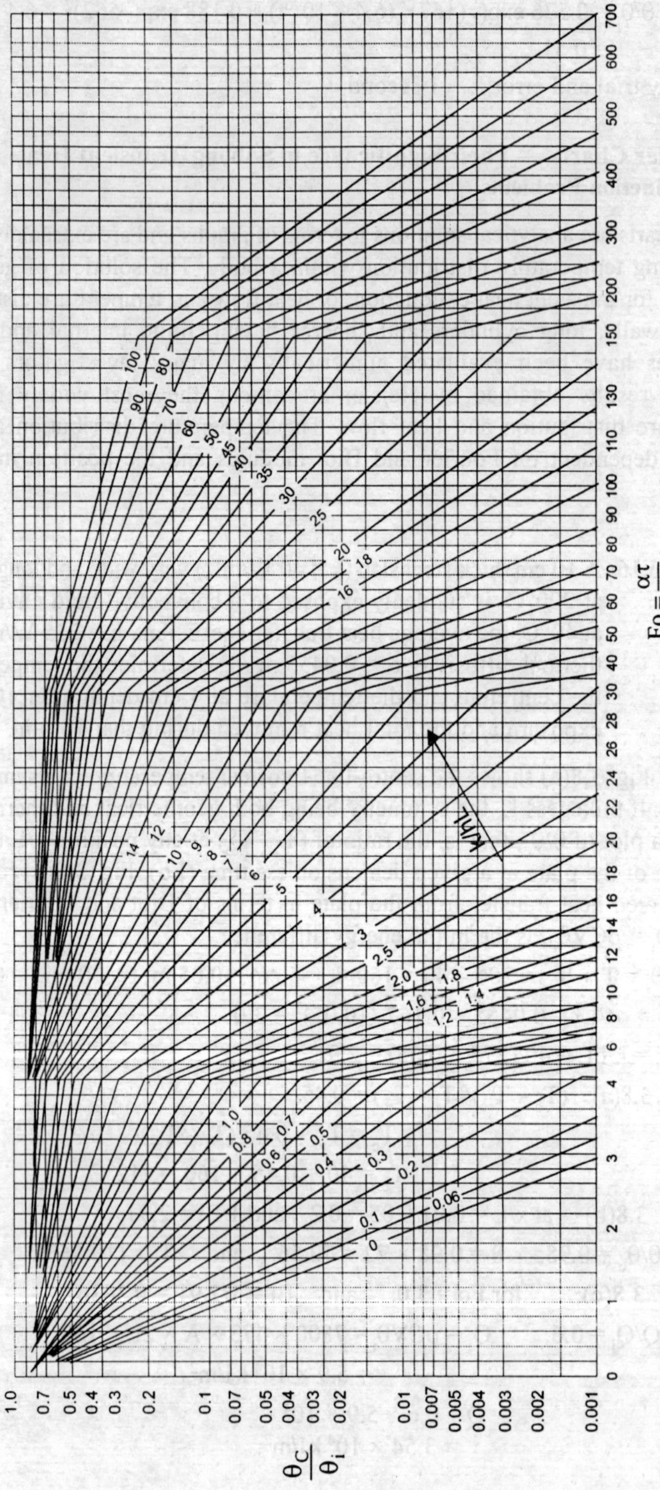

Fig. 3.8(a) Centre line temperature for a slab of thickness 2L

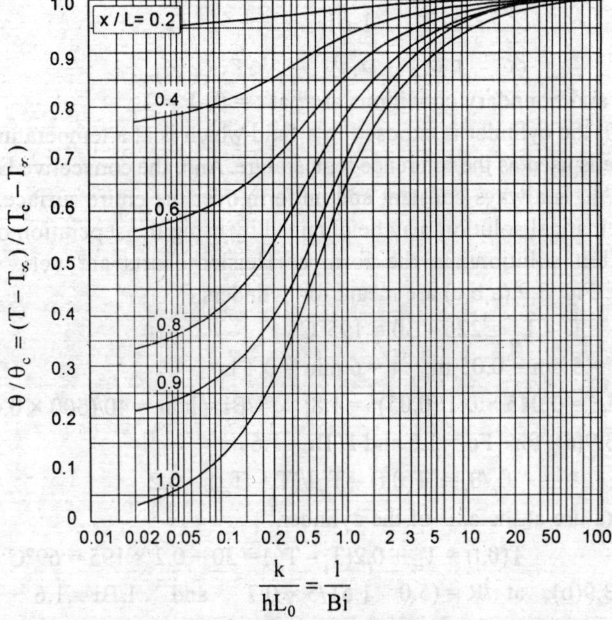

Fig. 3.8(b) Temperature as a function for centerline temperature in a slab of thickness 2*L*

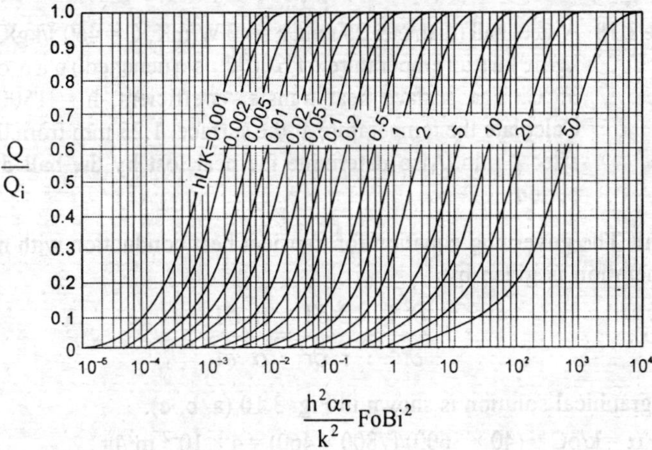

Fig. 3.8(c) Dimensionless heat loss of a slab of thickness 2L with time

Example 3.17 A very long cylinder (diameter 10 cm, k = 40 W/mK, $\alpha = 0.045$ m^2/h) is suddenly exposed to a convective fluid environment at 30°C. The convective heat transfer coefficient, h = 500 W/m^{2}K. Calculate the temperature at a depth of 1.5 cm from the surface after 6 min when the initial temperature is 225°C.

Solution: The basic differential equation governing radial flow in cylinders is given by the transient heat conduction equation with no internal heat generation,

or,
$$\frac{\partial^2 T}{\partial r^2} + \frac{1}{r}\frac{\partial T}{\partial r} = \frac{1}{\alpha}\frac{\partial T}{\partial t}$$

The initial and boundary conditions are : at $t = 0$, $T = T_i$;
and at $t = 0$, the cylinder is exposed to a fluid which is at a temperature T_∞. This temperature is used as the reference temperature. And, the convective heat transfer coefficient, h, is always constant and uniform over the entire surface.

The mathematical solution may be obtained by using the seperation of variables technique but solutions in the form of Heisler charts are very useful and convenient. Fig. 3.9 (a,b,c) are meant for cylinders.

$$\theta_i = T_i - T_\infty = 225 - 30 = 195°C$$

$$\text{radius} = 5 \text{ cm} = 0.05 \text{ m}; \quad t' = 6 \text{ min} = 0.1 \text{ hr}$$

$$\text{Fo} = \alpha t/L^2 = 0.045 \times 0.1/(0.05)^2 = 1.8; \quad 1/\text{Bi} = k/hr = 40/(500 \times 0.05) = 1.6$$

From Fig. 3.9(a): for   Fo = 1.8 and 1/ Bi = 1.6

$$\theta_c/\theta_i = [T(0,t) - T_\infty]/(T_i - T_\infty) = 0.2$$

The temperature at the axis of the cylinder,

$$T(0,t) = T_\infty + 0.2(T_i - T_\infty) = 30 + 0.2 \times 195 = 69°C$$

From Fig. 3.9(b):   at r/R = (5.0 – 1.5) /5 = 0.7   and   1/Bi = 1.6

$$\theta/\theta_c = 0.85$$

$\therefore$
$$T(r,t) = T_\infty + 0.85 \times 39 = 63.15°C.$$

Example 3.18 A steel ball (diameter 25 mm, k = 40 W/mK, C = 460 J/kgK, ρ = 7800 kg/m^3) at a temperature of 600°C is quenched in an oil bath at 35°C. The surface heat transfer coefficient, h = 1500 W/m^{2}K. Calculate the temperature at the surface, 1.25 mm from the surface after 1 min. Also determine the heat lost by the ball during that period.

Solution: The governing equation for transient heat conduction with no internal heat generation is given by

$$\frac{\partial^2 T}{\partial r^2} + \frac{2}{r}\frac{\partial T}{\partial r} = \frac{1}{\alpha}\frac{\partial T}{\partial t}$$

and the graphical solution is shown in Fig. 3.10 (a, b, c).

$$\alpha = k/\rho C = (40 \times 3600)/(7800 \times 460) = 4 \times 10^{-2} \text{ m}^2/\text{hr}$$

$$\text{Fo} = \alpha t/L^2 = (4 \times 10^{-2} \times 1/60)/(12.5 \times 10^{-3})^2 = 4.27$$

$$1/\text{Bi} = k/hR = 40/(1500 \times 12.5 \times 10^{-3}) = 2.13$$

From Fig. 3.10(a): for Fo = 4.27 and 1/Bi = 2.13, $\theta/\theta_c = 0.004$

$$\{T(0,t) - T_\infty\}/(T_i - T_\infty) = 0.004;$$

$\therefore$
$$T(0,t) = 35 + 0.004(600 - 35) = 37.26°C$$

The temperature at the surface 1.25 mm below the outer surface has the radius equal to = 12.5 – 1.25 = 11.25 mm, $\therefore$ r/R = 11.25/12.5 = 0.9

From Fig. 3.10(b): at r/R = 0.9,   1/Bi = 2.13, $\theta/\theta_c = 0.83$; $\therefore$ T = 36.87°C

The heat lost by sphere is obtained by Fig. 3.10(c).

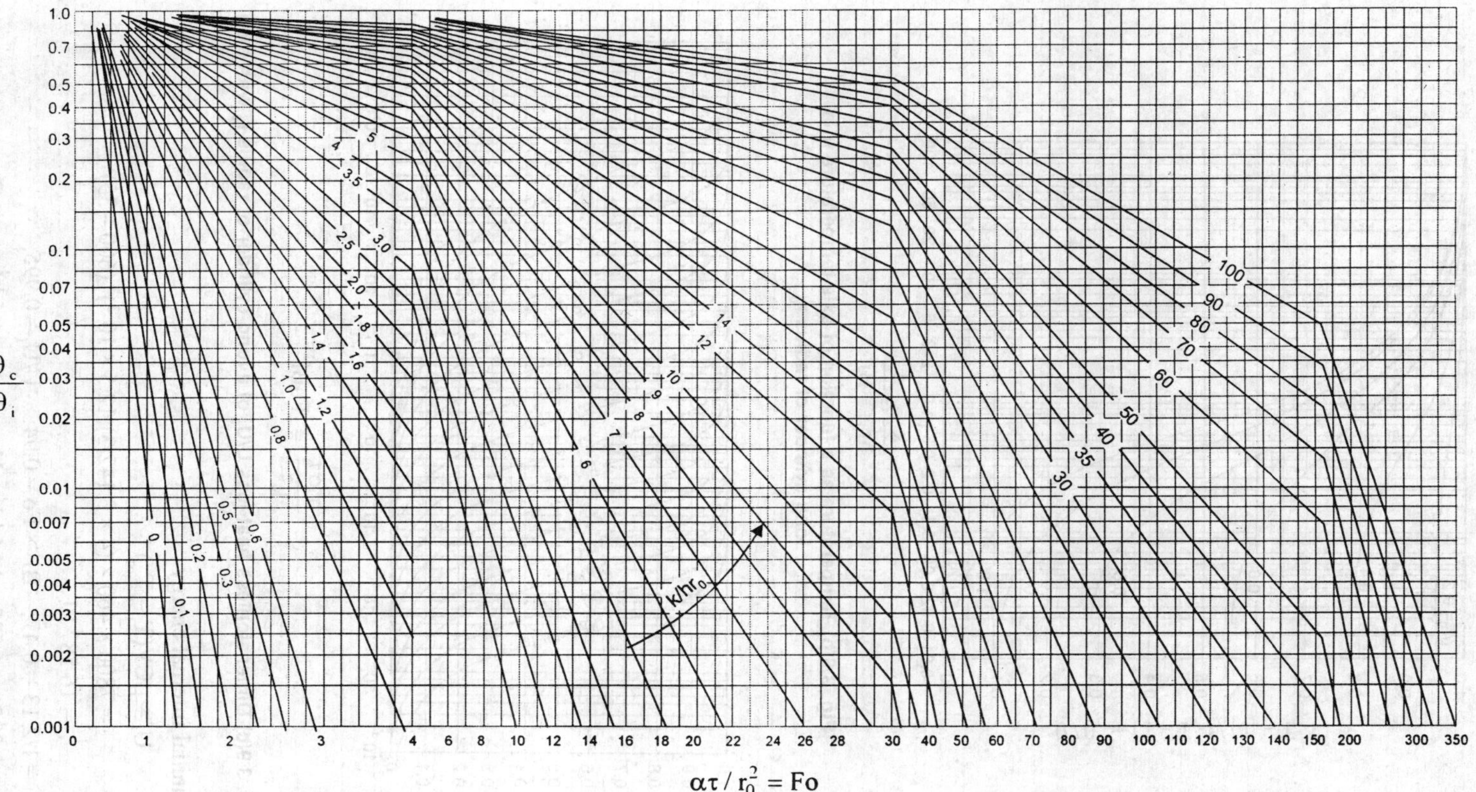

Fig. 3.9 (a) Axis temperature for a long cylinder of radius r_0, Bi = hr_0/k

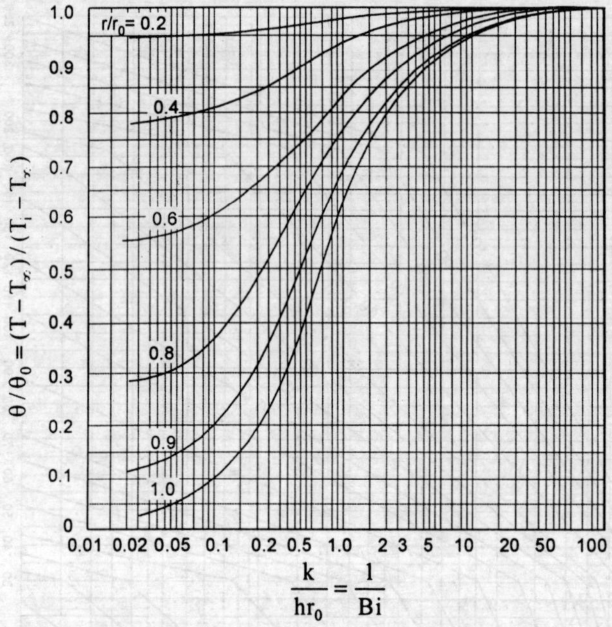

Fig. 3.9(b) Temperature as a function of axis temperature in a
long cylinder of radius r_0

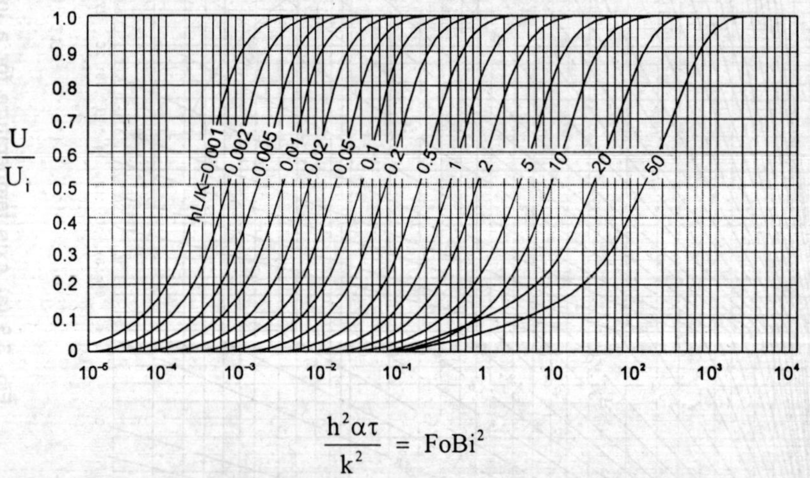

Fig. 3.9(c) Dimensionless heat loss U/U_0 of a long cylinder of radius r_0 with time

The initial internal energy

$$U_i = \rho CV(T_i - T_\infty)$$
$$= 7800 \times 460 \times (4 \times 3.142/3)(12.5 \times 10^{-3})^3 (600 - 35)$$
$$= 16.587 \text{ kJ}$$

$$\text{Bi} = 1/2.13 = 0.47; \quad \text{Bi}^2 \times \text{Fo} = 0.94; \quad U/U_i = 0.995,$$
$$\therefore U = 0.995 \times 16.587 = 16.5 \text{ kJ and heat lost} = 16.5 \text{ kJ}.$$

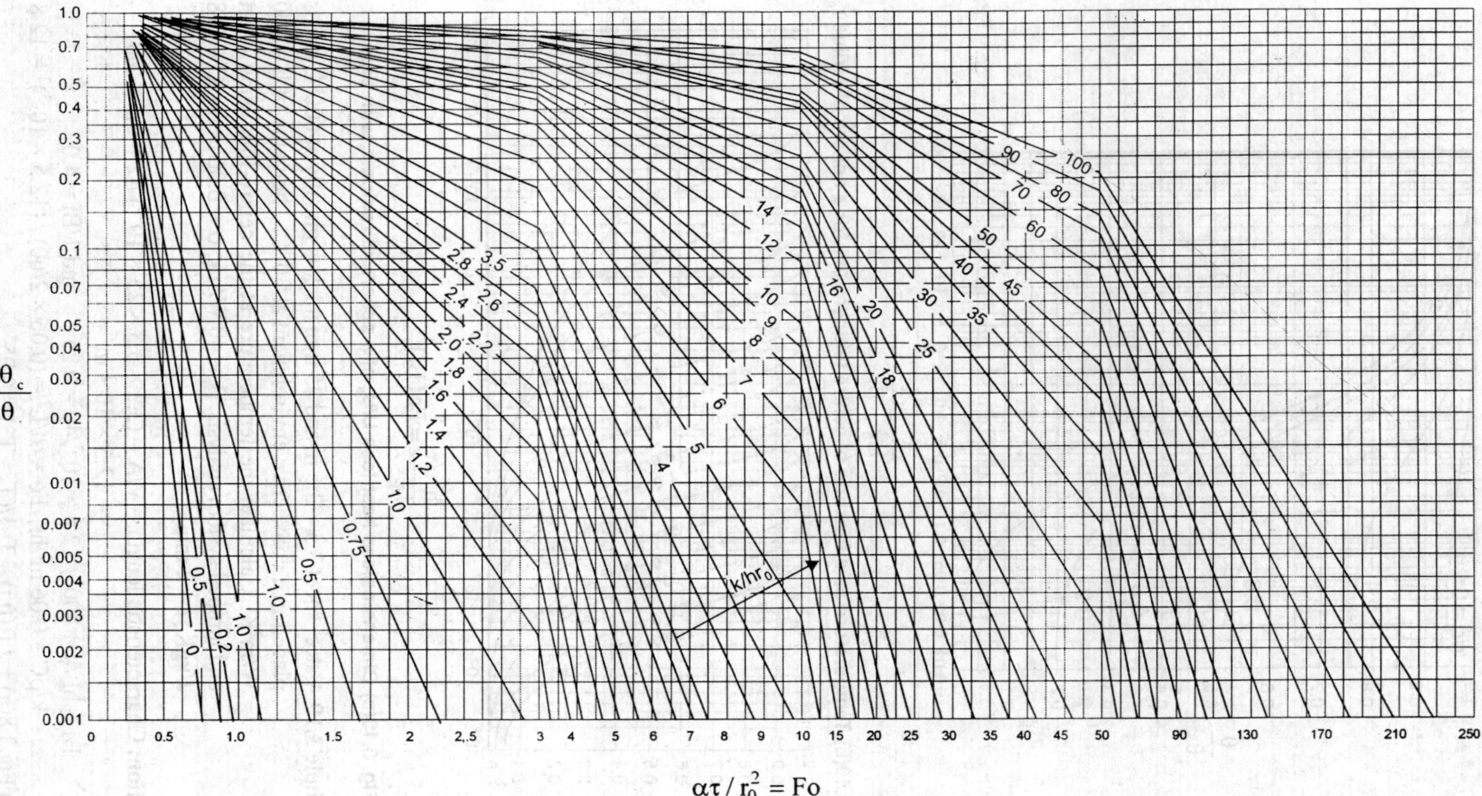

Fig. 3.10(a) Centre of temperature for a sphere of radius r_0

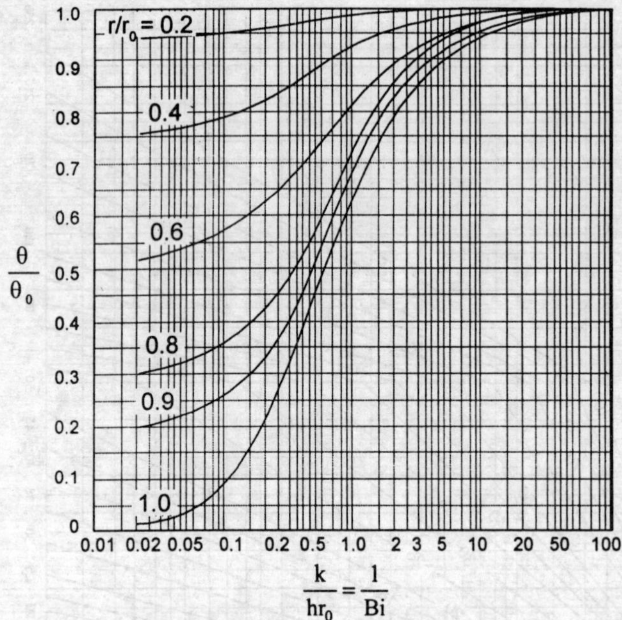

Fig. 3.10(b) Temperature as a function of centre temperature for a sphere of radius r_0

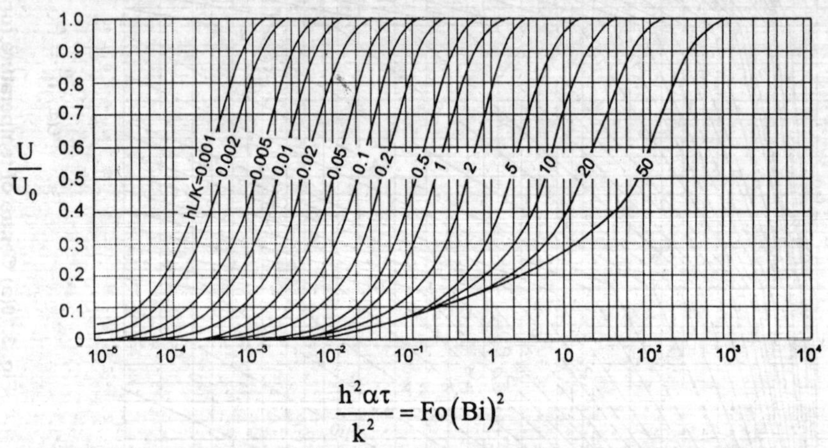

$$\frac{h^2 \alpha \tau}{k^2} = Fo\,(Bi)^2$$

Fig. 3.10(c) Dimensionless heat loss U/U_0 of a sphere of radius r_0 with time

Example 3.19 A steel sheet 3.5×3.5 m and 25 mm thick is initially at 600°C and is placed in a convective environment at 30°C. The value of h = 800 W/m²K. Calculate the temperature at the centre-line using chart and lumped system analysis. (k = 40 W/mk, α = 0.06 m²/hr, C = 418 J/kgK, t = 2 min).

Solution: Characteristic length = V/A = $(3.3 \times 3.5 \times 25 \times 10^{-3})$ /$(2 \times 3.5 \times 3.5)$
$$= 12.5 \times 10^{-3} \text{ m}$$

Bi = hL/k = $800 \times 12.5 \times 10^{-3}/40 = 0.25$ and 1/Bi = 4.0

α = k/ρC = 0.06 m²/hr, Fo = αt/L² = $(0.06 \times 2/60)$ / $(12.5 \times 10^{-3})^2 = 12.8$

From Fig 3.8(a) : $\{T\,(0,t) - T_\infty\}/(T_i - T_\infty) = 0.05$

∴ $T = 30 + 0.05\,(600 - 30) = 58.5°C$

By lumped system analysis:
$$(T - T_\infty)/(T_i - T_\infty) = \exp(-Bi\ Fo) = 0.041;$$
$$T = 30 + 0.041(600 - 30) = 53.37°C$$
an error of 5°C because Bi > 0.1.

11. Describe how One-dimensional Transient Solution can be Extended for Solving Two and Three-dimensional Problems

The solution to problems of transient conduction in two or three space dimensions are obtained by a simple product superposition of the solution of one-dimensional problems.

Let us consider a cylinder of height 2L, shown in Fig. 3.11(f). We can imagine that this body has been formed by the intersection of a large plane wall of thickness 2L and a long cylinder of radius r_0. We can also have different convective heat transfer coefficients, i.e., h_p at the top and bottom of the cylinder and h_c on the cylindrical faces of the body. The temperature distribution of this body is given by

$$[T(r, x, t) - T_\infty] / (T_0 - T_\infty) = P(x, t) \times C(r, t)$$

where T_0 is the initial temperature, and P (x, t) is the transient solution of a large plane wall of thickness 2L with convective heat transfer coefficient h_p, and C (r, t) is the transient solution for a long cylinder of radius r_0 with convective heat transfer coefficient h_c.

Similarly, the temperature-time history of a long rectangular bar (width 2a, thickness 2b) can be obtained by multiplying the dimensionaless temperature ratios for two infinite plates: one having a thickness 2a and the other having a thickness 2b. Symbolically, we write,

$$[(T - T_\infty) / (T_0 - T_\infty)]_{bar} = [(T - T_\infty) / (T_0 - T_\infty)]_{plate\ 2a} \times [(T - T_\infty) / (T_0 - T_\infty)]_{plate\ 2b}$$

Example 3.20 A short cylinder (outer diameter 100 mm, k = 40 W/mK, α = 0.045 m²/hr, length 12 cm) is at a uniform temperature of 250°C. The cylinder is suddenly exposed to convection environment where h = 400 W/m²K and T_∞ = 40°C. Calculate the temperature at the centre of the cylinder after 6 min.

Solution: A short cylinder is formed by the intersection of a slab (thickness 12 cm) and an infinitely long cylinder of diameter 10 cm. Thus the temperature distribution at the point (r, x, t) is written as

$$[T(r, x, t) - T_\infty] / (T_0 - T_\infty) = C(r,t) \times P(x,t)$$

For slab: the characteristic length, L = 12/2 = 6 cm = 0.06 m

$$Bi = hl/k = 400 \times 0.06/40 = 0.6 > 0.1$$
$$Fo = \alpha t/L^2 = (0.045 \times 6/60) / (0.06)^2 = 1.25$$

From Fig. 3.8(a): For Fo = 1.25 and 1/Bi = 1.67

$$\{T(0, t) - T_\infty\} / (T_0 - T_\infty) = 0.7$$

For cylinder : characteristic length = $r_0/2$ = 25 mm = 0.025 m.

$$Bi = hL/k = 400 \times 0.025/40 = 0.25 \quad \text{and} \quad 1/Bi = 4.0$$
$$Fo = \alpha t/L^2 = (0.045 \times 0.1) / (0.025)^2 = 7.2$$

From Fig. 3.9 (a): for $\qquad$ Fo = 7.2 and 1/Bi = 4.0

$$\{T(0,t) - T_\infty\}/(T_0 - T_\infty) = 0.04$$

$\therefore \qquad \{T(0,0,t) - T_\infty\}/(T_0 - T_\infty) = 0.7 \times 0.04 = 0.028$

and T at the centre of the cylinder $= 40 + 0.0228\,(25\text{--}40) = 45.9°C$

(It may be noted that if it were a long cylinder, the temperature at the centre-line (axis) of the cylinder would have been

$$T(0,t) = 0.04 \times 210 + 40 = 48.4°C$$

i.e., the axial heat transfer lowers the temperature.)

We can also calculate the loss in heat energy in multi-dimensional system. Langston (Int. J. of Heat and and Mass Transfer, vol 25, p 149, 1982) has shown that the heat loss for a multi-dimensional body, obtained by the intersection of two bodies, is given by

$$\left(\frac{U}{U_0}\right)_{total} = \left(\frac{U}{U_0}\right)_1 + \left(\frac{U}{U_0}\right)_2\left[1 - \left(\frac{U}{U_0}\right)_1\right] + \left(\frac{U}{U_0}\right)_3\left[1 - \left(\frac{U}{U_0}\right)_2\right]$$

Therefore, in order to compute the heat loss from the cylinder of Ex. 3.20 we calculate the dimensionless heat loss ratios for plates and infinitely long cylinders.

For plate : $\qquad$ Bi = 0.6; $\quad$ Bi^2 Fo = $0.36 \times 1.25 = 0.45$

From Fig 3.8(c); $\quad (U/U_0) = 0.4$

For cylinder: $\qquad$ Bi^2 Fo = $(0.25)^2 \times 7.2 = 0.45$

From Fig. 3.9(c) : $\quad (U/U_0) = 0.8$

$$(U/U_0)\ total = 0.4 + 0.8\,(1 - 0.4) = 0.88$$

$$U_0 = \rho CV(T_0 - T_\infty)$$
$$= 7800 \times 473 \times 3.142\,(0.05)^2\,(0.12)(250 - 40)$$
$$= 7.3 \times 10^5\ J$$

and heat loss $\qquad U = 0.88 \times 7.3 \times 10^5 = 642.66\ kJ.$

Example 3.21 A rectangular parallelopiped aluminium billet (100 cm × 200 cm × 300 cm, k = 200 W/mK, $\alpha = 8.4 \times 10^{-5}$ m²/s) at a temperature 30°C is heated in furnace to 650°C. Calculate the temperature at the geometric centre of the billet after one hour. The convective heat transfer coefficient, h = 200 W/m²K.

Solution: The object is a three-dimensional one. The characteristic dimensions of the three different infinite plates would be:

$\qquad L_1 = 100/2 = 50\ cm = 0.5\ m, \quad L_2 = 100\ cm = 1.0\ m \quad$ and $\quad L_3 = 1.5\ m.$

For plate 1: $\qquad$ Bi = hL/k = $200 \times 0.5/200 = 0.5$

$$Fo = \alpha t/L^2 = \frac{8.4 \times 10^{-5} \times 3600}{(0.5)^2} = 1.21$$

From Fig. 3.8(a): $\qquad \{T(0,\,t) - T_\infty\}\,/\,(T_i - T_\infty) = 0.7$

For plate 2: $\qquad$ Bi = hL/k = $200 \times 1.0/200 = 1.0$

$$Fo = \alpha t/L^2 = \frac{8.4 \times 10^{-5} \times 3600}{1} = 0.302$$

From Fig 3.8(a) $\Rightarrow \qquad [T(0,t) - T_\infty]\,/\,(T_i - T_\infty) = 0.9$

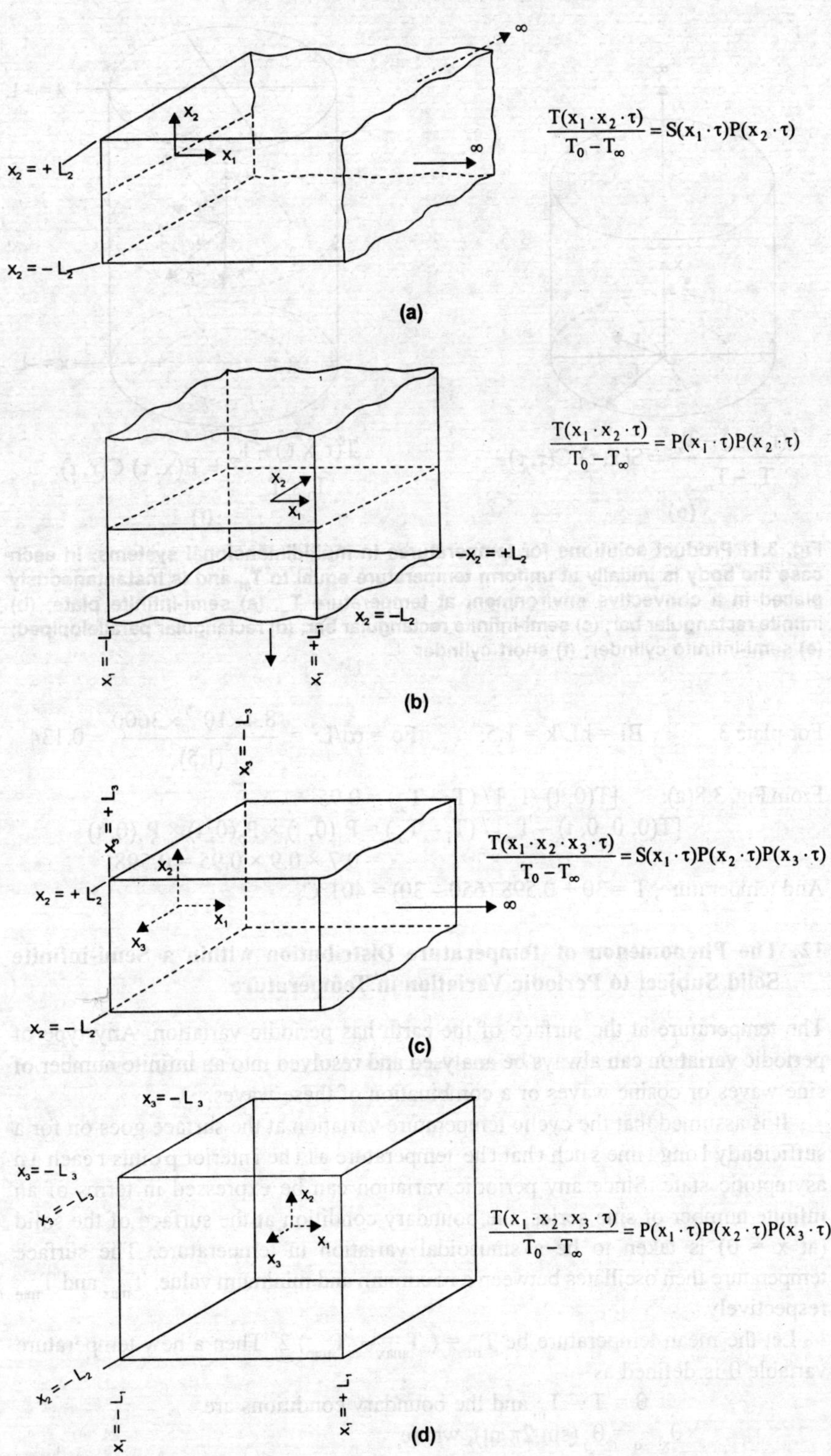

Fig. 3.16 Product solutions for... cases in multidimensional systems. In each case the body is initially at uniform temperature equal to T_0 and is instantaneously placed in a convective environment at temperature T_∞. (a) semi-infinite plate; (b) infinite rectangular bar; (c) semi-infinite rectangular bar; (d) rectangular parallelepiped; (e) semi-infinite cylinder; (f) short cylinder.

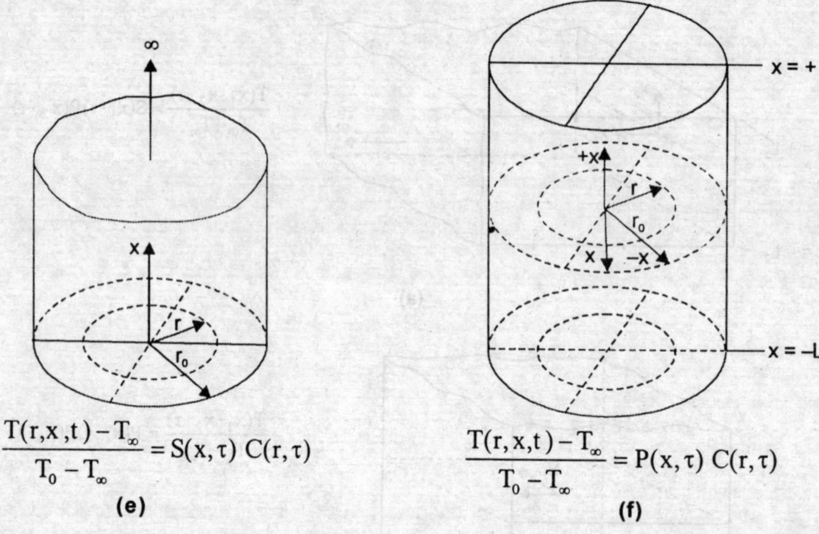

$$\frac{T(r,x,t)-T_{\infty}}{T_0-T_{\infty}} = S(x,\tau)\, C(r,\tau)$$

(e)

$$\frac{T(r,x,t)-T_{\infty}}{T_0-T_{\infty}} = P(x,\tau)\, C(r,\tau)$$

(f)

Fig. 3.11 Product solutions for temperatures in multidimensional systems. In each case the body is initially at uniform temperature equal to T_0, and is instantaneously placed in a convective environment at temperature T_{∞}. (a) semi-infinite plate; (b) infinite rectangular bar; (c) semi-infinite rectangular bar; (d) rectangular parallelopiped; (e) semi-infinite cylinder; (f) short cylinder

For plate 3: $\quad$ Bi = hL/k = 1.5; $\quad$ Fo = $\alpha t/L^2 = \dfrac{8.4 \times 10^{-5} \times 3600}{(1.5)^2} = 0.134$

From Fig. 3.8(a): $\quad [T(0,\,t)\!-\!T_{\infty}] / (T_i - T_{\infty}) = 0.95$

$\therefore \qquad [T(0,\,0,\,0,\,t) - T_{\infty}] / (T_i - T_{\infty}) = P_1(0,\,t) \times P_2(0,\,t) \times P_3(0,\,t)$
$$= 0.7 \times 0.9 \times 0.95 = 0.598$$

And temperature, T = 30 + 0.598 (650 – 30) = 401°C.

12. The Phenomenon of Temperature Distribution within a Semi-infinite Solid Subject to Periodic Variation in Temperature

The temperature at the surface of the earth has periodic variation. Any type of periodic variation can always be analysed and resolved into an infinite number of sine waves or cosine waves or a combination of these waves.

It is assumed that the cyclic temperature variation at the surface goes on for a sufficiently long time such that the temperature at the interior points reach an asymptotic state. Since any periodic variation can be expressed in terms of an infinite number of sine series, the boundary condition at the surface of the solid (at x = 0) is taken to be a sinusoidal variation in temperature. The surface temperature then oscillates between a maximum and minimum value, T_{max} and T_{min} respectively.

Let the mean temperature be $T_m = (T_{max} + T_{min})/2$. Then a new temperature variable θ is defined as

$$\theta = T - T_m \text{ and the boundary conditions are :}$$
$$\theta_{x\,=\,0} = \theta_0\,(\sin 2\pi\, nt), \text{ where}$$

n is the frequency of oscillation and θ_0 is amplitude = $(T_{max} - T_{min})/2$. The governing equation for one-dimensional, non-steady conduction heat transfer with no internal heat generation is given by

$$\frac{\partial^2 \theta}{\partial x^2} = \frac{1}{\alpha}\frac{\partial \theta}{\partial t} \quad \text{since } T_m \text{ is a constant.}$$

The solution of the above equation with the boundary conditions, using the separation of variable technique has been obtained as:

$$\frac{\theta}{\theta_0} = \exp(-x\sqrt{nx/\alpha}\,\sin\,(2\pi nt - x\sqrt{n\pi/\alpha}); \qquad \text{for } x > 0$$

and the temperature at a given distance $x = 0$, has the periodic variation of the same period but of an amplitude that decreases exponentially with distance. Or, the amplitude at a distance x is

$$(\theta_m)_x = \theta_0 \exp(-x\sqrt{nx/\alpha})$$

The periodicity at a distance x from the surface has the same period as that at the surface, but lags by a time increment equal to

$$\Delta t = \frac{x}{2}\sqrt{1/\alpha\pi n}\,, \qquad \text{shown in Fig. 3.12}$$

The instantaneous heat flow rate at the surface per unit area is

$$\left(\frac{\dot{Q}}{A}\right)_{x=0} = \left(-k\frac{\partial T}{\partial x}\right)_{x=0} = \left(-k\frac{\partial \theta}{\partial x}\right)_{x=0} = k\theta_0\sqrt{\frac{2n\pi}{\alpha}}\,\sin\,(2\pi nt + \pi/4)$$

i.e., the surface heat flow rate is a periodic function of time, leading the temperature variation by a time increment of n/8. The cumulative heat flow is then:

$$\left(\frac{\dot{Q}}{A}\right)_{x=0} = \int\left(\frac{\dot{Q}}{A}\right)_{x=0} dt = -k\theta_0\left(\frac{1}{2\pi n\alpha}\right)^{\frac{1}{2}}\cos\,(2\pi nt + \pi/4)$$

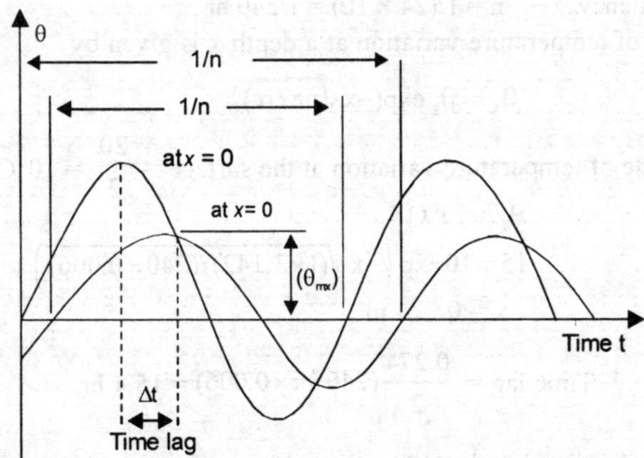

Fig. 3.12 Temperature distribution at two different depths in a semi-infinite solid subjected to a periodic surface temperature

Example 3.22 The earth is assumed to have an average thermal diffusivity $\alpha = 0.006$ m^2/hr. If the 24 hour variation in the temperature of a certain locality is from 20°C to 45°C, calculate the amplitude of the temperature variation at a depth of 15 cm. What would be the time lag of the temperature wave at this depth, assuming a sinusoidal variation at the surface.

Solution: The amplitude of temperature variation is $(T_{max} - T_{min})/2$,

or, amplitude $= (45 - 20)/2 = 12.5$°C; the frequency, $n = 1/24$

Thus at $x = 15$ cm,

$$(\theta_m)_x = 12.5 \exp[-0.15\{3.142/(24 \times 0.006)\}^{1/2}]$$

$$= 12.5 \times 0.496 = 6.2°C$$

Time lag, $\Delta t = (0.15/2)[24/(0.006 \times 3.142)]^{1/2} = 2.676$ hour $\equiv 2$ hour 40.56 min.

Example 3.23 A single cylinder 2 stroke engine runs at 4000 rpm. Calculate the depth where the temperature variation is 2% of its surface value. $\alpha = 0.045$ m^2/hr.

Solution: The frequency n for a 2-stroke engine $= 4000 \times 60 = 24000$ per hour

The amplitude at a distance x is : $(\theta_m)_x = \theta_0 \exp(-x(3.142\,n/\alpha)^{1/2})$

$$(\theta_m)_x/\theta_0 = 0.02 = \exp[-x(24000 \times 3.142/0.045)^{1/2}]$$

or, $x = -\ln(0.02)/(24000 \times 3.142/0.045)^{1/2} = 0.96$ mm.

Example 3.24 A cold wave for ten days duration causes a temperature drop of 20°C at the surface and the temperature variation can be assumed to follow a sinusoidal wave form. Calculate the depth at whcih the temperature was 15°C and the water pipes are laid below that depth.

Solution: Frequency, $n = 1/(24 \times 10) = 1/240$ hr
The amplitude of temperature variation at a depth x is given by

$$\theta_x = \theta_s \exp(-x\sqrt{n\pi/\alpha})$$

θ_s, the amplitude of temperature variation at the surface $= \dfrac{20}{2} = 10$°C

$$\theta_x = 15°C$$

$\therefore$ $15 = 10\exp\left(-x\sqrt{(1 \times 3.142)/(240 \times 0.006)}\right)$

or, $x = 0.274$ m

$$\text{Time lag} = \frac{0.274}{2}(240/\pi \times 0.006) = 15.4 \text{ hr.}$$

Example 3.25 The temperature distribution at a given instant of time through a 80 cm thick wall (k = 10 W/mK, $\alpha = 0.02$ m^2/hr) is given by
$T = 190 - 280x + 60\,x^2 + 40x^3$,

where T is C and x is in metre from its left face (x = 0). Calculate
  (i)   heat energy stored per unit area of the wall;
  (ii)   rate of change of temperature at x = 40 cm.

Solution: $\qquad T = 190 - 280x + 60x^2 + 40x^3$
$$dT/dx = -280 + 120x + 120x^2$$

Heat entering at the wall at x = 0 = $-k(dT/dx)$ = $10 \times 280 = 2800$ W/m^2

Heat leaving the wall at x = 0.8 m = $-k(dT/dx)$ at x = 0.8 m

$$Q_{out} = -10\,(-280 + 120 \times 0.8 + 120(0.8)^2)$$
$$= 972 \text{ W/m}^2$$

Energy storage rate = $2800 - 972 = 1828$ W/m^2

$$d^2T/dx^2 = 120 + 240x$$

at x = 0.4 $d^2T/dx^2 = 120 + 96 = 216$

Rate of temperature change, $dT/dt = \alpha d^2T/dx^2$
$$= 0.02 \times 216 = 4.32°\text{C/hr.}$$

Example 3.26 The temperature distribution at a given instant of time in a long
cylindrical fire tube (60 cm I.D., 100 cm O.D., k = 55 W/mK α =
0.005 m^2/hr) is given by T = 850 + 1100 r – 5100r^2 where T is in °C
and r is in metres.

Estimate (i) r ate o f heat e nergy s torage p er m etre length o f t he
tube; (ii) rate of change of temperature at the inner and outer
surface.

Solution : $\qquad T = 850 + 1100r - 5100r^2$

$$dT/dr = 1100 - 10200r; \qquad \text{and} \quad d^2T/dr^2 = -10200$$

Heat energy entering at the inner surface = $-kA_i dT/dr$ at r = r_i

$\therefore \qquad Q_{in} = -55(2\pi \times 0.3 \times 1)(1100 - 10200 \times 0.3) = 203.22$ kW/m.

Heat energy leaving at the outer radius,

$$Q_{out} = -55(2\pi \times 0.5 \times 1)(1100 - 10200 \times 0.5) = 691.24 \text{ kW/m}$$

Since $Q_{out} > Q_{in}$, the tube is losing heat at the rate of 488.02 kW per metre length
of the tube.

Again $\qquad \alpha\left[\dfrac{d^2T}{\partial r^2} + \dfrac{1}{r}\dfrac{\partial T}{\partial r}\right] = \dfrac{\partial T}{\partial t}$ for a cylinder

$\therefore$ Rate of change in temperature at the inner surface

$$\frac{\partial T}{\partial t} = 0.005\left[-10200 + \frac{1}{0.3}(1100 - 10200 \times 0.3)\right] = -60.8°\text{C}$$

Rate of change in temperature at the outer surface

$$\frac{\partial T}{\partial t} = 0.005\left[-10200 + \frac{1}{0.5}(1100 - 10200 \times 0.5)\right] = -91°\text{C.}$$

Example 3.27 Rework Ex. 3.25 when the wall has a unifrom heat generation rate of 1000 W/m³.

Solution: Since the temperature distribution remains the same as before, heat energy entering the wall at $x = 0 = 2800$ W/m² and heat energy leaving the wall at $x = 0.8$ m $= 972$ W/m². The rate of change of energy storage in the wall can be obtained by making an energy balance:

Energy coming in + Energy generated within the wall
$\qquad$ – Energy leaving the wall = Rate of energy storage

or, $\qquad 2800$ W/m² $+ 1000 \times 0.8$ W/m² $- 972$ W/m² = Rate of energy storage

$\therefore \qquad$ E storage $= 2628$ W/m²

The time rate of change of the temperature at any point in the wall can be obtained from the heat equation:

$$\frac{1}{\alpha}\frac{\partial T}{\partial t} = \frac{\partial^2 T}{\partial x^2} + \frac{\dot{q}}{k} ; \qquad \text{for } x = 0.4$$

or

$$\frac{\partial T}{\partial t} = \alpha \frac{\partial^2 T}{\partial x^2} + \frac{\alpha}{k}\dot{q}$$

$$= \frac{0.02 \times 216}{3600} + \frac{0.02 \times 1000}{10 \times 3600}$$

$$= 1.756 \times 10^{-3} \ {}^\circ\text{C/s}.$$

Example 3.28 For the aluminium slab of Ex. 3.12, calculate the time required for the temperature gradient at the surface to be 5°C/cm and the depth at which the rate of heating is maximum after 60 seconds.

Solution: The temperature distribution is given by the relation, Eq. (3.7)

$$\frac{(T - T_0)}{(T_i - T_0)} = \text{erf}\left(\frac{x}{2\sqrt{\alpha t}}\right) = \frac{2}{\sqrt{\pi}} \int_0^n e^{-n^2 d\eta}$$

The temperature gradient at the surface,

$$\left.\frac{\partial T}{\partial x}\right|_{x=0} = \frac{T_i - T_0}{\sqrt{\pi \alpha t}} = 5^\circ/\text{cm} = \frac{500^\circ\text{C}}{\text{m}}$$

or, $\qquad (T_i - T_o)^2 = (500)^2 \times \pi \alpha t$

$$t = (40 - 300)^2 / [(500)^2 \times 3.142 \times 8.4 \times 10^{-5}]$$

$$= 0.28459 \text{ hours}$$

The maximum heating rate requires $\partial T/\partial x$. to be maximum.

or when $t = x^2/2\alpha$ [Eq. (3.8)]

At $t = 60$ seconds, $x = \sqrt{2\alpha t}$

$$= \sqrt{2 \times 8.4 \times 10^{-5} \times 60}$$

$$= 0.1 \text{ m or } 10 \text{ cm}.$$

SUMMARY

1. While solving transient conduction problems, if Bi < 0.1, the internal thermal resistance is negligible and lumped capacity analysis should be applied.
2. For Bi > 0.1 and less than 100, solution should be obtained with the help of Heisler Charts.
3. Fourier Modulus is a dimensionless number and is defined as $Fo = \alpha t/L^2$ where α is the thermal diffusivity, t is the time and L is the characteristic length.
4. Response time is the time taken by a temperature measuring instrument to attain the source temperature and (ρ CV/hA) for the instrument should be very large.
5. A semi-infinite body extends from $-\infty$ to $+\infty$ in Y– and Z– directions and extends to $+\infty$ only in the X – direction.
6. The error function is defined as $\text{erf}(z) = \int_0^z \exp(-\eta^2)d\eta$

   where n is a dummy variable and erf (0) = 0;

   and as $z \to \infty$, erf (z) $\to$ 1.
7. Heisler Charts are analytical solutions in form of graphs for one-dimensional transient heat conduction temperature distribution and heat flow in plane wall, long cylinders and spheres.
8. Heisler Charts can be conveniently used for solving multi-dimensional transient heat conduction problems using product solutions.
9. In order to compute the overall energy conservation for the finite element of width x = 0 to x = δ in a semi-infinite body, we write the energy balance equation as

$$k\left(\frac{\partial T}{\partial x}\right)_{x=\delta} - k\left(\frac{\partial T}{\partial x}\right)_{x=0} = \rho C \frac{\partial}{\partial t}\int_0^\delta (T - T_0)dx$$

The above expression is known as 'heat balance integral'.

MULTIPLE CHOICE QUESTIONS

1. Assertion (A) : Lumped capacity type of analysis assumes a uniform temperature distribution throughout a solid body.
   Reasoning (R) : When the surface convection resistance is large compared with the internal conduction resistance.
   (a) Both A and R are false   (b) Both A and R are true
   (c) A is true, R is false   (d) A is false, R is true.   **(b)**
2. Biot number is defined as
   (a) k/hL   (b) kL/h   (c) hL/k   (d) h/kL   **(c)**
3. Unsteady-state heat transfer processes are:
   (a) When the temperature varies with time at a particular place
   (b) When the heat transfer rate varies with time at a particular place
   (c) When both the temperature and heat transfer rate vary with time at a particular place
   (d) All of the above   **(d)**

4. Thermal diffusivity is a combination of physical properties and is defined as
    (a) $\rho C/k$ (b) $k/\rho C$ (c) $C/\rho k$ (d) ρ/kC **(b)**

5. The time constant has the dimension of time and is defined as
    (a) $hA/\rho CV$ (b) $hV/\rho CA$ (c) $\rho CV/hA$ (d) $CV/\rho hA$ **(c)**

6. Assertion (A) : The temperature response of a thin hot copper wire is more in water than in air
    Reasoning (R) : Because the specific heat capacity of water is more than air
    (a) Both A and R are true
    (b) Both A and R are true but R is not the correct explanation of A
    (c) Both A and R are false
    (d) A is false and R is true **(b)**

7. The equation describing the temperature – time history of a system exposed to periodic temperature fluctuations and the equation derived for a constant ambient temperature are :
    (a) same (b) identical (c) different (d) dependent on Biot number **(b)**

8. The total heat added to a semi-infinite solid during time 't' is equal to:
    (a) $hA\,(T_0 - T_i)$ where T_0 and T_i are the final and initial temperature at the surface
    (b) the change in the internal energy of the solid
    (c) $-kA\,dT/dx$
    (d) none of the above **(b)**

9. Heisler Charts show the dimensionless temperature distribution as a function of
    (a) Fourier modulus and Biot modulus
    (b) Reciprocal of Fourier modulus and Biot modulus
    (c) Fourier modulus and reciprocal of Biot modulus
    (d) Reciprocal of Fourier modulus and reciprocal of Biot modulus **(c)**

10. Cooling and heating of the surface of the earth can be analysed by
    (a) infinite slab model (b) semi-infinite slab model
    (c) lumped parameter system (d) either (a) or (b) **(b)**

11. Thermal diffusivity is the highest for
    (a) Iron (b) Lead (c) Concrete (d) Wood **(b)**

12. Assertion (A) : Thermocouples are preferred over mercury in glass thermometers when high sensitivity and fast response are desired.
    Reasoning (R) : Because mercury thermometers have large thermal capacity.
    Code : (a) Both A and R are true and R is the only explanation of A
            (b) Both A and R are true but R is not the only reason for A
            (c) A is true, R is false
            (d) Both A and R are false. **(b)**

13. Fourier modulus is a dimensionless number and is defined as
    (a) $L^2/t\alpha$ (b) $L^2 t/\alpha$ (c) $\alpha t/L^2$ (d) $\alpha t/L$ **(c)**

14. Lumped capacity system of analysis can be applied when the Biot modulus is less than
    (a) 1 (b) 0.5 (c) 0.1 (d) Fourier modulus **(c)**

15. Match List I with List II and select the correct answer using the code given below

List I (Parameter)	List II (Definition)
A. Time constant of a thermometer of radius r_0	1. $hr_0/3k$
B. Biot number for a sphere of radius r_0	2. k/hL
C. Heisler Chart	3. $x/2\,(\alpha t)^{1/2}$
D. Transient conduction	4. $2\pi r_0 L\,h/\rho CV$

CODE

	A	B	C	D
(a)	4	1	3	2
(b)	4	3	2	1
(c)	4	1	2	3
(d)	1	2	3	4

(c)

16. 1. For Bi $<<$ 1, the temperature gradient within a solid is approximately equal to zero.
2. Fourier number increases with increasing thermal diffusivity.
3. Reynolds Analogy is valid for fluids having $1 < Pr < 50$
4. Hydraulic diameter for non-circular ducts is defined as 4A/P
   of the above statements
(a) 1 and 2 are only correct   (b) 1, 2 and 4 are correct
(b) 2, 3 and 4 are correct   (d) 1, 2, 3 and 4 are correct   **(b)**

NUMERICALS

1. A sphere made of aluminium mass 7 kg, density 2707 kg/m³, C = 900 J/kgK k = 204 W/mK is initially at 260°C. Determine the time required to cool the sphere to 90°C.

(28.93 min)

2. A metal sphere initially at uniform temperature, T_0, is immersed in a fluid which is being heated by an electric heater such that the temperature of the fluid, $T_\infty = T_0 + 10t$. Derive an expression for the temperature of the sphere as a function of the convective heat transfer coefficient and time. Assume that the sphere has negligible internal resistance.

$$\left(T - T_0 = \frac{10\rho cv}{hA}\left(\exp\left(-\frac{hA}{\rho cv}\right)t - 1\right) + 10t\right)$$

3. A steel cylindrical ingot (k = 50 W/mK, diameter 20 cm, length 25 cm) is initially at 90°C. It is treated in a furnace where the temperature is 1260°C and h = 100 W/m^{2}K. If the thermal diffusivity is 0.42 × 10^{-5} m^2/s, calculate the time required for the ingot to attain a temperature of 800°C.

4. A mettalic cube (ρ = 2700 kg/m^3, C = 900 J/kgK, 6 cm on a side) is initially at 145°C is suddenly immersed in a liquid at 25°C. The convective heat transfer coefficient is 120W/m^{2}K. The cube takes 198.5 seconds to reach a temperature of 70°C. Estimate the thermal conductivity of the material, using lumped capacity analysis.

(204 W/mK)

5. A copper plate (30 cm × 30 cm × 2 mm) initially at 400°C is dipped in water at 20°C. It reaches a temperature of 40°C in 106 seconds. Estimate the convective heat rransfer coefficient when k = 370 W/mK, ρ = 8800 kg/m^3 and specific heat capacity 381 J/kgK, using lumped capacity analysis.

(93 W/m^{2}K)

6. A mercury thermometer bulb idealized as a sphere of 2 mm radius is used for measuring the temperature of a fluid where h = 50 W/m^{2}K. The thermal conductivity k = 10 W/mK and thermal diffusivity being 0.017 m^2/h. Estimate the time constant.

7. A large slab of aluminium initially at a temperature of 200°C has its surface temperature lowered to 70°C. Estimate the time required for the temperature to be 120°C at a depth of 4 cm. Also estimate the quantity of heat removed per unit surface area during this period. Take α = 0.302m^2/h, and k = 215 W/mK.

(37.7 s, – 21.13 MJ/m^2)

8. A semi–infinite slab of copper is subjected to a constant heat flux at the surface of 0.3 MW/m^2. Neglecting convection at the surface, calculate the surface temperature after 10 minutes if the initial temperature of the slab is 30°C Calculate the temperature at a distance of 20 cm from the surface after 10 min.

(387.4°C, 187.6°C)

9. A large mass (k = 1.396 W/mK) at 400°C is suddenly exposed to 40°C at one of its surface for a long time. Calculate (i) the tmperature at a depth of 2.5 cm after 5 minutes (ii) the quantity of heat removed through the plane at a depth of 2.5 cm within the first 1 hour and (c) the time after which the temperature at that plane reaches 200°C.

(300°C, 14.6 kW/m^2, 16.2 min)

10. A 10 mm thick special alloy steel (k = 32 W/mK, α = 0.06 m^2/h, convective heat transfer coefficient on hot gas side 8.5 kW/m^{2}K) is initially at 40°C. One side if the plate is exposed to hot gases at 2000°C. Calculate the surface temperature of the plate after 1 second assuming that the other surface is insulated and the first two eigen values are considered.

11. A 5 cm thick aluminium plate (k = 215 W/mK, ρ = 2700 kg/m^3, C = 0.9 kJ/kgK α = 3.02 m^2/h), initially at 200°C, is immersed in a fluid at 70°C. The convective heat transfer coefficient is 525 W/m^{2}K. Determine the temperature at a depth

of 1.25 cm from one of the faces 1 minute after immersion. Also estimate the energy removed per unit area of the surface during that period.

$$(147°C, 6.5 \text{ MJ/m}^2)$$

12. A long cylinder, 12 cm in diameter and initially at 40^0C is suddenly exposed to hot gases in a furnace at 650°C. The convective heat transfer coefficient is 22 W/m^{2}K. Calculate the time required for the centre to reach a temperature of 255°C and the temperature of the surface at that time. The properties of the material are : $\rho = 580$ kg/m^3, k = 0.2 W/mK, C = 1.05 kJ/kgK.

$$(33 \text{ min, } 58°C)$$

13. A sphere (diameter 2.5 cm, k = 1.52 W/mK, $\alpha = 9.5 \times 10^{-7}$ m^2/s) initially at a uniform temperature of 25°C, is suddenly exposed to a convection environment at 200°C. The convective heat transfer coefficient is 110 W/m^{2}C. Calculate the temperatures at the centre and at a radius of 0.64 cm after a time of 4 minute.

14. A steel cylinder (k = 25 W/mK, $\rho = 7600$ kg/m^3, C = 0.46 kJ/kgK) 10 cm in diameter and 10 cm long is initially at 300°C. It is suddenly immersed in an oil bath at 40°C with h = 280 W/m^{2}C. Find (a) the temperature of the solid after 2 min and (b) the temperature at the centre of one of the circular faces after 2 min.

15. A long steel bar 5 cm × 10 cm is initially maintained at a uniform temperature of 250°C. It is suddenly exposed to an environment at 35°C with h = 23 W/m^{2}C. Using Heisler Charts, estimate the time required for the centre temperature to be 90°C. (The properties of steel may be taken from problem no. 14.)

16. An aluminium cube 5 cm on a side is initially at a uniform temperature of 100°C and is suddenly exposed to room air at 25°C. The convective heat transfr coefficient is 20 W/m^2 °C. Calculate the time required for the geometric centre temperature to be 50°C.

17. The temperature distribution across as 30 cm thick concrete slab is given by $T = 20 + 15x - 2x^2 + 5x^3$. Calculate (a) the heat entering and leaving each face (b) the heat gained or lost by the slab per unit area per unit time and (c) the change in temperature per unit time at the centre of the slab. Take $\rho = 2180$ kg/m^3, C = 0.84 kJ/kg C, k = 0.68 W/mK.

18. The temperature of the earth's surface varies over a period of 24 hours from – 20°C to 20°C. Assuming sinusoidal variation in temperature, calculate (a) the amplitude of temperature oscillations at a depth of 30 cm (b) the time lag (c) the temperature at that depth (d) the heat flowing into the surface per unit area every half cycle. The values are: k = 0.63 W/mK, $\alpha = 1.41 \times 10^{-3}$ m^2/h, time 5 hours after the surface temperature reaches the minimum.

$$(1.11°C, 1.1h, 0.022°C, 4723 \text{ kJ/m}^2)$$

CHAPTER 4

Two Dimensional Steady State Conduction

1. Governing Equation for Steady State Conduction

In the preceeding sections, we have considered cases where the temperature gradient was significant in only one direction and many practical problems could be solved. But when the boundaries of a system are irregular or when the temperature along a boundary is non-uniform, the temperature distribution within the system would be a function of more than one coordinates and therefore, some methods for analyzing conduction in two or three-dimension requires consideration.

For a steady state and constant thermal conductivity, the appropriate differential equation describing the temperature distribution is the Laplace's equation:

$$\nabla^2 T = \frac{\partial^2 T}{\partial x^2} + \frac{\partial^2 T}{\partial y^2} + \frac{\partial^2 T}{\partial z^2} = 0$$

We shall consider a few methods for analysing two-dimensional problem because they are less diffficult and at the same time, illustrate the basic method of analysis equally well. The various methods for solving two-dimensional problems have been divided into four categories: (i) analytical, (ii) graphical (iii) analogical and (iv) numerical. None of these methods consititutes a satisfactory approach for all engineering problems but the availability of high speed digital computers throughout industry allows the engineer to solve many conduction problems for which we are unable to obtain a closed form mathematical solution.

2. Analytical Solution—Separation of Variable Technique

The classical approach to obtain an exact solution to the 2-dimensional Laplace's equation (steady state heat conduction in a rectangular plate) is by separation of variables technique and is best illustrated with the following examples:

Example 4.1 A long rectangular bar (30 cm × 24 cm) with homogeneous composition (uniform properties) has one of its sides at 100°C and the other three edges are at 25°C. Calculate the temperature along the centerline of the bar.

Solution: The physical configuration with its boundary conditions are shown in Fig. 4.1. Since the bar is very long, it can be assumed that there is no variation in temperature in the Z-direction. Laplace's equation is written with the boundary conditions:

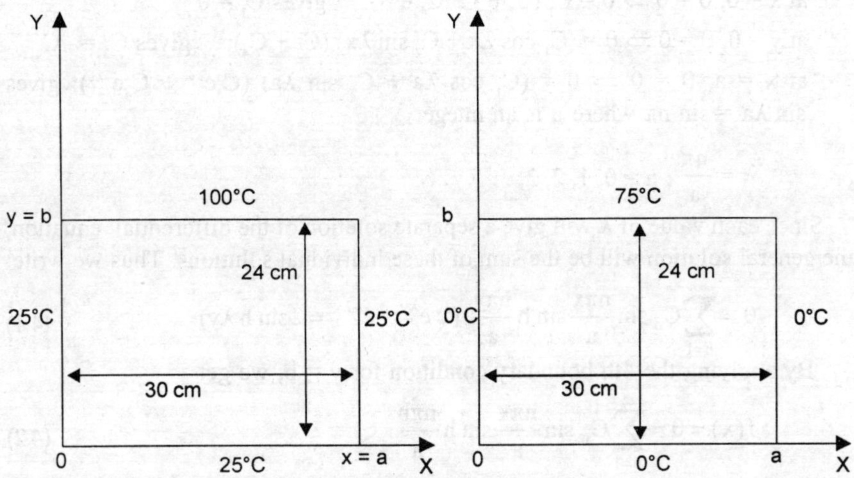

Fig. 4.1 A long rectangular bar with one nonhomogeneous boundary codition

$$\frac{\partial^2 T}{\partial x^2} + \frac{\partial^2 T}{\partial y^2} = 0$$

$$T = T_1 = 25°C \text{ at } x = 0, \quad T = T_1 = 25°C \text{ at } x = a = 0.3m$$
$$T = T_1 = 25°C \text{ at } y = 0, \quad T = T_1 + f(x) = 100 \text{ at } y = 0.24 \text{ m}$$

Let $\theta = T - T_1$, then the the equation reduces to

$$\frac{\partial^2 \theta}{\partial x^2} + \frac{\partial^2 \theta}{\partial y^2} = 0$$

$$\theta = 0, \quad \text{at } x = 0, \quad \theta = 0, \quad \text{at } x = a$$
$$\theta = 0, \quad \text{at } y = 0, \quad \theta = f(x) = 75 \quad \text{at } y = b$$

We seek a solution of the differential equation, using the separation of variables technique and assume that

$$\theta = X(x) \times Y(y); \quad \therefore \frac{\partial^2 \theta}{\partial x^2} = \frac{d^2 X}{dx^2} \cdot Y \text{ and } \frac{d^2 \theta}{dy^2} = X \frac{d^2 Y}{dy^2}$$

or,
$$-\frac{1}{X}\frac{d^2 X}{dx^2} = \frac{1}{Y}\frac{d^2 Y}{dy^2}$$

Since each side of the above equation involves only one of the independent variable, they would be equal only if they are equal to a constant, λ^2.

The solution of the above equation will depend on the sign of λ^2. The sign is so chosen that it s atisfies the b oundary conditions.

Assuming $\lambda^2 > 0$, we have

$$\frac{d^2 X}{dx^2} + \lambda^2 X = 0 \quad \text{and} \quad \frac{d^2 Y}{dy^2} - \lambda^2 Y = 0$$

The solution is: $X = C_1 \cos \lambda x + C_2 \sin \lambda x$ and $Y = C_3\, e^{\lambda y} + C_4\, e^{-\lambda y}$

or, $\quad \theta = XY = (C_1 \cos \lambda x + C_2 \sin \lambda x)(C_3\, e^{\lambda y} + C_4\, e^{-\lambda y})$

Substituting the boundary conditions:

at $x = 0$, $\theta = 0 \Rightarrow 0 = C_1\,(C_3\, e^{\lambda y} + C_4\, e^{-\lambda y})$; gives $C_1 = 0$

at $y = 0$, $\theta = 0 \Rightarrow 0 = (C_1 \cos \lambda x + C_2 \sin \lambda x)\,(C_3 + C_4)$; gives $C_3 = -C_4$

at $x = a$, $\theta = 0 \Rightarrow 0 = (C_1 \cos \lambda a + C_2 \sin \lambda a)\,(C_3 e^{\lambda y} + C_4 e^{-\lambda y})$; gives $\sin \lambda a = \sin n\pi$ where n is an integer

or, $\qquad \lambda = \dfrac{n\pi}{a}, n = 0, 1, 2, 3, \ldots$

Since each value of λ will give a separate solution of the differential equation, the general solution will be the sum of these individual solutions. Thus we write:

$$\theta = \sum_{n=1}^{\infty} C_n \sin \frac{n\pi x}{a} \sin h \frac{n\pi y}{a} \quad (\because e^{\lambda y} - e^{-\lambda y} = 2\sin h\,\lambda y) \tag{4.1}$$

By applying the 4th boundary condition for $y = b$, we get

$$f(x) = \theta = \sum_{n=1}^{\infty} C_n \sin \frac{n\pi x}{a} \sin h \frac{n\pi b}{a} \tag{4.2}$$

Since $C_n \sin h \left(\dfrac{n\pi b}{a}\right)$ is constant, we can express the constants C_n in terms of another constant B_n, such that

$$C_n \sin h \left(\frac{n\pi b}{a}\right) = B_n = \frac{2}{a} \int_0^a f(x) \sin \frac{n\pi}{a} x \tag{4.3}$$

and the above integration can be performed after f (x) is specified.

Since the edge at $y = b$ is maintained at a constant temperature ($\theta_1 = 75$) the solution is written as

$$\frac{\theta}{\theta_1} = \frac{2}{\pi} \sum_{n=1}^{\infty} \frac{1 - (-1)^n}{n} \sin \frac{n\pi x}{a} \frac{\sin h(n\pi\, y / a)}{\sin h(n\pi\, b / a)} \tag{4.4}$$

In order to evaluate the temperature at the centreline, we have $x = a/2$ and $y = b/2$. Therefore, the temperature is given by

$$\frac{\theta}{\theta_1} = \frac{2}{\pi} \sum_{n=1}^{\infty} \frac{(-1)^{n+1} + 1}{n} \sin \frac{n\pi}{2} \frac{\sin h(n\pi / 2 \cdot b / a)}{\sin h(n\pi\, b / a)} ; \text{ where } b/a = 0.8 \tag{4.5}$$

$n = 1 :\quad \theta/\theta_1 = 0.335;$

$n = 2 :\quad \theta/\theta_1 = 0$

$n = 3 :\quad \theta/\theta_1 = -0.00977$

Truncating the series at $n = 3$, $\qquad \theta/\theta_1 = (0.335 - 0.00977) = 0.325$

or, $\qquad (T - 25) / (100 - 25) = 0.325.$

$$\therefore \qquad T = 55.875°C.$$

From Example 4.1, one can conclude that the principle of superposition can be used to solve two-dimensional steady state conduction problems. It may be mentioned that:

(a) The principle of superposition can be applied to any linear, homogeneous differential equation such as Laplace equation.

(b) When the boundary conditions are non-homogeneous, the problem is to be divided into a number of simpler subproblems each having only one non-homogeneous boundary condition, like Ex. 4.1.

(c) A non-homogeneous boundary condition can be eliminated by selecting an appropriate reference temperature.

(d) The physical geometry of all the subproblems must remain the same as that of the original problem.

Example 4.2 Calculate the centreline temperature of the long bar of Ex. 4.1 when the upper edge is at 250°C, the right hand vertical edge is at 300°C and the other two edges are at 50°C.

Solution: The physical geometry of the problem is shown in Fig. 4.2. In order to remove two of the non-homogeneities, we define $\theta = (T - 50)$, and the complete problem is shown in Fig. 4.3. This can be subdivided into two simpler problems Fig. 4.4, such that each one has only one non-homogenous boundary condition. Making use of the solution giben by Eq. (4.5) we get

$$\frac{\theta_1}{200} = 0.325, \qquad \therefore \quad \theta_1 = 65°C$$

The solution for θ_2 will be

$$\frac{\theta_2}{250} = \frac{2}{\pi} \sum_{n=1}^{\infty} \frac{(-1)^{n+1} + 1}{n} \sin \frac{n\pi y}{b} \frac{\sin h(n\pi x/b)}{\sin h(\pi n a/b)}$$

$$250 = 0.1735 \text{ when we truncate the series in 3 terms.}$$

$$\therefore \qquad \theta_2 = 43.375$$

and $\qquad \theta = \theta_1 + \theta_2 = 65 + 43.375 = 108.375$ and $T = 158.375°C$

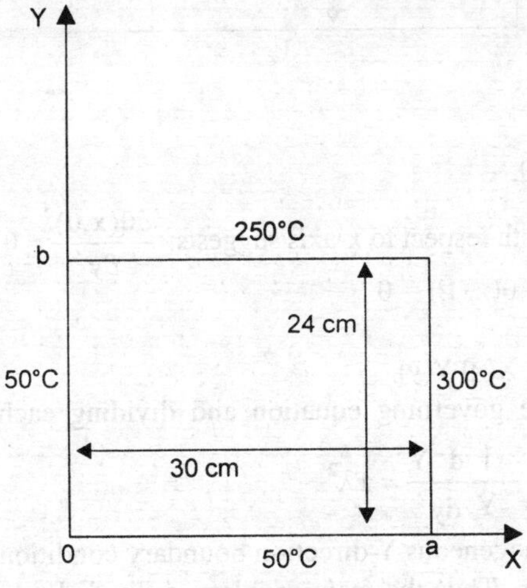

Fig. 4.2

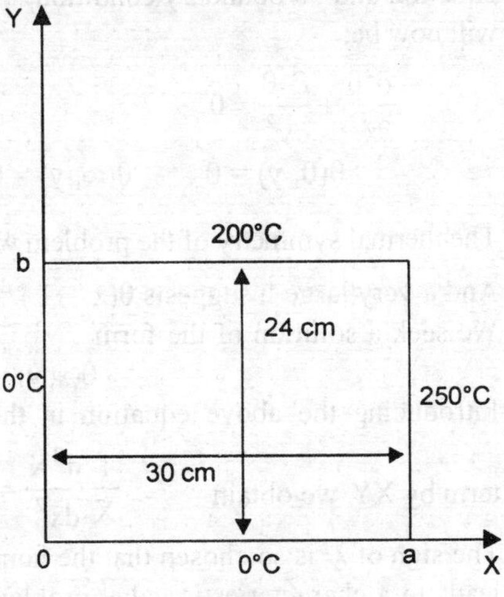

Fig. 4.3

Example 4.3 An infinitely long two-dimensional fin of thickness 2L has its base at temperature T_0. The ambient temperature is T_∞. The heat transfer coefficient is very large. Derive an expression for temperature distribution in the fin.

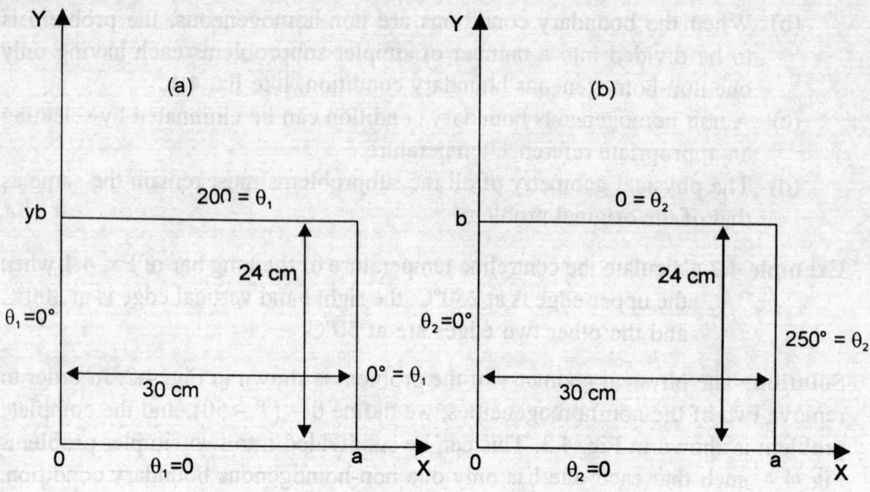

Fig. 4.4 Subproblem (a) and Subproblem (b)

Solution: The fin is shown in the accompanying figure. Since the method of separation of variables requires that the differential equation and three of the boundary conditions be homogeneous, we choose:

$$\theta = T - T_\infty$$

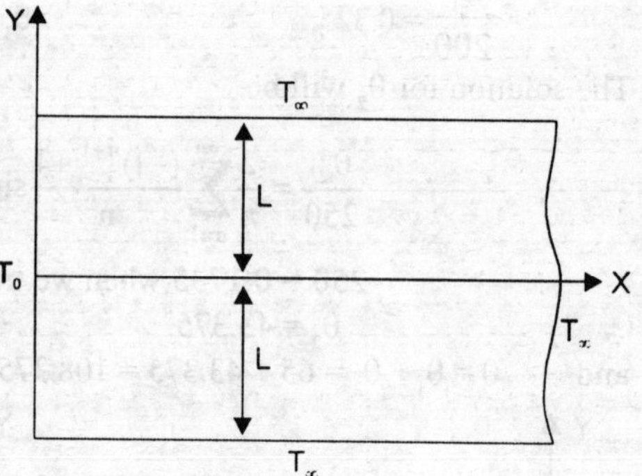

The governing differential equation and its boundary condition will now be:

$$\frac{\partial^2 \theta}{\partial x^2} + \frac{\partial^2 \theta}{\partial y^2} = 0$$

$$\theta(0, y) = \theta_o; \qquad \theta(\infty, y) = 0.$$

The thermal symmetry of the problem with respect to x-axis suggests: $\dfrac{\partial \theta(x,0)}{\partial y} = 0$

And a very large h suggests $\theta(x, -L) = \theta(x, L) = 0$

We seek a solution of the form

$$\theta(x, y) = X(x)\, Y(y)$$

Introducing the above equation in the governing equation and dividing each term by XY, we obtain

$$\frac{1}{X}\frac{d^2 X}{dx^2} = -\frac{1}{Y}\frac{d^2 Y}{dy^2} = \pm\lambda^2$$

The sign of λ^2 is so chosen that the homogeneous Y-direction boundary condition leads to a characteriestic value problem. Thus the main problem is divided into two subproblems expressed separately in the X-direction and Y- direction:

$$\frac{d^2 Y}{dy^2} + \lambda^2 y = 0; \qquad \frac{dY(0)}{dy} = 0; \qquad Y(L) = 0$$

$$\frac{d^2X}{dx^2} - \lambda^2 x = 0; \quad X(\infty) = 0; \quad X(0) = \theta_0 = T_0 - T_\infty$$

The solutions are:

$Y_n(y) = C_n \cos \lambda_n y$ where C_n is an arbitrary constant and
$\lambda_n = (2n+1)\pi/2L$ for n = 0, 1, 2, ...

The solution of the boundary value problem in x is

$$X_n(x) = B_n\, e^{-\lambda_n X}$$

And, the product solution is

$$\theta(x, y) = \sum_{n=0}^{\infty} a_n\, e^{-\lambda_n x} \cos \lambda_n y \text{ where } a_n = B_n C_n.$$

Using the non homogeneons boundary condition;

$$\theta_0 = \sum_{n=0}^{\infty} a_n \cos \lambda_n y$$

Since θ_o in uniform, $a_n = \dfrac{(-1)^n\, 2\theta_0}{\lambda_n L}$

and the solution of the problem is $\dfrac{\theta}{\theta_0} = \dfrac{2}{L} \displaystyle\sum_{n=0}^{\infty} \dfrac{(-1)^n\, e^{-\lambda_n x}}{\lambda_n} \cos \lambda_n y$

Example 4.4 Determine the temperature distribution for the infinitely long two dimensional fin of Ex. 4.3 for a finite heat transfer coefficient, h.

Solution: The fin with the boundary conditions is shown in the accompanying figure.

The governing equation and the boundary conditions can be written as:

$$\frac{\partial^2 \theta}{\partial x^2} + \frac{\partial^2 \theta}{\partial y^2} = 0$$

$$\theta(0, y) = \theta_0; \quad \theta(\infty, y) = 0$$

$$\frac{\partial \theta(x,0)}{\partial y} = 0 \text{ and } -k\frac{\partial \theta(x,L)}{\partial y} = h\theta(x, L)$$

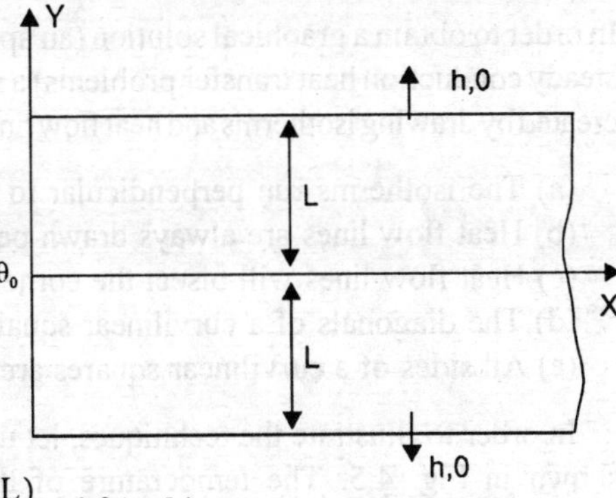

We seek a product solution as before. The two sub-problems with their boundary conditions are:

$$\frac{d^2Y}{dy^2} + \lambda^2 Y = 0; \quad \frac{dY(0)}{dy} = 0; \quad -K\frac{dY(L)}{dy} = hY(L)$$

$$\frac{d^2X}{dx^2} - \lambda^2 X = 0; \quad X(\infty) = 0; \quad X(0) = \theta_o$$

The characteristic value problem of the Y-direction has the solution

$$Y_n(y) = C_n \cos \lambda_n y$$

Where the characteristic values would be the roots of the transcendental equation

$$k\lambda_n \sin \lambda_n L = h \cos \lambda_n L$$

And, the above equation can be rearranged in either one of the form

$$\tan \lambda_n L = Bi/\lambda_n l ; \quad \text{or} \cot \lambda_n L = \lambda_n L/Bi \quad \text{where } Bi = hL/k$$

The boundary-value problem of the X-direction is the same as in Ex. 4.3

$$X_n(x) = B_n e^{-\lambda_n x}$$

and the product solution of the problem is

$$\theta_0 = \sum_{n=1}^{\infty} a_n e^{-\lambda_n x} \cos \lambda_n y$$

Using the non-homogeneous boundary condition, we have

$$\theta_0 = \sum_{n=1}^{\infty} a_n \cos \lambda_n y$$

The value of a_n is to be obtained from

$$a_n = \frac{2\theta_0 \lambda_n L}{\lambda_n L + \sin \lambda_n L \cos \lambda_n L}$$

The temperature distribution will then be given by

$$\frac{\theta}{\theta_0} = 2\sum_{n=1}^{\infty} \left(\frac{\sin \lambda_n L}{\lambda_n L + \sin \lambda_n L \cos \lambda_n L} \right) \exp(-\lambda_n x) \cos \lambda_n y$$

3. Graphical Analysis—Shape Factor

In order to obtain a graphical solution (an approximate solution to two-dimensional steady conduction heat transfer problems) a network of curvilinear squares has to be created by drawing isotherms and heat flow lines according to the following guidelines:

(a) The isotherms run perpendicular to insulated surfaces.
(b) Heat flow lines are always drawn perpendicular to isotherms.
(c) Heat flow lines will bisect the corners of isothermal boundaries.
(d) The diagonals of a curvilinear square intersect at right angles.
(e) All sides of a curvilinear squares are approximately equal.

In order to illustrate the techniques, let us consider a rectangular plate, ABCD, shown in Fig. 4.5. The temperature of the sides AB and CD are T_1 and T_2 respectively, i.e., they are isotherms. Therefore, heat flow lines will be at right angles to them (parallel to the sides AD or BC). Since the heat flow lines are tangent to the direction of heat flow (similar to streamline in fluid-flow field), no heat energy can flow across heat flow lines. We divide the side AD in five equal parts (say) i.e., we draw four isotherms (lines parallel to AB or CD). And, we draw heat flow lines (lines parallel to AD or BC) to complete the flownet. In this particular case, the squares have straight edges but this will not be always true. The flow net is to be drawn in such a way that the sides of the curvilinear square are almost equal and it is possible that one curvilinear square is smaller or larger than the other.

Let us examine the general case of a curvilinear square as shown in Fig. 4.6 Since the number of isotherms are chosen at the beginning of the graphical solution, the number of temperature increments, ΔT, across the body will be equal to the number of isotherms − 1, or,

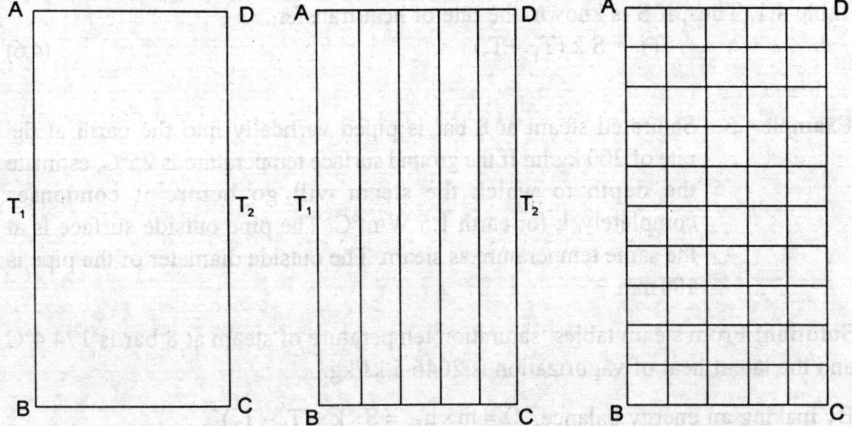

Fig. 4.5 A network of curvilinear squares for a rectangular plate

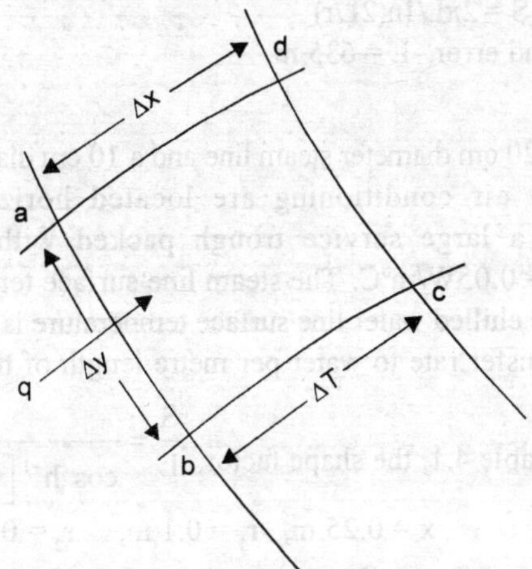

Fig. 4.6 A typical curvilinear square

N = number of isotherms -1

Let, M = number of heat flow lanes (the space between two adjacent heat flow lines is one heat flow lane)

$\Delta x \approx \Delta y$ for any curvilinear square

k = thermal conductivity

$\dot{q}$ = rate of heat flow due to the temperature gradient,

Δy = area of heat flow per unit depth

Therefore, $\dot{q} = k\,\Delta y \times \Delta T/\Delta x = k\,(\Delta T)$

Since there are N increments of ΔT, $\therefore \Delta T = (T_1 - T_2)/N$,

and, $\dot{q} = k\,(T_1 - T_2)/N$

The number of heat flow lanes are M, the total rate of heat flow is

$$\dot{Q} = M\dot{q} = M\,k\,(T_1 - T_2)/N.$$

The quantity M/N is often called the 'Conduction Shape Factor', S. The value of S for some of most common encountered configurations has been tabulated in

Table 4.1. Thus, if S is known, the rate of heat transfer,
$$\dot{Q} = S\,k\,(T_1 - T_2) \qquad\qquad (4.6)$$

Example 4.5 Saturated steam at 8 bar is piped vartically into the earth at the rate of 200 kg/hr. If the ground surface temperature is 25°C, estimate the depth to which the steam will go before it condenses completely. k for earth 1.5 W/m°C. The pipe outside surface is at the same temperature as steam. The outside diameter of the pipe is 100 mm.

Solution: From steam tables, saturation temperature of steam at 8 bar is 174.4°C and the latent heat of vaporization is 2046.5 kJ/kg.

By making an energy balance, $\dot{Q} = \dot{m} \times h_{fg} = S \times k \times (T_1 - T_2)$

$$S = (200/3600) \times 2046.5 \times 10^3 / [1.5 \times (174.4 - 25)] = 507.338$$

From Table 4.1, $S = 2\pi L / \ln(2L/r)$
Solving by trial and error, $L = 635$ m.

Example 4.6 A 20 cm diameter steam line and a 10 cm diameter chilled water line for air conditioning are located horizontally 25 cm apart in a large service trough packed with insulating material, k = 0.05W/m°C. The steam line surface temperature is 145°C and the chilled water line surface temperature is 5°C. Calculate the heat transfer rate to water per metre length of the pipe.

Solution: From Table 4.1, the shape factor, $\dfrac{S}{L} = \dfrac{2\pi}{\cos h^{-1}\left[(x^2 - r_1^2 - r_2^2)/2r_1 r_2\right]}$

$$x = 0.25 \text{ m}, \quad r_1 = 0.1 \text{ m}, \quad r_2 = 0.05 \text{ m}$$

$\therefore \qquad (x^2 - r_1^2 - r_2^2)/2r_1 r_2 = 5$

$\therefore \qquad\qquad S/L = 2\pi/\cos h^{-1}(5) = 2.74 \qquad \text{and} \quad S = 2.74$

$\dot{Q} = S\,K\,(T_1 - T_2) = 2.74 \times 0.05\,(145 - 5) = 19.18\text{W}.$

Example 4.7 A small cubical furnace constructed of 15 cm thick fire-brick walls has its inside dimensions as 40 cm by 40 cm by 40 cm. The inside and outside wall temperatures are 500°C and 50°C respectively. Compute the heat loss through the furnace if the thermal conductivity of the all material is 1 W/m°C.

Solution: From Table 4.1 we will compute the shape factor for the walls, edges and corners because the furnace is a three dimensional configuration.

For one wall: shape factor = area/thickness = 0.4 × 0.4/0.15 = 1.067

For one edge: shape factor = 0.54 × length = 0.54 × 0.4 = 0.216

For one corner: shape factor = 0.15 × thickness = 0.15 × 0.15 = 0.0225

Table 4.1 Conduction shape factors

Physical System	Schematic Diagram	Shape Factor	Restriction
Plane wall	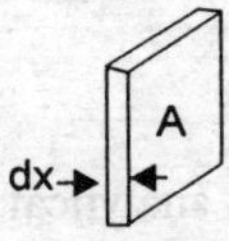	A/dx	1-dimensional heat-flow
Conduction through edge section of two walls, inner and outer surface temperature uniform	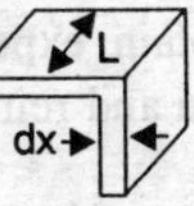	$0.54\,L$	Inside dimension must be greater than $(1/5)\,dx$
Conduction through corner section of three homogeneous walls, inner and outer surface temperature uniform	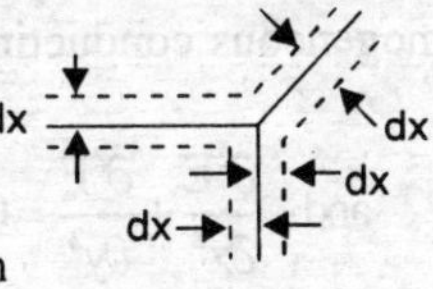	$0.15\,dx$	$dx \ll L$, the length of wall
Isothermal cylinder, radius r buried in semi-infinite medium having isothermal surface	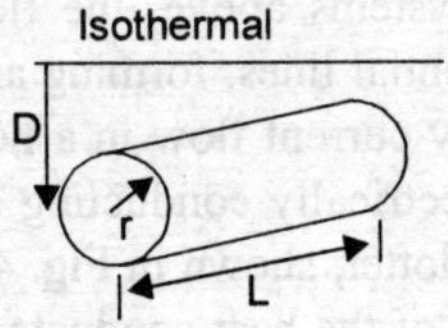	$2\pi L/\cos h^{-1}(D/r)$	$L \gg r$
Conduction between two isothermal cylinders buried in infinite medium	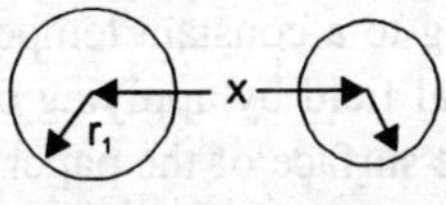	$\dfrac{2\pi L}{\cos h^{-1}\left(\dfrac{x^2 - r_1^2 - r_2^2}{2 r_1 r_2}\right)}$	$L \gg r$ $L \gg x$
Isothermal sphere of radius r buried in semi-infinite medium having isothermal surface	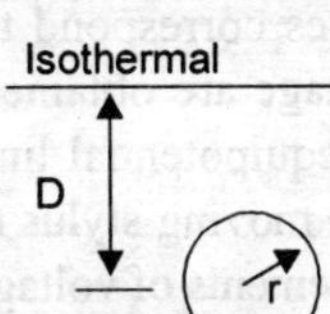	$4\pi r/(1 - r/2D)$	
Isothermal cylinder of radius r placed in semi-infinite medium as shown	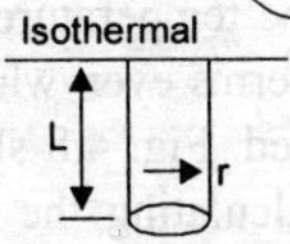	$2\pi L/\ln(2L/r)$	$L \gg 2r$
Cylinder centered in a square of length L	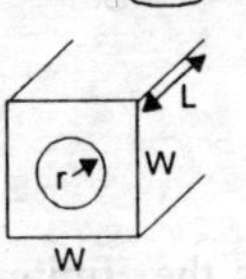	$2\pi L/\ln(0.54W/r)$	$L \gg W$
Hollow cylinder length, L	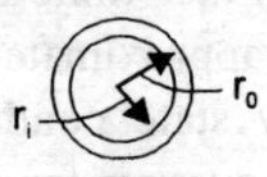	$2\pi L/\ln(r_o/r_i)$	$L \gg R$
Hollow sphere	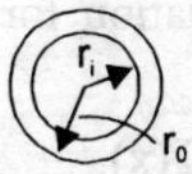	$4\pi r_o r_i/(r_o - r_i)$	
Hemisphere buried in semi-infinite medium		$2\pi r$	

Since there are six walls, twelve edges and eight corners, the total shape factor is:
$$S = 6 \times 1.067 + 12 \times 0.216 + 8 \times 0.0225 = 9.174 \text{ m}$$

Rate of heat transfer, $\dot{Q} = S\,k\,(T_1 - T_2) = 9.174 \times 1 \times 450 = 4.128$ kW.

4. Discuss the Utility of Analogical Method

There are many heat flow problems for which analytical solutions are extremely difficult and the experimental solutions are either too expensive or too time consuming. But, it is often possible to obtain experimental solution of such problems quite easily in an analogous system and reinterpret in terms of thermal problem.

An important class of analogical methods exists in two-dimensional steady state systems. The differential equation for the steady state heat conduction and the steady flow of current in a homogeneous conducting plane are analogous to each other. That is,

$$\frac{\partial^2 T}{\partial x^2} + \frac{\partial^2 T}{\partial y^2} = 0; \quad \text{and} \quad \frac{\partial^2 E}{\partial x^2} + \frac{\partial^2 E}{\partial y^2} = 0$$

In the thermal and electrical systems above, the flux lines are everywhere normal to their respective equipotential lines, forming an orthogonal net.

Heat conduction is simulated by current flow in a homogeneous, electrically conducting medium such as an electrically conducting sheet 'Teledeltos' paper, with the help of an Analog Field Plotter, shown in Fig. 4.7. The paper is cut to a shape geometrically similar to that of the heat conductor.

Boundary conditions corresponding to a constant temperature potential in the heat flow are obtained in the electrical field by applying copper wires or highly conductive areas of silver paint, to the surface of the paper and attaching them to a direct-voltage source. Insulated surfaces correspond to the plain edges of the conducting paper. Lines of constant voltage are obtained directly by selecting a voltage level on the 'null detector' and equipotential lines are traced by making small perforations in the paper while the moving stylus maintains a zero reading on the instrument. By selecting equal increments of voltage, adjacent lines become analogous to isotherms separated by the same temperature difference. Since the heat flow lines are perpendicular to the isotherms everywhere, they are sketched free-hand and curvilinear squares are obtained. Fig. 4.8 shows the arrangement for obtaining the curvilinear squares for calculating the shape factor for heat transfer through corner of thick-walled duct.

5. Numerical Methods

The most commonly used numerical method is the 'finite-difference' method in which a differential equation is replaced by an approximate algebraic expression using Taylor series expansion. Under steady state conduction with no heat generation within the system, the governing equation for heat transfer in one-dimension is written as:

$$d^2T/dx^2 = 0, \quad \text{where} \quad T = T(x).$$

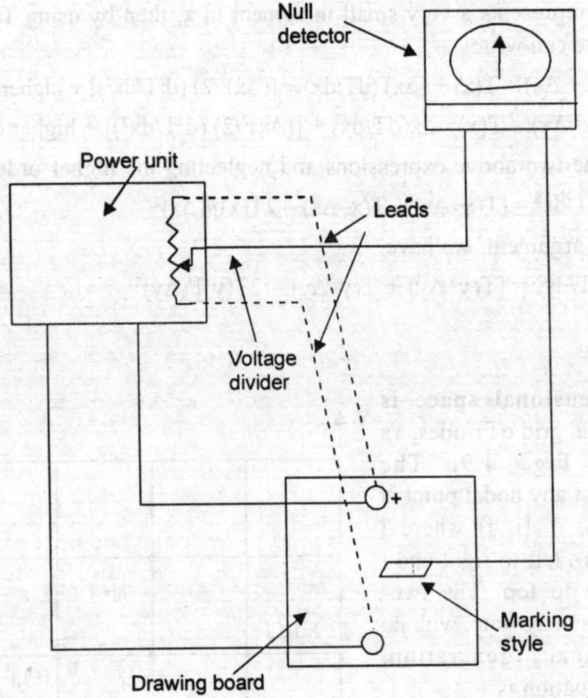

Fig. 4.7 Schematic Circuit Diagram for Analog Field Plotter

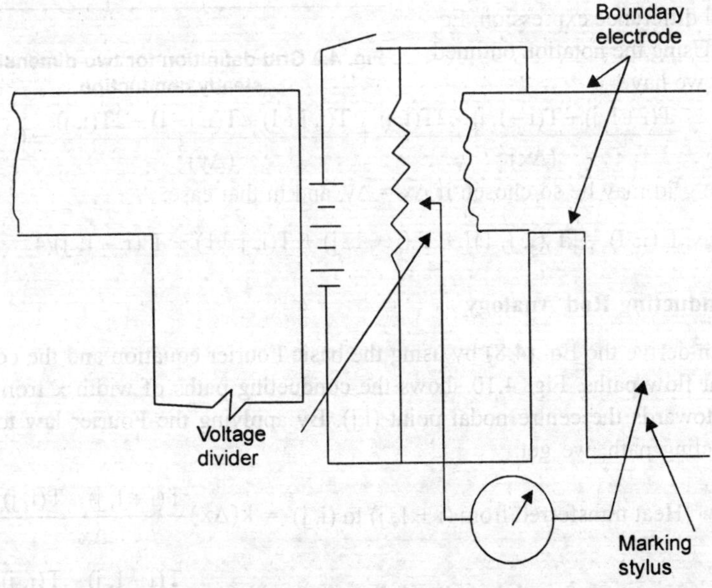

Fig. 4.8 Arrangement for analog field plotter for thick walled duct corner

Let Δx represents a very small increment in x, then by using Taylor series expansion, we can write:

$$T(x + \Delta x) = T(x) + (\Delta x)(dT/dx) + [(\Delta x)^2/2)(d^2T/dx^2)] + \text{higher order terms}$$

and $\quad T(x-\Delta x) = T(x) - \Delta x(dT/dx) + [(\Delta x)^2/2)(d^2T/dx^2)] + \text{higher order terms}$

By adding the two above expressions and neglecting the higher order terms,

$$d^2T/dx^2 = [T(x+\Delta x) + T(x-\Delta x) - 2T(x)]/(\Delta x)^2 \qquad (4.7a)$$

By the same argument, we have

$$d^2T/dy^2 = [T(y+\Delta y) + T(y-\Delta y) - 2T(y)]/(\Delta y)^2 \qquad (4.7b)$$

5A. Notations

A two-dimensional space is divided into a grid of nodes, as shown in Fig. 4.9. The temperature at any nodal point is designated as T (i, j) where i increases from left to right and j from bottom to top. The two-dimensional steady state, with no internal heat generation, governing equation is:

$$\frac{\partial^2 T}{\partial x^2} + \frac{\partial^2 T}{\partial y^2} = 0$$

and this equation can be put into finite difference form by using the central difference expression, Eq. (4.7). Using the notation outlined above we have

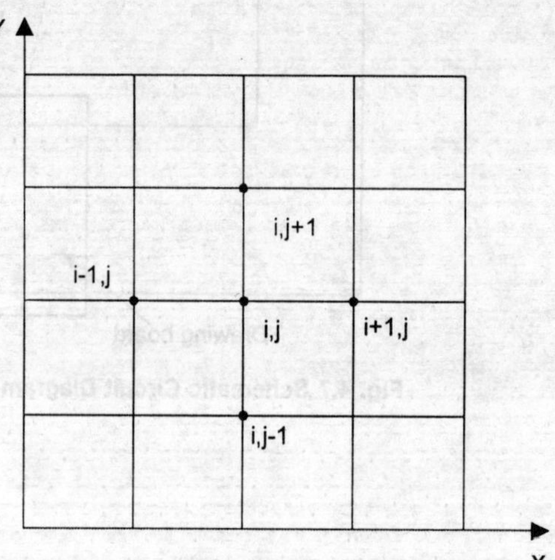

Fig. 4.9 Grid definition for two-dimensional steady conduction

$$\frac{T(i+1, j) + T(i-1, j) - 2T(i, j)}{(\Delta x)^2} + \frac{T(i, j+1) + T(i, j-1) - 2T(i, j)}{(\Delta y)^2} = 0$$

The grid may be so chosen if $\Delta x = \Delta y$, and in that case:

$$T(i, J) = [T(i, j-1)] + T(i+1, j) + T(i, j+1) + T(i-1, j)/4 \qquad (4.8)$$

6. Conducting Rod Analogy

We can derive the Eq. (4.8) by using the basic Fourier equation and the concept of heat flow paths. Fig. 4.10 shows the conducting paths of width x from each point towards the centre nodal point (i,j). By applying the Fourier law to each conducting path, we get:

$$\text{Heat transferred from } (i+1, j) \text{ to } (i, j) = k(\Delta x)\frac{T(i+1, j) - T(i, j)}{\Delta x}$$

$$\text{Heat transferred from } (i-1, j) \text{ to } (i, j) = k(\Delta x)\frac{T(i-1, j) - T(i, j)}{\Delta x}$$

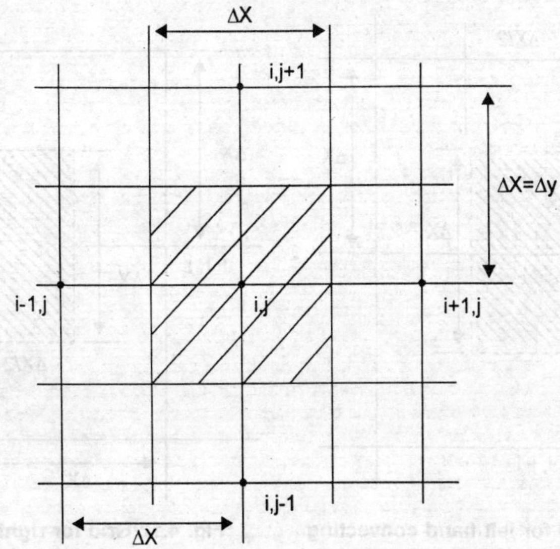

Fig. 4.10 Conducting rod analogy for two dimensional steady conduction

Using the same procedure, a simple energy balance gives

$$k(\Delta x)\frac{\left[T(i+1,j)-T(i,j)\right]}{\Delta x}+k(\Delta x)\frac{\left[T(i-1,j)-T(i,j)\right]}{\Delta x}+$$

$$k(\Delta x)\frac{\left[T(i,j+1)-T(i,j)\right]}{\Delta x}+k(\Delta x)\frac{\left[T(i,j-1)-T(i,j)\right]}{\Delta x}$$

This equation reduces to the same expression as in Eq. (4.8).

7. Conducting Rod Analogy for Different Boundary Conditions

The Eq. (4.8) is restricted mainly to points interior to the body in which the heat conduction is taking place. We can very easily develop such equations for imposed conditions on the exterior surfaces of system.

(a) *Surface Convecting to a Fluid:* Fig. 4.11 depicts the grid for a left hand face having convective heat transfer to a fluid. For the point (i,j) on the surface, we make an energy balance:

Heat received by conduction = Heat lost by convection.

or,
$$\frac{k\Delta x\,[T(i+1,j)-T(i,j)]}{\Delta x}+\frac{k\Delta x\,[T(i,j-1)-T(i,j)]}{2\Delta x}$$

$$+\frac{k\Delta x\,[T(i,j+1)-T(i,j)]}{2\Delta x}=h\Delta x\,[T(i,j)-T_\infty]$$

Upon simplification,

$$T(i,j)=\frac{2T(i+1,j)+T(i,j+1)+T(i,j-1)+2h\Delta xT_\infty/k}{4+2h\Delta x/k} \tag{4.9}$$

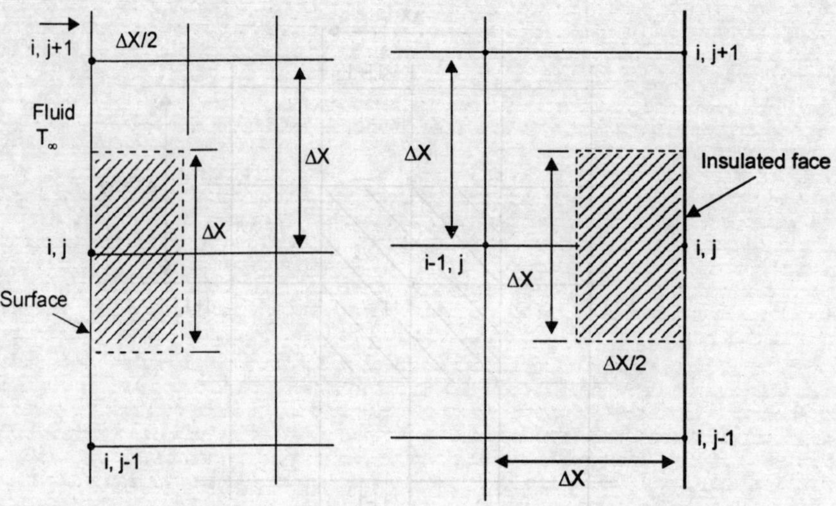

Fig. 4.11 Grid for left hand convecting surface   **Fig. 4.12 Grid for right hand face insulated**

(b) *Insulated Boundary:* Fig. 4.12 depicts the grid for the right hand face insulated. For the point (i, j) we make an energy balance:

$$\frac{k\Delta x[T(i-1,j)-T(i,j)]}{(\Delta x)} + \frac{k\Delta x[T(i,j+1)-T(i,j)]}{2(\Delta x)}$$

$$+ \frac{k\Delta x[T(i,j-1)-T(i,j)]}{2(\Delta x)} = 0$$

Upon simplification,

$$T(i,j) = \frac{2T(i-1,j)+T(i,j+1)+T(i,j-1)}{4} \tag{4.10}$$

(c) *Exterior Corner with Convective Boundary:* Fig. 4.13 depicts the grid and by making an energy balance for the point (i, j) we have

$$\frac{k\Delta x[T(i,j-1)-T(i,j)]}{2\Delta x} + \frac{k\Delta x[T(i-1,j)-T(i,j)]}{2\Delta x} = h\Delta x[T(i,j)-T_\infty]$$

or, $$T(i,j) = \frac{T(i,j-1)+T(i-1,j)+2h\Delta xT_\infty}{2(1+h\Delta x/k)} \tag{4.11}$$

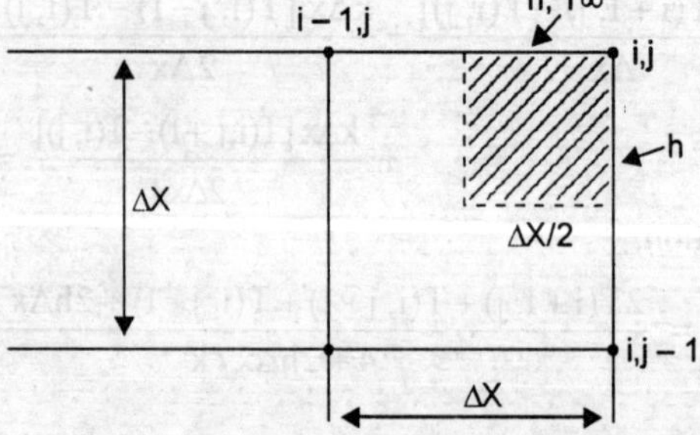

Fig. 4.13 Exterior corner with convective boundary

(d) *Reentrant Corner with Convective Boundary:* Fig. 4.14 depicts the grid and by making an energy balance for the point (i, j), we have

$$T(i, j) = \frac{2T(i-1, j) + 2T(i, j+1) + T(i+1, j) + T(i, j-1) + 2h\Delta xT_\infty / k}{2(3 + h\Delta x / k)} \qquad (4.12)$$

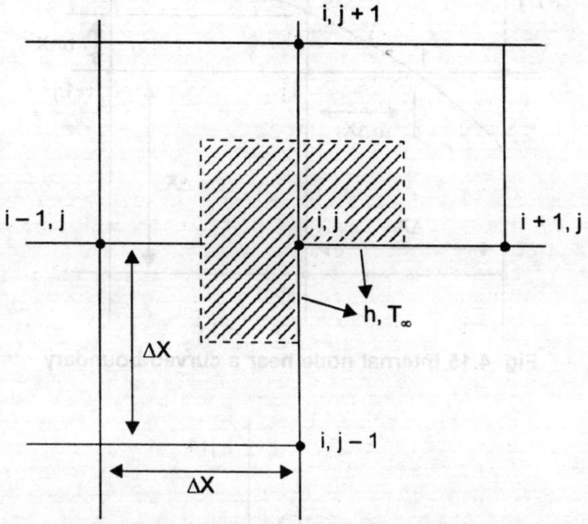

Fig. 4.14 Reentrant corner with convective boundary

(e) *Internal Node near Curved Boundary:* In many practical applications of numerical methods, the boundaries of the solid are irregular and it is either very inconvenient or is impossible to arrange the spacings so that the boundary points coincide with regular points of the net. For such cases special relations have been developed for the nodal resistances connecting the points on the boundary to points on the interior. Fig. 4.15 depicts the grids and T (i, j) is related by the expression:

$$T(i, j) = \frac{\left[\dfrac{2}{b(b+1)} T_2 + \dfrac{2}{a+1} T(i+1, j) + \dfrac{2}{b+1} T(i, j-1) + \dfrac{2}{a(a+1)} T_1 \right]}{\left[\dfrac{2(a+b)}{ab} \right]} \qquad (4.13)$$

(f) *A Nodal Point Located on a Plane, Insulated Surface with Uniform Heat Generation:* Fig. 4.15 (a) depects the grid and by making the energy balance for the point (i, j), we have:

$$k(T_{i-1,j} - T_{i,j}) + \frac{k}{2}(T_{i,j+1} - T_{i,j}) + \frac{k}{2}(T_{i,j-1} - T_{i,j}) + \dot{q}\left(\Delta y \cdot \frac{\Delta x}{2} \cdot 1 \right) = 0$$

or, $T (i-1, j) + T(i, j+1)/2 + T(i, j-1)/2 - 2T(i, j) + q (\Delta x)^2/2k = 0$

and, $2T (i-1, j) + T (i, j+1) + T (i, j-1) - 4T (i, j) + q(\Delta x)^2/k = 0.$

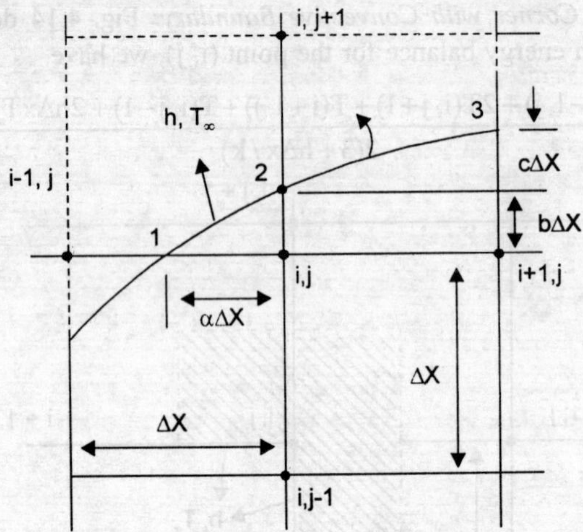

Fig. 4.15 Internal node near a curved boundary

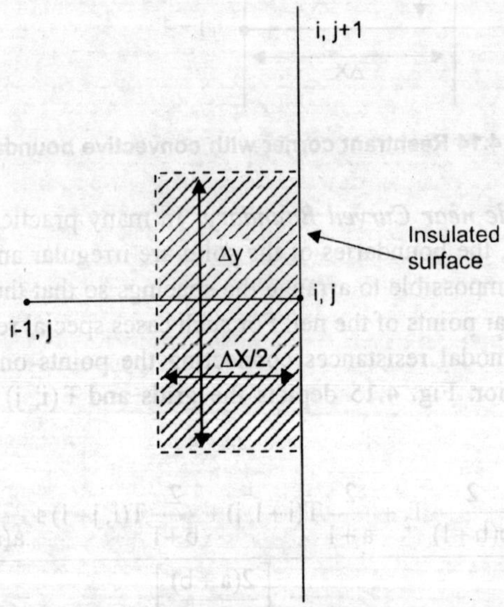

**Fig. 4.15 (a) Nodal point on a plane insulated surface with
uniform heat generation (Δy = Δx)**

Example 4.8   A straight circular steel fin (diameter 15 mm, length 100 mm, k = 25 W/mK) has its base temperature at 175°C. Set up the equation using finite difference method for determining the temperature distribution in the fin and for the rate of heat flow, $T_\infty = 25°C$.

Solution: Let us divide the fin into (M–1) equal parts of length Δx, giving M nodes, A value of M equal to 10 should be satisfactory. However, the accuracy of the solution will improve as the value of M increases. Fig. 4.16 (a) depicts the grid for the fin. In order to determine the temperature distribution in the fin, we perform an energy balance on the ith node, which is typical of all internal nodes from i = 2 to i = (M–1). Thus, for the ith node, we write, as shown in Fig. 4.16(b),

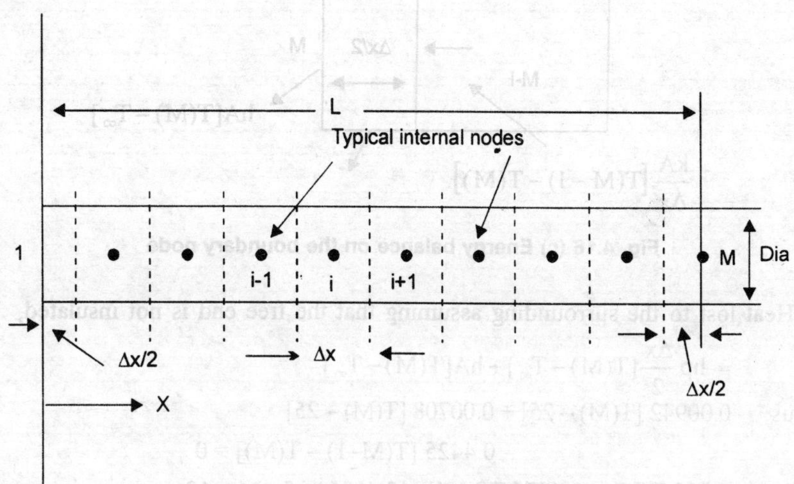

Fig. 4.16 (a) Grid for straight fin (Example 4.6)

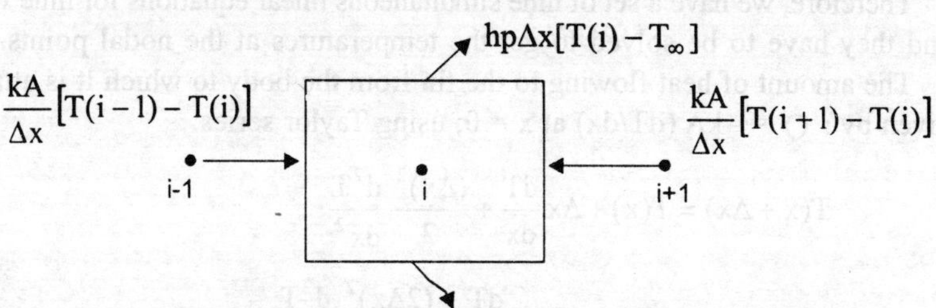

Fig. 4.16 (b) Energy balance for ith internal node of the fin

Heat conducted from (i–1) th node to ith node is: kA [T(i – 1) – T(i)]/Δx

Heat conducted from (i + 1) th node to ith node is: kA [T(i + 1) – T(i)]/Δx

Heat lost to the surroundings is : hp(Δx)(T(i) – T_∞)

or, $\dfrac{kA}{\Delta x}[T(i-1) - T(i)] + \dfrac{kA}{\Delta x}[T(i+1) - T(i)] - hp\Delta x[T(i) - T_\infty] = 0$

Δx = 0.1/M = 0.01m;   A = π/4 d² = 0.000177 m²;   p = πd = 0.0471 m

∴   kA/Δx = 0.4425,   hpΔx = 0.01884

Substituting the values:

0.4425 [T(i-1) – T(i)] + 0.4425 [T(i+1) – T(i)] – 0.01884 [T(i) – 25] = 0

or,   0.90384 T(i) = 0.4425 [T(i–1) + T(i+1) + 0.471], for i = 2 to (M–1) = 9

and   T(1) = 175

By making an energy balance for the node M, Fig. 4.16 (c) , we have

Heat conducted from (M–1) mode is : (kA/Δx) [T(M–1) – T(M)]

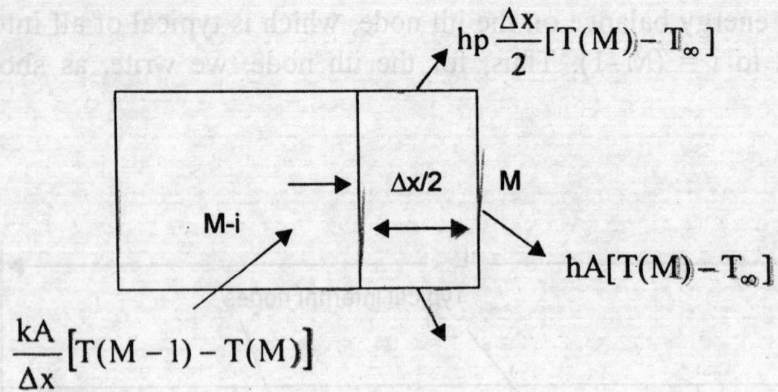

Fig. 4.16 (c) Energy balance on the boundary node

Heat lost to the surrounding assuming that the free end is not insulated,

$$= hp\frac{\Delta x}{2}[T(M)-T_\infty]+hA[T(M)-T_\infty]$$

Thus, $0.00942\,[T(M)-25]+0.00708\,[T(M)-25]$

$$-0.4425\,[T(M-1)-T(M)]=0$$

and, $0.459\,T(M)=0.4425\,T(M-1)+0.4125;\quad$ for $M=10.$

Therefore, we have a set of nine simultaneous linear equations for nine variables and they have to be solved to get the temperatures at the nodal points.

The amount of heat flowing to the fin from the body to which it is attached is given by : $\dot{Q}=-kA\,(dT/dx)$ at $x=0$; using Taylor series:

$$T(x+\Delta x)=T(x)+\Delta x\frac{dT}{dx}+\frac{(\Delta x)^2}{2}\frac{d^2T}{dx^2} \tag{a}$$

and $$T(x+2\Delta x)=T(x)+2\Delta x\frac{dT}{dx}+\frac{(2\Delta x)^2}{2}\frac{d^2T}{dx^2} \tag{b}$$

Multiplying the above equation (a) by 4 and subtracting the result from the equation (b), we get

$$T\,(x+2\Delta x)-4T\,(x+\Delta x)=-3T(x)-2\,\Delta x\,(dT/dx)+\text{higher order terms}$$

or, $$\left(\frac{dT}{dx}\right)=\frac{4T(x+\Delta x)-T(x+2\Delta x)-3T(x)}{2\Delta x}$$

and $$\left(\frac{dT}{dx}\right)_{x=0}=\frac{4T(2)-T(3)-3T(1)}{2\Delta x}$$

Therefore, $$\dot{Q}=kA\frac{3T(1)-T(3)-4T(2)}{2\Delta x}$$

$$=0.22125\,[3\times175-T(3)-4T(2)]$$

Example 4.9 A 1 metre square steel plate, $k=25\,W/m^\circ C$, has its left face maintaind at $200^\circ C$, top face maintained at $300^\circ C$. The other two face are exposed to an environment at $50^\circ C$, $h=25\,W/m^2\,^\circ C$. Set up the equations, using finite difference method, for calculating the temperature distribution in the plate.

Solution: We divide the square plate into 25 nodal points as shown in Fig. 4.9 where i varies from 1 to 5 and j also varies from 1 to 5. The left hand face, (i = 1) is maintained at 200°C, and the top face (j = 5) is maintained at 300°C. The nodes (i varying from 2 to 4 and J verying from 2 to 4) are internal nodes and we will get 9 equations for these 9 nodes by using Eq. (4.8), i.e.,

$$T (i, j) = [T (i, j\text{-}1) + T (i, j + 1) + T (+ 1, j) + T (i - 1, J)] / 4;$$

Using Eq. (4.9), we can write equations for 6 nodes: J = 1, i varying from 2 to 4 and i = 5, j varying from 2 to 4

Since $\qquad 2\,h\,\Delta x/k = 2 \times 25 \times 0.25 / 25 = 0.5$ and $T = 50$

$$\therefore \qquad T(i, j) = \frac{2T(i + 1, j) + T(i, j + 1) + T(i, j - 1) + 25}{4.5}$$

For the node i = 5, j = 1, a corner node, we make use of Eq. (4.11),

or, $\qquad T (5,1) = [T (4, 1) + T (5, 2) + 625] / 2.5.$

Several methods are available for the solution of a system of simultaneous algebraic equations. Certain methods particularly suited to the form of equations developed above have been discussed in detail in Schneider, P.J., *Conduction Heat Transfer*, Cambridge, Mass, Addison-Wesley 1955.

8. Transient Numerical Methods—Usefulness

Heisler Charts are very useful for calculating temperatures in certain regular shaped systems under transient heat flow conditions (discussed in chapter 3). There are many geometrical shapes of practical interest which do not fall under these categories and moreover, their boundary conditions vary with time. Such problems (geometrical shapes for which Heisler Charts are not available and time dependent boundary conditions) cannot be solved analytically and they are best handled by a numerical technique with computers.

Let us consider the one-dimensional transient heat transfer equation:

$$\frac{\partial^2 T}{\partial x^2} = \frac{1}{\alpha}\frac{\partial T}{\partial t} \text{ where } \alpha = \frac{k}{\rho C_p}$$

Forward Difference—the forward difference representation is obtained by using the forward-difference expression or the time derivative in the above equation and the central derivative for the term on the left side of the equation. Thus, we write,

$$\frac{[T(x + \Delta x, t) + T(x - \Delta x, t) - 2T(x, t)]}{(\Delta x)^2} = \frac{1}{\alpha}\frac{[T(x, t + \Delta t) - T(x, t)]}{\Delta t}$$

If we use the nodal notations for x, x + Δx, x − Δx, and denote T' for t + Δt while T denotes the temperature at a node at time t, we get

$$\frac{[T(i + 1, t) + T(i - 1, t) - 2T(i, t)]}{(\Delta x)^2} = \frac{1}{\alpha}\frac{[T'(i) - T(i)]}{\Delta t}$$

or, $\qquad T'(i) = \dfrac{\alpha\Delta t}{(\Delta x)^2}[T(i + 1, t) + T(i - 1, t)] + \left(1 - \dfrac{2\alpha\Delta t}{(\Delta x)^2}\right) T(i, t)$

and if the time and distance increments are so chosen that $\dfrac{(\Delta x)^2}{\alpha(\Delta t)} = 2$, the temperature at node i is given as

$$T'(i) = (T(i + 1, t) + T(i - 1, t))/2$$

i.e., the temperature at the node i after a time Δt is the arithmetic average of the two adjacent nodal temperatures at the beginning of the time increment.

If $\Delta x = \Delta y$, the equivalent two-dimensional forward difference (explicit) formulation would be:

$$\frac{[T(i+1,j,t) + T(i-1,j,t) + T(i,j+1,t) + T(i,j-1,t) - 4T(i,j,t]}{(\Delta x)^2}$$

$$= \frac{T'(i, j) - T(i, j)}{\alpha(\Delta t)}$$

and when $\dfrac{(\Delta x)^2}{\alpha(\Delta t)} = 4$, the temperature at the node (i, J) after a time increment Δt is given by:

$$T'(i, j) = [T(i + 1, j, t) + T(i-1, j, t) + T(i, j + 1, t) + T(i, j-1, t)]/4$$

Backward Difference – The backward difference or 'implicit' representation is obtained by using Taylor's expansion around $x = x$ and $t = t + \Delta t$. Thus, the one-dimensional heat conduction equation under transient conditions is written as

$$\frac{[T(x + \Delta x, t + \Delta t) + T(x - \Delta x, t + \Delta t) - 2T(x, t + \Delta t)]}{(\Delta x)^2} = \frac{[T(x, t + \Delta t) - T(x, t)]}{\alpha(\Delta t)}$$

Using the same notation as used in explicit formulation, we get

$$\frac{[T'(i+1) + T'(i-1) - 2T'(i)]}{(\Delta x)^2} = \frac{[T'(i) - T(i, t)]}{\alpha(\Delta t)}$$

Thus the implicit formulation gives the future temperature of node i, T' (i) in terms of the current temperature at node i, T (i), and the future temperatures of the neighbouring points. The two-dimensional case can be written as, after assuming $\Delta x = \Delta y$,

$$\frac{[T'(i+1, j) + T'(i-1, j) + T'(i, j+1) + T'(i, j-1) - 4T'(i, j)]}{(\Delta x)^2} = \frac{T'(i, j) - T(i, j)}{\alpha(\Delta t)}$$

It can be observed that the advantage of the forward-difference method over the backward difference method is that the former method gives the future temperature of a particular node in terms of current temperatures of that node and its neighbours. But, the magnitude of the time step is restricted by the spatial increment Δx, because for one-dimensional system $\dfrac{(\Delta x)^2}{\alpha(\Delta t)} \geq 2$ and for a two dimensional system $\dfrac{(\Delta x)^2}{\alpha(\Delta t)} \geq 4$, and t his r estriction m ay a ctually r equire m ore calculation time. In the backward method, there is no restriction in selecting the time or space increments but a set of simultaneous equations has to be solved because the implicit method expresses future nodal temperature in terms of its current value and the future values of its neighbours' temperatures.

Example 4.10 A 20 cm thick wall (thermal diffusivity $\alpha = 0.5 \times 10^{-6}$ m²/s) is initially at 100°C. The temperature of the left hand face linearly increases to 250°C in 1 hour. Estimate the temperature distribution in the wall at the end of this period.

Solution: The stability criteria for one-dimensional transient conduction problems using forward difference method is $(\Delta x)^2 / (\alpha \, \Delta t) \geq 2.0$

Let $\Delta x = 0.025$ m and $\Delta t = 10$ min $= 600$ seconds

$$(\Delta x)^2 / (0.5 \times 10^{-6} \times 600) = 2.08$$

Thus we will have eight space intervals, six temperature increments and the temperature at the node i after an interval of time $\Delta t = 10$ min can be summarised in the tabular form:

$$T_0 \quad T_1 \quad T_2 \quad T_4 \quad T_5 \quad T_6 \quad T_7 \quad T_8 \quad T_9$$

$$|\longleftarrow \quad 0.025 \times 8 = 0.20\text{m} \quad \longrightarrow|$$

Time in min	T_0	T_1	T_2	T_3	T_4	T_5	T_6	T_7	T_8	T_9
0	100	100	100	100	100	100	100	100	100	100
10	125	100	100	100	100	100	100	100	100	100
20	150	112	100	100	100	100	100	100	100	100
30	175	124.5	105.77	100	100	100	100	100	100	100
40	200	139.8	112	102.8	100	100	100	100	100	100
50	225	155.4	120.9	105.9	101.3	100	100	100	100	100
60	250	172.3	130.3	110.9	102.9	100.6	100	100	100	100

where $\quad T_i' = T(i, t) + (1/2.08) \cdot [T(i + 1, t) + T(i - 1, t) - 2T(i, t)]$

Example 4.11 A long power transmission cable is buried at a depth of 2m (ground to cable centreline distance). The cable is encased in a thin walled pipe 0.1 m in diameter. The space between the cable and the pipe is filled with liquid nitrogen at 77 K. The pipe is covered with an insulation (k = 0.005 W/mK, thickness 0.05 m). The surface of the earth is at 300 K and k = 1.2 W/mK. Calculate the cooling load on a refrigerating system which maintains the temperature of the liquid nitrogen.

Solution: Let the temperature at the outer surface of the insulation be T. The temperature at the inner surface of the insulation is 77K.

$$\therefore \quad \dot{Q}/L \text{ through insulation} = (T - 77) \Big/ \frac{\log_e 2}{2\pi k}$$

$$= (T - 77)/22.06$$

$\dot{Q}/L$ from the surface of the earth to the insulation

$$= Sk (T_\infty - T) = [2\pi/\cosh^{-1} (2/0.1)] \times 1.2 (300 - T)$$

$$= 2.045 (300 - T) = (T - 77)/22.06 \qquad \therefore T = 295.164 \text{ K}$$

and Q/L, the load on the refrigerator = 9.89 W/m.

SUMMARY

1. Multi–dimensional steady state conduction heat transfer problems can be solved analytically, graphicallly, analogically and numerically.
2. Separation of variables technique can be applied to solve linear partial differential equations provided the differential equation can be expressed in a product form satisfying the prescribed boundary conditions.
3. Two–dimensional Laplace's equation with non-homogeneous boundary conditions should be divided into a number of simpler subproblems each having only one non-homogeneous boundary condition.
4. A non-homogeneous boundary condition can be eliminated by selecting an appropriate surface temperature.
5. Graphical methods give an approximate solution to steady state, two–dimensional conduction heat transfer problems.
6. Isotherms are constant temperature lines and are perpendicular to insulated surfaces.
7. Heat flow lines are always perpendicular to isotherms and bisect the corners of isothermal boundaries.
8. In flux plot (graphical solutions), conduction shape factor is the ratio of the number of temperature increments (equal to number of isotherms − 1) to the number of heat flow lanes.
9. For two–dimensional steady state conduction heat transfer problems, electrical analogy to thermal systems are used to obtain the flow net (a set of curvilinear squares) for calculating the shape factor of different geometries.
10. Linear partial differential equation can be easily converted into a set of simultaneous equations by finite difference method using Taylor series expansion.
11. Forward finite difference method requires more calculation time because the magnitude of time step is limited by the spatial increments to ensure convergence in transient numerical method.
12. Implicit finite difference method expresses future nodal temperatures in terms of its current values and the future values of its neighbours' temperature in transient numerical methods.

MULTIPLE CHOICE QUESTIONS

1. In one-dimensional transient numerical method, the choice of the time interval Δt in the finite difference equations is limited by the condition that
$(\Delta x)^2/\alpha(\Delta t)$
   (a) is less than 2   (b) is equal to 2
   (c) is greater than 2   (d) either (b) or (c)   **(d)**
2. Assertion (A): The backward finite difference method when applied to solve problems involving a large number of nodes may require less computer time.
   Reasoning (R): Because very small times increments may be imposed in the forward difference method from stability requirements.
   Code: (a) A is true, R is false   (b) Both A and R are true
         (c) Both A and R are false   (d) A is false, R is true   **(b)**

3. Assertion (A): Electrical Analogs of thermal systems have found wide use.
   Reasoning (R):Because accurate measurement of heat transfer rates present
   rather formidable instrumentation difficulties.
   Code: (a) Both A and R are true and R is the correct explanation of A
       (b) Both A and R are true and R is not the only reason for A
       (c) A is false, R is true   (d) A is true, R is false   **(b)**

4. The conversion of a system to its analog involves the following:
   (a) All elements of the original system are functionally duplicated
   (b) Analog quantities must be scaled in relation to their corresponding thermal
       quantities
   (c) Either (a) or (b)   (d) both (a) and (b)   **(d)**

.5. The finite-difference method consists of
   (a) replacing the pertinent differential equation by a set of simultaneous
       linear equation
   (b) replacing a continuous system by a lumped parameter network
   (c) Either (a) or (b)   (d) both (a) and (b)   **(d)**

6. Assertion (A): The solution of 2-dimensional steady state heat conduction
   equation is obtained by assuming the temperature distribution to be
   expressible as the porduct of functions each of which involves only one of
   the independent variables.
   Reasoning (R): Because the governing equation is a linear differential
   equation.
   Code: (a) Both A and R are true   (b) Both A and R are false
          (c) A is true, R is false   (d) A is false, R is true   **(a)**

7. The solution of steady state conduction equation for a circular cylinder of
   finite length is a product of hyperbolic functions and Bessels functions of
   the order
   (a) zero   (b) one
   (c) zero and first kind   (d) zero and second kind   **(c)**

8. If the number of isotherms is M and the number of heat flow lanes is N, the
   conduction shape factor S is defined as
   (a) M/N   (b) N/M   (c) (M–1)/N   (d) (N–1)/M   **(c)**

9. Choose the correct answer:
   Isotherms are constant temperature lines and they are
   (a) parallel to insulated surfaces
   (b) parallel to heat flow lines
   (c) perpendicular to both insulated surfaces and heat flow lines
   (d) bisecting the corners of isothermal boundaries.   **(c)**

10. In the implicit finite difference method,
   (a) the magnitude of time step is restricted by the spatial increment
   (b) there is no restriction is selecting the time increment
   (c) there is no restriction in selecting the space increment
   (d) both (b) and (c)   **(d)**

11. Assertion (A): A nodal point in an exterior corner between isothermal surface has no residual equation
    Reasoning (R): Because no heat can flow to or from that point.
    Code: (a) Both A and R are false   (b) Both A and R are true
           (c) A is true and R is false   (d) A is false, R is true   **(b)**

12. Assertion (A): To facilitate the geometrical construction of a net, the net should consist of small squares.
    Reasoning (R): Because the rate of heat flow in each stream tube is irrespective of the size of the squares.
    Code: (a) A is true, R is false   (b) A is false, R is true
           (c) A is true and R is the only   (d) A is true and R is not the only
              correct reason for A   correct reason for A.   **(d)**

13. Choose the incorrect answer:
    (a) Diagonals of curvilinear squares in a flux plot bisect each other at 90 deg but do not bisect the corners
    (b) Flow lines are perpedicular to isothermal boundaries
    (c) Isotherms are perpendicular to insulated boundaries
    (d) Isotherms and flow lines form a network of curvilinear squares.   **(a)**

14. To ensure convergence of the explicit finite difference method for a two-dimensional case, the magnitude of the time step and the spatial increment should be
    (a) $(\Delta x)^2/\alpha(\Delta T) \leq 2$ (b) $\alpha(\Delta t)/(\Delta x)^2 \geq 4$
    (c) $(\Delta x)^2/\alpha(\Delta T) \leq 4$ (d) $(\Delta x)^2/\alpha(\Delta T) \geq 4$ **(d)**

15. 1. Numerical methods involve dividing the problem domain into discrete control volumes and writing an energy balance for each control volume
    2. The energy balance equations for boundary control volumes cannot be generalised.
    3. We can obtain a closed form solution for two-dimensional steady-state conductions problems with complex boundary conditions also.
    4. Unsteady state conductions problems are subdivided into problems that can be handled by the lumped-capacity methods where the temperature is a function of time and space coordinates.
    (a) only 1 and 2 are correct   (b) 1, 2, and 3 are correct
    (c) 1, 2, 3, and 4 are correct   (d) 1, 2, and 4 are correct   **(d)**

NUMERICALS

1. A thin flat plate has its three faces (x = 0, x = L, and y = 0) at 0°C and the fourth face (y = b) at a temperature $T = T_m \sin \pi x/L$
   Show that the temperature distribution is given by

$$T(x,y) = T_m \frac{\sin h\,(\pi y/L)}{\sin h\,(\pi b/L)} \sin \pi x/L$$

2. A square plate (30 cm by 30 cm) has its two adjacent sides maintained at 100°C, the upper surface and the right hand vertical surfaces are maintained at 200°C and 300°C respectively. Find the temperature at the centre of the plate.
             (175°C)

3. A semi–infinite plate having constant thermal conductivity has its dimensions
    $(0 \leq x \leq 60$ cm$)$ and $(0 \leq y \leq \infty)$. The temperature along the face $y = 0$ is
    $500°$ C and the three other faces are at $250°$C. Calculate the temperature at a
    point $x = 30$ cm and $y = 40$ cm.   **(289°C)**

4. A small laboratory furnace having inside dimensions 30 cm $\times$ 45 cm $\times$ 45 cm
    and a wall thickness of 10 cm is made of a material $k = 10$ W/mK.   Using
    Table 4.1, determine the heat lost through the furnace walls if the temperatures
    on the inside and outside surfaces ar $400°$C and $50°$C.

                                                                          (42.56 kW)

5. A sphere 1 m in diameter is buried such that its uppermost point is 2 m
    below the surface of the earth. The temperature at the outer surface of the
    sphere and the free surface of the earth are $450°$C and $25°$C respectively. If
    the loss of heat energy by the sphere is 6.12 kW, estimate the thermal
    conductivity of the soil.   **(2 W/mK)**

6. A long 1.5 cm diameter electric copper cable $(k = 375$ W/mK$)$ is embedded
    coaxially in a 15 cm square concrete block. If the outside temperature of the
    concrete is $30°$C and the rate of energy dissipation in the cable is 400 W/m
    length of the wire, determine the temperature at the outer surface and at the
    centre of the cable.

7. A square fin (15 cm long and 1 cm by 1 cm, $k = 40$ W/mK) has its base
    raised to temperature $200°$C. The convective heat transter coefficient =
    50 W/m^{2}K and the ambient temperature is $35°$C. Take $\alpha = 0.01$ m^2/h, Δx=0.5
    cm. Set up the finite difference equation for determining the temperature
    distribution in the fin choosing the value of Δt to ensure convergence.

8. A 10 cm by 20 cm metallic strip $(k = 25$ W/mK, $\rho = 7600$ kg/m^3, C = 0.9 kJ/
    kgK$)$ has its sides maintained at $500°$C. The bottom face is insulated and the
    top s urface h as surface h eat t ransfer c oefficient of 5 0 W/m^{2}K. S et up t he
    finite difference equation for temperature distribution within the strip choosing
    $\Delta x = \Delta y = 1$cm.

9. A fin (5 cm long), cross-sectional area 4.5 cm^2, perimeter 12 cm, $k = 25$ W/
    mK$)$ has its root temperature $500°$C. It is exposed to hot gases at $900°$C with
    convective heat transfer coefficient h = 500 W/m^{2}K. Using nodal network
    determine (a) the t emperature distributions in the fin, (b) the rate of heat
    transfer from the fin and compare the results with that calculated by the
    exact method.

[**Note :** Finite difference method gets complicated when either the arrangement of
    the nodes or the geometry is irregular. The finite element method simplifies
    many of these problems. The basic difference between the two techniques
    is that the former attacks the differential equation by replacing derivatives
    by differences whereas the finite element method casts the problem as an
    integral to be minimised and a numerical approximation of the integral is
    used to obtain a solution. This method requires the basics of Calculus of
    variation.]

CHAPTER 5

Heat Transfer by Natural Convection

1. Convection Heat Transfer—Requirements

The heat transfer by convection requires a solid-fluid interface, a temperature difference between the solid surface and the surrounding fluid and a motion of the fluid. The process of heat transfer by convection would occur when there is a movement of macro-particles of the fluid in space from a region of higher temperature to lower temperature.

2. Convection Heat Transfer Mechanism

Let us imagine a heated solid surface, say a plane wall at a temperature T_w, placed in an atmosphere at temperature T_∞, Fig. 5.1 Since all real fluids are viscous, the fluid particles adjacent to the solid surface will stick to the surface. The fluid particle at A, which is at a lower temperature, will receive heat energy from the plate by conduction. The internal energy of the particle would increase and when the particle moves away from the solid surface (wall or plate) and collides with another fluid particle at B which is at the ambient temperature, it will transfer a part of its stored energy to B. And, the temperature of the fluid particle at B would increase. This way the heat energy is transferred from the heated plate to the surrounding fluid. Therefore the process of heat transfer by convection involves a combined action of heat conduction, energy storage and transfer of energy by mixing motion of fluid particles.

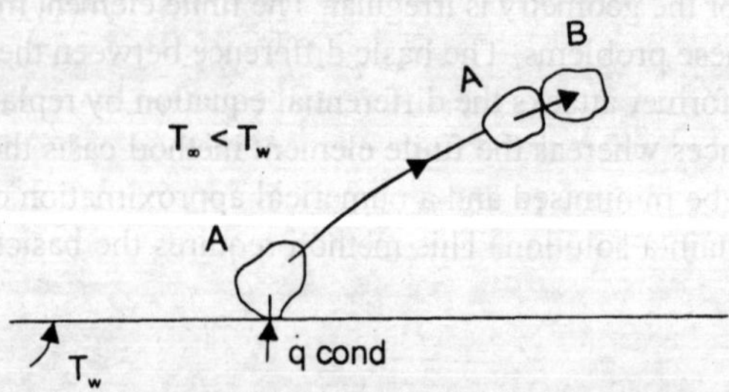

Fig. 5.1 Principle of heat transfer by convection

3. Free and Forced Convection

When the mixing motion of the fluid particles is the result of the density difference caused by a temperature gradient, the process of heat transfer is called natural or free convection. When the mixing motion is created by an artificial means (by some e xternal agent), t he process o f heat t ransfer is c alled forced c onvection. Since the effectiveness of heat transfer by convection depends largely on the mixing motion of the fluid particles, it is essential to have a knowledge of the characteristics of fluid flow.

4. Basic Difference between Laminar and Turbulent Flow

In laminar or streamline flow, the fluid particles move in layers such that each fluid p article follows a s mooth and c ontinuous path. T here i s no m acroscopic mixing of fluid particles between successive layers, and the order is maintained even when there is a turn around a corner or an obstacle is to be crossed. If a time dependent fluctuating motion is observed i n directions which are parallel and t ransverse to t he main f low, i.e., there i s a radom m acroscopic mixing o f fluid particles across successive layers of fluid flow, the motion of the fluid is called 'turbulent f low'. T he p ath o f a f luid p article w ould t hen b e z igzag a nd irregular, but on a statistical basis, the overall motion of the macroparticles would be regular and predictable.

5. Formation of a Boundary Layer

When a fluid flows over a surface, irrespective of whether the flow is laminar or turbulent, the fluid particles adjacent to the solid surface will always stick to it and their velocity at the solid surface will be zero, bacause of the viscosity of the fluid. Due to the shearing action of one fluid layer over the adjacent layer moving at the faster rate, there would be a velocity gradient in a direction normal to the flow.

Let us consider a two-dimensional flow of a real fluid about a solid (slender in cross-section) as shown in Fig. 5.2. Detailed investigations have revealed that

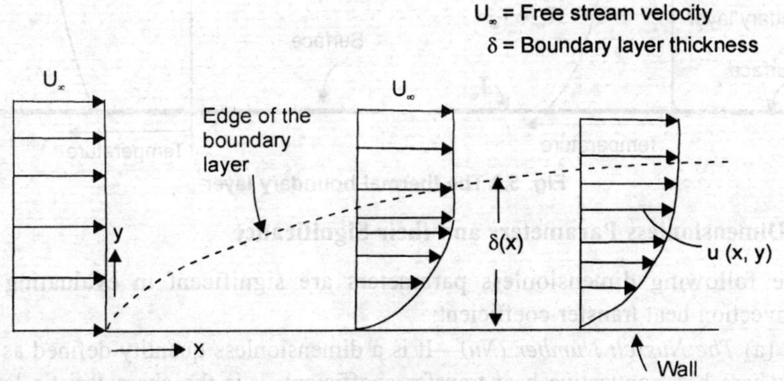

Fig. 5.2 Sketch of a boundary layer on a wall

the velocity of the fluid particles at the surface of the solid is zero. The transition from zero velocity at the surface of the solid to the free stream velocity at some distance away from the solid surface in the Y-direction (normal to the direction of flow) takes place in a very thin layer called 'momentum or hydrodynamic boundary layer'. The flow field can thus be divided in two regions:

(i) A very thin layer in the vicinity of the body where a velocity gradient normal to the direction of flow exists, the velocity gradient du/dy being large. In this thin region, even a very small viscosity μ of the fluid exerts a substantial influence and the shearing stress $\tau = \mu$ du/dy may assume large values. The thickness of the boundary layer is very small and decreases with decreasing viscosity.

(ii) In the remaining region, no such large velocity gradients exist and the influence of viscosity is unimportant. The flow can be considered frictionless and potential.

6. Thermal Boundary Layer

Since the heat transfer by convection involves the motion of fluid particles, we must superimpose the temperature field on the physical motion of fluid and the two fields are bound to interact. It is intuitively evident that the temperature distribution around a hot body in a fluid stream will often have the same character as the velocity distribution in the boundary layer flow. When a heated solid body is placed in a fluid stream, the temperature of the fluid stream will also vary within a thin layer in the immediate neighbourhood of the solid body. The variation in temperature of the fluid stream also takes place in a thin layer in the neighbourhood of the body and is termed 'thermal boundary layer'. Fig. 5.3 shows the temperature profiles inside a thermal boundary layer.

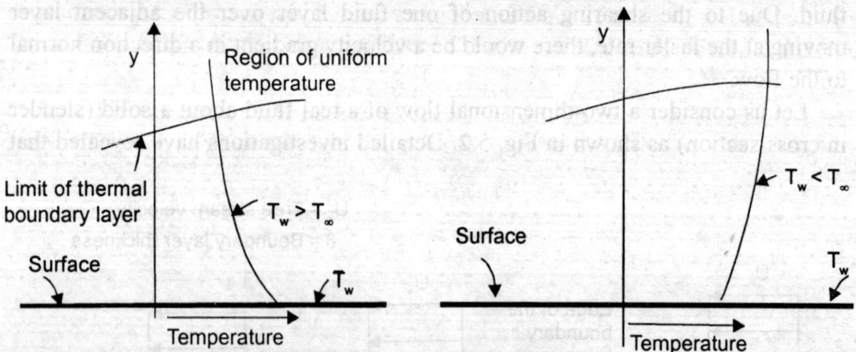

Fig. 5.3 The thermal boundary layer

7. Dimensionless Parameters and their Significance

The following dimensionless parameters are significant in evaluating the convection heat transfer coefficient:

(a) *The Nusselt Number (Nu)*—It is a dimensionless quantity defined as hL/k, where h = convective heat transfer coefficient, L is the characteristic length and k is the thermal conductivity of the fluid. The Nusselt number could be interpreted physically as the ratio of the temperature gradient in the fluid immediately in contact with the surface to a reference temperature gradient $(T_s - T_\infty)/L$. The convective heat transfer coefficient can easily be obtained if the Nusselt number, the thermal conductivity of the fluid in that temperature range and the characteristic dimension of the object is known.

Let us consider a hot flat plate (temperature T_w) placed in a free stream (temperature $T_\infty < T_w$). The temperature distribution is shown in Fig. 5.4. Newton's Law of Cooling says that the rate of heat transfer per unit area by convection is given by

$$\dot{Q}/A = h\,(T_w - T_\infty)$$

$$\frac{\dot{Q}}{A} = h(T_w - T_\infty)$$

$$= k\frac{T_w - T_\infty}{\delta_t}$$

$$h = \frac{k}{\delta_t}$$

$$Nu = \frac{hL}{k} = \frac{L}{\delta_t}$$

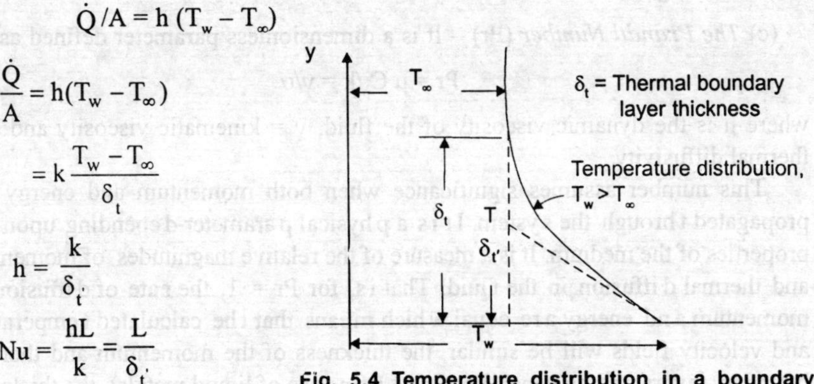

Fig. 5.4 Temperature distribution in a boundary layer: Nusselt modulus

The heat transfer by convection involves conduction and mixing motion of fluid particles. At the solid fluid interface $(y = 0)$, the heat flows by conduction only, and is given by

$$\frac{\dot{Q}}{A} = -k\left(\frac{dT}{dy}\right)_{Y=0} \qquad \therefore\ h = \frac{\left(-k\,dT/dy\right)_{y=0}}{(T_w - T_\infty)}$$

Since the magnitude of the temperature gradient in the fluid will remain the same, irrespective of the reference temperature, we can write $dT = d(T-T_w)$ and by introducing a characteristic length dimension L to indicate the geometry of the object from which the heat flows, we get

$$\frac{hL}{k} = -\frac{\left(dT/dy\right)_{y=0}}{(T_w - T_\infty)/L}\ ,\ \text{and in dimensionless form,}$$

$$= \left(\frac{d(T_w - T)/(T_w - T_\infty)}{d(y/L)}\right)_{y=0}$$

(b) *The Grashof Number (Gr)*—In natural or free convection heat trransfer, the motion of fluid particles is created due to buoyancy effects. The driving force for fluid motion is the body force arising from the temperature gradient. If a body with a constant wall temperature T_w is exposed to a quiscent ambient fluid at T_∞, the force per unit volume can be written as $\rho g \beta (t_w - T_\infty)$ where ρ = mass density of the fluid, β = volume coefficient of expansion and g is the acceleration due to gravity.

The ratio of Inertia force × Buoyancy force/(viscous force)^2 can be written as

$$Gr = \frac{(\rho V^2 L^2) \times \rho g \beta (T_w - T_\infty) L^3}{(\mu V L)^2}$$

$$= \frac{\rho^2 g \beta (T_w - T_\infty) L^3}{\mu^2} = g \beta L^3 (T_w - T_\infty)/\nu^2$$

The magnitude of Grashof number indicates whether the flow is laminar or turbulent. If the Grashof number is greater than 10^9, the flow is turbulent and for Grashof number less than 10^8, the flow is laminar. For $10^8 < Gr < 10^9$, it is the transition range.

(c) *The Prandtl Number* (Pr) – It is a dimensionless parameter defined as

$$Pr = \mu\, C_p/k = \nu/\alpha$$

where μ is the dynamic viscosity of the fluid, ν = kinematic viscosity and α = thermal diffusivity.

This number assumes significance when both momentum and energy are propagated through the system. It is a physical parameter depending upon the properties of the medium. It is a measure of the relative magnitudes of momentum and thermal diffusion in the fluid. That is, for Pr = 1, the rate of diffusion of momentum and energy are equal which means that the calculated temperature and velocity fields will be similar, the thickness of the momentum and thermal boundary layers will be equal. For Pr <<1 (in case of liquid metals), the thickness of the thermal boundary layer will be much more than the thickness of the momentum boundary layer and vice versa. The product of Grashof and Prandtl number is called Rayleigh number. Or, Ra = Gr × Pr.

8. Evaluation of Convective Heat Transfer Coefficient

The convective heat transfer coefficient in free or natural convection can be evaluated by two methods:

(a) Dimensional Analysis combined with experimental investigations
(b) Analytical solution of momentum and energy equations in the boundary layer.

Dimensional Analysis and Its Limitations

Since the evaluation of convective heat transfer coefficient is quite complex, it is based on a combination of physical analysis and experimental studies. Experimental observations become necessary to study the influence of pertinent variables on the physical phenomena.

Dimensional analysis is a mathematical technique used in reducing the number of experiments to a minimum by determining an empirical relation connecting the relevant variables and in grouping the variables together in terms of dimensionless numbers. And, the method can only be applied after the pertinent variables controlling the phenomenon are identified and expressed in terms of the primary dimensions. (Table 1.1)

In natural convection heat transfer, the pertinent variables are: h, ρ, k, μ, C_p, L, (ΔT), β and g. Buckingham π's method provides a systematic technique for arranging the variables in dimensionless numbers. It states that the number of dimensionless groups, π's, required to describe a phenomenon involving 'n' variables is equal to the number of variables minus the number of primary dimensions 'm' in the problem.

In SI system of units, the number of primary dimensions are 4 and the number of variables for free convection heat transfer phenomenon are 9 and therefore, we should expect $(9 - 4) = 5$ dimensionless numbers. Since the dimension of the coefficient of volume expansion, β, is θ^{-1}, one dimensionless number is obviously $\beta(\Delta T)$. The remaining variables are written in a functional form:

$$\phi(h, \rho, k, \mu, C_p, L, g) = 0.$$

Since the number of primary dimensions are 4, we arbitrarily choose 4 independent variables as primary variables such that all the four dimensions are represented. The selected primary variables are : ρ, g, k, L. Thus the dimensionless group,

$$\pi_1 = \rho^a\, g^b\, k^c\, L^d\, h = (ML^{-3})^a\, (LT^{-2})^b \cdot (MLT^{-3}\theta^{-1})^c\, (L)^d\, (MT^{-3}\theta^{-1}) = M^0L^0T^0\theta^0$$

Equating the powers of M, L, T, θ on both sides, we have

$$\left.\begin{array}{lll} M & : & a + c + 1 = 0 \\ L & : & -3a + b + c + d = 0 \\ T & : & -2b - 3c - 3 = 0 \\ \theta & : & -c - 1 = 0 \end{array}\right\}$$

Upon solving them,
$c = -1$, $b = a = 0$ and $d = 1$.

and $\pi_1 = hL/k$, the Nusselt number.

The other dimensionless number

$$\pi_2 = \rho^a\, g^b\, k^c\, L^d\, C_p = (ML^{-3})^a\, (LT^{-2})^b\, (MLT^{-3}\theta^{-1})^c\, (L)^d\, (L^2T^{-2}\theta^{-1}) = M^0L^0T^0\theta^0$$

Equating the powers of M,L,T and θ and upon solving, we get

$$\pi_2 = \rho^2\, g\, L^3\, C_p^2\, /k^2$$

again, $\quad \pi_3 = \rho^a\, g^b\, k^c\, L^d\, \mu = (ML^{-3})^a \cdot (LT^{-2})^b \cdot (MLT^{-3}\theta^{-1})^c\, (L)^d\, (MT^{-1}\theta^{-1}) = M^0L^0T^0\theta^0$

Equating the powers of M,L,T and θ and upon solving, we get

$$\pi_3 = \mu^2/\rho^2 g\, L^3$$

By combining π_2 and π_3, we write $\quad \pi_4 = \left[\pi_2 \times \pi_3\right]^{1/2}$

$$= \left[\rho^2 g L^3 C_p^2 / k^2 \times \mu^2 / \rho^2 g L^3\right]^{1/2} = \frac{\mu C_p}{k} \quad \text{(the Prandtl number)}$$

By combining π_3 with $(\beta\,\Delta T)$, we have $\pi_5 = \beta(\Delta T) * \dfrac{1}{\pi_3}$

$$= \beta(\Delta T) \times \frac{\rho^2 g L^3}{\mu^2} = g\beta(\Delta T)L^3 / v^2 \quad \text{(the Grashof number)}$$

Therefore, the functional relationship is expressed as:

$$\phi(Nu, Pr, Gr) = 0; \text{ Or, } Nu = \phi_1(Gr\, Pr) = Const \times (Gr \times Pr)^m \qquad (5.1)$$

and values of the constant and 'm' are determined experimentally.

Table 5.1 gives the values of constants for use with Eq. (5.1) for isothermal surfaces.

Table 5.1 Constants for use with Eq. 5.1 for Isothermal Surfaces

Geometry	$Gr_f Pr_f$	C	m
Vertical planes and cylinders	$10^4 - 10^9$	0.59	1/4
	$10^9 - 10^{13}$	0.021	2/5
	$10^9 - 10^{13}$	0.10	1/3
Horizontal cylinders	$0 - 10^{-5}$	0.4	0
	$10^4 - 10^9$	0.53	¼
	$10^9 - 10^{12}$	0.13	1/3
	$10^{-10} - 10^{-2}$	0.675	0.058
	$10^{-2} - 10^2$	1.02	0.148
	$10^2 - 10^4$	0.85	0.188
	$10^4 - 10^7$	0.48	1/4
	$10^7 - 10^{12}$	0.125	1/3
Upper surface of heated plates or lower surface of cooled plates	$8 \times 10^6 - 10^{11}$	0.15	1/3
- do -	$2 \times 10^4 - 8 \times 10^6$	0.54	1/4
Lower surface of heated plates or upper surface of cooled plates	$10^5 - 10^{11}$	0.27	1/4
Vertical cylinder height = diameter characteristic length = diameter	$10^4 - 10^6$	0.775	0.21
Irregular solids, characteristic length = distance the fluid particle travels in boundary layer	$10^4 - 10^9$	0.52	1/4

Analytical Solution—Flow over a Heated Vertical Plate in Air

Let us consider a heated vertical plate in air, shown in Fig. 5.5. The plate is maintained at uniform temperature T_w. The coordinates are chosen in such a way that x - is in the streamwise direction and y - is in the transverse direction. There will be a thin layer of fluid adjacent to the hot surface of the vertical plate within

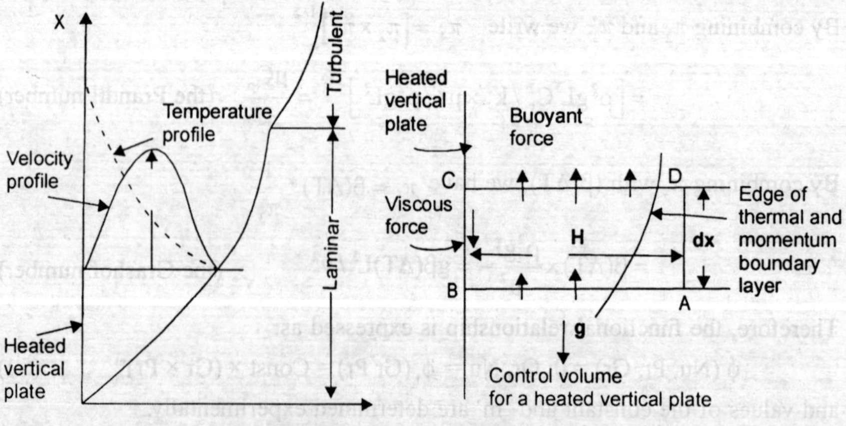

Fig. 5.5 Boundary layer on a heated vertical plate

which the variations in velocity and temperature would remain confined. The relative thickness of the momentum and the thermal boundary layer strongly depends upon the Prandtl number. Since in natural convection heat transfer, the motion of the fluid particles is caused by the temperature difference between the temperatures of the wall and the ambient fluid, the thickness of the two boundary layers are expected to be equal. When the temperature of the vertical plate is less than the fluid temperature, the boundary layer will form from top to bottom but the mathematical analysis will remain the same.

The boundary layer will remain laminar upto a certain length of the plate (Gr < 10^8) and beyond which it will become turbulent (Gr > 10^9). In order to obtain the analytical solution, the integral approach, suggested by von-Karman is preferred.

We choose a control volume ABCD, having a height H, length dx and unit thickness normal to the plane of paper, as shown in Fig. 5.5. We have:

(i) Conservation of Mass:

$$\text{Mass of fluid entering through face AB} = \dot{m}_{AB} = \int_0^H \rho u\, dy$$

$$\text{Mass of fluid leaving face CD} = \dot{m}_{CD} = \int_0^H \rho u\, dy + \frac{d}{dx}\left[\int_0^H \rho u\, dy\right] dx$$

$$\therefore \quad \text{Mass of fluid entering the face } DA = \frac{d}{dx}\left[\int_0^H \rho u\, dy\right] dx$$

(ii) Conservation of Momentum :

$$\text{Momentum entering face } AB = \int_0^H \rho u^2\, dy$$

$$\text{Momentum leaving face } CD = \int_0^H \rho u^2\, dy + \frac{d}{dx}\left[\int_0^H \rho u^2\, dy\right] dx$$

$$\therefore \quad \text{Net efflux of momentum in the } + x\text{–direction} = \frac{d}{dx}\left[\int_0^H \rho u^2\, dy\right] dx$$

The external forces acting on the control volume are:

$$\text{(a) Viscous force} = \mu \left.\frac{du}{dy}\right|_{y=0} dx \quad \text{acting in the } -ve\ x\text{-direction}$$

$$\text{(b) Buoyant force approximated as } \left[\int_0^H \rho g \beta (T - T_\infty)\, dy\right] dx$$

From Newton's law, the equation of motion can be written as:

$$\frac{d}{dx}\left[\int_0^\delta \rho u^2\, dy\right] = -\mu \left.\frac{du}{dy}\right|_{y=0} + \int_0^\delta \rho g \beta (T - T_\infty)\, dy \qquad (5.2)$$

because the value of the integrand between δ and H would be zero.

(iii) Conservation of Energy:

$$\dot{Q}_{AB},\ \text{convection} + \dot{Q}_{AD},\ \text{convection} + \dot{Q}_{BC},\ \text{conduction} = \dot{Q}_{CD},\ \text{convection}$$

or,
$$\int_0^H \rho uCTdy + CT_\infty \left[\frac{d}{dx}\int_0^H \rho udy\right]dx - k\frac{dT}{dy}\bigg|_{y=0} dx$$

$$= \int_0^H \rho uTCdy + \frac{d}{dx}\left[\int_0^H \rho uTCdy\right]dx$$

or,
$$\frac{d}{dx}\int_0^\delta \rho u(T_\infty - T)dy = \frac{k}{\rho C}\frac{dT}{dy}\bigg|_{y=0} = \alpha\frac{dT}{dy}\bigg|_{y=0} \tag{5.3}$$

The boundary conditions are:

Velocity profile	Temperature profile
$u = 0$ at $y = 0$	$T = T_w$ at $y = 0$
$u = 0$ at $y = \delta$	$T = T_\infty$ at $y = \delta_t \equiv \delta$
$du/dy = 0$ at $y = \delta$	$dT/dy \equiv 0$ at $y = \delta_t \equiv \delta$

Since the equations (5.2) and (5.3) are coupled equations, it is essential that the functional form of both the velocity and temperature distribution are known in order to arrive at a solution.

The functional relationship for velocity and temperature profiles which satisfy the above boundary conditions are assumed of the form:

$$\frac{u}{u_*} = \frac{y}{\delta}\left(1 - \frac{y}{\delta}\right)^2 \tag{5.4}$$

where u_* is a fictitious velocity which is a function of x; and

$$\frac{(T - T_\infty)}{(T_w - T_\infty)} = \left(1 - \frac{y}{\delta}\right)^2 \tag{5.5}$$

After the Eqs. (5.4) and (5.5) are inserted in Eqs. (5.2) and (5.3) and the operations are performed (details of the solution are given in Chapman, A.J. Heat Transfer, Macmillan Company, New York), we get the expression for boundary layer thickness as:

$$\delta/x = 3.93 \ Pr^{-0.5} (0.952 + Pr)^{0.25} \ Gr_x^{-0.25} \tag{5.6}$$

where Gr_x is the local Grashof number $= g\beta x^3(T_w - T_\infty)/v^2$
The heat transfer coefficient can be evaluated from:

$$\dot{q}_w = -k\frac{dT}{dy}\bigg|_{y=0} = h(T_w - T_\infty)$$

Using Eq. (5.5) which gives the temperature distribution, we have

$$h = 2k/\delta \text{ or, } hx/k = Nu_x = 2x/\delta$$

The non-dimensional equation for the heat transfer coefficient is

$$Nu_x = 0.508 \ Pr^{0.5} (0.952 + Pr)^{-0.25} \ Gr_x^{0.25} \tag{5.7}$$

The average heat transfer coefficient, $\bar{h} = \frac{1}{L}\int_0^L h_x dx = 4/3 h_{x=L}$

$$Nu_L = 0.677 \ Pr^{0.5} (0.952 + Pr)^{-0.25} \ Gr^{0.25} \tag{5.8}$$

Limitations of Analytical Solution: Except for the analytical solution for flow over a flat plate, experimental measurements are required to evaluate the heat transfer coefficient. Since in free convection systems, the velocity at the surface of the wall and at the edge of the boundary layer is zero and its magnitude within the boundary layer is so small, it is very difficult to measure them. therefore, velocity measurements require hydrogen-bubble technique or sensitive hot wire anemometers. The temperatuire field measurement is obtained by interferometer.

Expression for 'h' for a Heated Vertical Cylinder in Air

The characteristic length used in evaluating the Nusselt number and Grashof number for vertical surfaces is the height of the surface. If the boundary layer thickness is not to large compared with the diameter of the cylinder, the convective heat transfer coefficient can be evaluated by the equation used for vertical plane surfaces. That is, when $D/L \geq 35/(Gr_L)^{0.25}$

Example 5.1   A large vertical flat plate 3 m high and 2 m wide is maintained at 75°C and is exposed to atmosphere at 25°C. Calculate the rate of heat transfer.

Solution: The physical properties of air are evaluated at the mean temperature, i.e, $T = (75 + 25)/2 = 50°C$
From the data book, the values are:

$$\rho = 1.088 \text{ kg/m}^3; \qquad C_p = 1.00 \text{ kJ/kg.K};$$
$$\mu = 1.96 \times 10^{-5} \text{ Pa–s} \qquad k = 0.028 \text{ W/mK}.$$

$$Pr = \mu C_p/k = 1.96 \times 10^{-5} \times 1.0 \times 10^3 / 0.028 = 0.7$$

$$\beta = \frac{1}{T} = \frac{1}{323}$$

$$Gr = \rho^2 g\beta \, (\Delta T) \, L^3/\mu^2$$

$$= \frac{(1.088)^2 \times 9.81 \times 1 \times (3)^3 \times 50}{323 \times (1.96 \times 10^{-5})^2}$$

$$= 12.62 \times 10^{10}$$

$$Gr.Pr = 8.834 \times 10^{10}$$

Since Gr.Pr lies between 10^9 and 10^{13}
We have from Table 5.1

$$Nu = \frac{hL}{k} = 0.1(Gr.Pr)^{1/3} = 441.64$$

$$\therefore \qquad h = 441.64 \times 0.028/3 = 4.122 \, W/m^2 K$$

$$\dot{Q} = hA(\Delta T) = 4.122 \times 6 \times 50 = 1236.6 \, W$$

We can also compute the boundary layer thickness at x = 3m

$$\delta = \frac{2x}{Nu_x} = \frac{2 \times 3}{441.64} = 1.4 \text{ cm}$$

Example 5.2 A vertical flat plate at 90°C, 0.6 m long and 0.3 m wide, rests in air at 30°C. Estimate the rate of heat transfer from the plate. If the plate is immersed in water at 30°C, calculate the rate of heat transfer.

Solution: (a) *Plate in Air*: Properties of air at mean temperature 60°C

$Pr = 0.7$, $k = 0.02864$ W/mK , $v = 19.036 \times 10^{-6}$ m²/s

$Gr = 9.81 \times (90 - 30) (0.6)^3 / [333 (19.036 \times 10^{-6})^2]$

$\quad = 1.054 \times 10^9$; $Gr \times Pr = 1.054 \times 10^9 \times 0.7 = 7.37 \times 10^8 < 10^9$

From Table 5.1: for $Gr \times Pr < 10^9$, $Nu = 0.59 (Gr. Pr)^{1/4}$

∴ $h = 0.02864 \times 0.59 (7.37 \times 10^8)^{1/4}/0.6 = 4.64$ W/m²K

The boundary layer thickness, $\delta = 2 k/h = 2 \times 0.02864/4.64 = 1.23$ cm

and $\dot{Q} = hA (\Delta T) = 4.64 \times (2 \times 0.6 \times 0.3) \times 60 = 100$W.

Using Eq (5.8), $Nu = 0.677 (0.7)^{0.5} (0.952 + 0.7)^{-0.25} (1.054 \times 10^9)^{0.25}$,

which gives $h = 4.297$ W/m²K and heat transfer rate, $\dot{Q} = 92.81$ W

Churchill and Chu have demonstrated that the following relations fit very well with experimental data for all Prandtl numbers.

For $Ra_L < 10^9$, $Nu = 0.68 + (0.67 \, Ra_L^{0.25})/[1 + (0.492/Pr)^{9/16}]^{4/9}$ $\qquad$ (5.9)

$\quad Ra_L > 10^9$, $Nu = 0.825 + (0.387 \, Ra_L^{1/6})/[1 + (0.492/Pr)^{9/16}]^{8/27}$ $\qquad$ (5.10)

Using Eq(5.9): $Nu = 0.68 + [0.67(7.37 \times 10^8)^{0.25}]/[1 + (0.492/0.7)^{9/16}]^{4/9}$

$\qquad\qquad = 58.277$ and $h = 4.07$ W/m²k; $\dot{Q} = 87.9$ W

(b) *Plate in Water*: Properties of water from the Table

$\quad Pr = 3.01$, $\rho^2 g \beta C_p/\mu k = 6.48 \times 10^{10}$;

$\quad Gr.Pr = \rho^2 g \beta C_p L^3 (\Delta T)/\mu k = 6.48 \times 10^{10} \times (0.6)^3 \times 60 = 8.4 \times 10^{11}$

Using Eq (5.10) : $Nu = 0.825 + [0.387 (8.4 \times 10^{11})^{1/6}]/[1 + (0.492/3.01)^{9/16}]^{8/27}$

$\qquad\qquad = 89.48$ which gives $h = 97.533$ and $\dot{Q} = 2.107$ kW.

9. Modified Grashof Number

When a surface is being heated by an external source like solar radiation incident on a wall, a surface heated by an electric heater or a wall near a furnace, there is a uniform heat flux distribution along the surface. The wall surface will not be an isothermal one. Extensive experiments have been performed by many research workers for free convection from vertical and inclined surfaces to water under constant heat flux conditions. Since the temperature difference (ΔT) is not known beforehand, the Grashof number is modified by multiplying it by Nusselt number. That is,

$$Gr_x^* = Gr_x . Nu_x = (g \beta x^3 \Delta T/v^2) \times (hx/k) = g \beta x^4 q/kv^2 \qquad (5.11)$$

where q is the wall heat flux in Wm². (q = h (ΔT))

It has been observed that the boundary layer remains laminar when the modified Rayleigh number, $Ra^* = Gr_x^*$. Pr is less than 3×10^{12} and fully turbulent flow appears for $Ra^* > 10^{14}$. The local heat rransfer coefficient can be calculated from:

$\quad$ q constant and $\qquad 10^5 < Gr_x^* < 10^{11}$: $Nu_x = 0.60 (Gr_x^*. Pr)^{0.2}$ $\qquad$ (5.12)

$\quad$ q constant and $2 \times 10^{13} < Gr_x^* < 10^{16}$: $Nu_x = 0.17 (Gr_x^* . Pr)^{0.25}$ $\qquad$ (5.13)

Although these results are based on experiments for water, they are applicable to air as well. The physical properties are to be evaluated at the local film temperature.

Example 5.3   Solar radiation of intensity $700\,W/m^2$ is incident on a vertical wall, 3 m high and 3 m wide. Assuming that the wall does not transfer energy to the inside surface and all the incident energy is lost by free convection to the ambient air at 30°C, calculate the average temperature of the wall.

Solution: Since the surface temperature of the wall is not known, we assume a value for h = 7 W/m^2K.

$$\Delta T = \dot{q}/h = 700/7 = 100°C \text{ and the film temperature} = (30 + 130)/2 = 80°C$$

The properties of air at 273 + 80 = 353 are : $\beta = 1/353$, Pr = 0.697

$$k = 0.03 \text{ W/mK}, \nu = 20.76 \times 10^{-6} \text{ m}^2/s.$$

Modified Grashof number, $Gr_x^* = 9.81 \cdot (1/353) \cdot (3)^4 \times 700/[0.03 \times (20.76 \times 10^{-6})^2]$

$$= 1.15 \times 10^{14}$$

From Eq. (5.13), $h = (k/x)\,(0.17)\,(Gr_x^* \, Pr)^{0.25}$

$$= (0.03/3)\,(0.17)\,(1.15 \times 10^{14} \times 0.697)^{1/4}$$

$$\doteq 5.087 \text{ W/m}^2K, \text{ a different value from the assumed value.}$$

Second Trial: $\Delta T = \dot{q}/h = 700/5.087 = 137.66$ and film temperature = 98.8°C

The properties of air at (273 + 98.8) °C are: $\beta = 1/372$,   k = 0.0318 W/mK

$$Pr = 0.693, \qquad \nu = 23.3 \times 10^{-6} \text{ m}^2/s$$

$$Gr_x^* = 9.81 \cdot (1/372) \cdot (3)^4 \times 700/[0.318\,(23.3 \times 10^{-6})^2] = 8.6 \times 10^{13}$$

Using Eq (5.13), $h = (k/x)\,(0.17)\,(Gr_x^* \, Pr)^{1/4} = 5.015$ W/m^2k, an acceptable value.

In tubulent heat transfer by convection, Eq. (5.13) tells us that the local heat transfer coefficient h_x does not vary with x and therefore, the average and local heat transfer coefficients are the same.

$$\Delta T = \dot{q}/h = 700/5.015 = 139.8°C$$

and the average wall temperature = 139.8 + 30 = 169.8°C.

10. Free Convection from Horizontal Cylinders—Empirical Relations

Natural convection near horizontal round tubes is of great practical importance. The pattern of free convection flow near hot horizontal cylinders is shown in Fig. 5.6(a). The flow is laminar over the entire surface and the plume is likely to be turbulent much earlier with increasing diameter of the cylinder. The local heat transfer coeffecient is maximum at $\theta = 0$ (thickness of the boundary layer is minimum) and it increases gradually as θ increases. Hermann has shown local value of 'h' can be calculated by:

$$Nu(\theta) = \frac{hD}{k}(\theta) = 0.604\,Gr^{0.25}\,\phi\,(\theta)$$

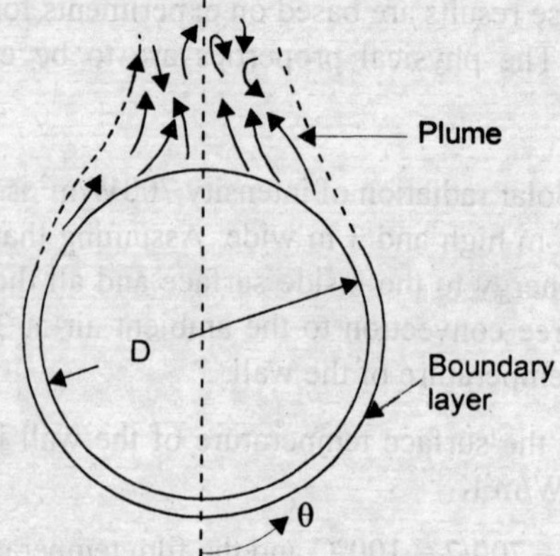

(a) Free flow across heated horizontal tubes

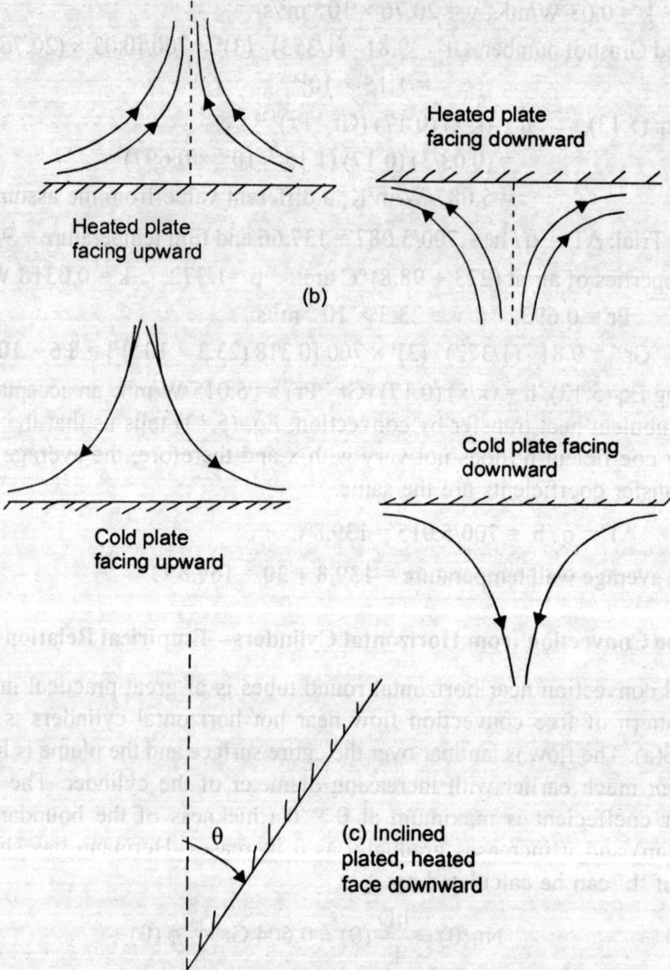

Fig. 5.6 Natural convection flow pattern

and the numerical values of $\phi(\theta)$ are :

θ	0°	30°	60°	90°	120°	150°	165°	180°
$\phi(\theta)$	0.76	0.75	0.72	0.66	0.58	0.46	0.36	0

If the cylinder is placed in a hot fluid, the boundary layer starts developing at $\theta = \pi$ and the local Nu is maximum at that location and the plume descends from the cylinder.

Churchill and Chu have suggested a complicated relation for use over a wide range of Gr.Pr for evaluating the average heat transfer coefficient:

For $10^{-5} < $ Gr . Pr $ < 10^{12}$:

$$Nu = [0.60 + 0.387\{Gr.Pr/(1 + (0.559/Pr)^{9/16})^{16/9}\}^{1/6}]^2 \; ; \text{and} \qquad (5.14a)$$

For laminar range : $10^{-6} < $ Gr.Pr $ < 10^{9}$

$$Nu = 0.36 + 0.518 \, (Gr. \, Pr)^{1/4}/[1 + (0.559 / Pr)^{9/16}]^{4/9} \qquad (5.14b)$$

Heat transfer from horizontal cylinders to liquid metals may be calculated by:

$$Nu = 0.53 \, (Gr. \, Pr^2)^{0.25} \qquad (5.15)$$

Example 5.4   A nichrome wire 0.6 mm in diameter has its surface temperature maintained at 230°C. The wire is exposed to ambient air at 25°C. Calculate the electric power necessary to maintain the wire temperature per metre length of the wire.

Solution: The mean temperature = $(230 + 25)/2 = 127.5C = 400$ K
The physical properties are: $\beta = 1/400$,   Pr $= 0.689$
       $k = 0.0336$ W/m°C; $v = 25.9 \times 10^{-6}$ m²/s
The characteristic length to be used for calculating the Gr and Nu numbers is the diameter of the wire.

       Gr $= 9.81 \, (1/400) \, (205) \, (0.6 \times 10^{-3})^3/(25.9 \times 10^{-6})^2$
          $= 1.62$;
    Gr.Pr $= 1.115$
From Eq. (5.14b), Nu $= 0.36 + 0.518 \, (1.115)^{1/4}/ \, [1 + (0.559/0.689)^{9/16}]^{4/9}$
            $= 0.7612$
$\therefore$ $h = 0.7612 \, (k/d) = 42.628$ W/m²°C

and the electrical power required $=$ hA $(\Delta T) = 42.68 \times 3.142 \, (0.6 \times 10^{-3})^2 \times 1 \times 205$
                   $= 16.47$ W $= I^2R$

(Thus for a given resistance, the current carrying capacity can be obtained.)

11. Film Convection—Defined

The conditions of heat transfer for thin wirPes (d = 0.2 to 2.0 mm) are of special character. The quantity of heat transferred is insignificant, because the surface of the wire is small and at small temperature differences, a stationary film of heated air forms around the wire. The value of h is obtained by: Nu = 1.18 $(Ra)^{1/8}$ for Ra <500.

Using the above relation in Ex. 5.4 Nu = 1.18 $\times (1.115)^{1/8} = 1.196$ and h = 66.98 W/m²K. (a much higher value because ΔT is very large)

Example 5.5   A horizontal steam pipe, outside diameter 15 cm, carries saturated steam at 115°C in a room having air temperature 30°C. Calculate the quantity of steam condensed for a 10m length of the pipe.

Solution: Assuming that the temperature of the outer surface of the pipe is also 115°C, the mean temperature would be $(115 + 30)/2 = 72.5$°C

The physical properties of air at 72.5°C (345.5K) are:

$$k = 0.0336 \text{ W/mK}, \quad Pr = 0.689, \quad \beta = 1/345.5, \quad \nu = 25.9 \times 10^{-6} \text{ m}^2/\text{s}$$

$$Gr.Pr = 9.81(1/345.5)(115 - 30)(0.15)^3 \times 0.689/(25.9 \times 10^{-6})^2 = 7.3 \times 10^6$$

From Table 5.1: $Nu = 0.53 (Gr.Pr)^{0.25} = 0.53(7.3 \times 10^6)^{0.25} = 27.55$

$\therefore \qquad h = 27.55 \times k/d = 6.17 \text{ W/m}^2\text{°C}$

$$Q = hA(\Delta T) = 6.17 \times 3.142 \times 0.15 \times 10 \times 85 = 2.47 \text{ kW}$$

Latent heat of vaporization of steam at 115°C = 2216 kJ/kg

Mass of steam condensed $= (2.47/2216) \times 3600 = 4$ kg/hr.

Example 5.6   A 2.5 cm diameter immersion heater (length 20 cm) is placed horizontally in a pool of water at 30°C Calculate the power required if the surface temperature of the heater is maintained at 60°C.

Solution: The mean temperature $= (60 + 30)/2 = 45$°C

The properties of water are: $k = 0.64$ W/m C, $Pr = 3.84$, $g\beta\rho^2 C/\mu k = 4.54 \times 10^{10}$

$$Gr.Pr = 4.54 \times 10^{10}(60 - 30)(2.5 \times 10^{-2})^3 = 2.13 \times 10^7$$

From Table 5.1 : $Nu = 0.53 (Gr.Pr)^{0.25} = 0.53 (2.13 \times 10^7)^{0.25} = 36$

$$\therefore \qquad h = \frac{36 \times 0.64}{0.025} = 921.6 \text{ W/m}^2\text{K}; \qquad \dot{Q} = hA(\Delta T) = 434.35 \text{ W}.$$

Example 5.7   An electric bulb. 500 W capacity, can be approximated as a sphere of 8 cm in diameter. Estimate the percentage of power dissipated by convection when the surface temperature is 550°C and the ambient air is at 30°C.

Solution: The mean temperature $= (550 + 30)/2 = 290$°C $= 563$ K

The properties of air are: $k = 0.0443$ W/mK. $Pr = 0.68$, $\nu = 44 \times 10^{-6}$ m²/s

$$Gr. Pr = [9.81 \cdot (1/563) \cdot (550 - 30) \cdot (0.08)^3/(46 \times 10^{-6})^2] \times 0.68 = 1.49 \times 10^6$$

Based on experiments, the recommended relation for free convection from a sphere is given by:

$$Nu = hd/k = 2 + 0.43 (Ra)^{0.25}, \tag{5.16}$$

$$h = (0.0443/0.08) (2 + 0.43 (1.49 \times 10^6)^{0.25}) = 1.94 \text{ W/m}^2\text{C}.$$

$$Q = hA (\Delta T) = 1.94 \times 4 \times 3.142 (0.04)^2 (550 - 30) = 20.28 \text{ W}$$

From Eq. (5.16), it is clear that at very small values of Rayleigh number, the heat transfer will be by pure conduction and $Nu = 2.0$. For higher ranges of Rayleigh numbers. (experiments conducted with water) the following relation is recommended:

For $\qquad 3 \times 10^5 < Gr.Pr < 8 \times 10^8, \quad Nu = 2 + 0.5 (Gr. Pr)^{1/4}$

12. Mechanism of Free Convection from Horizontal Plates

The heat transfer coefficients for horizontal surfaces depend on whether the plate is hot or cold in relation to the ambient fluid. When the direction of heat flow and the direction of buoyant forces are the same, the value of h will be greater than that when the two directions are opposite to each other. The flow pattern is shown in Fig. 5.6 (b). Goldstein. R.J; et al and other investigators suggest that the characteristic length should be calculated as:

$$L = \text{Surface area of the plate/Perimeter of the plate}$$

The average heat transfer coefficient from isothermal horizontal plates is calculated with Eq. (5.1) and the value of the constants are given in Table 5.1.

13. Horizontal and Inclined Flat Plates with Uniform Heat Flux—Empirical Relations

The following empirical relations are being used for evaluating convecive heat transfer coefficients from horizontal plates having constant heat flux (flat roofs receiving solar radiation, or plates being heated from below):

   (a) Heated surface facing upwards (cold surface facing downward)

for $\text{Ra} = \text{Gr.Pr} < 2 \times 10^8$: $\text{Nu} = 0.13\,(\text{Ra})^{1/3}$

for $2 \times 10^8 < \text{Ra} < 10^{11}$: $\text{Nu} \quad = 0.16\,(\text{Ra})^{1/3}$ (5.17)

   (b) Heated surfacee facing downward (cold surface facing upward)

for $10^6 < \text{Ra} < 10^{11}$: $\text{Nu} = 0.58\,(\text{Ra})^{1/5}$ (5.18)

In the above equations, all properties, except β, have to be evaluated at a temperature, $T_e = T_w - 0.25\,(T_w - T_\infty)$

The average wall temperature is obtained from the relation $h = \dot{q}\,/(T_w - T_\infty)$,

and the Nusselt number, $\text{Nu} = hL/k = \dot{q}\,L/k\,(T_w - T_\infty)$.

Inclined Surfaces – Fuzi and Imura have performed experiments on natural convection heat transfer from a plate of length L with pure water. The inclination angle θ was measured from the vertical, Fig. 5.6(c), and it was varied from 0°C to 89°C. The correlation for a downward facing hot plate is:

$$10^5 < \text{Gr.Pr} < 10^{11};\ \theta < 89°;\ \text{Nu} = 0.59\,(\text{Gr.Pr}\cos\theta)^{1/4} \qquad (5.19)$$

where all properties are evaluated at a mean temperature (except for β)

$$T_e = T_w - 0.25\,(T_w - T_\infty).$$

T_w is the mean wall temperature. T_∞ = Free stream temperature. β is evaluated at temperature of $T_\infty + 0.5\,(T_w - T_\infty)$.

Example 5.8   A thin 20 cm diameter horizontal heated plate, maintained at 136°C is immersed in a pool of water at 50°C. The plate loses heat by convection from both bottom and top surfaces. Calculate the rate of heat input to maintain the temperature of the plate.

Solution: The characteristic length. $L = A/P = (\pi/4)d^2/\pi d = d/4 = 5$ cm.
Mean temperature, $T = (136 + 50)/2 = 93°C$,

The properties of water at 93°C, $\rho = 963.2$ kg/m³, $k = 0.678$ W/mK

$$g\beta\rho^2 C/\mu k = 9.81 \times 0.21 \times 10^{-3} \times (963.2)^2 \times 4.2 \times 10^3/(3.06 \times 10^{-4} \times 0.678)$$

$$= 3.32 \times 10^{11}$$

$$Gr.Pr = g\beta\rho^2 C/\mu k \times (\Delta T) \, L^3 = 3.32 \times 10^{11} \times 86 \times (0.05)^3 = 3.57 \times 10^9$$

From Table 5.1, for heated surface upward:

$$Nu = 0.27 \, (3.57 \times 10^9)^{1/3} = 2.292 \times 10^2; \quad \text{and h} = 3108 \text{ W/m}^2\text{K}$$

For heated surface downward: $Nu = 0.27(3.57 \times 10^9)^{1/3} = 66$ and $h = 894.9$ W/m²k.
The rate of heat transfer from the plate, $Q = h \, A \, (\Delta T)$

$$Q = (3108 + 894.9) \, (\pi/4) \, (0.2)^2 \times 86 = 10.816 \text{ kW}.$$

Example 5.9 A horizontal duct (35 cm × 25 cm) passes through a room where the air temperature is 40°C. If the temperature of the duct surface is manitained at 10°C, calculate the heat transfer by free convection per metre length of the duct.

Solution: The mean temperature $= (40 + 10) /2 = 25°C$
The properties of air are: $\nu = 15.69 \times 10^{-6}$ m²/s, $k = 0.026$ W/mK, $Pr = 0.708$
Let the width of the duct be 35 cm and height 25 cm.

$$Gr.Pr = 9.81 \times (1/298) \times L^3 \times 0.708/(15.69 \times 10^{-6})^2 = 2.84 \times 10^9 \, L^3$$

For vertical surface: $Gr. \, Pr = 2.84 \times 10^9 \times (0.25)^3 = 4.44 \times 10^7$
For horizontal surface, $L = $ surface area/ perimeter $=$ width/2 $= 0.175$ m
$\therefore \; Gr. \, Pr = 2.84 \times 10^9 \times (0.175)^3 = 1.52 \times 10^7$
For horizontal cold surface upward: $Nu = 0.27(1.57 \times 10^7)^{1/4} = 17$

$$h = 17 \times 0.026/0.175 = 2.525 \text{ W/m}^2\text{C}$$

For horizontal cold surface downward : $Nu = 0.15 \, (1.52 \times 10^7)^{1/3} = 37.0$

$$h = 37.0 \times 0.026/0.175 = 5.5 \text{ W/m}^2\text{C}$$

For vertical surfaces : $Nu = 0.59 \, (4.44 \times 10^7) = 48.16$ and $h = 5.0$ W/m²C
Rate of heat transfer, $Q = (2.525 + 5.0) \, (0.35 \times 1) \, (30) + (5.0) \, (2 \times 0.25 \times 1) \, (30)$

$$= 154 \text{ W}.$$

Example 5.10 A shopkeeper can place coke cans (200 mm long, 60 mm in diameter) in a bottle cooler either vetically or horizontally. The initial temperature is 40°C and the ambient temperature inside the cooler is 5°C. Estimate the cooling rate.

Solution: The mean temperature $= (40 + 5)/2 = 22.5°C = 295.5$ K
The properties of air are: $\nu = 15.3 \times 10^{-6}$ m²/s, $Pr = 0.7025$, $k = 0.0225$ W/mK

(a) When the can is placed horizontally:

$$Gr.Pr = 9.81 \, (1/295.5) \times 35 \times (0.06)^3 \times 0.7025/(15.3 \times 10^{-6})^2 = 7.53 \times 10^5$$

Using Eq. (5.14b): $Nu = 0.36 + 0.515 \, (7.53 \times 10^5)^{1/4}/[(1 + 0.559/0.7025)^{9/16}]^{4/9}$

$$= 11.89 \text{ and h} = 4.46 \text{ W/m}^2 \, °C.$$

Rate of heat removal = $4.46 \times 3.42 \times 0.06 \times 0.2 \times 35 = 5.89$ W/can.

(b) When the can is placed vertically:

$$Gr.Pr = 9.81(1/295.5) \times 35 \times (0.2)^3 \times 0.7025/(15.3 \times 10^{-6})^2 = 2.79 \times 10^7$$

Since $35/(Gr_L)^{1/4} = 35/(93.97 \times 10^7)^{1/4} \ll (D/L = 0.06/0.2)$, the equation for vertical surfaces can be applied for vertical cylinders.

$Nu = 0.59(2.79 \times 10^7)^{1/4} = 42.88$, and $h = 4.83$ W/m²°C.

Rate of heat removal = $4.83 \times 3.142 \times 0.06 \times 0.2 \times 35 = 6.36$ W/can.

Example 5.11 A metal piece 6 cm × 9 cm × 11 cm, at a temperature of 60°C is being cooled in ambient air at 30°C. If the height is 11 cm, calculate the convective heat transfer coefficient.

Solution: For rectangular parallelepiped, the characteristic length is obtained from the relation, $L = (L_h \times L_v)/(L_h + L_v)$ where L_v is the height of the parallelepiped , L_h is the longer of the two horizontal dimensions.

$\therefore$ $L = 11 \times 9/(11 + 9) = 4.995$ cm

The mean temperature = $(60 + 30)/2 = 45$°C = 318 K

The properties of air are: $v = 17.46 \times 10^{-6}$ m²/s, Pr = 0.7, k = 0.024 W/mK

$$Gr.\ Pr = 9.81(1/318) \times 30 \times (0.04995)^3 \times 0.7/(17.46 \times 10^{-6})^2$$
$$= 2.85 \times 10^5, \text{ a laminar flow.}$$

From Table 5.1: $Nu = 0.52\ (Gr.Pr)^{1/4} = 0.52(2.86 \times 10^5)^{1/4} = 12.02$, and

$h = 5.8$ W/m²°C and the heat transfer rate,

$Q = 5.8 \times (2 \times 6 \times 9 + 2 \times 6 \times 11 + 2 \times 9 \times 11) \times 10^{-4} \times 30$
$= 7.62$ W.

Example 5.12 A hot plate (0.5 m x 0.6 m) receives heat energy at a constant rate of 1000W/m² and is placed in a room at temperature 25°C. Calculate the value of convective heat transfer coefficient and the surface temperature when the hot side is facing (a) upward (b) downward and (c) hot side making an angle of 45° with the vertical, heated surface downward.

Solution: *(a) Hot side facing upward*

Characteristic length = Area/perimeter = $(0.5 \times 0.6)/2(0.5 + 0.6)$
                                                     $= 0.1364$ m.

The property has to be evaluated at a temperature, $T_e = T_w - 0.25\ (T_w - T_\infty)$

Assuming a value of $h = 8$ W/m² C, $(T_w - T_\infty) = 1000/8 = 125$C

$T_e = 150 - 0.25(125) = 118.75$ C

Properties of air at 118.75°C are:

$v = 25.4 \times 10^{-6}$ m²/s. $k = 0.0285$ W/mK, Pr = 0.686.

β has to be evaluated at temperature = $T_\infty + 0.5(T_w + T_\infty) = 112.5 \equiv 385.5$K

$Gr.Pr = 9.81 \times (1/385.5) \times 125 \times (0.1364)^3 \times 0.686/(25.4 \times 10^{-6})^2 = 8.58 \times 10^6$

$Nu = 0.13(Gr.Pr)^{1/3} = 0.13(8.58 \times 10^6)^{1/3} = 26.1$

and $h = 26.1 \times 0.0285/0.1364 = 5.56$, a little lower value

With this new value of h, $(T_w - T_\infty) = 1000/5.56 = 179.8$ and $T_w = 204.8$

$T_e = 204.8 - 0.25(179.8) = 159.85$ or $160°C$

Properties of air at $160°C$: $\nu = 30.09 \times 10^{-6}$ m²/s, $k = 0.0313$, $Pr = 0.682$

For β : $T = [T_\infty + 0.5(T_w + T_\infty)] = [25 + 0.5(204.8 + 25)] \equiv 413$ K

$Gr.Pr = 9.81 \times (1/413) \times 179.8 \times (0.1364)^3 \times 0.682/(30.09 \times 10^{-6})^2$

$= 8.16 \times 10^6$

$Nu = 0.13(8.16 \times 10^6)^{1/3} = 26.17$, and $h = 6.0$ W/m² °C, value closer to second assumption.

$\therefore$ $T_w = 204.8°C$.

(b) Hot side facing downward:

Let $h = 4$ W/m² °C, $T_w - T_\infty = 1000/4 = 250°C$; $T_w = 275°C$

$T_e = 275 - 0.25(250) = 212.5$ and $1/\beta = 273 + 25 + 0.5 (273 + 25) = 428$ K

$Gr.Pr = 9.81 (1/428) \times 250 \times (0.1364)^3 \times 0.681/(36.3 \times 10^{-6})^2 = 1.926 \times 10^6$

$Nu = 0.58(1.926 \times 10^6)^{1/5} = 10.48$ and $h = 10.48 \times 0.0346/0.1364 = 2.66$

Second trial: $h = 2.66$, $T_w - T_\infty = 1000/2.66 = 376°C$, $T_w = 401°C$

$T_e = 401 - 0.25(376) = 307°C$, $1/\beta = 273 + 25 + 0.5 (401 +25) = 511K$

Properties of air at $307°C$: $\nu = 49.2 \times 10^{-6}$ m²/s, $k = 0.0399$ W/mK, $Pr = 0.674$

$Gr.Pr = 9.81 \times (1/511) \times 376 \times (0.1364)^3 \times 0.674/ (49.2 \times 10^{-6})^2$

$= 5.1 \times 10^6$

$Nu = 0.58 (5.1 \times 10^6)^{1/5} = 12.73$ and $h = 3.72$ W/m²C

Following the same procedure, a third trial gives

$Gr.Pr = 6.95 \times 10^6$; $Nu = 13.54$ and $h = 2.45$ W/m² C.

and $T_w = 25 + 1000 / 2.45 = 433.16°C$

i.e., the temperature of the surface is higher when the heated surface faces downward.

(c) Inclined surface, heated surface facing downward:

$Nu = 0.56 (Gr.Pr \cos \theta)^{1/4} = 0.56(6.95 \times 10^6 \times \cos 45)^{1/4} = 26.36$

$h = 26.36 \times 0.02465/0.1364 = 4.77$ W/m² °C and $T_w = 234.6°C$.

(A second trial will give more accurate temperature)

14. Simplified Free Convection Relations for Air

In most cases, free convection heat transfer takes place with air as ambient fluid. And, therefore, we have simplified relations for the evaluation of the convective heat transfer coefficient. From dimensional analysis

$Nu = \phi (Gr.Pr)$ and we have an empirical relation as

$Nu = C (g\beta(\Delta T)L^3/\nu^2 \times \mu C_p/k)^m$

or, $h = C_1 [k(g\beta C_p/\nu^2)^m] [(\Delta T)^m/L^{1-3m}]$, where C_1 is a constant.

For a quick estimate of heat transfer coefficient in air at normal atmospheric conditions, we can write

$h = C_2 (\Delta T)^m/L^{1-3m}$, where C_2 is another constant.

In laminar range, m = ¼ and the expression for 'h' is
$$h = C_2 (\Delta T/L)^{\frac{1}{4}} \tag{5.20}$$
and in the trubulent range, m = 1/3 and the characteristic length vanishes.

or, $h = C_2 (\Delta T)^{1/3}$ (5.21)

Thus, we have the following simplified equations for free convection from vertical surfaces to air at atmospheric pressure, Table 5.2.

Table 5.2

Surface	Laminar	Turbulent
Horizontal cylinder	$10^4 <$ Gr.Pr. $< 10^9$	Gr. Pr $> 10^9$
	$h = 1.32\ (\Delta T/D)^{\frac{1}{4}}$	$h = 1.24\ (\Delta T)^{1/3}$
Vertical plate/cylinder	$h = 1.42\ (\Delta T/L)^{\frac{1}{4}}$	$h = 1.31\ (\Delta T)^{1/3}$
Horizontal plate		
(i) heated plate facing upward or cooled plate facing downward	$h = 1.32\ (\Delta T/L)^{\frac{1}{4}}$	$h = 1.52\ (\Delta T)^{1/3}$
(ii) heated plate facing downward or cooled plate facing upward	—	$h = 0.59\ (\Delta T/L)^{\frac{1}{4}}$

Example 5.13 Compute the rate of heat transfer by vertical plate of Ex. 5.1 using the simplified relations for air.

Solution: The characteristic dimension for the vertical plate of Ex 5.1 was 3 m, and Gr. Pr was greater than 10^9, hence a turbulent flow.

$$h = 1.31\ (\Delta T)^{1/3} = 1.31\ (75 - 25)^{1/3} = 4.8\ \text{W/m}^2\text{C};\quad \text{a very high value.}$$

Example 5.14 Calculate the convective heat transfer heat transfer coefficient for the steam pipe of Ex. 5.5 using the simplified relation.

Solution: From Table 5.2: $h = 1.32\ (\Delta T/D)^{0.25} = 1.32((115 - 30)/0.15)^{0.25}$
$$= 6.44\ \text{W/m}^2\,°\text{C, a very good approximation.}$$

Example 5.15 A 10 cm pipe carries hot gases at 200°C. The ambient air is at 30°C. Calculate the rate of heat transfer per metre length of the pipe by natural convection using the relation $h = 1.32\ (\Delta T/D)^{\frac{1}{4}}$. What would be the rate of heat transfer if the pipe is provided with a 3 cm thick insulation (k = 0.07 W/mK)?

Solution: Assuming that the temperature of the pipe surface is 200°C,
$$h = 1.32\ [200 - 30)/0.1]^{0.25} = 8.48\ \text{W/m}^2\text{C},$$

and $\dot{Q} =$ hA $(\Delta T) = 8.48 \times 3.142 \times (0.1) \times 1 \times 170 = 452.95$ W/m

when the pipe is insulated, the inner diameter of the insulation is 10cm and the outer diameter of the insulation 16 cm. Let T be the temperature at the outside surface of the insulation. Under steady state condition, rate of heat transfer by conduction through the insulation must equal to the rate of heat transfer by convection at the outer surface.

or, $\quad \dot{Q}/L = (200 - T)/\ln(16/10)/2\pi \times 0.07 = h(\pi D)(T - 30)$

$\qquad = 1.32((T - 30)/0.16)^{1/4} \cdot \pi D(T - 30)$

or, $\quad (200 - T)1.0685 = 1.049(T - 30)^{1.25}$; and

$\qquad (200 - T) = 1.12(T - 30)^{1.25}$

solving by trial and error, $T = 73.5°C$

$\qquad \dot{h} = 1.32[(73.5 - 30)/0.16]^{0.25} = 5.36$

and $\quad \dot{Q} = 117.2$ W; (74% reduction).

15. Heat Transfer by Free Convection in a Limited Volume

Free convection currents occuring at and near a hot or cold body are seriously affected if that body is placed in an enclosure where the volume of the ambient fluid is small. In that case, the rate of heat transfer depends on the properties of fluid, its temperature and temperature difference and on the shape and size of the space.

Let us consider two horizontal infinite flat planes, Fig. 5.7 (a) separated by a small distance δ. When the upper plate is at a higher temperature than of the lower plate, there would be no free flow of fluid and heat energy be transferred by pure conduction ($Nu_\delta = h\delta/k = 1.0$). If the lower plate is at a higher temperature, pure conduction occurs upto $Gr_\delta < 1700$ and the convection currents would occur for $Gr_\delta > 1700$. Hot fluid particles (of smaller density) would rise upward and alternate rising and falling currents would appear in the enclosed space. Fig. 5.7 (b). These patterns are called 'Benard Cells' and this cellular pattern is destroyed when turbulence begins at $Gr_\delta > 50,000$.

Circulation also develops in vertical slots. If the two vertical plates are separated by a large distance, rising and descending flows do not interfere, Fig. 5.7(c). When δ is small, interference between two currents give rise to inner circuits, Fig. 5.7(d). The height of the inner circuit is determined by the width of the enclosed space, the property of fluid involved and the intensity of the process. In spherical and horizontal cylindrical spaces, fluid circulates as shown in Fig. 5.7 (e, f, g) depending on diameter ratio. The emprical relations for free convection in enclosures have been expressed in the form:

$$k_e/k = \text{const}\,(Gr_\delta \cdot Pr)^n (L/\delta)^m \tag{5.22}$$

where k_e is the effective thermal conductivity and $Nu = k_e/k$. The summary of empirical relations for free convection in enclosures are given in Table 5.3.

Example 5.16 Two vertical parallel walls (4 m high and 5 m wide) are separated by a distance of 10 cm. The inner surfaces of the inner and outer walls are at 25°C and 15°C respectively. Calculate the heat loss by convection and conduction. If the air space is divided in half by a sheet of metal 0 025 mm thick and parallel to the walls, calculate the heat loss.

Solution: The mean temperature $= (25 + 15)/2 = 20°C = 293$K
Properties of air: $\nu = 15.06 \times 10^{-6} \cdot$ $Pr = 0.703$, $\quad k = 0.0223$ W/mK, $\delta = 0.1$ m,

$\qquad Gr_\delta = 9.81 \times (1/293)(25 - 15)(0.1)^3/(15.06 \times 10^{-6})^2$

$\qquad = 1.476 \times 10^6$

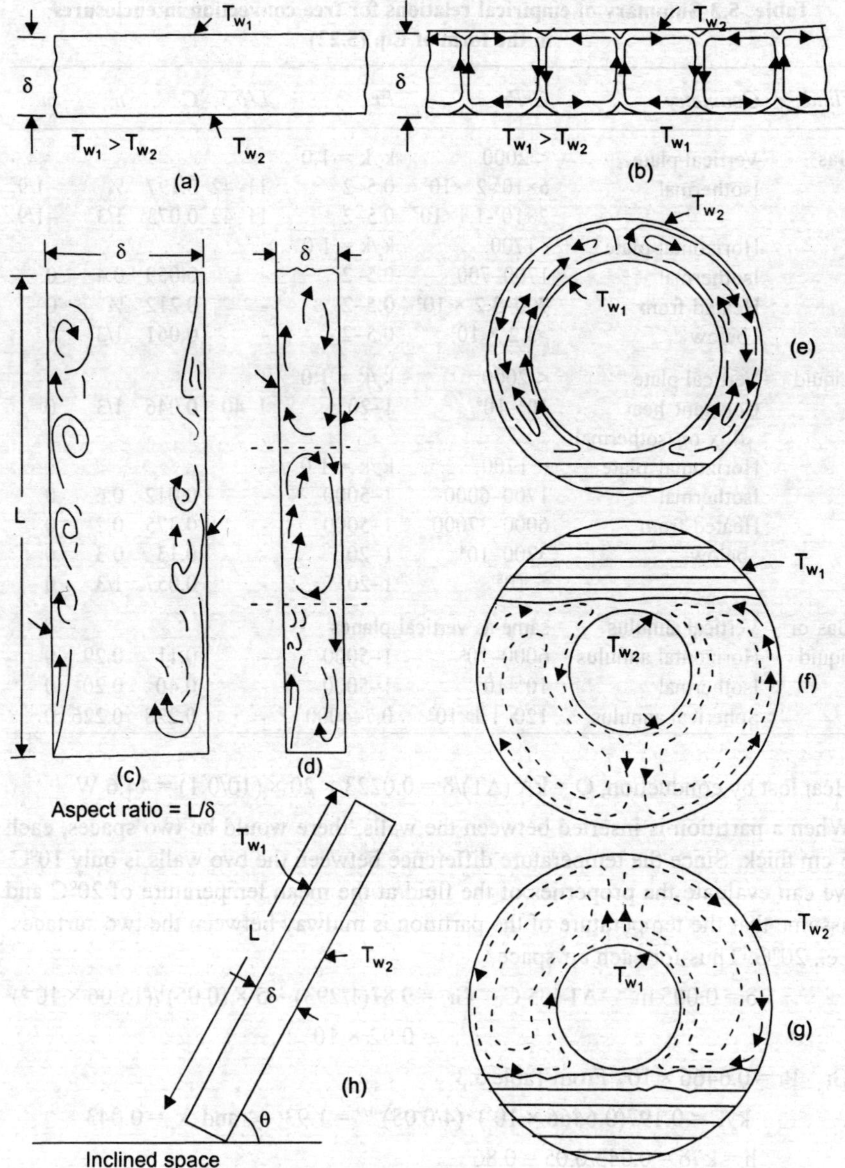

Fig. 5.7 Free convection in enclosed space

$$Gr_\delta \cdot Pr = 1.038 \times 10^6$$

From Table 5.3, for Gr·Pr = 1.038×10^6 and Pr = 0.703,

$$k_e/k = 0.073 \, (1.038 \times 10^6)^{1/3} \, (L/\delta)^{-1/9}$$

where L = 4 m and δ = 0.1m = 4.9; and the equivalent thermal conductivity, k_e = 0.1094

$$Nu = h\delta/k = k_e/k; \quad h = k_e/\delta = 0.1094/0.1 = 1.094$$

Heat lost by convection, Q = hA (ΔT) = 1.094 (4 × 5) × 10 = 218.8 W.

Table 5.3 Summary of empirical relations for free convection in enclosures in the form of Eq. (5.22)

Fluid	Geometry	GrPr	Pr	L/δ	C	n	m
Gas	Vertical plate	< 2000	$k_c/k = 1.0$				
	Isothermal	6×10^3-2×10^5	0.5–2	11–42	0.197	¼	–1/9
		2×10^5-1.1×10^7	0.5–2	11–42	0.073	1/3	–1/9
	Horizontal plate	<1700	$k_c/k = 1.0$				
	Isothermal	1700–700	0.5–2	-	0.059	0.4	0
	Heated from	700–3.2×10^5	0.5–2	-	0.212	¼	0
	below	$>3.2\times10^5$	0.5–2	-	0.061	1/3	0
Liquid	Vertical plate	< 2000	$k_c/k = 1.0$				
	Constant heat flux or isothermal	10^6–10^9	1–20	1–40	0.046	1/3	0
	Horizontal plate	< 1700	$k_c/k = 1.0$	-			
	Isothermal	1700–6000	1–5000	-	0.012	0.6	0
	Heated from	6000–37000	1–5000	-	0.375	0.2	0
	below	3700–10^8	1–20	-	0.13	0.3	0
		$> 10^8$	1–20	-	0.057	1/3	0
Gas or	Vertical annulus	same as vertical planes					
liquid	Horizontal annulus	6000–10^6	1–5000	-	0.11	0.29	0
	Isothermal	10^6–10^8	1–5000	-	0.40	0.20	0
	Spherical annulus	120–1.1×10^9	0.7–4000	-	0.228	0.226	0

Heat lost by conduction, $Q = kA\,(\Delta T)/\delta = 0.0223 \times 20 \times (10/0.1) = 44.6$ W

When a partition is inserted between the walls, there would be two spaces, each 5 cm thick, Since the temperature difference between the two walls is only 10°C, we can evaluate the properties of the fluid at the mean temperature of 20°C and assume that the temperature of the partition is midway between the two surfaces, i.e., 20°C. Thus for each air space,

$$\delta = 0.005 \text{ m}, \quad \Delta T = 5 \text{ C}, \quad Gr_\delta = 9.81(1/293) \times 5 \times (0.05)^3/(15.06 \times 10^{-6})^2$$
$$= 0.92 \times 10^5$$

$Gr_\delta \cdot Pr = 0.6466 \times 10^5$. From Table 5.3

$$\therefore \quad k_c/k = 0.197(0.6466 \times 10^5)^{¼}(4/0.05)^{-1/9} = 1.93; \quad \text{and } k_c = 0.043$$
$$h = k_c/\delta = 0.043/0.05 = 0.86$$

Heat lost by convection between the hotter wall and the partition

$$Q = hA\,(\Delta T) = 0.86 \times 20 \times 5 = 86 \text{ W and the same energy will flow from}$$
the pertition to the colder wall.

Heat lost by conduction will be again 44.6 W.

Example 5.17 Air at atmosheric pressure is contained between horizontal plates (5 m × 4 m), separated by a distance of 5 cm. The temperature of the lower and upper plates are respectively 100°C and 40°C. Calculate the rate of heat transfer.

Solution: The mean temperature = 70°C = 343 K and the properties of air:
$$\nu = 20.02 \times 10^{-6} \text{ m}^2/\text{s}, \quad k = 0.0255 \text{ W/mK}, \quad Pr = 0.694$$
$$Gr_\delta \cdot Pr = 9.18 \times (1/343) \times 60 \times (0.05)^3/(20.02 \times 10^{-6})^2 \times 0.694$$
$$= 3.7 \times 10^5$$
$$\frac{k_e}{k} = 0.061(3.7 \times 10^5)^{1/3} = 4.378$$
$$\therefore \quad k_e = 0.1116, \quad h = \frac{k_e}{\delta} = 2.233 \text{ and}$$
$$\dot{Q} = hA(\Delta T) = 2.233 \times 20 \times 60 = 2.68 \text{ kW}.$$

Example 5.18 Calculate the heat lost by the lower plate of Ex. 5.17 when the space between the two plates is occupied by water.

Solution: The properties of water at the mean temperature 70°C are:
$$\nu = 0.421 \times 10^{-6} \text{ m}^2/\text{s}, \quad \beta = 0.21 \times 10^{-3}, \quad Pr = 2.62, \quad k = 0.066$$
$$Gr_\delta \cdot Pr = 9.81 \times 0.21 \times 10^{-3} (60) (0.05)^3 \times 2.62/(0.421 \times 10^{-6})^2$$
$$= 1.61 \times 10^8$$
$$k_e/k = 0.057 (1.61 \times 10^8)^{1/3} = 31.01; \quad \text{and} \quad k_e = 20.47; \quad h = 20.47/0.05 = 809.4$$
$$\dot{Q} = h A (\Delta T) = 809.4 \times 20 \times 60 = 491.28 \text{ kW}.$$

Example 5.19 Air at 1 atmosphere is contained between two concentric spheres having radii 100 mm and 125 mm. Calculate the rate of heat transfer when the temperature at the inner and outer surfaces are 75°C and 45°C.

Solution: The mean temperature is (75 + 45)/2 = 60°C = 333 K
The properties of air: $\nu = 18.97 \times 10^{-6} \text{ m}^2/\text{s}, \quad Pr = 0.696, \quad k = 0.02896$

Characteristic length $= (r_o - r_i) = 25 \text{ mm} = 0.025 \text{ m}$

$Gr_\delta \cdot Pr = 9.18(1/333) \cdot (75 - 45) \cdot (0.025)^3 \times 0.696/(18.97 \times 10^{-6})^2 = 2.67 \times 10^4$

From Table 5.3 $k_e/k = 0.228(2.67 \times 10^4)^{0.226} = 2.28; \quad \text{and} \quad k_e = 0.066$

$$\dot{Q} = \frac{4\pi k_e r_i r_0}{r_0 - r_i}(\Delta T) = \frac{4 \times 3.142 \times 0.066 \times 0.1 \times 0.125 \times 30}{0.025} = 12.44 \text{ W} .$$

Example 5.20 A cylindrical vessel (20 cm in diameter) contains water at 20°C. The bottom of the vessel is at 100°C (placed on a gas stove) and the depth of water is 8 cm. Calculate the initial rate of heat transfer from the stove to the water.

Solution: The mean temperature is (100 + 20)/2 = 60°C. The properties of water are:
$$Pr = 3.02, \quad \nu = 0.478 \times 10^{-6} \text{ m}^2/\text{s}, \quad \beta = 5.18 \times 10^{-4} \text{ K}^{-1}, \quad k = 0.657 \text{ W/mK}$$
$$Gr \times Pr = \frac{(9.8) \times (5.18 \times 10^{-4})(80) \times (0.08)^3 (3.02)}{(0.478 \times 10^{-6})^2} = 2.75 \times 10^9$$

From Table 5.3: Nu = 0.057 (Gr·Pr)$^{1/3}$ = 0.057(2.75 × 10^9)$^{1/3}$ = 79.85

$\qquad$ h = Nu × k/δ = 79.85 × 0.657/0.08 = 655.77 W/m^{2}K

Therefore the initial rate of heat transfer,

$$\dot{Q} = hA\,(\Delta T) = 655.77 \times (3.142/4)(0.20)^2\,(80) = 1648 \text{ W.}$$

16. Heat Transfer from Inclined Surfaces

Natural convection in a cavity formed between two inclined planes, Fig. 5.7(h), is encountered in flat plate solar collectors and the heat transfer rate can be calculated by substituting (g cos θ) in place of (g) in Grashof number where θ is the angle which the heater surface makes with the horizontal. This transformation is expected to hold up to an inclination angle of 60°C and applies only to those cases where the hotter surface is facing upward. For angles between 70°C and 90°C, Catton (Natural convection in enclosures, Proceedings, Sixth Internatonal Heat Transfer Conference, Toronto, Vol. 6, pp 13-31, Hampshire, 1978) has recommended that the Nusselt number for a vertical enclosure be multiplied by sin θ$^{1/4}$.

17. Heat Transfer by Natural Convection from Rotating Cylinders and Disks

Convection heat transfer is of importance in the thermal analysis of various machines having rotating components, such as turbine rotors, flywheels, rotating shafts, etc. When a cylindrical shaft rotates, the temperature and velocity field near the periphery of the surface changes. And, the heat transfer rate depends upon the peripheral speed Reynolds number = $\pi D^2 \omega / \nu$. For Re$_w$<8000 the rate of heat transfer is controlled by conventional Grashof number. And for Re$_w$>8000 in air, the peripheral speed Reynolds number becomes the controlling parameter. For a horizontal cylinder rotating in air, the combined effects of Reynolds, Prandtl and Grashof numbers on the Nusselt number is given by

$$\text{Nu} = 0.11 \,(0.5\,\text{Re}_w^2 + \text{Gr} \cdot \text{Pr})^{0.35} \qquad\qquad (5.23)$$

Example 5.21 A 20 cm diameter steel shaft initially at 100°C is cooled in air at 20°C by rotating the shaft about its own horizontal axis. Calculate the rate of convection heat transfer when the shaft is rotating at 4 rpm and 16rpm.

Solution: The mean temperature is (100 + 20)/2 = 60°C. The properties of air are:

$\qquad$ ν = 1.94 × 10^{-5} m^2/s, β = 1/333, k = 0.0279 W/mK, Pr = 0.71

(a) Rotational speed = 4 rpm, ω = 2 × 3.142 × 4/60 = 0.419 rad/s

$\qquad$ Re$_w$ = 3.142 × (0.2)^2 × 0.419/(1.94 × 10^{-5}) = 2714.4 < 8000

$\qquad$ Ra = Gr·Pr

$\qquad\qquad$ = 9.8 × (1/333) (80)(0.2)^3 × 0.71/(1.94 × 10^{-5})^2 = 3.55 × 10^7

(b) Rotational speed = 16 rpm, ω = 1.676 and Re$_w$ = 10857.6

$\qquad$ When the shaft is stationary, from Table 5.1, Eq. (5.1)

$$Nu = 0.125 \ (3.55 \times 10^7)^{1/3} = 41.06; \ \text{and}$$

$$h = 41.06 \times 0.0279/0.2 = 5.73 \ W/m^2K$$

For $\quad Re_w = 2714.4$, Eq (5.23)

$$Nu = 0.125 \ (0.5(2714.4)^2 + 3.55 \times 10^7)^{0.35} = 50$$

and $\quad h = 6.975 \ W/m^2K$

For $\quad Re_w = 10857.6$

$$Nu = 0.11(0.5(10857.6)^2 + 3.55 \times 10^7)^{0.35} = 68 \ \text{and} \ h = 9.5 \ W/m^2K.$$

Thus, the rotation of shaft increases the heat transfer coefficient and for $Re_w < 8000$, the gravity induced natural convection dominates, but at $Re_w > 8000$, the peripheral speed Reynolds number dominates.

Serveral research workers have investigated (experimentally and theoretically) the heat transfer from a rotating disk. They have observed that the boundary layer on the disk is laminar and of uniform thickness at rotational Reynolds number, $\omega D^2/\nu$, below 10^6. At higher Reynolds number, the flow becomes turbulent near the outer edge and with increasing Re_w, the transition point moves radially inward, and the thickness of the boundary layer increases with increasing radius. The empirical relations for evaluating the heat transfer coefficient are:

a) $\qquad Nu = 0.36 \ (Re_w)^{1/2}$ for $Re_w < 10^6$ $\hfill$ (5.24)

b) $\qquad Nu = hr/k = 0.0195 \ (\omega r^2/\nu)^{0.8}$ for turbulent flow $\hfill$ (5.25)

c) the average value of Nusselt number for Laminar flow between $r = 0$ and $r = r_e$ and turbutent flow between $r = r_e$ and $r = r_o$,

$$Nu = \frac{hr_0}{k} = 0.36 \left(\frac{\omega r_0^2}{\nu} \right)^{1/2} \cdot \left(\frac{r_c}{r_0} \right)^2 + 0.015 \left(\frac{\omega r_0^2}{\nu} \right)^{0.8} \left[1 - \left(\frac{r_c}{r_0} \right)^{2.6} \right] \tag{5.26}$$

Example 5.22 A horizontal disk 1 m in diameter rotates in air at 20°C. If the disk is at 100°C, estimate the rpm of disk at which the natural convection heat transfer for a stationary disk is less than 10% of the heat transfer for a rotating disk.

Solution: Properties of air at mean temperature (60°C) are:

$$\nu = 1.94 \times 10^{-5} \ m^2/s, \quad \beta = 1/333, \quad k = 0.0279 \ W/mK, \quad Pr = 0.71,$$

Characteristic length $= D/4 = 0.25$ m

$$Gr \cdot Pr = 9.8 \times (1/333) \ (80) \times (0.25)^3 \times 0.71/(1.94 \times 10^{-5})^2 = 6.94 \times 10^7$$

For natural convection from a heated horizontal disk, from Eq. (5.1) Table 5.1

$$Nu = hL/k = 0.15(Ra)^{1/3} = 0.15(6.49 \times 10^7)^{1/3} = 57.5$$

$$h = 57.5 \times 0.0279/0.25 = 6.417 \ W/m^2K$$

Assuming laminar flow, h for rotating disc $= 6.417/0.9 = 7.13 \ W/m^2K$

$$Nu = h \ D/k = 7.13 \times 1.0/0.0279 = 255.55 = 0.36 \ (Re_w)^{1/2};$$

$\therefore \qquad Re_w = 0.504 \times 10^6 < 10^6$

$$\omega = 0.504 \times 10^6 \times 1.94 \times 10^{-5}/(1)^2 = 9.8 \ rad/s; \ \text{and the rpm}, \ N = 93.3.$$

SUMMARY

1. Natural convection heat transfer requires: (i) a solid fluid interface, (ii) a temperature difference between the temperature of the solid and the surrounding fluid, and (iii) mixing motion of fluid particles due to the density difference created by the temperature gradient.

2. In laminar flow, the fluid particles follow a smooth and continuous path and do not have a macroscopic mixing between successive layers. In a turbulent flow, there is a random macroscopic mixing of fluid particles across successive layers of fluid flow.

3. All real fluids are viscous. Fluids past a solid surface lead to the formation of a boundary layer. The thickness of the boundary layer is very small in comparison with the characteristic dimension of the solid object. The velocity of fluid particles is zero at the solid surface and is equal to the free stream velocity at the edge of the boundary layer. The boundary layer may be laminar or turbulent.

4. In natural convection, the heat transfer coefficient is evaluated from the functional relationship

$$\text{Nu} = C \, (\text{Gr} \times \text{Pr})^m$$

   and the values of C and m are determined experimentally.

5. The significant dimensionless parameters in natural convection heat transfer are:

   Nusselt Number, $\text{Nu} = h \, L/k$, ratio of the temperature gradient in the fluid immediately in contact with the surface to a reference temperature gradient.

   Grashof Number, $\text{Gr} = g\beta(\Delta T)L^3/v^2 = \text{Intertia force} \times \text{Buoyancy forces}/(\text{Viscous forces})^2$

   Prandtl Number, $\text{Pr} = \mu \, C/k = v/\alpha$ a measure of the relative magnitudes of momentum and thermal diffusion in fluid.

   Rayleigh Number, $\text{Ra} = \text{Gr} \times \text{Pr}$; the boundary layer is laminar for $\text{Ra} < 10^9$

   Modified Grashof Number, $\text{Gr}^* = \text{Gr} \times \text{Nu}$, and is significant in evaluating 'h' under constant heat flux condition.

   Also, $\text{Gr}^* = (g\beta x^4/kv)q$; where $q = h \, (\Delta T)$

6. Film convection is the rate of heat transfer from the surface of a thin wire when a stationary film of hot air is formed around the wire at small temperature difference (Ra < 500).

7. When two horizontal infinite planes are separated by a small distance, δ and the upper plate is at a higher temperature or when the lower plate is at a higher temperature and $\text{Gr}_\delta < 1700$, the heat transfer between the two plates takes place by pure conduction and $h\delta/k = 1.0$.

8. Benard cells or cellular patterns are formed in enclosed spaces between two horizontal infinite planes when the lower plate is at a higher temperature than the upper plate and $\text{Gr}_\delta > 1700$. These cells get destroyed when turbulence sets in and $\text{Gr}_\delta > 50,000$.

9. The rate of heat transfer by free convection in a limited space depends upon the temperatures and properties of fluid, the temperature difference between

the space and the shape and size of the space. The heat transfer coefficient 'h' is given by the relation:

$$h = k_e/\delta \text{ where } k_e \text{ is the effective or equivalent thermal}$$

conductivity and is obtained from the relation

$$k_e/k = const (Gr_\delta \times Pr)^n (L/\delta)^m$$

10. Natural convection heat transfer coefficient from horizontal circular cylinders and horizontal plates can be increased by imparting rotary motion to cylinders about their axis and by rotating plates about their vertical axis, Fig. 5.8.

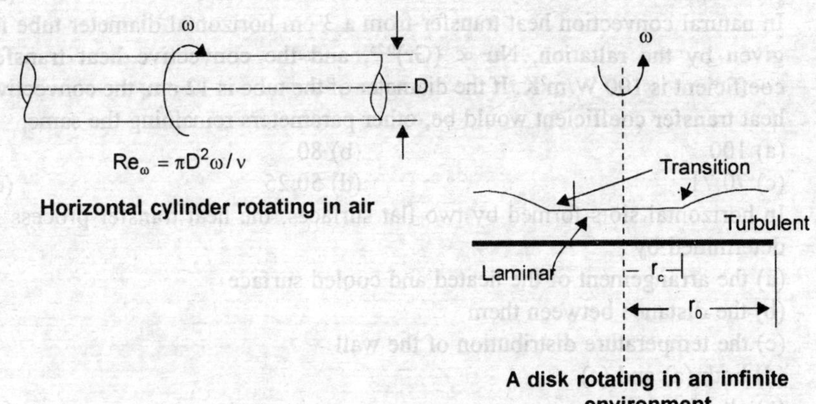

Fig. 5.8 Rotating cylinders and disks

MULTIPLE CHOICE QUESTIONS

1. Dimensional analysis is applied when
    (a) mathematical analysis is impractical
    (b) some rapid, qualitative answers are required
    (c) either (a) or (b)   (d) both (a) and (b)   **(d)**
2. Assertions (A): Dimensional analysis cannot be a substitute for the exact or the approximate mathematical solution.
    Reasoning (R): Because it leads to the formulation of semi-empirical relation.
    Code: (a) Both A and R are false   (b) A is false, R is true
            (c) A is true, R is false   (d) Both A and R are true   **(d)**
3. The advantages of dimensional analysis are:
    (a) lays the foundation of an efficient experimental programme
    (b) indicates a possible form of semi-empirical correlation
    (c) suggests logical grouping of quantities for presenting the results
    (d) all of the above   **(d)**
4. Grashof number is defined as
    (a) $g\beta(\Delta T)L/\nu$ (b) $g\beta(\Delta T)L^2/\nu^2$
    (c) $g\beta(\Delta T)L^3/\nu^2$ (d) inertia forces/buoyancy forces   **(c)**

5. The Nusselt number in natural convection heat transfer is a function of
   (a) Re & Pr
   (b) Re & Gr
   (c) Gr & Pr
   (d) Gr & Bi
   where   Re is Reynolds number, Gr is Grashof number,
           Pr is Prandtl number and Bi is Biot number   **(c)**

6. The characteristic length for computing Grashof number in the case of a vertical cylinder is
   (a) diameter of the cylinder
   (b) length of the cylinder
   (c) the perimeter of the cylinder
   (d) either a or b   **(b)**

7. In natural convection heat transfer from a 3 cm horizontal diameter tube is, given by the raltation, $Nu \propto (Gr)^{0.25}$, and the convective heat transfer coefficient is 100 W/m²K. If the diameter of the tube is 12 cm, the convective heat transfer coefficient would be, other parameters remaining the same,
   (a) 100
   (b) 80
   (c) 70.71
   (d) 50.25   **(c)**

8. In horizontal slots formed by two flat surfaces, the heat transfer process is determined by
   (a) the arrangement of the heated and cooled surface
   (b) the distance between them
   (c) the temperature distribution of the wall
   (d) both (a) and (c)
   (e) all of the above   **(e)**

9. In natural convection heat transfer under uniform heat flux, the modified Grashof number is defined as:
   (a) Pr × Gr
   (b) Nu × Gr
   (c) Nu × Gr/Pr
   (d) Gr × Pr/Nu   **(b)**

10. Prandtl number is defined as:
   (a) k/µc
   (b) α/ν
   (c) µc/k
   (d) kc/µ   **(c)**

11. Assertion (A) : At small temperature potentials a stationary film of heated air forms around a thin wire and this type of heat transfer and free flow is called film convection.
   Reasoning (R) : Because the surface of the wire (diameter 0.2 – 2.0 mm) is very small and the quantity of heat transferred is insignificant.
   Code : (a) Both A and R are false   (b) A is true and R is false
          (c) Both A and R are true   (d) A is false, R is true.   **(c)**

12. In natural convection heat transfer, the boundary layer becomes turbulent when Gr × Pr is
   (a) equal to 10^9
   (b) equal to 10^{10}
   (c) greater than 10^9
   (d) greater than 6×10^{10} **(d)**

13. The assumptions made in the analysis of lamnar boundary layers over a heated vertical plate in natural convection are:
   (a) the physical parameters are constant
   (b) transport of heat by conduction and convection along the moving layer of the fluid is negligible

(c) inertia forces are very small in comparison with gravity and viscous forces

(d) only b and c   (e) all of the above   **(b)**

14. In natural convection heat transfer, the maximum velocity in the laminar boundary layer occurs when y is approximately equal to

(a) 0   (b) $\delta/3$

(c) $\delta/2$ (d) δ **(b)**

where δ is the boundary layer thickners

15. Heat transfer by natural convection takes place in the following system:

(a) hot gases flowing through a tube

(b) cooling of nuclear reactors

(c) cooling of a glass of milk placed on a table

(d) cooling of an internal combustion engine   **(c)**

16. The average Nusselt number in laminar free convection from a vertical wall at 180°C with still air at 20°C is found to be 48. If the wall temperature becomes 30°C all other parameters remaining the same, the average Nusselt number will be

(a) 8   (b) 16   (c) 24   (d) 32   **(c)**

17. A 320 cm high vertical pipe at 175⁰C wall temperature is in a room with still air at 25°C. This pipe supplies heat at the rate of 8 kW into the room air by natural convection. Assuming laminar flow, the height of the pipe required to supply 1 kW only is

(a) 80   (b) 40 cm   (c) 20 cm   (d) 10 cm   **(c)**

18. Assertion (A) : In free convection turbulent flow over a heated vertical plate. h is independent of the characteristic length.

Reasoning (R) : because $Nu = C(Gr.Pr)^{1/3}$

Code : (a) Both A and R are false   (b) both A and R are true

(c) A is true, R is false   (d) A is false, R is true   **(b)**

19. In natural convection heat transfer from a heated horizontal circular disc, diameter D, rotating about its axis:

1. the characteristic length is equal to D,

2. the heat transfer rate is controlled by peripheral speed Reynolds number Re_w

3. the boundary layer is laminar and is of uniform thickness, if $Re_w < 10^6$

4. the boundary layer thickness increases with increasing radius for $Re_w > 10^8$. Of these statements:

(a) 1 and 2 are correct   (b) 3 and 4 are correct

(c) 1, 3 and 4 are correct   (d) 2, 3 and 4 are correct   **(b)**

NUMERICALS

1. The warm air in a heating system is circulated through a vertical duct 60 cm by 40 cm by 6 m. The duct carries warm air at 85°C and the air surrounding the duct is at 15°C. Calculate the heat lost by convection from the duct to the surroundings.

   (1185W)

2. Determine the rate of heat input into a vertical steel plate that loses heat by natural convection to surrounding air. The plate is 30 cm by 50 cm and is maintained at 150°C. The surrounding air is at 18°C. If the plate is immersed in water at 30°C, calculate the rate of heat transfer.

   (2023 W in air)

3. Solar energy at a rate of 280 W/m² is incident on a roof inclined at an angle of 40°C with the horizontal. Assuming that the surface of the roof behaves like a back body, determine the equilibrium temperature of the roof if the ambient air is at 0°C and all the energy received by the roof is lost by free convection.

   (94°C)

4. A large circular duct, 1.5m in diameter carries hot gases at 250°C. The duct passes through a room where the air temperature is 20°C. Calculate the heat loss per metre length of the duct.

5. A cube, 20 cm on a side, is maintained at 70°C and is exposed to ambient air at 20°C. Calculate the heat transfer by free convection. (hint: the characteristic length may be taken as 0.4 m which is the distance a particle travels in the boundary layer)

   (109 W)

6. Calculate the rate of heat transfer from a 100 W electric bulb at 150°C in air at 25°C. Assume the bulb as a sphere of 75 mm diameter and use the relation.
   $$Nu = 0.60 \, (Gr \times Pr)^{1/4}$$
   For air: $v = 28.22 \times 10^{-6}$ m²/s, $k = 0.03$ W/mK, $Pr = 0.69$

7. Calculate the rate of heat loss by free convection per metre length from a horizontal pipe 150 mm diameter, surface temperature 277°C. The room temperature is 17°C. Use the relation
   $$Nu = 0.53 \, (Gr \times Pr)^{1/4}$$
   For air: $v = 28.22 \times 10^{-6}$ m²/s, $Pr = 0.69$, $k = 0.15$ W/mK

   (1.119 kW)

8. A vertical surface l m high and at a temperature of 900K loses heat by natural convection to air at 300 K. Estimate the loss per metre width by using the relation
   $$h = 1.42 \, (\Delta T/L)^{0.25} \quad \text{for } 10^4 < Gr < 10^9$$
   $$\text{or} \quad h = 1.31 \, (\Delta T)^{1/3} \qquad \text{for } 10^9 < Gr < 10^{12}$$
   For air $v = 51.128 \times 10^{-6}$ m²/s

   (5250 W)

9. A tube of 2.5 cm diameter and surface temperature of 50°C is losing heat by free convection to air at 15°C. In order to estimate the heat loss, model tests with a wire heated electrically to 270°C are to be carried out in compressed air

at 15°C. The diameter of the wire is 0.25 cm. Calculate the pressure that would be required. Assume $Nu = \phi(Gr.Pr)$.

(24.0 bar)

10. A 2000 W immersion heater is placed in a 5 cm diameter sealed horizontal tube, and is used to heat water to 60°C. If the surface temperature of tube is not to exceed 95°C, calculate the length of the tube required. For free convection
$$Nu = 0.53 \ (Gr \times Pr)^{1/4}$$
For water $\beta = 6.21 \times 10^{-4} \ K^{-1}$; $\quad C = 4191/kgK$; $\quad \rho = 973.7 \ kg/m^3$
$$\mu = 3.72 \times 10^{-4} \ Pa\text{--}s, \quad k = 0.668 \ W/mK.$$

(0.355m)

11. A hot square plate 50 cm by 50 cm at 100°C is placed in ambient air at 25°C. Calculate the heat loss from both surfaces of the plate if (a) the plate is vertical and (b) the plate is horizontal.

12. Two flat plates (6 m by 11 m) are separated by an air gap width 2.5 cm. One plate is at 45°C and the other is at 35°C. Find the heat transfer by natural convection and by conduction.

13. The surfaces of two long horizontal concentric thin walled tubes having radii 100 m and 120 mm are maintained at 120°C and 35°C, respectively. Estimate the rate of heat transfer per unit length of the tube if air is trapped in the annular space.
Use the relations:
$k_e/k = 0.11(Gr.Pr)^{0.29} \quad$ for $Gr_\delta Pr < 10^6$
$ = 0.40 \ (Gr.Pr)^{0.20} \quad$ for $Gr_\delta \cdot Pr$ between 10^6 and 10^8

14. Two 50 cm horizontal square plates are separated by a distance of 1 cm. The lower plate is maintained at 38°C and the upper plate is maintaned at 27°C. Water at atmospheric pressure occupies the space between the plates. Calclate the heat lost by the lower plate.
$$k_e/k = 0.13 \ (Gr.Pr)^{0.3}$$
For water: $k = 0.623 \ W/mK$ $\ C = 4174 \ J/kgK$, $\quad \rho = 994.9 \ kg/m^3$,
$$\mu = 7.65 \times 10^{-4} \ Pa\text{ -s}, \qquad\qquad \beta = 6.2 \times 10^{-4} \ K^{-1}.$$

(965 W)

15. A 0.25m long vertical plate is at 70°C. Estimate the thickness of the boundary layer at the trailing edge if the temperature of air is 25°C.
For air: $\qquad v = 17.95 \times 10^{-6} \ m^2/s,$ $\qquad Pr = 0.7,$
Use the relation: $Nu = 0.59 \ (Gr.Pr)^{0.25}$

(1.0 cm)

16. A 20cm diameter steel shaft at 100°C is to be cooled in air at 20°C by rotating about its own horizontal axis. Estimate the speed of the shaft in rpm if a 20% increase in the heat transfer coefficient is required.
For air: $\quad v = 1.94 \times 10^{-5} \ m^2/s,$ $\quad k = 0.0279 \ W/mK,$ $\qquad Pr = 0.71,$
Use the following relations:
$$Nu = 0.125 \ (Ra)^{0.333}; \text{ and } Nu = 0.125 \ [0.5 \ Re_w^2 + Ra]^{0.35}$$

(2.98 rpm)

CHAPTER 6

Laminar Flow Forced Convection Heat Transfer

1. Forced Convection Heat Transfer Principles

The mechanism of heat transfer by convection requires mixing of one portion of fluid with another portion due to gross movement of the mass of the fluid. The transfer of heat energy from one fluid particle or a molecule to another one is still by conduction but the energy is transported from one point in space to another by the displacement of fluid.

When the motion of fluid is created by the imposition of external forces in the form of pressure differences, the process of heat transfer is called 'forced convection'. And, the motion of fluid particles may be either laminar or turbulent and that depends upon the relative magnitude of inertia and viscous forces, determined by the dimensionless parameter Reynolds number. In free convection, the velocity of fluid particle is very small in comparison with the velocity of fluid particles in forced convection, whether laminar or turbulent. In forced convection heat transfer, $Gr/Re^2 << 1$, in free convection heat transfer, $Gr/Re^2 >> 1$ and we have combined free and forced convection when $Gr/Re^2 \approx 1$.

2. Methods for Determining Heat Transfer Coefficient

The convective heat transfer coefficient in forced flow can be evaluated by:

    (a) Dimensional Analysis combined with experiments;

    (b) Reynolds Analogy – an analogy between heat and momentum transfer;

    (c) Analytical Methods – exact and approximate analyses of boundary layer equations.

3. Method of Dimensional Analysis

As pointed out in Chapter 5, dimensional analysis does not yield equations which can be solved. It simply combines the pertinent variables into non-dimensional numbers which facilitate the interpretation and extend the range of application of experimental data. The relevant variables for forced convection heat transfer phenomenon whether laminar or turbulent, are

(i) the properties of the fluid – density ρ, specific heat capacity C_p, dynamic or absolute viscosity μ, thermal conductivity k.

(ii) the properties of flow – flow velocity V, and the characteristic dimension of the system L.

As such, the convective heat transfer coefficient, h, is written as

$$h = f(\rho, V, L, \mu, Cp, k)$$

or, $$f_1(h, \rho, V, L, \mu, Cp, k) = 0 \qquad (6.1)$$

Since there are seven variables and four primary dimensions, we would expect three dimensionless numbers. As before, we choose four independent or core variables as ρ, V, L, k, and calculate the dimensionless numbers by applying Buckingham π's method:

$$\pi_1 = \rho^a\, V^b\, L^c\, K^d\, h = (ML^{-3})^a\, (LT^{-1})^b\, (L)^c\, (MLT^{-3}\,\theta^{-1})^d\, (MT^{-3}\,\theta^{-1})$$
$$= M^\circ\, L^\circ\, T^\circ\, \theta^\circ .$$

Equating the powers of M, L, T and θ on both sides, we get

$\quad$ M : a + d + 1 = 0

$\quad$ L : $-$ 3a + b + c + d = 0

$\quad$ T : $-$ b $-$ 3d $-$ 3 = 0 $\qquad$ By solving them, we have

$\quad$ θ : $-$ d $-$ 1 = 0. $\qquad$ d = $-$ 1, a = 0, b = 0, c = 1.

Therefore, π_1 = hL/k is the Nusselt number.

$$\pi_2 = \rho^a\, V^b\, L^c\, k^d\, \mu = (ML^{-3})^a\, (LT^{-1})^b\, (L)^c\, (MLT^{-3}\,\theta^{-1})^d\, (ML^{-1}\, T^{-1})$$
$$= M^\circ\, L^\circ\, T^\circ\, \theta^\circ .$$

Equating the powers of M, L, T and on both sides, we get

$\quad$ M : a + d + 1 = 0

$\quad$ L : $-$ 3a + b + c + d = 1 = 0

$\quad$ T : $-$ b $-$ 3d $-$ 1 = 0

$\quad$ θ : $-$ d = 0.

By solving them, d = 0, b = $-$ 1, a = $-$ 1, c = $-$ 1

and $\quad$ $\pi_2 = \mu/\rho VL$; or, $\pi_3 = \dfrac{1}{\pi_2} = \dfrac{\rho VL}{\mu}$ is the Reynolds number.

(Reynolds number is a flow parameter of greatest significance. It is the ratio of inertia forces to viscous forces and is of prime importance to ascertain the conditions under which a flow is laminar or turbulent. It also compares one flow with another provided the corresponding length and velocities are comparable in two flows. There would be a similarity in flow between two flows when the Reynolds numbers are equal and the geometrical similarities are taken into consideration.)

$$\pi_4 = \rho^a\, V^b\, L^c\, k^d\, C_p = (ML^{-3})^a\, (LT^{-1})^b\, (L)^c\, (MLT^{-3}\,\theta^{-1})^d\, (L^2\, T^{-2}\,\theta^{-1})$$
$$= M^\circ\, L^\circ\, T^\circ\, \theta^\circ$$

Equating the powers of M, L, T, on both sides, we get

$\quad$ M : a + d = 0; $\qquad\qquad$ L : $-$ 3a + b + c + d + 2 = 0

$\quad$ T : $-$ b $-$ 3d $-$ 2 = 0; $\qquad$ θ : $-$ d $-$ 1 = 0

By solving them,

$$d = -1, a = 1, b = 1, \quad c = 1,$$

$$\pi_4 = \frac{\rho VL}{k} C_p; \quad \pi_5 = \pi_4 \times \pi_2$$

$$= \frac{\rho VL}{k} C_p \times \frac{\mu}{\rho VL} = \frac{\mu C_p}{k}$$

$\therefore \quad \pi_5$ is Prandtl number.

Therefore, the functional relationship is expressed as:

$$Nu = f (Re, Pr); \quad \text{or} \quad Nu = C \, Re^m \, Pr^n \tag{6.2}$$

where the values of c, m and n are determined experimentally.

4. Principles of Reynolds Analogy

Reynolds was the first person to observe that there exists a similarity between the exchange of momentum and the exchange of heat energy in laminar motion and for that reason it has been termed 'Reynolds analogy'. Let us consider the motion of a fluid where the fluid is flowing over a plane wall. The X-coordinate is measured parallel to the surface and the Y-coordinate is measured normal to it. Since all fluids are real and viscous, there would be a thin layer, called momentum boundary layer, in the vicinity of the wall where a velocity gradient normal to the direction of flow exists. When the temperature of the surface of the wall is different than the temperature of the fluid stream, there would also be a thin layer, called thermal boundary layer, where there is a variation in temperature normal to the direction of flow. Fig. 6.1 depicts the velocity distribution and temperature profile for the laminar motion of the fluid flowing past a plane wall.

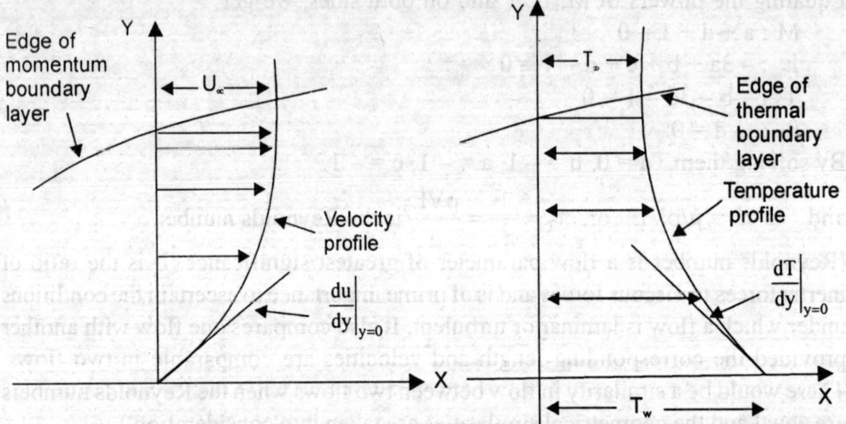

Fig. 6.1 Velocity distribution and temperature profile for laminar motion
of the fluid over a plane surface

In a two-dimensional flow, the shearing stress is given by $\quad \tau_w = \mu \left.\dfrac{du}{dy}\right|_{y=0}$

and the rate of heat transfer per unit area is given by $\quad \dfrac{\dot{Q}}{A} = -k \left.\dfrac{dT}{dy}\right|_{y=0}$

Combining the two equations, we have $\quad \dfrac{\dot{Q}}{A} = -\dfrac{\tau_w \, k}{\mu} \dfrac{dT}{du}$

For $Pr = \mu C_p/k = 1$, we have $k/\mu = C_p$ and therefore, we can write after separating the variables,

$$\frac{\dot{Q}}{A\tau_w C_p} du = -\, dT \tag{6.3}$$

Assuming that Q and τ_w are constant at any station x, we integrate equation (6.3) between the limits: $u = 0$ when $T = T_w$, and $u = U_\infty$ when $T = T_\infty$ and we get,

$$\dot{Q}/(A\tau_w C_p) \times U_\infty = (T_w - T_\infty)$$

Since by definition, $\dot{Q}/A = h_x (T_w - T_\infty)$, and $\tau_w = C_{fx} \times \rho U_\infty^2/2$,

where C_{fx} is the skin friction coefficient at the station x. We have

$$C_{fx}/2 = h_x / (C_p \rho \, U_\infty) \tag{6.4}$$

Since $\quad h_x/C_p\rho U_\infty = (h_x . x/k) \times (\mu/\rho \times U_\infty) \times (k/\mu . C_p) = Nu_x /(Re . Pr)$,

$$Nu_x/Re . Pr = C_{fx}/2 = \text{Stanton numer}, \quad St. \tag{6.5}$$

Equation (6.5) is satisfactory for gases in which Pr is approximately equal to unity. Colburn has shown that Eq. (6.5) can also be used for fluids having Prandtl numbers ranging from 0.6 to about 50 if it is modified in accordance with experimental results.

Or, $\quad \dfrac{Nu_x}{Re_x \, Pr} . Pr^{2/3} = St_x \, Pr^{2/3} = C_{fx}/2 \tag{6.6}$

Eq. (6.6) expresses the relation between fluid friction and heat transfer for laminar flow over a plane wall. The heat transfer coefficient could thus be determined by making measurements of the frictional drag on a plate under conditions in which no heat transfer is involved.

Example 6.1   Glycerine at 35°C flows over a 30 cm by 30 cm flat plate at a velocity of 1.25 m/s. The drag force is measured as 9.8 N (both side of the plate). Calculate the heat transfer for such a flow system.

Solution: From tables, the properties of glycerine at 35°C are:

$\rho = 1256 \text{ kg/m}^3$, $\quad C_p = 2.5$ kJ/kgK, $\quad \mu = 0.28$ kg/m–s, $\quad k = 0.286$ W/mK, $\quad Pr = 2.4$
$Re = \rho VL/\mu = 1256 \times 1.25 \times 0.30/0.28 = 1682.14$, a laminar flow.[*]

Average shear stress on one side of the plate = drag force/area

$$= 9.8/(2 \times 0.3 \times 0.3) = 54.4$$

and shear stress = $C_f \, \rho \, U^2/2$

$\therefore$ The average skin friction coefficient, $C_f / 2 = \dfrac{\tau}{\rho U^2}$

$$= 54.4/(1256 \times 1.25 \times 1.25) = 0.0277$$

From Reynolds analogy, $\quad C_f/2 = St. \, Pr^{2/3}$

or, $\quad h = \rho \, C_p \, U \times C_f/2 \times Pr^{-2/3} = \dfrac{1256 \times 2.5 \times 1.25 \times 0.0277}{(2.45)^{0.667}} = 59.8$ kW/m²K.

---

[*] (For flow over a flat surface, the flow remains laminar upto $Re \le 5\times10^5$, and since the properties of fluids vary with temperature, it is necessary to calculate the properties of the fluid at the mean temperature when evaluating Nu, Re and Pr.)

5. Analytical Evaluation of 'h' for Laminar Flow over a Flat Plate—Assumptions

As pointed out earlier, when the motion of the fluid is caused by the imposition of external forces, such as pressure differences, and the fluid flows over a solid surface, at a temperature different from the temperature of the fluid, the mechanism of heat transfer is called 'forced convection'. Therefore, any analytical approach to determine the convective heat transfer coefficient would require the temperature distribution in the flow field surrounding the body. That is, the theoretical analysis would require the use of the equation of motion of the viscous fluid flowing over the body along with the application of the principles of conservation of mass and energy in order to relate the heat energy that is convected away by the fluid from the solid surface.

For the sake of simplicity, we will consider the motion of the fluid in 2 space dimension, and a steady flow. Further, the fluid properties like viscosity, density, specific heat, etc are constant in the flow field, the viscous shear forces in the Y-direction is negligible and there are no variations in pressure also in the Y-direction.

6. Derivation of the Equation of Continuity—Conservation of Mass

We choose a control volume within the laminar boundary layer as shown in Fig. 6.2. The mass will enter the control volume from the left and bottom face and will leave the control volume from the right and top face. As such, for unit depth in the Z-direction,

$$\dot{m}_{AD} = \rho \, u dy \quad ; \qquad \dot{m}_{BC} = \rho(u + \frac{\partial u}{\partial x} \cdot dx)dy;$$

$$\dot{m}_{AB} = \rho \, v dx \quad ; \qquad \dot{m}_{CD} = \rho(v + \frac{\partial v}{\partial y} \cdot dy)dx;$$

For steady flow conditions, the net efflux of mass from the control volume is zero, therefore,

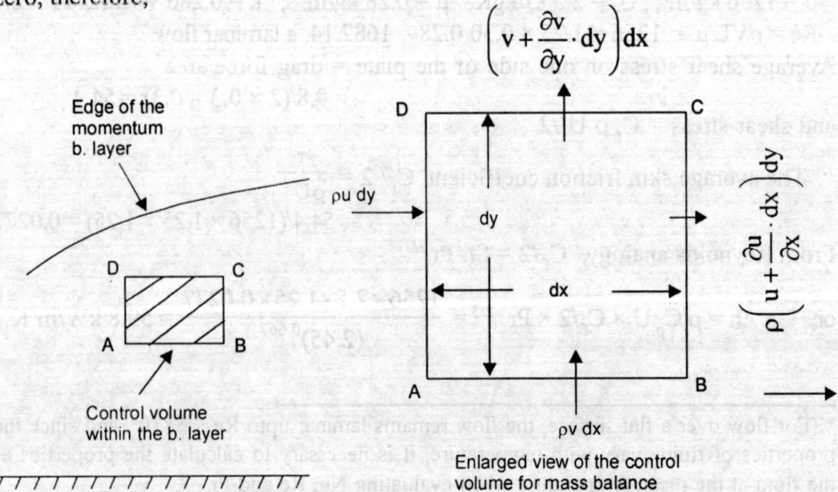

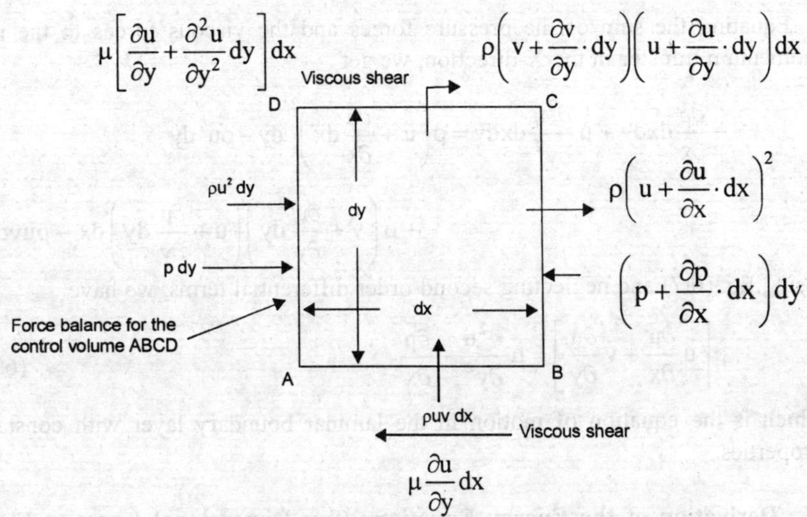

Fig. 6.2 A differential control volume within the boundary layer for laminar flow over a plane wall

$$\rho u\,dy + \rho v\,dx = \rho u\,dy + \rho\frac{\partial u}{\partial x}\,dxdy + \rho v\,dx + \rho\frac{\partial v}{\partial x}\,.dxdy$$

or, $\quad \partial u/\partial x + \partial v/\partial y = 0$, the equation of continuity. $\hfill (6.7)$

7.   Derivation of Equation of Motion—Two-Dimensional Laminar Flow over a Flat Plate

We make a force balance in the X-direction (in the flow direction) with reference to the control volume within the laminar boundary layer as shown in Fig. 6.2.

$$\Sigma F_x = \text{increase in the momentum flux in the X-direction.}$$

Momentum entering the face AD $= \rho\, u^2\, dy$

Momentum leaving the face BC $= \rho\, (u + \partial u/\partial x \cdot dx)^2\, dy$

Momentum in the X-direction which enters the bottom face AB $= \rho\, v\, u\, dx$ and the momentum in X-direction which leaves the top face CD is

$$\rho(v + (\partial v/\partial y)dy)\,(u + (\partial u/\partial y)dy)dx$$

The external forces acting on the control volume in the X-direction are: pressure

force on the left face $= p\, dy$, and that on the right face $= -\left[p + \dfrac{\partial p}{\partial x}dx\right]dy$ and

therefore, the net pressure force in the direction of motion is: $-(\partial p/\partial x)dxdy$

The viscous shear force on the bottom face is: $-\mu\,(\partial u/\partial y)\,dx$,

and the shear force on the top face is: $\mu dx\left[\dfrac{\partial u}{\partial y}+\dfrac{\partial}{\partial y}\left(\dfrac{\partial u}{\partial y}\right)dy\right]$

$\therefore$ The net viscous-shear force in the direction of motion is $= \mu\dfrac{\partial^2 u}{\partial y^2}dxdy$

Equating the sum of the pressure forces and the viscous forces to the net momentum transfer in the X-direction, we get

$$-\frac{\partial p}{\partial x}\,dxdy + \mu\frac{\partial^2 u}{\partial y^2}\,dxdy = \rho\left(u + \frac{\partial u}{\partial x}\,dx\right)^2 dy - \rho u^2 dy$$

$$+ \rho\left(v + \frac{\partial u}{\partial y}\,dy\right)\left(u + \frac{\partial u}{\partial y}\,dy\right)dx - \rho uvdx$$

Using Eq. (6.7) and neglecting second-order differential terms, we have

$$\rho\left[u\frac{\partial u}{\partial x} + v\frac{\partial u}{\partial y}\right] = \mu\frac{\partial^2 u}{\partial y^2} - \frac{\partial p}{\partial x} \tag{6.8}$$

which is the equation of motion in the laminar boundary layer with constant properties.

8. Derivation of the Energy Equation—Two-Dimensional Laminar Flow over a Flat Plate

We choose an elemental control volume shown in Fig. 6.3 for energy analysis of laminar boundary layer. The assumptions are: incompressible steady flow, constant physical properties and negligible heat conduction in the direction of flow (X-direction). Applying the principles of conservation of energy, we have:

Energy convected in left face + energy convected in bottom face + heat conducted in bottom face + net viscous work done on the element = energy convected out right face + energy convected out top face + heat conducted out top face.

With reference to Fig. 6.3, the different terms are as follows:

Energy convected in left face = $\rho C_p\, u\, T\, dy$

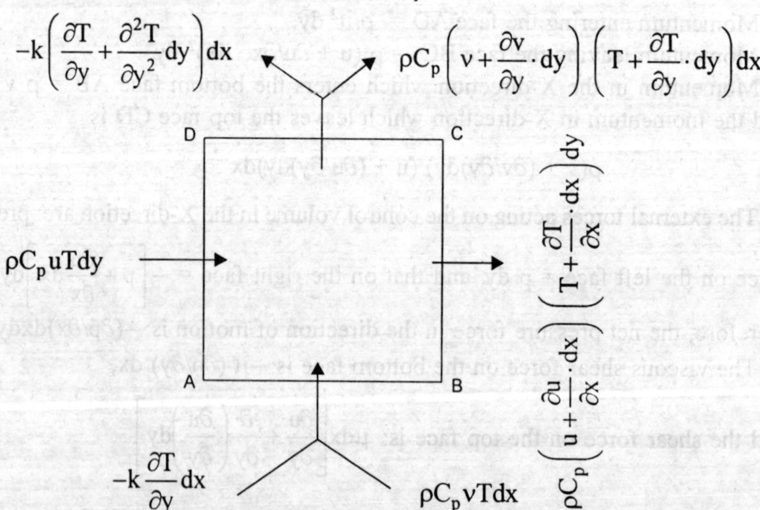

Fig. 6.3 A differential control volume within the thermal boundary layer for laminar flow over a plane surface for energy analysis

Energy convected out right face $= \rho C_p \left(u + \dfrac{\partial u}{\partial x} dx \right) \left(T + \dfrac{\partial T}{\partial x} dx \right) dy$

Energy convected in bottom face $= \rho C_p \, v \, T \, dx$

Energy convected out top face $= \rho C_p \left(v + \dfrac{\partial v}{\partial y} dy \right) \left(T + \dfrac{\partial T}{\partial y} . dy \right) dx$

Energy conducted in bottom face $= -k dx \dfrac{\partial T}{\partial y}$

Energy conducted out top face $= -k dx \left[\dfrac{\partial T}{\partial y} + \dfrac{\partial}{\partial y} \left(\dfrac{\partial T}{\partial y} \right) dy \right]$

The viscous-shear force is the product of the shear-stress and the

$$\text{area} \quad dx = \mu \dfrac{\partial u}{\partial y} dx$$

and the distance through which it moves per unit time in respect to the elemental

control volume $(dx.dy) = \left(\dfrac{\partial u}{\partial y} \right) dy$

so that the net viscous energy delivered to the element is $\mu \left(\dfrac{\partial u}{\partial y} \right)^2 dx dy$.

Substituting these values in the energy balance equation and after neglecting the second-order differentials, we get

$$\rho C_p \left[u \dfrac{\partial T}{\partial x} + v \dfrac{\partial T}{\partial y} + T \left(\dfrac{\partial u}{\partial x} + \dfrac{\partial v}{\partial y} \right) \right] dx dy = k \dfrac{\partial^2 T}{\partial y^2} dx dy + \mu \left(\dfrac{\partial u}{\partial y} \right)^2 dx dy$$

Using Eq. (6.7), the energy equation reduces to

$$u \dfrac{\partial T}{\partial x} + v \dfrac{\partial T}{\partial y} = \alpha \dfrac{\partial^2 T}{\partial y^2} + \dfrac{\mu}{\rho C_p} \left(\dfrac{\partial u}{\partial y} \right)^2 \tag{6.9}$$

Equation (6.9) is the energy equation of the laminar thermal boundary layer.

Comparison of Equation of Motion and Energy Equation

If the pressure remains constant, the Eq. (6.8) reduces to

$$u \dfrac{\partial u}{\partial x} + v \dfrac{\partial u}{\partial y} = v \dfrac{\partial^2 u}{\partial y^2} \tag{6.10}$$

The viscous-work term is of importance only at high velocities because its magnitude will be small in comparison with the other terms when low–velocity flow is under consideration. Therefore, for low velocity incompressible flow, we have

$$u \dfrac{\partial T}{\partial x} + v \dfrac{\partial T}{\partial y} = \alpha \dfrac{\partial^2 T}{\partial y^2} \tag{6.11}$$

Thus there is a striking similarity between the two equations (6.10 & 6.11) The solutions to the two equations will have the same form if $\upsilon = \alpha$.

9. Significance of Prandtl Number

The ratio υ/α is the Prandtl number and represents the relative magnitudes of diffusion of momentum and heat. It takes into account the heat storage capacity of the fluid, heat conducting capacity and the viscous nature of the fluid and therefore is a function of temperature.

10. Solution of the Momentum and Energy Equation

The solution of the momentum equation Eq. (6.10) along with the equation of continuity (Eq. 6.7) has to be obtained with the following boundary conditions:
   (i) condition of no slip at the surface $u = v = 0$ at $y = 0$.

   (ii) from Eq. (6.10) it is evident that at $y = 0$, $\dfrac{\partial^2 u}{\partial y^2} = 0$

   (iii) at the edge of the boundary layer, $u = U_\infty$ at $y = \delta$
   The left hand side and the right-hand side of Eq. (6.10) represent the inertia terms and viscous terms respectively. If both the terms are of the same order of magnitude, then

$$u\frac{\partial u}{\partial x} = 0(1), \qquad v\frac{\partial u}{\partial y} = 0(1) \quad \text{and} \quad \upsilon\frac{\partial^2 u}{\partial y^2} = 0(1)$$

Since $\qquad \dfrac{\partial^2 u}{\partial y^2} = 0\left(\dfrac{U_\infty}{\delta^2}\right), \quad \upsilon = 0(\delta^2), \text{ i.e., } \quad \delta = 0(\upsilon^{1/2})$

The Reynolds number for the flow, defined as Re $= U_\infty x / \upsilon$ must be of $0(1/\upsilon)$ in magnitude and therefore,

$$\delta/x = 0\,(\text{Re}^{-1/2}) \tag{6.12}$$

   The Eq. (6.12) tells us that for large Reynolds number, i.e., when the relative magnitude of the viscous forces is smaller than the inertia forces, the viscous forces will remain confined to a relatively thin layer adjacent to the solid wall or boundary of the flow.

   H. Blassius, a German engineer, in 1908, succeeded in solving the Eq. (6.10) with the help of continuity equation and the boundary conditions mentioned above. He introduced a dimensionless parameter

$$\eta = y\,(U_\infty/\upsilon x)^{1/2} \text{ and a stream function } \psi(x, y) = (\upsilon \times U_\infty)^{1/2}\, f(\eta),$$
   (6.12a)

where $f(\eta)$ denotes the dimensionless stream function. Thus the velocity components become:

$$u = \frac{\partial \psi}{\partial y} = \frac{\partial \psi}{\partial \eta} \times \frac{\partial \eta}{\partial y} = U_\infty f'(\eta); \quad \text{and } v = -\frac{\partial \psi}{\partial x} = \frac{1}{2}\left(\frac{\upsilon U_\infty}{x}\right)^{1/2}(\eta f' - f)$$

Writing down the further terms of Eq. (6.10) and after simplification, we get

$$2f''' + ff'' = 0.$$

The boundary conditions then become:

at $\quad\eta = 0,\quad f = 0,\quad f' = 0;\quad$ at $\eta = \infty,\quad f' = 1.$

Thus, the partial differential equations (6.10) and (6.11) were transformed into an ordinary differential equation and Blassius obtained

$$\eta = 5.0 \quad \text{when} \quad u/U_\infty = 0.9915;$$

Thus, $\quad y = \delta = 5.0\,(vx/U_\infty)^{1/2}$

or, $\qquad \delta/x = 5.0\,\mathrm{Re}_x^{-1/2}$ $\hfill$ (6.13)

The skin friction coefficient was defined as: $\quad C_{fx} = \tau_w \Big/ \dfrac{1}{2}\rho U_\infty^{\;2} = \left(\dfrac{\partial u}{\partial y}\right)_{y=0} \Big/ \dfrac{1}{2}\rho U_\infty^{\;2}$

and therefore, $C_{fx} = 0.664/\mathrm{Re}_x^{1/2}$, and the average value, $C_f = 1.328/\mathrm{Re}_L^{1/2}$

The boundary conditions for the simplified energy equation of the thermal boundary layer (Eq. 6.11) are:

$$T = T_w \quad \text{at} \quad y = 0 \quad \text{and} \quad T = T_\infty \text{ at } y = \infty$$

The temperature T can be non-dimensionalized by assuming

$$\theta = (T - T_w)/(T_\infty - T_w),\hfill (6.14)$$

Then the Eq. (6.11) becomes $\quad u\dfrac{\partial\theta}{\partial x} + v\dfrac{\partial\theta}{\partial y} = \alpha\dfrac{\partial^2\theta}{\partial y^2}$ $\hfill$ (6.15)

With boundary conditions: at $y = 0,\ \theta = 0\quad$ and at $y = \infty,\ \theta = 1.$ $\hfill$ (6.16)

Pohlhausen obtained a solution to Eq. (6.15) subjected to the boundary conditions given by Eq. (6.16) for fluids with arbitrary Prandtl number. Since there is a striking similarity in the two equations (6.10 and 6.11), he assumed the similarity parameters η and the stream function ψ defined by Eq. (6.12a) and obtained

$$\left(\frac{d\theta}{d\eta}\right)_{n=0} = \frac{1}{\displaystyle\int_0^\infty \exp\left(-\frac{\mathrm{Pr}}{2}\int_0^\eta f\,d\eta\right)d\eta}\hfill (6.17)$$

Pohlhausen numerically integrated the Eq. (6.17) and his results were

$$\left(\frac{d\theta}{d\eta}\right)_{n=0} = 0.332\mathrm{Pr}^{0.343} \quad \text{for fluids with } 0.5 < \mathrm{Pr} < 10.\hfill (6.18)$$

For simplicity, the exponent of Pr is taken as 1/3, and the equation (6.18) is written as

$$\left(\frac{d\theta}{d\eta}\right)_{n=0} = 0.332\mathrm{Pr}^{1/3}\hfill (6.19)$$

The local heat transfer coefficient, h_x, is defined as

$$h_x = \frac{\dot{Q}/A}{(T_w - T_\infty)} = -\frac{k(T_\infty - T_w)\left(\dfrac{d\theta}{dy}\right)_{y=0}}{(T_w - T_\infty)} \quad \text{where } \frac{\dot{Q}}{A} \text{ is the heat flux per}$$

unit area and therefore, $\quad h_x = k\left(\dfrac{U_\infty}{\upsilon x}\right)^{1/2}\left(\dfrac{d\theta}{d\eta}\right)_{\eta=0}$

Combining the above relation with Eq. (6.19), we have

$$h_x = 0.332k\sqrt{\frac{U_\infty}{\nu.x}} \cdot Pr^{1/3} \tag{6.20}$$

which can be expressed in terms of the dimensionless Nusselt number

$$Nu_x = \frac{h_x x}{k} = 0.332 Re_x^{0.5} Pr^{1/3} \; ; \; (or \; h_x \propto x^{-0.5}) \tag{6.21}$$

When we consider the approximation to the non-dimensional velocity and thermal gradients as

$$\left(\frac{d\theta}{d\eta}\right)_{\eta=0} = \frac{1}{\delta_t} ; \quad and \quad \left(\frac{d(u/U_\infty)}{d\eta}\right)_{\eta=0} \cong \frac{1}{\delta}$$

where $1/\delta_t = 0.332 \, Pr^{1/3}$ and $1/\delta = 0.332$,

The Pohlhausen results indicate that

$$\delta_t = \delta/Pr^{1/3} \tag{6.22}$$

In order to determine the average value of 'h' for a plate of length L and unit

width, we have $\;h_{av} \equiv h = \dfrac{C\displaystyle\int_0^L x^{-1/2}dx}{L} = 2CL^{-1/2}$ which shows that the average

value of the convective heat transfer coefficient for a plate at constant temperature over its entire length is exactly twice of the local value at the end of the plate.
Thus, $\quad Nu_L = 0.664 \, Re_L^{1/2} . Pr^{1/3}$ $\tag{6.23}$

Example 6.2   Air at pressure 1 bar, temperature 25°C flows over a horizontal flat plate, temperature 75°C and 1m long. Estimate the maximum drag experienced by the plate and the rate of heat transfer per unit width of the plate assuming laminar flow. The properties of air may be taken as:

$$\rho = 1.2 \text{ kg/m}^3, \quad \upsilon = 15 \times 10^{-6} \text{ m}^2/\text{s}, \; Pr = 0.7, \; k = 0.027 \text{ W/mK}$$

Solution: Since the flow remains laminar, the maximum Reynolds number occuring at the trailing edge will have the value $= 5 \times 10^5$.

$$Re_L = U_\infty L/\nu = 5 \times 10^5 ; \quad \therefore U_\infty = 5 \times 10^5 \times 15 \times 10^{-6}/1.0 = 7.5 \text{ m/s}$$

The average skin friction coefficient, $C_f = 1.328/Re_L^{1/2}$

$$= 1.328/(5 \times 10^5)^{1/2} = 1.878 \times 10^{-3}$$

Shear stress at the wall $= C_f \times \rho \times U_\infty^2/2$

$$= 1.878 \times 10^{-3} \times 1.2 \times (7.5)^2/2 = 0.0638$$

Drag = shear stress × area = 0.0638 × (1× 1) = 0.0638 N, on each side of the plate.

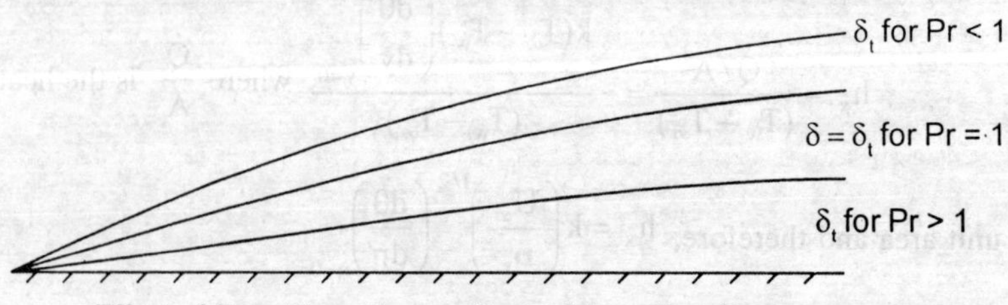

Effect of Prandtl number on thermal boundary layer

Further, $\qquad Nu = 0.664 \times Re^{1/2}\, Pr^{1/3}$

$$= 0.664 \times (5 \times 10^5)^{1/2}\ (0.7)^{1/3} = 416.937$$

and, the average heat transfer coefficient, $h = 416.937 \times 0.027/1.0$

$$= 11.257\ W/m^2K$$

The rate of heat transfer, $\dot{Q} = h\,A\,(\Delta T) = 11.25 \times (1 \times 1)\,(75 - 25)$

$$= 562.86\ W\ \text{per unit width of the plate.}$$

The thickness of the momentum boundary layer at the trailing edge would be,

$$\delta = 5 \times L/Re_L^{1/2} = 5 \times 1.0/(5 \times 10^5)^{1/2} = 7.07\ mm,$$

and the thickness of the thermal boundary layer at the trailing edge would be,

$$\delta_t = \delta\,/Pr^{1/3} = 7.07/(0.7)^{1/3} = 7.96\ mm.$$

[i.e. for Pr < 1, the thermal boundary layer thickness is greater than momentum boundary layer thickness.]

11. Derivation of the Integral Momentum Equation—Its Use and Solution

The closed form solution of the simplified equation of motion for laminar boundary layer is quite difficult. As such, von Karman has suggested an approximate method to analyse boundary layer flows. The method attempts to describe, approximately, the overall behaviour of the boundary layer and although, the results obtained by the integral approach are not as complete and detailed as the results obtained by the solution of the differential equations, a greater variety of problems can be handled by this method.

Let us consider the boundary-layer flow system shown in Fig. 6.4. It is assumed that the boundary layer thickness has a definite value at a station x and it varies with the coordinate X. The free stream velocity outside the boundary layer is U_∞.

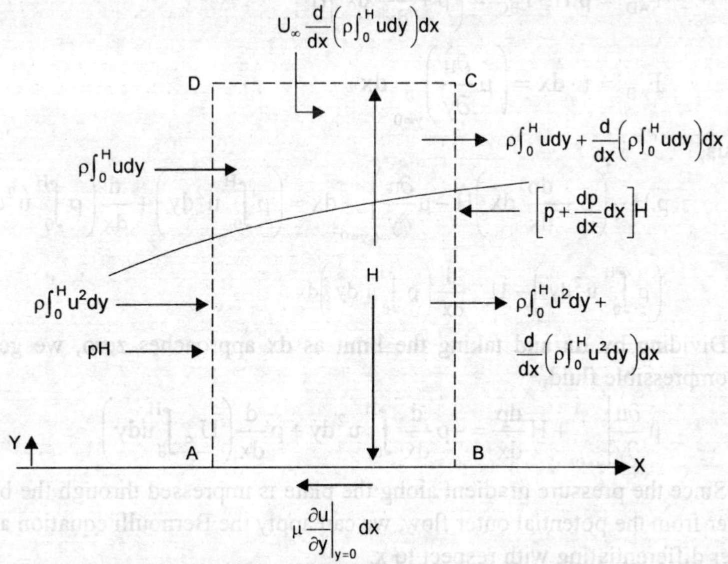

Fig. 6.4 Control volume for approximate mass and momentum
analysis of a boundary layer

We choose a control volume ABCD such that the upper face CD lies outside the domain of the boundary layer, for unit depth in the Z-direction.

$$\dot{m}_{AD} = \left(\rho \int_0^H u \, dy \right)_x ; \qquad \dot{m}_{BC} = \left(\rho \int_0^H u \, dy \right)_{x+dx}$$

$$= \left[\rho \int_0^H u \, dy \right]_x + \frac{d}{dx} \left[\rho \int_0^H u \, dy \right]_x dx$$

$$\dot{m}_{AB} = 0 \text{ (solid wall)}; \quad \dot{m}_{DC} = \dot{m}_{BC} - \dot{m}_{AD} - \dot{m}_{AB} = \frac{d}{dx} \left[\rho \int_0^H u \, dy \right] dx$$

A Taylor series expansion has been written above, to express the mass flow at $x + dx$ in terms of the mass flow at x, with only first two terms considered since dx will subsequently be allowed to approach zero.

The momentum flux in the X-direction across each of the sides of the control volume is:

$$\dot{M}_{AD} = \left(\int_0^H (\rho u) \, u \, dy \right)_x = \left(\rho \int_0^H u^2 dy \right)_x , \dot{M}_{AB} = 0$$

$$\dot{M}_{BC} = \left(\rho \int_0^H u^2 dy \right)_x + \frac{d}{dx} \left(\rho \int_0^H u^2 dy \right)_x ;$$

$$M_{DC} = \dot{m}_{DC} \cdot U_\infty = U_\infty \cdot \frac{d}{dx} \left(\rho \int_0^H u \, dy \right)_x dx$$

The forces acting on the control volume are: the viscous force at the bottom face only, because the top face is outside the boundary layer. The pressure forces will act on the left and right face of the control volume. Thus, the external forces are:

$$F_{AD} = p.H; \quad F_{BC} = \left(p + \frac{dp}{dx} .dx \right) H ;$$

$$F_{AB} = \tau \cdot dx = \left(\mu \frac{\partial u}{\partial y} \right)_{y=0} dx$$

Thus,

$$p.H - \left(p + \frac{dp}{dx} .dx \right) H - \mu \frac{\partial u}{\partial y} \bigg)_{y=0} dx = \left(\rho \int_0^H u^2 dy \right) + \frac{d}{dx} \left(\rho \int_0^H u^2 dy \right) dx -$$

$$\left(\rho \int_0^H u^2 dy \right) - U_\infty \frac{d}{dx} \left(\rho \int_0^H u \, dy \right) dx \qquad (6.24)$$

Dividing by dx and taking the limit as dx approaches zero, we get for an incompressible fluid,

$$\mu \frac{\partial u}{\partial y} \bigg|_{y=0} + H \frac{dp}{dx} = -\rho \frac{d}{dx} \int_0^H u^2 dy + \rho \frac{d}{dx} \left(U_\infty \int_0^H u \, dy \right) \qquad (6.25)$$

Since the pressure gradient along the plate is impressed through the boundary layer from the potential outer flow, we can apply the Bernoulli equation and write after differentiating with respect to x,

$$\frac{dp}{dx} + \rho U_\infty \frac{dU_\infty}{dx} = 0$$

And therefore,

$$H \frac{dp}{dx} = -H\rho U_\infty \frac{dU_\infty}{dx} = -\rho \frac{dU_\infty}{dx} \int_0^H U_\infty dy$$

Since U_∞ is independent of y, substituting into Eq. (6.25), we get

$$\mu \frac{\partial u}{\partial y}\bigg|_{y=0} = \rho \frac{d}{dx}\left[\int_0^\delta (U_\infty - u)u dy\right] + \rho \frac{dU_\infty}{dx} \int_0^\delta (U_\infty - u) dy \quad (6.26)$$

where the upper limit of integration has been changed from H to δ, the boundary layer thickness because $(U_\infty - u)$ vanishes outside the boundary layer.

Solution of the Integral Momentum Equation

For laminar incompressible flow along a flat plate with zero pressure gradient, the free-stream velocity U_∞ is a constant, and Eq. (6.26) reduces to

$$\rho \frac{d}{dx}\left[\int_0^\delta (U_\infty - u)u dy\right] = \mu \frac{\partial u}{\partial y}\bigg|_{y=0} \quad (6.27)$$

The relevant boundary conditions are:
  (i) no slip condition: u = 0 at y = 0.

  (ii) from Eq. (6.10), at the surface of the plate, i.e., at y = 0, $\dfrac{\partial^2 u}{\partial y^2} = 0$

  (iii) at the edge of the boundary layer, y = δ , u = U_∞ and du/dy = 0

The general procedure followed in solving the equation (6.27) is the assumption of a reasonable velocity profile, u = u(y/δ), within the boundary layer such that it satisfies the boundary conditions and the shear stress at the wall. A polynomial that will satisfy the conditions is chosen as

$$\frac{U}{U_\infty} = C_0 + C_1 \frac{y}{\delta} + C_2 \left(\frac{y}{\delta}\right)^2 + C_3 \left(\frac{y}{\delta}\right)^3 \quad (6.28)$$

Substituting the four boundary conditions, Eq. (6.28) reduces to

$$\frac{U}{U_\infty} = \frac{3}{2}\left(\frac{y}{\delta}\right) - \frac{1}{2}\left(\frac{y}{\delta}\right)^3 \quad (6.29)$$

Applying this velocity profile to Eq. (6.27) we have

$$\rho U_\infty^2 \frac{d}{dx}\left[\int_0^\delta \left(\frac{3}{2}\frac{y}{\delta} - \frac{1}{2}\left(\frac{y}{\delta}\right)^3\right)\left(1 - \frac{3}{2}\frac{y}{\delta} + \frac{1}{2}\left(\frac{y}{\delta}\right)^3\right) dy\right] = \frac{3}{2}\mu \frac{U_\infty}{\delta}$$

which upon integration is

$$\frac{39}{280}\rho U_\infty^2 \frac{d\delta}{dx} = \frac{3}{2}\mu \frac{U_\infty}{\delta}$$

Seperating the variables, we obtain

$$(140/13)\,(\mu/\delta U_\infty)\,dx = \delta \cdot d\delta$$

which can be readily integrated to give the boundary layer thickness by

$$\delta/x = 4.64/(Re_x)^{1/2} \quad (6.30)$$

The agreement between this and Blasius result is quite good. The present value is approximately 7 percent lower than that obtained by Blasius. The local skin friction coefficient, using the Integral Momentum Method, is given by

$$C_{fx} = 0.646/(Re_x)^{1/2} \tag{6.31}$$

12. Derivation of Integral Boundary-layer Energy Equation—Assumptions and Solution

The Integral Energy Equation can be derived in a manner similar to that employed in developing the Integral Momentum Equation. We select a control volume as shown in Fig. 6.5. In order to simplify the problem, it is assumed that the physical properties are independent of the temperature and shear work due to the frictional forces along the wall are very small because the velocity of the flow field is very small.

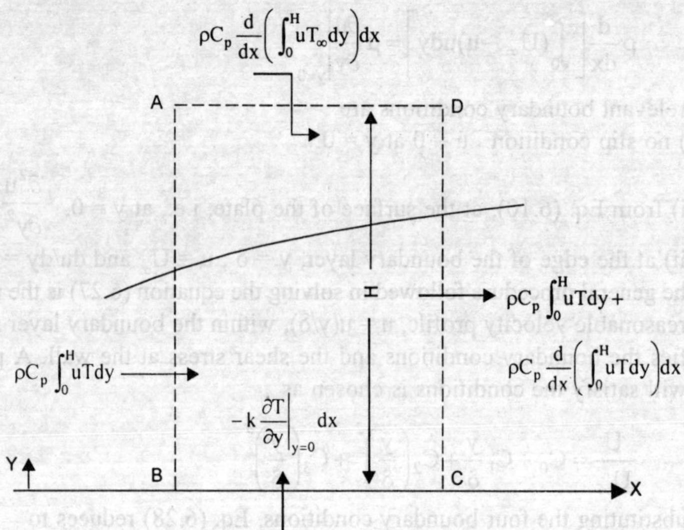

Fig. 6.5 Control volume for approximate energy analysis of a boundary layer

For the control volume ABCD, the energy flow rates are:

Face AB: $\rho C_p \displaystyle\int_0^H uTdy$

Face CD: $\rho C_p \displaystyle\int_0^H uTdy + \rho C_p \dfrac{d}{dx} \left(\displaystyle\int_0^H uTdy \right) dx$

Face AD: $\rho C_p \dfrac{d}{dx} \left(\displaystyle\int_0^H uT_\infty dy \right) dx$

Face BC: $-k \dfrac{\partial T}{\partial y} \Big|_{y=0} dx$

In the above expressions, the conductive fluxes along the surfaces AB and CD have been neglected because of the very small temperature gradient in the X-direction and the conductive heat flux along the surface AD is zero because the temperature of the fluid is constant outside the thermal boundary layer. Under steady state system, the net rate of heat flow by conduction to the control volume must be equal to the net efflux of energy by convection, and therefore we write

$$\frac{k}{\rho C_p}\frac{\partial T}{\partial y}\bigg|_{y=0} = \frac{d}{dx}\left(\int_0^H (T_\infty - T)u\,dy\right) = \frac{d}{dx}\left(\int_0^{\delta_t}(T_\infty - T)u\,dy\right) \tag{6.32}$$

The upper limit of integration can be changed to δ_t from H because $(T_\infty - T)$ vanishes for $y > \delta_t$, where δ_t is the thickness of thermal boundary layer. The boundary conditions on temperature are:

at $\qquad y = 0, \quad T = T_w \quad$ and $\dfrac{\partial^2 T}{\partial y^2} = 0;\qquad$ at $y = \delta_t, \quad T = T_\infty \quad$ and $\dfrac{\partial T}{\partial y} = 0$

Since there are four boundary conditions for temperature, we assume a temperature profile of the form:

$$\theta = (T - T_w)/(T_\infty - T_w) = a_0 + a_1\eta + a_2\eta^2 + a_3\eta^3, \qquad \text{where } \eta = y/\delta_t$$

Substituting the boundary conditions for temperature, the temperature profile is given by

$$\frac{(T - T_w)}{(T_\infty - T_w)} = \frac{3}{2}\cdot\left(\frac{y}{\delta_t}\right) - \frac{1}{2}\left(\frac{y}{\delta_t}\right)^3 \tag{6.33}$$

The velocity profile for the integral momentum equation is given by Eq. (6.29) Substituting Eq. (6.29) and Eq. (6.33) in Eq. (6.32) we get the right hand side as

$$(T_\infty - T_w)U_\infty \int_0^{\delta_t}\left(1 - \frac{3}{2}\left(\frac{y}{\delta_t}\right) + \frac{1}{2}\left(\frac{y}{\delta_t}\right)^3\right)\left(\frac{3y}{2\delta} - \frac{1}{2}\left(\frac{y}{\delta}\right)^3\right)dy$$

which can be integrated to obtain

$$(T_\infty - T_w)U_\infty\left[\frac{3}{4}\left(\frac{\delta_t^2}{\delta}\right) - \frac{3}{4}\left(\frac{\delta_t^2}{\delta}\right) - \frac{1}{8}\left(\frac{\delta_t^4}{\delta^3}\right) + \frac{3}{20}\left(\frac{\delta_t^2}{\delta} + \frac{\delta_t^4}{\delta^3}\right) - \frac{1}{28}\left(\frac{\delta_t^4}{\delta^3}\right)\right]$$

Let $\zeta = \dfrac{\delta_t}{\delta}$ the ratio of thermal to velocity boundary-layer thickness, the above expression can be written as

$$U_\infty(T_\infty - T_w)\delta\left(\frac{3\zeta^2}{20} - \frac{3\zeta^4}{280}\right) \tag{6.34}$$

Except for liquid metals which have Prandtl number <<1, it is reasonable to assume ζ equal to 1 or less than 1. In that case, the term involving ζ^4 in Eq. (6.34) can be neglected and Eq. (6.32) can be written as

$$\frac{3}{20}U_\infty(T_w - T_\infty)\zeta^2\frac{d\delta}{dx} = -\alpha\frac{dT}{dy}\bigg|_{y=0} = \frac{3}{2}\alpha\frac{T_w - T_\infty}{\delta\zeta}$$

or, $\qquad \dfrac{1}{10}U_\infty\zeta^3\delta\dfrac{d\delta}{dx} = \alpha$

since $\quad \delta d\delta/dx = (140/13)\,(\mu/\rho\,U_\infty),$

we get $\quad \zeta^3 = (10/10.75)\, \alpha/\upsilon$; and, $\delta_t = 0.976\, \delta\, Pr^{-1/3}$ $\qquad\qquad$ (6.35)

The rate of heat transfer by convection from the plate per unit area is

$$\dot{Q}/A = -k\,(dT/dy) \text{ at } y = 0 = -(3/2)\,(k/\delta_t)\,(T_\infty - T_w)$$

Substituting the values for δ_t we have

$$\dot{Q}/A = -(3/2)\,(k/x)\,(Pr^{1/3} \cdot Re_x^{1/2}) / (0.976 \times 4.64) \cdot (T_\infty - T_w)$$

$$= 0.33\, k/x \cdot Re_x^{1/2}\, Pr^{1/3}\, (T_w - T_\infty)$$

and, $\quad Nu_x = h_x\, x/k = 0.33\, Re_x^{1/2}\, Pr^{1/3}$ $\qquad\qquad$ (6.36)

This result is in agreement with the exact analysis by Pohlhausen except for the mathematical constant.

Example 6.3 $\quad$ Air at 25°C and 1 bar is flowing over a flat plate 25 cm wide with a velocity of 2.5 m/s. Calculate the following:
(a) boundary layer thickness, (b) local coefficient of friction, (c) local shearing stress, (d) thermal boundary layer thickness, and (e) the rate of heat transfer, when the length of the plate is 40 cm and the plate is at 80°C.

Solution: The properties of air at the mean temperature are:
$$\rho = 1.37 \text{ kg/m}^3, \quad \mu = 19 \times 10^{-6} \text{ Pa-s}, \quad C_p = 1.005 \text{ kJ /kgK}, \quad Pr = 0.7$$

Reynolds number, Re_x at $x = 0.4$ m $= \rho U_\infty\, x/\mu = 1.37 \times 2.5 \times 0.4/19 \times 10^{-6}$

$\therefore \qquad Re = 7.2 \times 10^4$, a laminar flow

Using the Blasius solution, the boundary layer thickness,
$$\delta = 0.4 \times 5.0/(7.2 \times 10^4)^{1/2} = 7.45 \text{ mm}$$

Local friction coefficient is, $C_{fx} = 0.664/(Re_x)^{1/2} = 0.664/(7.2 \times 10^4)^{1/2}$
$$= 0.00247$$

Local shearing stress, $= C_{fx} \times \rho\, U^2_\infty/2 = 0.00247 \times 1.37 \times (2.5)^2/2$
$$= 0.0106 \text{ N/m}^2.$$

Thermal boundary layer thickness $\delta_t = \delta / (Pr)^{1/3} = 7.45/(0.7)^{0.333}$
$$= 8.39 \text{ mm}.$$

Local heat transfer coefficient at $x = 0.4$ m is given by
$$Nu_x = 0.332\, Re_x^{1/2}\, Pr^{1/3} = 0.332 \times (7.2 \times 10^4)^{1/2}\, (0.7)^{0.333}$$
$$= 79.1$$

The thermal conductivity, $k = \mu C_p/Pr = 19 \times 10^{-6} \times 1005/0.7$
$$= 0.027 \text{ W/mK}$$

Therefore, the local heat transfer coefficient, $h_x = 5.34 \text{ W/m}^2\text{K}$

The average heat transfer coefficient, $\quad h = 2h_x = 10.68 \text{ W/m}^2\text{K}$

The rate of heat transfer, $Q = hA\,(T_w - T_\infty) = 10.68 \times (0.1)\,(80 - 25) = 58.74\,\text{W}.$

Example 6.4 $\quad$ Air at 25°C and 1 bar flows over a flat plate with a velocity 2.5 m/s. Calculate the thickness of the boundary layer at distances 18 and 32 cm from the leading edge of the plate. What would be the mass flow entering the boundary layer between $x = 18$ cm and $x = 32$ cm. μ at 25°C $= 19 \times 10^{-6}$ Pa-s.

Solution: The density of air at 25°C and 1 bar is, $\rho = p/RT$

$$\rho = 10^5/287 \times 298 = 1.169 \text{ kg/m}^3.$$

Reynolds number at 18 cm and 32 cm will be:

$$\text{Re} = 1.169 \times 2.5 \times 0.18/19 \times 10^{-6}; \quad \text{Re} = 1.169 \times 2.5 \times 0.32/19 \times 10^{-6}$$

$$= 2.77 \times 10^4, \text{ a laminar flow.} \qquad = 4.92 \times 10^4, \text{ a laminar flow}$$

From Fig. 6.4, the mass rate of flow at any station x is equal to

$$\dot{m} = \rho \int_0^\delta u\, dy \text{ where the velocity profile can be taken as}$$

$$\frac{u}{U_\infty} = \frac{3}{2}\left(\frac{y}{\delta}\right) - \frac{1}{2}\left(\frac{y}{\delta}\right)^3$$

The boundary layer thickness at the two stations are:

$$\delta_1 = 4.64 \times 0.18/(2.77 \times 10^4)^{1/2} = 5 \times 10^{-3} \text{ m, and}$$

$$\delta_2 = 4.64 \times 0.32/(4.92 \times 10^4)^{1/2} = 6.69 \text{ mm}$$

Now $\quad \dot{m} = \int_0^\delta \rho u\, dy = \int_0^\delta \rho U_\infty \left((3/2)\,(y/\delta) - \frac{1}{2}(y/\delta)^3 \right) dy = \delta 5 \rho U_\infty /8$

$\therefore \qquad \dot{m}_1 = 5 \times 1.169 \times 2.5 \times 5 \times 10^{-3}/8 = 7.3 \times 10^{-2}$ kg/s per unit width of plate

$$\dot{m}_2 = 5 \times 1.169 \times 2.5 \times 6.69 \times 10^{-3}/8 = 9.77 \times 10^{-2} \text{ kg/s}$$

and the mass entering the boundary layer between the two sections is

$$\dot{m}_2 - \dot{m}_1 = (9.77 - 7.3) \times 10^{-2} = 2.47 \times 10^{-2} \text{ kg/s.}$$

13. Empirical Relation for Flat Plates having (a) Unheated Initial Length and (b) Constant Heat Flux

The solution of the Prandtl boundary-layer energy equation (6.15) and given by Eq. (6.21) assumes that the plate is heated over its entire length and is maintained at constant wall temperature T_w. When the plate has an unheated initial length, as shown in Fig. 6.6, the Eq. (6.21) is modified as

$$\text{Nu}_x = 0.332\, \text{Re}_x^{1/2}\, \text{Pr}^{1/3}\, \{1 - (x_i/x)^{3/4}\}^{-1/3} \tag{6.37}$$

There are many practical situations where the surface heat flux is essentially constant over the flat plate, such as heating by solar energy. In such cases the local Nusslet number is obtained from:

$$\text{Nu}_x = 0.453\, \text{Re}^{1/2}\, \text{Pr}^{1/3} \tag{6.38}$$

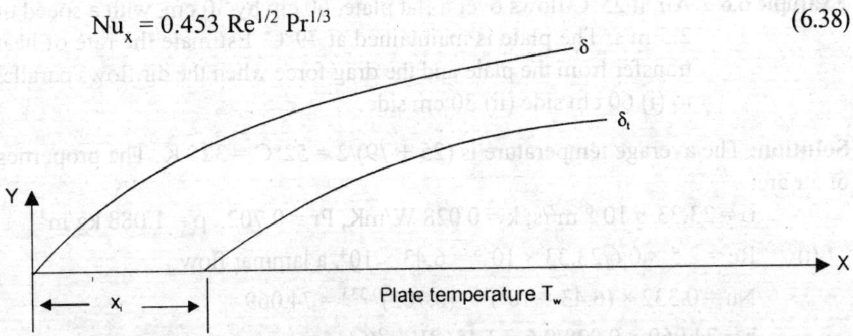

**Fig. 6.6 Thermal and velocity boundary layer on a flat plate
with an unheated initial length**

Example 6.5   A flat plate 50 cm by 50 cm is placed in an air stream at 27°C and 1 bar with air speed of 3 m/s. Calculate the local heat transfer coefficient at x = 25 cm and x = 50 cm when it receives solar energy at a rate of 1100 W/m².

Solution: The physical properties of the fluid are to be evaluated at the mean temperature. Since the temperature of the plate is not known, a trial and error solution is used.

First Trial: The temperature of air is 27 + 273 = 300K. The physical properties at this temperature are:

$\upsilon = 15.69 \times 10^{-6}$, $k = 0.02674$ W/mK,   Pr = 0.708.

at x = 25 cm,   Re = 3 × 0.25/15.69 × 10^{-6} = 4.78 × 10^4

$Nu = 0.453\ Re^{1/2}\ Pr^{1/3} = 0.453\ (4.78 \times 10^4)^{1/2}\ (0.708)^{0.333} = 88.273$

$T_w - T_\infty = (Q/A)/h = 1100 \times 0.25/(88.273 \times 0.02674) = 116.5$K or °C

Second Trial: Let the mean temperature be (116.5 + 27)/2 = 71.75°C = 345K. The physical properties at this temperature are:

$\upsilon = 20.16 \times 10^{-6}$, k = 0.03 W/mK, Pr = 0.698

$Re = 3 \times 0.25/20.16 \times 10^{-6} = 3.72 \times 10^4$

and $T_w - T_\infty = 1100 \times 0.25/[0.03 \times 0.0453 \times (3.72 \times 10^4)^{1/2}\ (0.698)^{0.333}]$
$= 118.26°C$

Thus taking the temperature difference as (116.5 + 118.26)/2 ≡ 117.4, local convective heat transfer coefficient at x = 0.25 m,

$h = 100/117.4 = 9.37$ W/m²K.

Repeating the same procedure, the convective heat transfer coefficient at the trailing edge, x = 0.50m, h = 6.67 W/m²K.
Since the temperature of the surface is varying the average convective heat transfer coefficient for the entire plate be obtained by using the relation

$$Nu_L = 0.6795\ Re_L^{1/2}\ Pr^{1/3} \tag{6.39}$$

and the properties should be evaluated at the average temperature given by

$$T_w - T_\infty = (\dot{Q}/A)/h_{av}.$$

Example 6.6   Air at 25°C flows over a flat plate, 60 cm by 30 cm, with a speed of 2.5 m/s. The plate is maintained at 79°C. Estimate the rate of heat transfer from the plate and the drag force when the air flows parallel to (i) 60 cm side (ii) 30 cm side.

Solution: The average temperature is (25 + 79)/2 = 52°C = 325 K. The properties of air are:

$\upsilon = 23.33 \times 10^{-6}$ m²/s; k = 0.028 W/mK, Pr = 0.702, $\rho = 1.088$ kg/m³

(i)   Re = 2.5 × 0.6/23.33 × 10^{-6} = 6.43 × 10^4, a laminar flow

$Nu = 0.332 \times (6.43 \times 10^4)^{1/2}\ (0.702)^{0.333} = 74.069$

$h = 74.069 \times 0.028/0.6 = 3.456$ W/m²K

The average convective heat transfer coefficient, $h_{av} = 2h = 6.912$ W/m^{2}K
Rate of heat loss = h × area × temperature difference

$$= 6.912 \times (0.6 \times 0.3) \times (79 - 25) = 67.18 \text{ W}$$

Average skin friction coefficient = $1.328/(6.43 \times 10^4)^{1/2} = 5.237 \times 10^{-3}$

Drag force = shear stress × area = $5.237 \times 10^{-3} \times 1.088 \times 6.25 \times (2 \times 0.6 \times 0.3)/2$
$$= 6.41 \times 10^{-3} \text{ N.}$$

(ii)   Re = $2.5 \times 0.3/23.33 \times 10^{-6} = 3.215 \times 10^4$, a laminar flow

$$\text{Nu} = 0.332 \times (3.215 \times 10^4)^{1/2} \ (0.702)^{0.333} = 52.91$$

The average heat transfer coefficient, $h_{av} = 2 \times 52.91 \times 0.028/0.3$
$$= 9.877 \text{W/m}^2\text{K}$$

Rate of heat transfer = $9.877 \times (0.6 \times 0.3) \times 54 = 96$ W

Average skin friction coefficient = $1.328/(3.215 \times 10^4)$
$$= 7.406 \times 10^{-3}$$

Drag force = $7.406 \times 10^{-3} \times 1.088 \times 6.25 \times (2 \times 0.6 \times 0.3)/2 = 9.065 \times 10^{-3}$ N.

Pohlhausen solution for the evaluation of rate of heat transfer from the laminar flow of a fluid over a plane surface, Eq. (6.21) is applicable to fluids having Prandtl number lying between 0.5 and 50. Churchill and Ozoe have correlated a large amount of data to give the following relation for laminar flow on an isothermal flat plate:

$$\text{Nu}_x = (0.3387 \ \text{Re}_x^{1/2} \ \text{Pr}^{1/3})/[1 + (0.0468/\text{Pr})^{2/3}]^{1/4} \tag{6.40}$$

When the plane surface receives heat energy at a constant rate, the constant 0.3387 should be changed to 0.4637 and 0.0468 be changed to 0.0207. The fluid properties are to be evaluated at the mean temperature as usual.

Example 6.7   Engine oil at 25°C is flowing over a 25 cm square plate at a velocity of 1 m/s. The plate is being maintained at a uniform temperature of 55°C. Calculate the heat transfer from the plate.

Solution: The properties of the engine oil at the mean temperature $(25 + 55)/2$ = 40°C are:

$$\rho = 875 \text{ kg/m}^3, \quad \nu = 0.00024 \text{ m}^2/\text{s}, \quad k = 0.144 \text{ W/mK}, \quad \text{Pr} = 2870$$

The Reynolds number,   Re = $1.0 \times 0.25/0.00024 = 1041.67$, a laminar flow

Since the Prandtl number is very large, we make use of Eq. (6.40)

$$\text{Nu} = (0.3387 \times (1041.67) \ (2870)^{1/3}/(1 + (0.0468/2870)^{2/3})^{1/4}$$
$$= 154.91$$

Local convective heat transfer coefficient, h = $154.91 \times 0.144/0.25$
$$= 89.23 \text{ W/m}^2\text{K}$$

It is evident from Eq (6.40) that the local heat transfer coefficient h_x varies as $x^{-1/2}$, the average heat transfer coefficient can be written as twice of the local value.

Therefore, $h_{av} = 2 \times 89.23 = 178.46$ W/m^{2}K, and the rate of heat transfer,

$$\dot{Q} = 178.46 \times (0.25 \times 0.25) \ (55 - 25) = 334.61 \text{ W.}$$

Example 6.8   Air at 25°C flows with a velocity of 3 m/s over a plate 1 m long 80 cm wide and 30 cm thick. The top surface of the plate is being maintained at 95°C. If the thermal conductivity of the plate is 10 W/mK, calculate the following: (a) rate of heat loss from the plate, and (b) temperature of the bottom surface of the plate. The properties of air at the mean temperature are:

$$\rho = 1.1 \text{ kg/m}^3, k = 0.029 \text{ W/mK}, Pr = 0.7, v = 18.9 \times 10^{-6} \text{ m}^2\text{/s}.$$

Solution: Reynolds number at the trailing edge,
$$Re_L = 3 \times 1/18.9 \times 10^{-6} = 1.587 \times 10^5$$
The flow over the plate is laminar, as such, the average Nusselt number is

$$Nu = 0.664 \; Re_L^{\frac{1}{2}} \; Pr^{1/3} = 0.664 \; (1.587 \times 10^5)^{\frac{1}{2}} \; (0.7)^{1/3} = 234.9$$
$$\therefore \qquad h = 234.9 \times 0.029/1.0 = 6.812 \text{ W/m}^2\text{K}$$
Rate of heat transfer, $\dot{Q} = 6.812 \times (1.0 \times 0.8)\,(95 - 25) = 381.47$ W
The heat lost by the plate from the top surface must be coming from the bottom surface by conduction, and therefore,

$$\dot{Q} = k \, A \, (T_o - 95)/\text{thickness; where } T_o \text{ is the temperature of the}$$
bottom surface of the plate.

$$T_o = 95 + 381.47 \times 0.3/(10 \times 1.0 \times 0.8) = 109.3°C.$$

Example 6.9   Air at 25°C flows over a flat plate, 1 m long and 0.5 m wide with a velocity of 3 m/s. The velocity profile in the boundary is approximated by the relation $u/U_\infty = \sin(\pi y/2\delta)$ If $\rho = 1.25 \text{ kg/m}^3$ and $v = 15 \times 10^{-6} \text{ m}^2\text{/s}$, estimate the boundary layer thickness and the shear stress at the trailing edge. Also estimate the drag force exerted by air on one side of the plate.

Solution: The momentum integral equation is

$$U_\infty^2 \frac{d}{dx} \int_0^\delta \frac{u}{U_\infty}\left(1 - \frac{u}{U_\infty}\right) = \frac{\tau_0}{\rho} = \frac{\mu(du/dy)}{\rho} \quad \text{at } y = 0$$

Substituting the given expression for the velocity distribution in the momentum integral equation, we have

$$U_\infty^2 \frac{d}{dx} \int_0^\delta \sin(\pi/2)\cdot(y/\delta)(1 - \sin \pi y/2\delta)dy = \frac{\mu d}{dy}[U_\infty \sin(\pi y/2\delta)] \text{ at } y = 0$$

$$\int_0^\delta \sin(\pi/2)\cdot(y/\delta)(1 - \sin \pi y/2\delta)dy \text{ can be written as:}$$

$$\int_0^\delta \sin(\pi y/2\delta)dy - \int_0^\delta [\sin^2(\pi y/2\delta)]dy$$

$$= \left[-\frac{2\delta}{\pi}\cdot\cos(\pi y/2\delta)\right]_0^\delta - \frac{1}{2}\int_0^\delta (1 - \cos \pi y/\delta)dy \text{ which upon integration}$$

gives: $2\delta/\pi - \frac{1}{2}[\delta - (\delta/\pi)\sin\pi] = 0.137\delta$

The R.H.S. of the equation is $\mu U_\infty \pi / 2\delta\rho$

Thus, $U_\infty^2 \dfrac{d}{dx}(0.137\delta) = \mu U_\infty \pi / 2\delta\rho$

or, $\delta \dfrac{d\delta}{dx} = \pi\mu /(2 \times 0.137\rho U_\infty) = \pi\mu / 0.274\rho U_\infty$

Integrating both sides we get, $\delta^2 \left[\dfrac{\pi}{0.137} \cdot \dfrac{\mu x}{\rho U_\infty} \right] + C$

Since at $x = 0$, $\delta = 0$,   we have $C = 0$.
and, $(\delta/x)^2 = 23\ \mu/\rho x U_\infty = 23/Re_x$

or, $\dfrac{\delta}{x} = \dfrac{4.8}{Re_x^{0.5}}$

Shear stress at the surface, $\tau_0 = \mu\ (du/dy)y = 0 = \mu U_\infty \pi / 2\delta$.

Since, $\delta = 4.8x / Re_x^{1/2}$

    $\tau_0 = (\mu\ U_\infty \pi\ Re^{1/2})/(2 \times 4.8x) = C_{fx} \cdot \rho U_\infty^2/2$

Thus, $C_{fx} = 0.654/Re^{1/2}$ and $\tau_0 = C_{fx}\ \rho\ U^2/2$

Reynolds number, Re, at the trailing edge $= 3 \times 1/15 \times 10^{-6} = 2 \times 10^5$

and $C_{fx} = 0.654/(2 \times 10^5)^{1/2} = 1.46 \times 10^{-3}$

Boundary layer thickness at the trailing edge,

    $\delta = 4.8 \times 1.0/(2 \times 10^5)^{1/2} = 0.01\ m = 10\ mm$

Shear stress $= C_{fx}\ \rho\ U_\infty^2/2 = 1.46 \times 10^{-3} \times 1.25 \times 9/2 = 8.21 \times 10^{-3}\ N/m^2$

Since the flow is laminar, the average shear stress $= 16.42 \times 10^{-3}\ N/m^2$

and the drag force on one side of the plate $= 16.42 \times 10^{-3} \times (1.0 \times 0.5)$

    $= 8.21 \times 10^{-3}\ N.$

(The thickness of the boundary layer at a distance x from the leading edge and the average skin friction coefficient for flow over a flat plate can be calculated for different velocity profiles. The results are tabulated in Table 6.1.)

Table 6.1 Boundary layer parameters for different velocity profiles

Velocity Profile	Boundary Condition		δ/x	C_f
	at y = 0	*at y = δ*		
1. $u/U_\infty = y / \delta$	$u = 0$	$u = U_\infty$	$3.64/Re_x^{1/2}$	$1.155/Re_L^{1/2}$
2. $u/U_\infty = \sin(\pi y/2\ \delta)$	$u = 0$	$u = U_\infty$	$4.8/Re_x^{1/2}$	$1.31/Re_L^{1/2}$
3. $u/U_\infty = 2(y/\delta)-(y/\delta)^2$	$u = 0$	$u = U_\infty$ $\partial u/\partial y = 0$	$5.48/Re_x^{1/2}$	$1.46/Re_L^{1/2}$
4. $u/U_\infty = (3/2)(y/\delta)-\tfrac{1}{2}(y/\delta)^3$	$u = 0$ $\partial^2 u/\partial y^2 = 0$	$u = U_\infty$ $\partial u/\partial y = 0$	$4.64/Re_x^{1/2}$	$1.292/Re_L^{1/2}$
5. Blasius exact solution			$5.0/Re_x^{1/2}$	$1.328/Re_L^{1/2}$

14. Laminar Flow through Tubes—Entry Length and Fully Developed Flow

Osborne Reynolds, in 1889, performed a series of experiments with fluids flowing through tubes and he observed that when Reynolds number, $Re = \rho VD/\mu$, (ρ, the density of the fluid, V, the average velocity of flow, D, the internal diameter of the tube and μ, the viscosity of the fluid) is less than 2300, the flow inside the circular tubes is always laminar.

When a fluid enters a tube, it has a uniform velocity at entrance and as the fluid flows along the tube, the velocity profile changes and is a function of the tube radius, r, shown in Fig. 6.7. After a distance, L_e, the velocity profile does not change, i.e., we have a 'fully-developed flow'. This distance L_e, is called the 'entry length' and is given by

$$L_e/D = 0.058\ Re \tag{6.41}$$

whereas, the distance from the inlet at which the temperature profile approaches its fully developed shape is given by

$$L_e/D = 0.05\ Re\ Pr$$

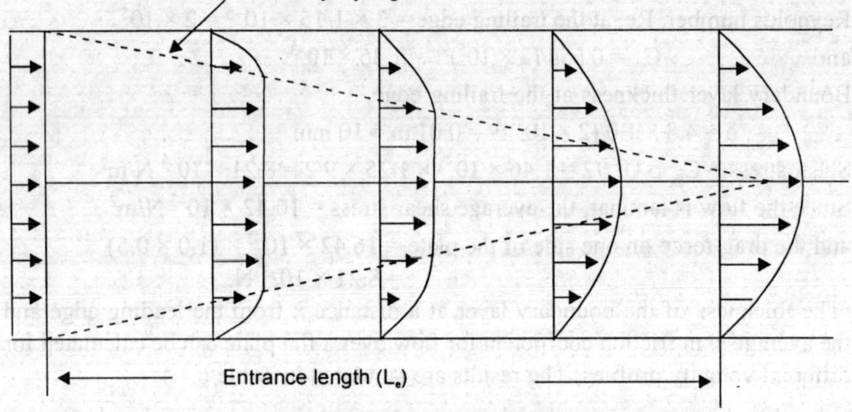

Fig. 6.7 Development of a laminar flow through a tube

15. An Expression for Pressure Loss in Hagen-Poiseuille Flow

Let us consider a fully developed laminar flow (known as Hagen-Poiseuille flow) through a tube of internal radius R. We choose a cylindrical control volume, shown in Fig. 6.8, such that the axis of the tube is the center-line of the control volume and U is the maximum velocity of the fluid along the axis of the tube. Since there is no acceleration in the flow, the pressure forces are balanced by the viscous-shear forces.

or, $p(\pi r^2) - (p + dp)\pi r^2 = -\mu\ 2\pi r\ (du/dr)\ dx,$

and $-dp\ \pi\ r^2 = -2\pi\mu\ (du/dr)\ r\ dx$

$$du = \frac{1}{2\mu}(dp/dx)\ r\ dr$$

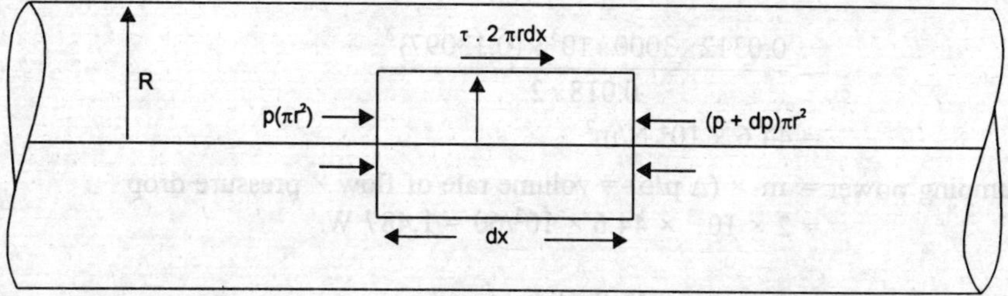

Fig. 6.8 Cylindrical control volume for laminar flow in a tube

Upon integration, $u = \dfrac{1}{4\mu}(dp/dx)\,r^2 + C$

when r = 0, u = U, therefore, C = U

and $u = U + \dfrac{1}{4\mu}(dp/dx)\,r^2$

at r = R, u = 0; therefore $u = -\dfrac{1}{4\mu}(dp/dx)\,(r^2 - R^2)$

and, $u/U = 1 - r^2/R^2$ (6.42)

Thus, the velocity profile for a fully developed laminar flow in a circular tube is parabolic.

The average velocity, $V = \dfrac{1}{\pi r^2}\displaystyle\int_0^R u(2\pi r)dr$

$= U/2 = R^2/8\mu\,(-dp/dx);$

and $-(dp/dx) = 8\,\mu\,V/R^2$

If the pressure gradient (dp/dx) is constant over a length L and we define a friction factor 'f' as: $f = D\,(-dp/dx)/(\tfrac{1}{2}\,\rho\,V^2)$

or, $f = D\,(8\mu\,V/R^2)\,/\,\tfrac{1}{2}\,\rho V^2 = 64/Re$ (6.43)

and, $(p_1 - p_2)/\rho\,g = f\,L\,V^2/2\,g.D$ (6.44)

Example 6.10 Water flows through a 18 mm inside diameter pipe at a rate of 2 litres/min at room temperature. Estimate the entry length, the required pressure drop and the pumping power to maintain the flow if the length of the pipe is 3000 metres.

Solution: The cross-sectional area of flow = $3.142\,(0.018)^2/4 = 2.545 \times 10^{-4}$ m²

Average velocity, V = volume rate of flow/cross-sectional area

$= 2 \times 10^{-3}/(60 \times 2.545 \times 10^{-4}) = 0.13097$ m/s

The dynamic viscosity of water at 20°C = 1.15×10^{-3} Pa-s

Reynolds number, Re $= 10^3 \times 0.13097 \times 0.018/1.15 \times 10^{-3}$

$= 2050$, a laminar flow

Entry length, $L_e = 0.058 \times (Re) \times D = 0.058 \times 2050 \times 0.018 = 2.14$ m.

For laminar flow, the friction factor, f = 64/Re = 64/2050 = 0.0312

Pressure drop, $\Delta p = f \times L/D \times (\rho\,V^2/2)$

$$= \frac{0.0312 \times 3000 \times 10^3 \times (0.13097)^2}{0.018 \times 2}$$

$$= 44.6 \times 10^3 \text{ N/m}^2$$

Pumping power $= \dot{m} \times (\Delta p/\rho) = $ volume rate of flow $\times$ pressure drop

$$= 2 \times 10^{-3} \times 44.6 \times 10^3/60 = 1.487 \text{ W.}$$

16. The Bulk Temperature—Definition

The bulk temperature T_b is the mean temperature of the fluid at a given cross-section of the tube, i.e., if the fluid from the cross-section is collected in a vessel for a short period of time and thoroughly mixed without allowing it to exchange energy with the surroundings, the resulting temperature of the fluid is the 'bulk temperature'.

or, $\quad T_b = \dfrac{\text{(Total thermal energy crossing a section of the tube per unit time)}}{\text{(Heat capacity of the fluid crossing the same section per unit time)}}$

$$= \int_0^R \rho(C_p T)(2\pi\, r\, u\, dr) / \int_0^R (\rho\, C_p\, 2\pi\, r\, u\, dr) \qquad (6.45)$$

17. Fully Developed Temperature Profile—Effect of Temperature Difference

When a fluid at a uniform temperature T is flowing through a tube whose walls receive a uniform heat flux q_w, its bulk temperature increases and the temperature profile of the fluid will undergo a change as shown in Fig. 6.9. After a certain distance from the entrance is reached, the temperature profile does not change and a fully developed temperature profile is said to be established. This distance is called the 'thermal entry length'. It should be mentioned that after the thermal entry length, the difference between the temperature of the fluid at the tube wall T_w and the bulk temperature T_b will remain constant, as shown in Fig. 6.9. Thus,

$$\frac{\partial}{\partial x}\left[(T_w - T)/(T_w - T_b)\right] = 0, \text{ for fully developed temperature profile.}$$

(When there is a large temperature difference in the flow, there would be an appreciable change in the fluid properties between the wall of the tube and the central flow. These property variations may be observed by a change in the velocity profile as shown in Fig. 6.10. The velocity profile for isothermal flow will be different from that when the liquids are being heated or cooled because of the change in the viscosity of liquid with temperature.)

18. Energy Analysis for Flow through a Tube—Evaluation of 'h'

We choose an annular control volume, Fig 6.11, at radius r, thickness dr and length dx. The principle of conservation of energy requires that under steady state conditions, heat entering the control volume from the radial and axial directions must be equal to the net heat leaving in the axial direction. Since the velocity is purely axial, there would be no convection in the radial direction.

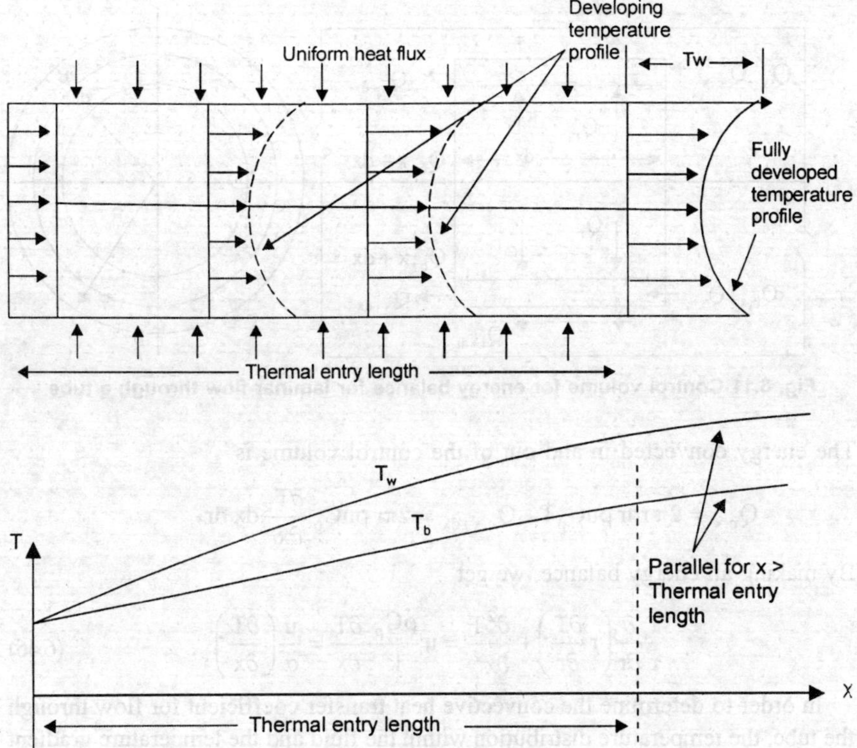

**Fig. 6.9 Development of temperature profile for laminar flow in a tube
with uniform heat flux**

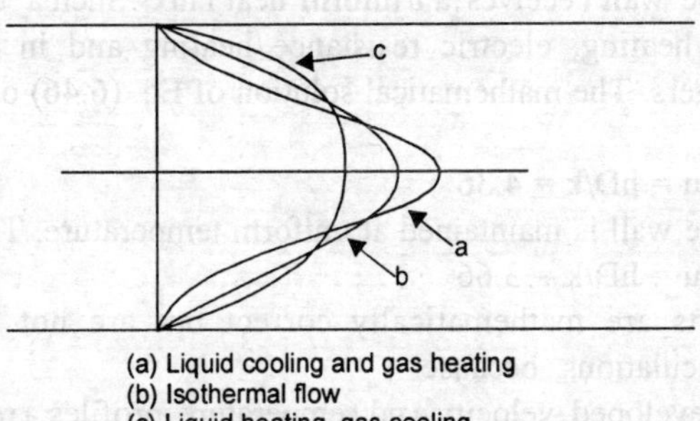

(a) Liquid cooling and gas heating
(b) Isothermal flow
(c) Liquid heating, gas cooling

Fig. 6.10 Influence of heating on velocity profile in laminar tube flow

Heat conducted in the radial direction $= = -k(2\pi\, r\, dx)\dfrac{\partial T}{\partial r} = \dot{Q}_r$

Heat conducted out in the radial direction,

$$\dot{Q}_{r+dr} = -k(2\pi\, r\, dx)\frac{\partial T}{\partial r} - 2\pi k\frac{\partial}{\partial r}\left(r\frac{\partial T}{\partial r}\right)dr\, dx$$

Rate of axial conduction of heat into and out of the control volume is

$$\dot{Q}_x = -k(2\pi\, r\, dr)\frac{\partial T}{\partial x}\, ; \quad \dot{Q}_{x+dx} = -k(2\pi\, r\, dr)\frac{\partial T}{\partial x} - 2\pi r k\frac{\partial^2 T}{\partial x^2}\cdot dr\, dx$$

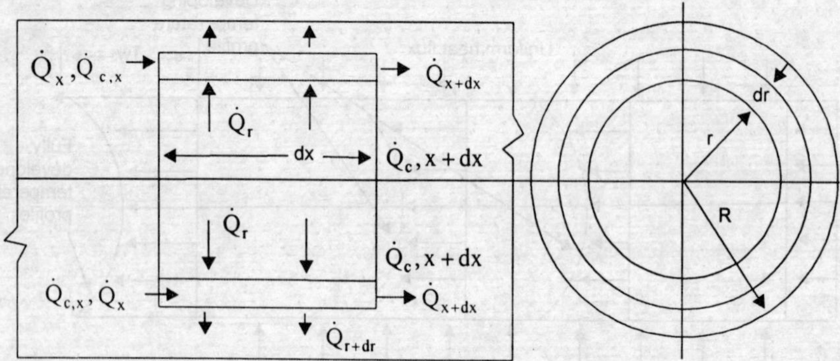

Fig. 6.11 Control volume for energy balance for laminar flow through a tube

The energy convected in and out of the control volume is

$$\dot{Q}_{c,x} = 2\pi\,rdr\,\rho uC_pT; \quad \dot{Q}_{c,x+dx} = 2\pi r\,\rho uC_p\,\frac{\partial T}{\partial x}\cdot dx\,dr$$

By making an energy balance, we get

$$\frac{1}{r}\frac{\partial}{\partial r}\left(r\frac{\partial T}{\partial r}\right) + \frac{\partial^2 T}{\partial x^2} = u\frac{\rho C_p}{k}\frac{\partial T}{\partial x} = \frac{u}{\alpha}\left(\frac{\partial T}{\partial x}\right) \tag{6.46}$$

In order to determine the convective heat transfer coefficient for flow through the tube, the temperature distribution within the fluid and the temperature gradient at the wall has to be obtained using the relevant boundary conditions. The possible thermal boundary conditions are:

(a) The tube wall receives a uniform heat flux. Such a condition arises in radiant heating, electric resistance heating and in counter-flow heat exchangers. The mathematical solution of Eq. (6.46) obtained by Nusselt gives:

$$Nu = hD/k = 4.36 \tag{6.47}$$

(b) The tube wall is maintained at uniform temperature. The solution is

$$Nu = hD/k = 3.66 \tag{6.48}$$

These results are mathematically correct but are not suitable for most engineering calculations, because:

(i) Fully developed velocity and temperature profiles are seldom found in laminar tube flow and the thermal entry length is given by

$$(L_e)_t = 0.05\,Re\,Pr\,D \tag{6.49}$$

(ii) The flow through the tube is laminar with liquids having high viscosities. Since viscosity of a liquid is highly dependent on temperature, the parabolic velocity profile is seldom obtained, Fig. 6.10.

(iii) It cannot be said that the temperature and velocity profiles develop simultaneously and are fully developed because highly viscous liquids have high Prandtl numbers and the thermal profile would develop at a slower rate than the velocity profile.

Hausen has recommended that for fully developed laminar flow through tubes at constant wall temperature, the heat transfer coefficient can be obtained by the relation

$$Nu = hD/k = 3.66 + (0.0668 \, (D/L)Re \, Pr)/(1 + 0.04 \, (D/L)Re \, Pr)^{2/3}) \qquad (6.50)$$

where h is the average heat transfer coefficient over the entire length of the tubes.

Sieder and Tate have proposed a simpler empirical relation for laminar heat transfer in tubes,

$$Nu = 1.86 \, (Re \, Pr)^{1/3} \, (D/L)^{1/3} \, (\mu/\mu_w)^{0.14} \text{ for } Re \, Pr \, D/L > 10 \qquad (6.51)$$

Colburn has suggested that heat transfer coefficient can be obtained by making an analogy between fluid friction and heat transfer and by the relation

$$St \, Pr^{2/3} = f/8 \qquad (6.52)$$

where the friction factor 'f' is evaluated by using Eq. (6.44) or Eq. (6.43).

If the channel through which the fluid is flowing is not circular in cross-section, the heat transfer correlation should be based on the hydraulic diameter D_H, defined as

$$D_H = 4 \, A/P$$

where A is the cross-sectional area of flow and P is the wetted perimeter.

Shah and London have compiled the heat transfer and fluid friction data for fully developed laminar flow in ducts with a variety of flow cross-sections as tabulated in Table 6.2.

Kays and Sellars, Tribus and Klein have calculated the local and average Nusselt number for laminar entrance regions of circular tubes when the velocity profile is fully developed. Their results can be approximated by the relation

$$Nu = 1.30 \, (Re \, Pr/ \, x/D)^{1/3}, \text{ for constant wall heat flux} \qquad (6.53a)$$

$$\text{and, } Nu = 1.077 \, (Re \, Pr/ \, x/D)^{1/3}, \text{ for constant wall temperature} \qquad (6.53b)$$

Eq. (6.53) is applicable if Graetz Number, Gz = Re. Pr.D/x > 200.
(x is the length of the tube)

Example 6.11 The velocity and temperature distribution for laminar flow through a circular tube, inner radius 5 cm, are given by
$u = 2(1 - 0.04 \, r^2)$ cm/s, and $T = 60(1 + 2r)°C$
where the distance r is measured from the axis. Estimate the average velocity and the bulk temperature of the fluid.

Solution: The average velocity is given by

$$V = (2/R^2) \int_0^R u \, r \, dr = (2/R^2) \int_0^R 2(1 - 0.04 \, r^2) \, r \, dr$$

$$= (4/R^2)(R^2/2 - 0.04 \, R^4/4) = 2 - 0.04R^2 = 1.0 \, cm/s$$

Table 6.2 Heat transfer and fluid friction for fully developed laminar Flow in ducts of various cross-sections

Nu_{H1} = Average Nusselt number for uniform heat flux in flow direction and uniform wall temperature at particular flow cross section

Nu_{H2} = Average Nusselt number for uniform heat flux both in flow direction and around periphery

Nu_T = Average Nusselt number for uniform wall temperature

$f \cdot Re$ = Product of friction factor and Reynolds number

Geometry ($L/D_h > 100$)		Nu_{H1}	Nu_{H2}	Nu_T	$f\,Re$
Triangle ($60°$), $2a$, $2b$	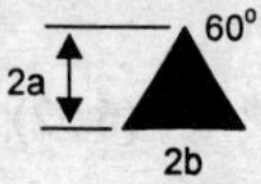 $\dfrac{2b}{2a} = \dfrac{\sqrt{3}}{2}$	3.111	1.892	2.47	13.333
Square, $2a$, $2b$	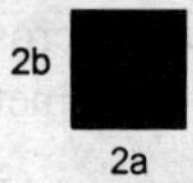 $\dfrac{2b}{2a} = 1$	3.608	3.091	2.976	14.227
Hexagon		4.002	3.862	3.34	15.054
Rectangle, $2a$, $2b$	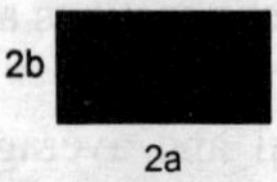 $\dfrac{2b}{2a} = \dfrac{1}{2}$	4.123	3.017	3.391	15.548
Circle		4.364	4.364	3.657	16.000
Rectangle, $2a$, $2b$	 $\dfrac{2b}{2a} = \dfrac{1}{4}$	5.099	4.35	3.66	18.700
Ellipse, $2a$, $2b$	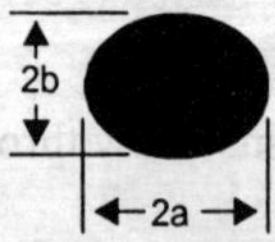 $\dfrac{2b}{2a} = 0.9$	5.331	2.930	4.439	18.233
Rectangle, $2a$, $2b$	 $\dfrac{2b}{2a} = \dfrac{1}{8}$	6.490	2.904	5.597	20.585
Parallel plates	$\dfrac{2b}{2a} = 0$	8.235	8.235	7.541	24.000
Insulated	 $\dfrac{b}{a} = 0$	5.385	-	4.861	24.000

From Eq. (6.45), the bulk temperature is

$$T_b = \left(\int_0^R u\, T\, r\, dr \right) / \int_0^R u\, r\, dr \right), \text{ assuming that } \rho \text{ and } C_p \text{ are const.}$$

$$= \left(\int_0^R 2(1 - 0.04r^2)60[(1 + 2r)\, r\, dr] \right) / \left(\int_0^R 2[1 - 0.04r^2)\, r\, dr] \right)$$

$$= 253.33^\circ C.$$

Example 6.12 Water, flowing through a 10 mm inner diameter tube, is heated from 2 5°C to 5 0°C. Calculate the mass flow rate of water such that the flow remains laminar. If the tube receives solar heat energy at a rate of 400 W/m length of the tube, estimate the length of the tube and the average heat transfer coefficient.

Solution: The average bulk temperature = (25 + 50)/2 = 37.5°C.
Properties of water at this temperature are:

$C_p = 4.174$ kJ/kgK; $\rho = 993$ kg/m³;
$k = 0.63$ W/mK;   Pr = 4.53, $\mu = 6.82 \times 10^{-4}$ Pa-s

Assuming that the flow remains laminar for Re = 2000, the average velocity,

$$V = 2000 \times 6.82 \times 10^{-4}/(993 \times 0.01) = 0.137 \text{ m/s}$$

Mass flow rate = density × area of flow × velocity

$$= 993 \times 3.142 \times 0.005 \times 0.005 \times 0.137 \times 3600$$

$$\dot{m} = 38.47 \text{ kg/hour}$$

Let L be the length of the tube. The quantity of heat energy received is 400 L watt. By making an energy balance,

$$400\, L = \dot{m}\, C_p\, (50 - 25) = 38.47 \times 4174 \times 25 / 3600$$

Therefore,   L = 2.79 m.
For constant heat flux condition, the average heat transfer coefficient is given by Eq. (6.47), or,

$$h = 4.36 \times 0.63/0.01 = 274.68 \text{ W/m}^2\text{K}.$$

Example 6.13 What would be the temperature of the inner wall of the tube of Ex. 6.12 if the temperature of the wall is maintained at a constant value.

Solution: Since the water is flowing at the rate of 38.47 kg/hour and is being heated from 25°C to 50°C, the quantity of heat supplied is

$$\dot{m} \times C_p \times (\Delta T) = 400 \times 2.79 = 1116 \text{ W}.$$

Using Eq. (6.48), the average heat transfer coefficient is,

$$h = 3.36 \times 0.63/0.01 = 211.68 \text{ W/m}^2\text{K},$$

and, $1116 = h \times \text{area} \times (\Delta T) = 211.68 \times 3.142 \times 0.01 \times 2.79 \times (\Delta T)$

or $(T_w - 37.5) = 167.795; \ T_w = 167.795 + 37.5$

$$= 207.29^\circ C.$$

The convective heat transfer coefficient at the exit end of the tube is,

$$h = 1116/ [3.142 \times 0.01 \times 2.79 \times (207.29 - 50)]$$
$$= 80.94 \text{ W/m}^2\text{K}.$$

Example 6.14 Water at 50°C flows through a 25 mm inner diameter tube. Calculate the exit water temperature if the length of the tube is (i) 2m (ii) 0.35 and the wall temperature is maintained at 70°C, and the mean velocity of flow is 2 cm/s.

Solution: The properties of the fluid are to be evaluated at the mean temperature. Since the temperature at exit is not known, as an approximation, we evaluate the properties at 50°C.

$$\rho = 990 \text{ kg/m}^3, \quad k = 0.642, \quad \mu = 5.53 \times 10^{-4} \text{ Pa-s},$$

$$Pr = 3.55, \qquad\qquad C_p = 4181 \text{ J/kgK},$$

$$Re = \rho \, V \, D/\mu = 990 \times 0.02 \times 0.025/5.53 \times 10^{-4}$$
$$= 895.1, \text{ a laminar flow}$$

When the length of the tube is 2 metre, Graetz number = Re.Pr.D/x
$$= 895.1 \times 3.55 \times 0.025/2$$
$$= 39.72 > 10$$

Therefore, we can use Eq. (6.51)

or, $\quad$ Nu = $(1.86) (39.72)^{1/3} (5.53 \times 10^{-4}/4.05 \times 10^{-4})$ $\quad \because \mu_w$ at 70°C = 4.05×10^{-4}

$$= 6.63 = hD/k$$

$\therefore \qquad\qquad$ h = $6.63 \times 0.642/0.025 = 170.21$ W/m^{2}K

Mass flow rate = $990 \times 3.142 \times (0.0125)^2 \times 0.02 = 9.72 \times 10^{-3}$ kg/s

By making an energy balance, h A $(\Delta T) = \dot{m} \, C_p \, (T_o - 50)$, where T_o is the bulk temperature at outlet.

Since $\quad$ T = $[70 - (T_i + T_o)/2]$ $\quad$ and $\quad T_i = 50$°C, $\quad$ we get $T_o = 60$°C

Because the temperature rise is not significant, another iteration will not make any significant change in T_o.

When the length of the tube is 0.35 m,

Graetz number $= $ Re. Pr. D/x = $895.1 \times 3.55 \times 0.025/0.35$

$$= 226.97 > 200$$

and we make use of Eq. (6.53 b),

$$Nu = 1.077 \, (Gz)^{1/3} = 1.077 \, (226.97)^{1/3} = 6.558 = h \, D/k$$

$\therefore \quad$ PPh = $6.558 \times 0.642/0.025 = 168.41$ W/m^{2}K

By making an energy balance, h A $(\Delta T) = m \, C_p \, (T_o - 50)$

which gives $T_o = 51.20$°C.

Example 6.15 Air at 25°C enters a tube, inner diameter 15 mm and length 2.5 m, with an average velocity of 1 m/s. If the exit temperature of air is 45°C, estimate the solar energy required for heating the air.

Solution: The mean temperature of air is $(25 + 45)/2 = 35°C$. The properties of air at this temperature are:

$$\rho = 1.15 \text{ kg/m}^3, \quad C_p = 1.007 \text{ kJ/kgK}, \quad k = 0.027 \text{ W/mK},$$
$$Pr = 0.70, \quad \upsilon = 16.89 \times 10^{-6}.$$

Reynolds number $= V \, D/\upsilon = 1 \times 0.015/16.89 \times 10^{-6}$
$$= 888, \text{ a laminar flow.}$$

Using Reynolds analogy, $St.Pr^{2/3} = f/8$, where $f = 64/Re = 64/888$
$$= 0.072$$

and, $\quad h/(\rho \, C_p \, V) \times Pr^{2/3} = 0.072/8$
$$h = 1.15 \times 0.009 \times 1007 \times 1/(0.7)^{0.667} = 13.22 \text{ W/m}^2\text{K}$$

Quantity of heat energy required $= \dot{m} \, C_p \, (45 - 25)$
$$= 1.15 \times 3.142 \times (0.0075)^2 \times 1 \times 1007 \times 20$$
$$= 4.09 \text{ W}$$
$$= h \, A \, (\Delta T)$$

$\Delta T = 4.09/(13.22 \times 3.142 \times 0.015 \times 2.5) = 2.63°C$

Temperature at the tube exit $= 45 + 2.63 = 47.63°C$

Solar insolation $= 4.09/(3.142 \times 0.015 \times 2.5) = 34.71 \text{ W/m}^2$

If the length of the tube is reduced to 1 m, the required solar insolation would be
$$= 4.09/ (3.142 \times 0.015 \times 1.0) = 86.78 \text{ W/m}^2.$$

Example 6.16 Freon-12 enters a rectangular duct, cross–section 10 mm by 20 mm, as a liquid at 7°C and leaves the duct as liquid at 17°C. The walls of the duct are maintained at –5°C. Assuming fully developed laminar flow, estimate the rate of heat transfer per metre length of the duct.

Solution: From Table 6.2, the Nusselt number for fully developed flow through a rectangular cross-section having sides in the ratio 1:2, with constant wall temperature is given as 3.391.

Hydraulic diameter for the duct, $\quad D_H = 4 \, A/P = 4(0.01 \times 0.02)/0.06$
$$= 0.01333 \text{ m}$$

The mean temperature of the refrigerant is $12°C \equiv 285K$

The thermal conductivity at that temperature is $= 0.073 \text{ W/mK}$

Therefore, the convective heat transfer coefficient, $h = 3.391 \times 0.073/0.0133$
$$= 18.61 \text{ W/m}^2\text{K}$$

Rate of heat transfer, $Q = h \, A \, (\Delta T) = 18.61(0.06 \times 1) \, (12 + 5)$
$$= 19.186 \text{ W}.$$

Example 6.17 Air is flowing through a circular tube of diameter 10 mm with a speed of 1 m/s. The mean bulk temperature of air is 40°C. Calculate the values of the convective heat transfer coefficient when air flows through a square duct of the same cross–sectional area. Assume that walls of the tube and duct are maintained at constant temperature.

Solution: The properties of air at 40°C are:

$$\rho = 1.128 \text{ kg/m}^3, \quad \mu = 19 \times 10^{-6} \text{ Pa-s}, \quad k = 0.028 \text{ W/mK}, \quad C_p = 1005 \text{ J/kgK}$$

When air is flowing through the circular tube,

$$\text{Re} = 1.128 \times 1 \times 0.01/19 \times 10^{-6} = 593.68, \quad \text{a laminar flow}$$

From Eq. (6.48), Nu = 3.66, therefore, h = 3.66 × 0.028/0.01 = 10.248 W/m²K

For the same cross-sectional area, the side of a square = $(3.142)^{1/2}$ (0.005)

$$= 8.86 \times 10^{-3} \text{m}$$

Hydraulic diameter, $D_H = 4A/P = 8.86 \times 10^{-3}$m

Reynolds number, Re = $1.128 \times 1 \times 8.86 \times 10^{-3}/19 \times 10^{-6}$ = 526, a laminar flow

Therefore, Nu = 3.66, and h $= 3.66 \times 0.028/8.86 \times 10^{-3}$

$$= 11.57 \text{ W/m}^2\text{K, an increased value.}$$

Example 6.18 A fluid is flowing through a pipe 30 cm in diameter and whose surface is maintained at a constant temperature. The length of the pipe is 3 m. At the inlet section of the pipe, the wall surface temperature exceeds the fluid temperature by 30°C. Calculate the increase in the fluid temperature at the exit section of the pipe, using Reynolds Analogy, St = f/8 where f = 0.02.

Solution: Let T_0 be the temperature of the fluid at the exit of the pipe.

By making an energy balance: $\dot{m} C_p (T_0 - T_i) = h A (\Delta T)$

The fluid enters at T_i and comes out at T_0. The temperature difference ΔT can be approximated as $[T_w - (T_0 + T_i)/2]$

$$(\pi/4) VD^2\rho C_p (T_0 - T_i) = h \times \pi DL [T_w - (T_0 + T_i)/2]$$

or, $(hL/ \rho V C_p) ((T_w - T_0) + (T_w - T_i)/2 = (D/4) (T_0 - T_i)$

$$= \frac{D}{4}\left[(T_w - T_i) - (T_w - T_0)\right]$$

Since $\quad$ St = h/ ρVC_p = f/8 = 0.0025, $\quad$ and $\quad T_w - T_i$ = 30, $\quad$ we have

$$(0.0025L/2) [(T_w - T_0) + 30] = (D/4) [30 - (T_w - T_0)]$$
$$= (0.3/4) [30 - (T_w - T_0)]$$

Collecting the term $(T_w - T_0)$ we get,

$$0.005 (T_w - T_0) + 0.15 = 3.0 - 0.1 (T_w - T_0); \quad \text{or, } 0.1050 (T_w - T_0) = 2.85$$

and $\quad T_w - T_0$ = 2.85/0.105 = 27.14°C. $\quad$ The temperature rise at exit = 2.86°C.

Example 6.19 Water at 25°C enters a thick walled tube (inner and outer diameter 25 and 50 mm). Water comes out at 65°C. The outer surface of the tube is well insulated and the electrical heating within the wall provides for a uniform generation rate of 10^6 W/m³. Estimate the length of the tube for a mass flow rate of 0.1 kg/s. If the inner surface temperature of the tube T_w is 75°C, find the local heat transfer coefficient at the outlet.

Solution: Since the outer surface of the tube is insulated, the energy generated within the walls of the tube is taken in by water.

Thus, $10^6 \times \pi(D_o{}^2 - D_i{}^2) \, L/4 = \dot{m} \, C_p \, (T_o - T_i)$

$\qquad L = 4 \times 0.1 \times 4182(65 - 25)/(3.142 \, (50^2 - 25^2) \times 10^{-6} \times 10^6) = 11.36 \text{ m.}$

Assuming that a uniform heat generation within the wall provides a constant heat flux,

$$\dot{q} = \text{Energy generated}/ \, \pi D_i \, L = 10^6 \times (50^2 - 25^2) \times 10^{-6}/(4 \times 25 \times 10^{-3})$$
$$= 18750 \text{ W/m}^2$$

Therefore, the local heat transfer coefficient, $h = \dot{q}/(T_w - T_o)$

or, $\qquad h = 18750/(75 - 65) = 1875 \text{ W/m}^2\text{K.}$

Example 6.20 Water, mass flow rate 1 kg/min, flows through a 40 mm inner diameter pipe. It is heated from 20°C to 60°C while the tube wall temperature is maintained at 80°C. Assuming fully developed flow, estimate the length of the tube required. The properties of water are:

$\rho = 990 \text{ kg/m}^3, \qquad\qquad C_p = 4182 \text{ J/kgK,}$
$\upsilon = 0.48 \times 10^{-6} \text{ m}^2\text{/s,} \qquad\quad k = 0.66 \text{ W/mK}$

Solution: Mass rate of flow = density × cross–sectional area × velocity

$\therefore \qquad$ Velocity, $V = (1/60)/(990 \times 3.142 \times (0.02)^2) = 0.0134 \text{ m/s}$

Reynolds number, $\text{Re} = V \, D/\upsilon = 0.0134 \times 0.04/0.48 \times 10^{-6}$
$\qquad\qquad\qquad\qquad = 1117$, a laminar flow.

Since the tube temperature is constant and the flow is fully developed, we use Eq. (6.48),

$\qquad\qquad \text{Nu} = h \, D/k = 3.66; \quad \therefore \ h = 3.66 \times 0.66/0.04 = 60.39 \text{ W/m}^2\text{K}$

Energy gained by water $= m \, C_p \, (T_o - T_i) = 1.0 \times 4182 \times 40/60 = 2788 \text{ W}$

By making an energy balance, $2788 = h \, A \, (\Delta T). = h \, (\pi D \, L) \, (\Delta T)$

The temperature of water increases from 20° to 60° and the difference in temperature at inlet to the tube, $\theta_1 = (80 - 20) = 60°$;

The difference in temperature at the tube outlet, $\theta_2 = (80 - 60) = 20°$

If we take the arithmetic mean, $\Delta T = (60 + 20)/2 = 40°$

Length of the tube, $L = 2788/(60.39 \times 3.142 \times 0.04 \times 40) = 9.18 \text{ m}$

If we take the logarithmic mean temperature difference,

$$\text{LMTD} = (\theta_1 - \theta_2)/\ln(\theta_1/\theta_2) = 40/\ln(3) = 36.41$$

and the length of the tube, $L = 2788/(60.39 \times 3.142 \times 0.04 \times 36.41)$
$$= 10 \text{ metres.}$$

19. Methods to Improve 'h' in Laminar Flow through Tubes

The convection heat transfer coefficient, h, may be increased by

    (a) Introducing surface roughness so that the flow is turbulent. This can be achieved by inserting a coil-spring wire which will provide a helical roughness element in contact with the tube inner surface.

(b) Introducing swirl by inserting a twisted tape (a thin strip periodically twisted through 360°) because the introduction of a tangential velocity component will increase the speed of the flow particularly near the tube wall.

(c) Using a coiled tube, because centrifugal forces induce a secondary flow consisting of a pair of vortices that increase 'h'.

20. Laminar flow through Concentric Tube Annulus—Evaluation of 'h'

Many practical problems involve heat transfer in a concentric tube annulus. Fluid flows through the annulus formed by the concentric tubes, and convection heat transfer may occur to or from both the outer and inner surfaces of the tube. The convective heat transfer coefficients at the inner and outer surfaces of the tube are likely to be different and the heat flux from each surface may be expressed as:

$$Q_i = h_i \, (T_{wi} - T_m), \qquad \text{and} \qquad Q_o = h_o \, (T_{wo} - T_m)$$

where T_{wi} and T_{wo} are the temperatures at the inside and outside surfaces of the annulus, and T_m is the mean temperature of the fluid. The corresponding Nusselt numbers can be written as:

$$Nu_i = h_i \, D_h/k; \quad \text{and} \quad Nu_o = h_o \, D_h/k$$

where D_h is the hydraulic diameter and will be equal to $D_o - D_i$.

For fully developed laminar flow through the annulus with one surface insulated and the other surface at constant temperature, the Nu_i or Nu_o may be obtained from Table 6.3.

Table 6.3 Nusselt number for fully developed laminar flow in a circular tube annulus with one surface insulated and the other at a constant temperature

D_i/D_o	Nu_i	Nu_o	D_i/D_o	Nu_i	Nu_o
0		3.66	0.25	7.37	4.23
0.05	17.46	4.06	0.50	5.74	4.43
0.10	11.56	4.11	1.00	4.86	4.86

Example 6.21 Water at $\dot{m}$ = 0.02 kg/s flows through an annular region formed by an inner tube of diameter 25 mm and an outer tube of diameter 100 mm. The surface temperature of the inner tube is constant at 100°C and the outer surface of the outer tube is well insulated. Estimate the length of the tube when water is heated from 20°C to 75°C, assuming fully developed flow. What would be the heat flux from the inner tube at outlet?

Solution: The mean temperature = [100 + (20 + 75)/2]/2 = 74°C. At this temperature, μ for water = 389 × 10⁻⁶ Pa-s; hydraulic diameter = 75mm, k = 0.668 W/mK

$\dot{m} = \rho AV \therefore \rho V = \dot{m}/A = 0.02/[\pi/4(D_o^2 - D_i^2)] = 2.7159;$
$Re = \rho VD/\mu = 2.7159 \times 75 \times 10^{-3}/389 \times 10^{-6} \equiv 524,$ A laminar flow
Mean temperature of water = 47.5°C;
From Table 6.3; for $D_i/D_o = 0.25,$
$Nu_i = 7.37 = h_iD_h/k;$ $\qquad\qquad \therefore h_i = 7.37 \times 0.668/0.075 = 65.64$ W/m^{2}K

$$\dot{Q} = \dot{m}C_p(\Delta T) = h_iD_h \times \pi L(T_w - T_m)$$

$$\therefore \quad L = \frac{0.02 \times 4187 \times (75-20)}{\pi \times 0.075 \times 65.64(100 - 47.5)} = 5.67 \text{ m}$$

From Table 6.3, for $D_i/D_0 = 0.25,$

$$Nu_0 = 4.23$$

$$h_0 = \frac{4.23 \times 0.668}{0.075} = 37.68 \text{ W/m}^2\text{K}.$$

SUMMARY

1. Reynolds number is a dimensionless parameter. It is the ratio of the inertia forces to viscous forces and ascertains whether the flow is laminar or turbulent.
2. Using dimensional analysis, the functional relationship for the evaluation of convective heat transfer coefficient in forced convection has the form
$$Nu = f(Re, Pr) = C\,Re^m\,Pr^n.$$
3. Reynolds analogy tells us that the local convective heat transfer coefficient in laminar flow can be obtained by measuring local skin friction coefficient because of similarity between the exchange of momentum and the exchange of energy. And for $Pr = 1$, Stanton Number, $St = C_{fx}/2$.
4. Colburn has analysed that for $0.6 \le Pr \le 50$, Reynolds analogy should be modified as $St.\,Pr^{2/3} = C_{fx}/2$.
5. Hydrodynamic boundary layer is a very thin layer close to the solid surface. It is automatically formed when a real fluid flows over a solid surface. The fluid velocity at the solid surface is zero and at the edge of the boundary layer, the velocity is equal to 99% of the free stream velocity.
6. Thermal boundary layer will develop only when the fluid temperature is different from the temperature of the solid surface. The thermal and the momentum boundary layers are identical for $Pr = 1$.
7. Prandtl number, $Pr\ (= \mu C_p/k)$ is a function of temperature and takes into account the heat storage capacity of the fluid, its conducting ability and the viscous nature. The thickness of the thermal boundary layer, $\delta_t = \delta/Pr^{1/3}$.
8. Blasius transformed the non-linear partial differential equation of motion (known as 2 – D Prandtl boundary layer equation for flow over a flat plate) into an ordinary differential equation with the help of a dimensionless similarity parameter, $\eta = y\,(U_\infty/vx)^{1/2}$ and obtained the thickness of the momentum boundary layer at a distance x from the leading edge as: $\delta/x = 5.0/(Re_x)^{1/2}$, and the local skin friction coefficient, $C_{fx} = 0.664/Re_x^{1/2}$. The local heat transfer coefficient can be evaluated from: $Nu_x = 0.332\,Re_x^{1.2}\,Pr^{1/3}$, for laminar flow.

9. The average heat transfer coefficient from a flat plate in laminar flow is twice the local value at the trailing edge of the plate. The convective heat transfer coefficient, h, is proportional to
$$k^{2/3} \times \rho^{1/2} \times U^{1/2} \times C_p^{1/3} \times L^{-1/2} \times \mu^{-1/6} \times x^{-1/2}$$

10. Von – Karman's Integral method gives an approximate but fairly accurate solution to momentum and energy equation for a 2 – D flow over a flat plate when the velocity and temperature profiles within the boundary layer are represented by a cubic polynomial.

11. For a fluid flowing through a tube, the velocity profile is fully developed after an entry length, $L_e = 0.058$ (Re) $\times$ inner dia of the tube.

12. The bulk temperature is the resulting temperature when the fluid flowing through a cross-section is collected in a vessel for a short period of time and thoroughly mixed without allowing it to exchange energy with the surroundings.

13. Colburn analogy can also be applied to evaluate the convective heat transfer coefficient in laminar flow through tubes by the relation St. $Pr^{2/3} = f/8$, where $f = 64/Re$.

14. While evaluating heat transfer coefficients for fluids flowing through non-circular cross-section or through concentric circular annulus, the characteristic dimension is called 'hydraulic diameter' $D_h = 4$ A/P.

15. For fully developed velocity profiles, the local and average Nusselt number for laminar entrance region of circular tubes is given by Nu = constant $\times$ $(Gz)^{1/3}$, where Gz, is the Graetz number, = (Re Pr D/x) should be greater than 200.

MULTIPLE CHOICE QUESTIONS

1. There would be a similarity in flow between two flows if
   (a) the two Reynolds number are equal
   (b) the geometrical similarity is taken into consideration
   (c) either a or b
   (d) both a and b   **(d)**

2. Reynolds analogy states that convective heat transfer coefficient for a laminar flow over a flat plate can be obtained by the relation:
   (a) Nu = Re $(C_f/2)$ (b) St = $C_f/2$
   (c) St = $Pr^{-2/3}$ $C_f/2$ (d) Nu = $C_f/2$ **(b)**
   where Nu is Nusselt number, St is Stanton number, Pr is Prandtl number

3. The boundary layer thickness at a distance x from the leading edge in a laminar boundary layer is given by
   (a) 5.0 $Re_x^{-1/2}$ (b) 4.64 $Re_x^{-1/2}$
   (c) 5.0 $(x^{1/2} \, v^{1/2}) U_\infty^{-1/2}$ (d) none of the above   **(c)**

4. The average value of heat transfer coefficient can be evaluated as:

   (a) $\int_0^L h_x \, d_x$, where h_x is the local heat transfer coefficient and L is the length of the plate
   (b) the arithmetic mean of the coefficients at the leading edge and the trailing edge.

(c) $1/L \int_0^L h_x \, d_x$,

(d) $1/3 \, h_L$, where h_L is the heat transfer coefficient at $x = L$. **(c)**

5. The local heat transfer coefficient in laminar flow over a flat plate
   (a) increases with the distance from the leading edge
   (b) increases with the viscosity of the fluid
   (c) increases with the velocity of the free stream
   (d) decreases with the density of the fluid **(c)**

6. Choose the correct statement:
   For a given value of Nusselt number, the convective heat transfer coefficient, h
   (a) decreases with increasing thermal conductivity of fluid
   (b) increases with distance x from the leading edge
   (c) increases with the thermal conductivity of fluid
   (d) is independent of the distance x **(c)**

7. In laminar boundary layers over a flat plate, the ratio of the δ_t/δ is given by:
   (a) $Pr^{1/2}$ (b) $Pr^{1/3}$ (c) $Pr^{-1/3}$ (d) $Pr^{1/4}$ **(c)**

8. Assertion (A): In heat transfer by convection through a fluid at a given rate, a larger temperature gradient is required in laminar flow.
   Reasoning (R): Because the transport of energy is aided materially by mixing currents on a macroscopic scale.
   Code: (a) Both A and R are true   (b) Both A and are false
         (c) A is false, R is true   (d) A is true, R is false **(d)**

9. Assertion (A): Von Karman's method is being used to obtain simple equations which d escribe t he v elocity a nd temperature d istributions i n t he b oundary layer.
   Reasoning (R): Because it yields solutions to problems which cannot be treated by an exact mathematical analysis.
   Code: (a) Both A and R are true and R is the only reason for A
   (b) Both A and R are true and R is not the only reason for A
   (c) A is true, R is false
   (d) A is false, R is true **(b)**

10. When a hot fluid is flowing over a cold flat plate, the temperature gradient
    (a) is zero at the surface
    (b) is negative at the surface
    (c) is zero at the edge of the thermal boundary layer
    (d) none of the above **(c)**

11. When all conditions are identical for fluids flowing through a pipe with heat transfer, the velocity profile will be identical for
    (a) liquid heating and liquid cooling
    (b) gas heating and gas cooling
    (c) gas heating and liquid cooling
    (d) cooling and heating of any fluid. **(c)**

12. For fully developed laminar flow and heat transfer in a uniformly heated long tube, if the flow velocity is doubled and the tube diameter is halved, the

heat transfer coefficient will be
   (a) double of the original value
   (b) half of the original value
   (c) same as before
   (d) four times the original value **(a)**

13. Graetz number is defined as:
   (a) Re × Pr   (b) Re × Pr (x/D)   (c) Re × Pr (D/x)   (d) Gr × Pr **(c)**

14. Assertion (A): Heat transfer coefficient in laminar flow through a tube should be obtained by making an analogy between fluid friction and heat transfer
   Reasoning (R): Because fully developed velocity and temperature profiles are seldom found in laminar tube flow
   Code: (a) Both A and R are true   (b) Both A and R are false
          (c) A is true, R is false   (d) A is false, R is true **(a)**

15. With increasing D_i/D_o, the Nusselt number ($h_o D_h/k$) for fully developed laminar flow in a circular tube annulus with one surface insulated and the other at constant temperature,
   (a) decreases (b) remains the same
   (c) increases (d) depends upon the Reynolds number **(c)**

16. Assertion (A) : Heat transfer may be enhanced without inducing turbulence or additional heat transfer area by coiling a tube.
   Reasoning (R): Because centrifugal forces induce a secondary flow
   Code (a) Both A and R are true and R is the correct explanation of A
         (b) Both A and R are true and R is not the only reason for A
         (c) Both A and R are false
         (d) A is false and R is true **(b)**

17. In a Laminar fully developed flow of a fluid with constant properties the local convection heat transfer coefficient in a tube is
   (a) a constant   (b) a constant and independent of x
   (c) dependent on Re even if the flow is laminar
   (d) none of the above **(b)**

18. In laminar flow through a tube, under fully developed conditions:
   (1) $(\partial u/\partial x) = 0$ (2) $(\partial T/\partial x)$ at any radius r is not zero
   (3) the temperature profile T(r) continuously changes with x
   (4) for constant tube wall temperature, surface heat flux is constant
   Out of the above statements:
   (a) only 1 and 2 are correct   (b) 1, 2 and 3 are correct
   (c) 1, 2, 3 and 4 are correct   (d) 1 and 4 are correct **(b)**

19. For laminar flow of a liquid (k = 0.4 W/mK, Pr = 1.2), the value of h at a distance x from the leading edge is 10 W/m²K. When another fluid (k = 0.8 W/mK, Pr = 0.6) flows over the same plate, the value of h would be, when other conditions remain the same,
   (a) 10   (b) 12   (c) 15.88   (d) 20 **(c)**

NUMERICALS

1. A flat plate 15 cm by 24 cm is exposed to an air stream flowing parallel to its surface. The air is at atmospheric pressure and 300K and is flowing at 30 m/s. A total drag force (for both sides) of 1.1 N is observed to act on the plate. Using Reynolds Analogy, estimate the convective heat transfer coefficient between the plate and air. C_p = 1.005 kJ/kgK

   (511.9 W/m^{2}K)

2. Calculate the heat transfer from a 30 cm square plate over which air at 1 atm pressure and 25°C flows with a velocity of 5 m/s when the plate is maintained at 100°C. Also estimate the drag force on the plate (both sides) (ρ = 1.1 kg/m^3, C_p = 1005 J/kgK, Pr = 0.7, k = 0.028 W/mK, υ = 20x10^{-6} m^2/s)

   (101.72 W, 0.013 N using Reynolds analogy)

3. Liquid ammonia at 10°C is forced across a square plate 20 cm on a side at a velocity of 80 cm/s. The plate is maintained at 50°C. Calculate the heat lost by the plate.

   (ρ = 626.16 kg/m^3, υ = 0.368 × 10^{-6} m^2/s, k = 0.531 W/mK, Pr = 2.04)

   (466 W)

4. Engine oil at 30°C flows over a flat plate at a velocity of 0.08 m/s. The length of the plate is 5 m. The plate is maintained at a constant temperature of 90°C. Calculate (i) the thickness of the momentum and thermal boundary layer at x = 5m, (ii) the drag force on each side of the plate assuming unit width (iii) the local heat transfer coefficient at the trailing edge and (iv) the heat transfer rate. The properties are:

   ρ = 956.8 kg/m^3, k = 0.213 W/mK, α = 7.2 × 10^{-8} m^2/s, υ = 6.5 × 10^{-5} m^2/s

   (31.87 cm, 3.3 cm, 0.26 N/m, 6430 W)

5. The velocity distribution in the laminar boundary layer can be approximated by the relation $u/U_\infty = 2(y/\delta) - (y/\delta)^2$. Using the momentum integral equation show that the expression for the boundary layer thickness is given by $\delta/x = 5.48/Re_x^{1/2}$.

6. Air at 25°C and 1 atm pressure flows over a flat plate at a speed of 4 m/s. Calculate the boundary layer thickness at a distance of 25 cm and 40 cm from the leading edge. Determine the mass flow which enters the boundary layer between x = 25 cm and x = 40 cm. The plate is maintained at 60°C. Also calculate the rate of heat transfer by the first 40 cm of the plate. The properties are: ρ = 1.1 kg/m^3 , υ = 17.4 × 10^{-6} m^2/s, k = 0.0275 W/mK, Pr = 0.7

   (1.043 mm 1.32 mm, 0.762x10^{-3} kg/s, 172 W/m)

7. Water flows through a tube of 2.5 cm inside diameter pipe at a rate of 90 litres per hour. Estimate the entry length the required pressure drop and the pumping power to maintain the flow if the length of the pipe is 2 km.; for water υ = 1 × 10^{-6} m^2/s.

   (1.812m, 1280 N/m^2, 0.32 W)

8. The velocity and temperature distribution for laminar flow through a tube, inner diameter 16 cm, are approximated as u = 1.8 (1 − 0.0156 r^2) and T = 55 (1 + 0.25 r)°C where r is the distance measured from the axis in cm. Estimate the average velocity and the bulk temperature of the fluid.

   (0.9 cm/s, 103.12°C)

9. Glycerine flows through a 40 mm diameter tube at a rate that the Reynolds number is 10. The glycerine enters at 25°C and comes out at 30°C The tube wall is maintained at 45°C. Estimate the length of the tube.

    Take: $\rho = 1258$ kg/m³, $\upsilon = 5 \times 10^{-4}$ m²/s, $k = 0.286$ W/mK

    Pr = 5.18, $C_p = 2445$ J/kgK.

    (42.27 m)

10. Water enters a 5 cm diameter tube at 25°C and leaves at 40°C. The flow is laminar and the Reynolds number is 1800. The tube length is 18 m and is maintained at 60°C. Estimate the heat transfer coefficient.

    $\rho = 993$ kg/m³, $C_p = 4182$ J/kgK, $\mu = 6.82 \times 10^{-4}$ Pa-s

    (59.7 W/m²K)

11. In a 25 mm diameter tube the pressure drop per metre length is 0.0002 bar at a section where the mean velocity is 24 m/s, and the mean specific heat capacity of the gas is 1.13 kJ/kgK. Calculate the heat transfer coefficient.

    (5.88 W/m²K)

12. Atmospheric air at 25°C flows over a flat plate with a velocity of 4 m/s. The plate is maintained at a uniform surface temperature of 50°C. Assuming that the transition occurs at Re $= 5 \times 10^5$. Estimate the length of the plate from the leading edge where the boundary layer changes from laminar to turbulent. At that location, calculate the following:

    (a) thickness of the momentum and thermal boundary layer

    (b) local and average heat transfer coefficient and heat transfer rate

    (c) skin-friction coefficient.

    Take: $\rho = 1.128$ kg/m³, $\upsilon = 16.96 \times 10^{-6}$ m²/s, $k = 0.0275$ W/mK, Pr = 0.699

    (2.12 m, 14.98 mm, 16.88 mm, 2.625 W/m²K,
    5.25 W/m²K, 445.2 W/m, 9.39 × 10⁻⁴)

13. Water at 25°C enters a triangular duct (each side 1.5 cm) and leaves at 45°C. Assuming fully developed laminar flow, estimate the rate of heat transfer if the tube wall temperature is maintained at 60°C and the length is 2m.
    (k = 0.63 W/mK)

    (381.66 W)

14. Water is being cooled in the annular section of a double pipe heat exchanger. The outer diameter of the inner tube is 2.5 cm and the inner diameter of the outer tube is 5 cm. Estimate the convective heat transfer coefficient at the inner and the outer surface of the annulus when the mass flow rate of water is 0.055 kg/s and the outside surface of the annulus is insulated and the inside surface is maintained at constant temperature. Take the properties of water as:

    $\rho = 1000$ kg/m³, $\upsilon = 1.397 \times 10^{-6}$ m²/s,   Pr = 10.31, $k = 0.575$ W/mK

    (132.02 and 101.89 W/m²K)

15. The velocity and temperature distribution for laminar flow through a circular tube of diameter 20 cm, is given by

    $u = 20r(9 - 10r)$ and $T = 100(1 - 2r)$ where r is the distance in metres measured from the inside surface of the tube. Calculate the average velocity and bulk temperature.

    (85°C)

CHAPTER 7

Turbulent Flow Forced Convection Heat Transfer

1. Laminar and Turbulent Flow—Basic Difference

In laminar flow, the fluid streamlines are observed to run in a well ordered manner with adjacent layers of fluid sliding relative to one another and without any motion or exchange taking place normal to the streamlines of the main flow. However, it has been observed that most flows occurring in nature are turbulent. Hinze. J.O. has defined the turbulent fluid motion as 'an irregular condition of flow in which various quantities (such as velocity and pressure) show a random variation with time and space, so that statistically distinct average values can be discerned'.

Since the velocity at any point fluctuates at random and with high frequency, the solution of the laws of motion in a turbulent flow is extremely difficult. The macroscopic mixing of fluid particles and the physical characteristics at any point in the flow field spread much more rapidly to another point. As such, in order to compute the flow properties one has to depend on experimental measurements rather than basic laws.

2. Turbulent Flow—Basic Definitions

A turbulent flow under certain conditions appears to be a steady one and can be represented by a steady mean flow proceeding in a specified direction with a superimposed time dependent fluctuating motion Fig. 7.1 (a). These fluctuations are observed to occur in directions parallel and transverse to the main flow. For such a two dimensional turbulent flow, we write

$$u = \bar{u} + u'; \qquad v = \bar{v} + v'$$

where $\bar{u}$ and $\bar{v}$ are the time average values of u and v and are constant with time, and u' and v' are fluctuations with time of u and v. Then for steady turbulent motion,

$$\bar{u} = \frac{1}{(t_2 - t_1)} \int_{t_1}^{t_2} u\,dt; \qquad \bar{v} = \frac{1}{(t_2 - t_1)} \int_{t_1}^{t_2} v\,dt \qquad (7.1)$$

where $\bar{u}$ and $\bar{v}$ are independent of time.

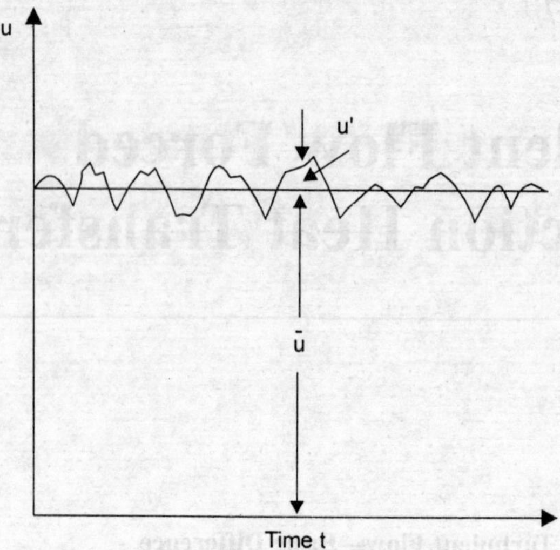

Fig. 7.1(a) Fluctuations in turbulent flow

The time interval $(t_2 - t_1)$ is chosen in such a way that $\bar{u}$ and $\bar{v}$ are constant. The time average of fluctuations is then zero, i.e.,

$$\bar{u}' = \frac{1}{(t_2 - t_1)} \int_{t_1}^{t_2} u'dt = 0, \quad \bar{v}' = \frac{1}{(t_2 - t_1)} \int_{t_1}^{t_2} v'dt = 0, \tag{7.2}$$

It should be mentioned that although the time averages of the velocity fluctuations ($\bar{u}'$, $\bar{v}'$) are defined as zero, the time averages of their products ($\overline{u'v'}$) and the square of the velocity fluctuations ($\overline{u'^2}, \overline{v'^2}$) are not necessarily zero.

The mean kinetic energy per unit volume can be written as:

$$KE = \frac{1}{2}\rho(\bar{u}^2 + \bar{v}^2 + \bar{w}^2 + u'^2 + v'^2 + w'^2)$$ and the level or intensity of turbulence

is defined as $I = \left((\overline{u'^2} + \overline{v'^2} + \overline{w'^2})/3 \right)^{\frac{1}{2}} \Big/ u_m$ where u_m is the mean velocity of

flow. The intensity of turbulence affects the separation and boundary layer transition, heat and mass transfer rates.

3. Shear Stress and Eddy Viscosity—Defined

Let us consider a two-dimensional flow in X-Y plane as shown in Fig. 7.1(b). It is assumed that the mean flow is only in X-direction (parallel to the wall), i.e., $u = f(y)$ only and $v = 0$. But this mean flow has velocity fluctuations in X- and Y-directions, and the time dependent turbulent velocity in X- and Y-direction is then

$$u = \bar{u} + u' \quad \text{and} \quad v = 0 + v' \tag{7.3}$$

At a distance y from the wall (sectional plane A – B), the mean velocity is $\bar{u}$ and for particles crossing the plane AB in the upward direction, the time rate of transport of mass per unit area is $\rho v'$. The change in momentum experienced in

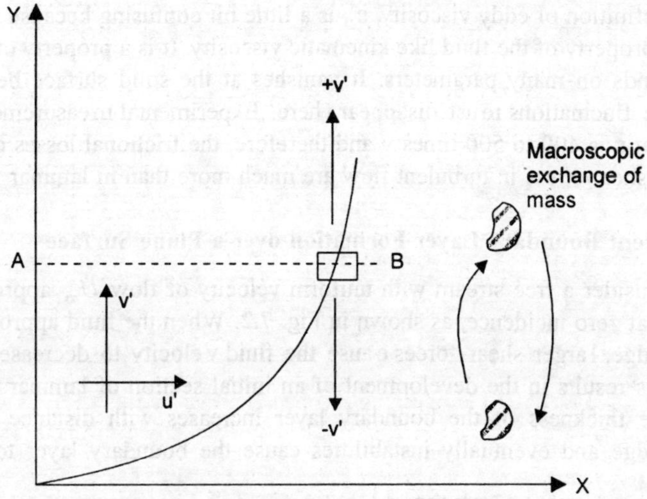

Fig. 7.1 (b) Momentum transport due to turbulent fluctuation

the X-direction by this mass will be dependent on the change in u that it experiences. The particles crossing the plane AB in the upward direction are coming from a region where a smaller mean velocity $\bar{u}$ prevails. Therefore, this will tend to produce a negative u' in the layer above the surface AB. The change in the X-wise momentum per unit area experienced by such particles will be $(-\rho v'u')$. Similarly, the particles which are crossing the plane AB in the downward direction (v' negative) are coming from a region where a larger mean velocity $\bar{u}$ prevails and this will give rise to $+ u'$. Thus, on the average, a positive v' is associated with a negative u' and a negative v' is associated with a positive u'. And the time average of the product $u'v'$ is not only different from zero but is negative. Thus exchange of momentum per unit area on the macroscopic scale gives rise to a 'turbulent shear stress' in the X-direction.

Or $$\tau' = -\rho \overline{u'v'} \tag{7.4}$$

This must be added to the stresses caused by laminar flow. The total stress acting on the plane AB is written as,

$$\tau = +\mu \frac{\partial \bar{u}}{\partial y} - \overline{\rho u'v'} \tag{7.5}$$

The Eq. (7.5) cannot be used because the exact nature of the fluctuations u' and v' are not known. Therefore, the apparent stress is defined in terms of apparent kinematic viscosity or eddy viscosity, ε , as

$$\tau' = -\overline{\rho u'v'} = \rho\varepsilon \frac{\partial \bar{u}}{\partial y} \tag{7.6}*$$

and the total shear stress is then,

$$\tau = \rho(v + \varepsilon) \frac{\partial \bar{u}}{\partial y} \tag{7.7}$$

---

* The apparent stress is proportional to $\left(\left|\dfrac{d\bar{u}}{dy}\right|\right)^2$, according to Parndtl mixing length hypothesis.

The definition of eddy viscosity, ε , is a little bit confusing because it is not a physical property of the fluid like kinematic viscosity. It is a property of the flow and depends on many parameters. It vanishes at the solid surface because all transverse fluctuations must disappear there. Experimental measurements have shown that ε is 400 to 500 times ν and therefore, the frictional losses caused by turbulent shear stress in turbulent flow are much more than in laminar flow.

4. Turbulent Boundary Layer Formation over a Plane Surface

Let us consider a free stream with uniform velocity of flow U_∞ approaching a flat plate at zero incidence, as shown in Fig. 7.2. When the fluid approaches the leading edge, larger shear forces cause the fluid velocity to decrease near the plate. This results in the development of an initial section of laminar boundary layer. The thickness of the boundary layer increases with distance from the leading edge and eventually instabilities cause the boundary layer to become

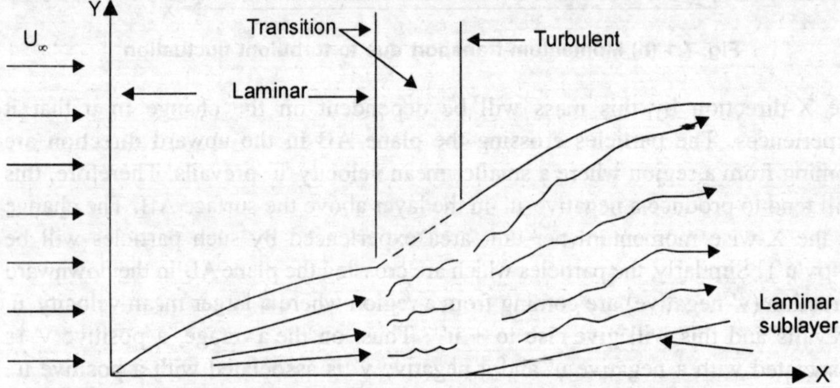

Fig. 7.2 Developing turbulent boundary layer on a flat plate

turbulent. The thickness of the turbulent boundary layer is much more because of the velocity fluctuations in the Y-direction and this leads to a much flatter velocity profile than laminar flow as shown in Fig. 7.3. In the laminar sublayer, however, there is a very steep gradient. And therefore, the shear stress at the wall is much greater for the turbulent boundary layer than for the laminar boundary layer. There is a zone of transition between the laminar and turbulent boundary layer and the total drag on a plate is highly dependent upon the location of the transition zone.

The transition to turbulent boundary layer flow depends upon (a) the critical

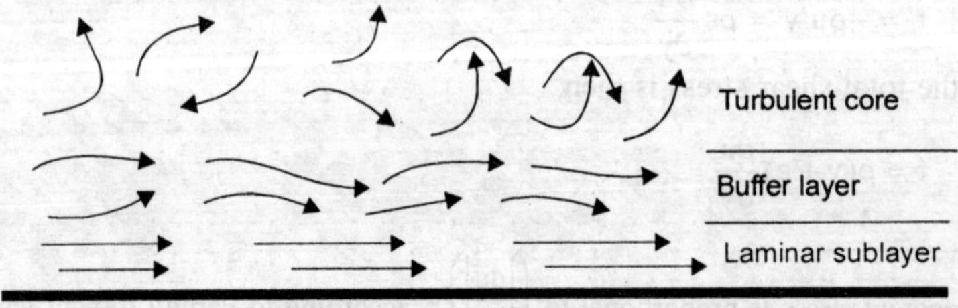

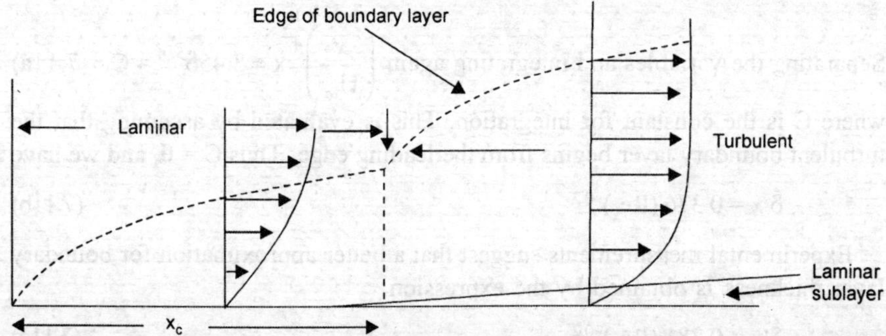

Fig. 7.3 Structure of a turbulent flow field near a solid surface and velocity profiles for laminar and turbulent layer in flow over a flat plate

Reynolds number $U_\infty x_c /v$, (b) the wall roughness, (c) the free stream turbulence, and (d) the external flow pressure-gradient.

5. Estimating Turbulent Boundary Layer Thickness—One-seventh Power Law

Since it is extremely difficult to ascertain the exact location for the beginning of a turbulent boundary layer and the turbulent boundary layer begins with a finite thickness, Prandtl has suggested that we should imagine that the turbulent boundary layer begins at the leading edge of the plate and the results obtained in this way are quite good for distances beyond x_c (Fig. 7.3) but are invalid for the laminar leading section.

The Von Kármán momentum integral equation (Eq. 6.27) can be written for a

turbulent boundary layer as $\tau_w = \rho \dfrac{d}{dx}\left[\int_o^\delta (U_\infty - \bar{u})\bar{u}\,dy \right]$ (7.8)

In order to integrate Eq. (7.8), a velocity profile is chosen in the form

$$u/U_\infty = (y/\delta)^{1/7} \qquad (7.9)$$

This one-seventh power law for the velocity profile is a good approximation for the bulk of the boundary layer but it totally fails near the wall. Therefore it cannot be used to obtain the velocity gradient for determining the wall shear stress.

Blasius has determined experimentally that wall shear stress in a turbulent boundary layer can be evaluated by

$$\tau_w = 0.0225\rho U_\infty^2 (v/U_\infty \delta)^{\frac{1}{4}} \qquad (7.10)$$

Substituting Eq (7.9) and Eq. (7.10) into Eq (7.8), we get

$$0.0225\, U_\infty^2 (v/U_\infty \delta)^{\frac{1}{4}} = \frac{d}{dx}\int_o^\delta U_\infty^2 [(y/\delta)^{1/7} - (y/\delta)^{2/7}]dy \qquad \text{and} \qquad \text{on}$$

integration, we have $0.0225\,(v/U_\infty \delta)^{\frac{1}{4}} = \dfrac{7}{72}\dfrac{d\delta}{dx}$.

Separating the variables and integrating again $\left(\dfrac{v}{U_\infty}\right)^{\frac{1}{4}} x = 3.45\delta^{5/4} + C$ (7.11a)

where C is the constant for integration. This is evaluated by assuming that the turbulent boundary layer begins from the leading edge. Thus C = 0, and we have

$$\delta/x = 0.376\,(Re_x)^{-0.2} \qquad (7.11b)$$

Experimental measurements suggest that a better approximation for boundary layer thickness is obtained by the expression

$$\delta/x = 0.381\,(Re_x)^{-0.2} \qquad (7.11c)$$

Assuming that the boundary layer is laminar up to $Re_c = 5 \times 10^5$, the thickness of the turbulent boundary layer should be obtained by:

$$\delta/x = 0.381\,(Re_x)^{-0.2} - 10256/Re_x \qquad (7.11d)$$

Thus it can be said that the turbulent boundary layer thickness grows with the 4/5 power of x, whereas, the thickness of the laminar boundary layer grows with 1/2 power of x. Or, a turbulent boundary layer will grow faster or will be thicker than a laminar boundary layer at a given distance from the leading edge.

Schlichting has surveyed experimental measurements of friction coefficient for turbulent flow on flat plates. By making the fluid friction – heat transfer analogy, the local skin friction coefficient is given by

$$C_{fx} = 0.0592\,Re_x^{-0.2} \qquad (7.12)$$

for Reynolds number between 5×10^5 and 10^7. At higher Reynolds number, the relation by Schultz – Grunow may be used:

$$C_{fx} = 0.37\,(\log Re_x)^{-2.584} \qquad (7.13)$$

The average friction coefficient for a plate of length L and unit width is obtained by integrating the Eq. (7.12) over the length of the plate.

or, $C_f = 0.072/Re_L^{1/5}$; C_f decreases with increasing Re

Experimental investigations reveal that the constant be increased to 0.074 for a better result, i.e.,

$$C_f = 0.074/Re_L^{1/5} \qquad (7.14)$$

Since the leading edge of the plane surface will always have a laminar boundary layer, the drag should be calculated by considering laminar boundary layer up to a distance x_c and turbulent boundary layer from this point downstream. That is, the following relation should be used:

$$C_f = 0.074/Re_L^{0.2} - A.Re_L^{-1}; \qquad Re_c < Re_L < 10^7 \qquad (7.15)$$

where values of A are:

Re_c	10^5	3×10^5	5×10^5	10^6	3×10^6
A	360	1050	1700	3300	8700

6. Turbulent Prandtl Number—Defined

Turbulent momentum transfer is analogous to turbulent heat transfer. And therefore, there would be a turbulent energy fluctuations, analogous to turbulent momentum flux given by Eq. (7.6), which can be written as

$(\dot{Q}/A)_{turb} = -\rho\, C_p\, \varepsilon_H\, \partial T/\partial y$, where ε_H is called eddy diffusivity and, when both molecular and turbulent energy transport are important, we

$$\dot{Q}/A = -\rho\, C_p\, (\alpha + \varepsilon_H)\, \partial T/\partial y \tag{7.16}$$

In the turbulent flow region, $\varepsilon_H \gg \alpha$ and $\varepsilon \gg \nu$, and as such, a turbulent Prandtl number is defined as:

$$Pr_t = \varepsilon/\varepsilon_H$$

Experimental results show that both eddy viscosity and eddy diffusivity increase in the same proportion as compared with their molecular values and therefore, the two Prandtl numbers, $Pr_t = Pr$, are assumed to be equal.

7. Fluid Friction—Heat Transfer Analogy in Turbulent Flow

Applying the fluid-friction analogy given by Eq. (6.6), we can obtain the local turbulent heat transfer coefficient using the relation:

$$St_x\, Pr^{2/3} = 0.0296\, Re_x^{-1/5}\,; \quad 5\times10^5 < Re_x < 10^7 \tag{7.17}$$

or $\quad St_x\, Pr^{2/3} = 0.185\, (\log Re_x)^{-2.584}; \quad 10^7 < Re_x < 10^9 \tag{7.18}$

The average heat transfer coefficient over the entire laminar-turbulent boundary layer is given by: (for $Re_c = 5\times10^5$ and $Re_L < 10^7$)

$$Nu_L = \frac{hL}{k} = (0.037\, Re_L^{0.8} - 850)\, Pr^{1/3} \tag{7.19}$$

Whitaker has suggested that the average heat transfer coefficient for liquids flowing over plane surfaces be evaluated from:

$$Nu_L = \frac{hL}{k} = 0.036\, (Re_L^{0.8} - 9200)\, Pr^{0.43} \left(\frac{\mu_\infty}{\mu_w}\right) \tag{7.20}$$

The viscosity-ratio term should be dropped when the equation is used for gases. All properties except μ_∞ and μ_w are to evaluated at the mean temperature.

Example 7.1 Air at 25°C and 1 bar flows over a flat plate with a speed of 45 m/s. The plate is 60cm long and is maintained at 65°C. Calculate the rate of heat transfer from the plate per unit width of plate.

Solution: The mean temperature is $(25 + 65)/2 = 45°C$. The properties are:

$$\rho = p/RT = 10^5/(287 \times 318) = 1.1 \text{ kg/m}^3; \quad Pr = 0.7, \quad k = 0.0296 \text{ W/mK}$$
$$\mu = 1.9 \times 10^{-5} \text{ Pa-s}$$

The Reynolds number, $Re = 1.1 \times 45 \times 0.60/1.9 \times 10^{-5} = 1.56 \times 10^6 > 5 \times 10^5$

Using Eq. (7.19),

$$Nu = hL/k = (0.037 \, Re_L^{0.8} - 850) \, Pr^{1/3}$$
$$= (0.037 \times (1.56 \times 10^6)^{0.8} - 850) \, (0.7)^{0.333}$$
$$= 2204$$

Therefore, $h = 2204 \times 0.0296/0.6 = 108.73 \text{ W/m}^2\text{K}$

Rate of heat transfer, $\dot{Q} = hA(\Delta T) = 108.73 \times 0.60 \times 1 \times (65 - 25)$
$$= 2609.5 \text{W}.$$

Example 7.2   A plane surface 40 cm × 30 cm at 95°C is placed in an air stream of 50 m/s. The temperature of air is 25°C. The critical Reynolds number where transition from laminar to turbulent flow occurs is 2×10^5. Calculate the heat transfer rate from the upper surface.

Solution: The mean temperature is $(95 + 25)/2 = 60°C$. The properties of air are:

$$\rho = 1.1 \text{ kg/m}^3, \nu = 18.7 \times 10^{-6} \text{ m}^2/\text{s}, Pr = 0.7, k = 0.028 \text{ W/mK}$$

Since the critical Reynolds number is 2×10^5, the critical length,
$$x_c = 2 \times 10^5 \times 18.7 \times 10^{-6}/50 = 0.0748 \text{ m} = 7.48 \text{ cm}.$$

For laminar flow, the average heat transfer coefficient is given by

$$h = (k/x_c) \, 0.664 \, (Re)^{1/2} \, (Pr)^{1/3} = (0.028/0.748) \times 0.664 \times 397.13$$
$$= 98.71 \text{ W/m}^2\text{K}$$

Rate of heat transfer from the first 7.48 cm of the surface is

$$\dot{Q}_1 = 98.71 \times (0.0748 \times 0.3) (95 - 25) = 155 \text{ W}$$

Turbulent flow exists between $x = 0.0748$ m and 0.40 m. The Reynolds number at the trailing edge of the plate is

$$Re_L = 50 \times 0.4/18.7 \times 10^{-6} = 1.07 \times 10^6 < 10^7$$

From Eq (7.17),

$$h_x = 0.0296 \, (k/x) \, (Re_x)^{0.8} \, (Pr)^{1/3}$$

The average value of h for the turbulent portion is given by

$$h = \frac{1}{L - x_c} \int_{x_c}^{L} h_x dx$$

$$= [1/\{(0.4 - 0.0748)\} \, \{0.0296 \, k \, (Pr)^{1/3} \, (U_\infty/\nu)^{0.8} \int_{x_c}^{L} dx/x^{0.2}\}]$$
$$= (1/0.3252) \, [100 \times (0.4^{0.8} - 0.0748^{0.8}) \, /0.8] = 136.38 \text{ W/m}^2\text{K}$$

Rate of heat transfer from the turbulent portion is

$$\dot{Q}_2 = 138.38 \times (0.3252 \times 0.3) (95 - 25) = 945 \text{ W}$$

and the total heat transfer rate $= 945 + 155 = 1100 \text{ W}$.

Example 7.3 A flat plate 1.5 m × 1.5 m is maintained at 95°C. The rate of heat transfer from the plate is 4.5 kW when an air stream at 25°C flows over the plate. Estimate the speed of air.

Solution: The mean temperature is $(95 + 25)/2 = 60°C$. The properties are:

$$\nu = 18.7 \times 10^{-6} \text{ m}^2/\text{s}, \ Pr = 0.7, \ k = 0.028 \text{ W/mK}$$

The average heat transfer coefficient, $h = Q/(A(\Delta T))$

$$h = 4.5 \times 10^3/(1.5 \times 1.5)70 = 28.57 \text{ W/m}^2\text{K}$$

Assuming a laminar-turbulent boundary layer over the plate, with the help of Eq. (7.19), we have,

$$h \, L/k = (0.037 \ Re_L^{0.8} - 850) \ Pr^{1/3}$$

or, $28.57 \times 1.5/0.028 = (0.037 \ Re_L^{0.8} - 850) \ (0.7)^{1/3}$

and, $Re_L^{0.8} = [(1530.535)/0.888 + 850]/0.037 = 6.9556 \times 10^4$

and the Reynolds number $= 1.129 \times 10^6$

Free stream velocity, $U_\infty = 1.129 \times 10^6 \times 18.7 \times 10^{-6}/1.5$
$$= 14.07 \text{ m/s}.$$

Example 7.4 Water at 25°C flows over a flat plate, 1m × 1m, with a speed of 4 m/s. Determine the thickness of the boundary layers, convective heat transfer coefficient and the drag force when the plate is maintained at 65°C.

Solution: The mean temperature is $(65 + 25)/2 = 45°C$.
The properties of water are:

$$\mu = 5.77 \times 10^{-4} \text{ Pa-s}, \ Pr = 3.77, \ k = 0.64 \text{ W/mK}, \ \mu_\infty = 8.55 \times 10^{-4} \text{ Pa-s},$$

$$\rho = 1000 \text{ kg/m}^3, \ \mu_w = 4.2 \times 10^{-4} \text{ Pa-s}.$$

Reynolds number at the trailing edge, $Re = 10^3 \times 4 \times 1/5.77 \times 10^{-4}$
$$= 6.93 \times 10^6, \text{ a turbulent flow.}$$

Assuming that the boundary layer is laminar upto $Re = 5 \times 10^5$, the thickness of the momentum boundary layer is given by Eq. (7.11c)

or, $\delta = 1 \times 0.381 (6.93 \times 10^6)^{-0.2} - 10256/6.93 \times 10^6$
$$= 0.0148 \text{ m} \equiv 1.48 \text{ cm}$$

Since in view of the intensive turbulent transport, the thicknesses of the thermal and momentum boundary layers practically coincide, the thickness of the thermal boundary layer will also be equal to 1.48 cm.

From Eq. (7.15), the average friction coefficient for $Re_c = 5 \times 10^5$, is given by:

$$C_f = 0.074 / Re_L^{0.2} - 1700/Re_L$$
$$= 0.074 / (6.93 \times 10^6)^{0.2} - 1700/6.93 \times 10^6 = 2.925 \times 10^{-3}$$

Drag on one side of the plate, $D = C_f (\rho U^2/2)$ area

$$= 2.925 \times 10^{-3} \times 10^3 \times 16/2 = 23.4 \text{ N}$$

From Eq. (7.20), the average heat transfer coefficient is

$$h = k/L \times 0.036 \ (Re_L^{0.8} - 9200) \ Pr^{0.43} \ (\mu_\infty/\mu_w)$$
$$= 0.64/1 \times 0.036 \ [(6.93 \times 10^6)^{0.8} - 9200] \ (3.77)^{0.43} \ (8.55 \times 10^{-4}/4.2 \times 10^{-4})$$
$$= 23.875 \text{ kW/m}^2\text{K}.$$

Example 7.5   A flat plate 2 m long is maintained at 95°C. An air stream at 25°C flows over the plate with a speed of 50 m/s. Estimate the rate of heat transfer and the drag force per unit width of the plate when (i) boundary layer is assumed to be turbulent from the leading edge (ii) boundary layer is partly laminar and partly turbulent.

Solution: The physical properties at the mean temperature is taken from Ex. 7.3
Reynolds number at the trailing edge, $Re_L = 50 \times 2/18.7 \times 10^{-6}$
$$= 5.348 \times 10^6, \text{ a turbulent}$$
For a turbulent boundary layer from the leading edge,

$$Nu = 0.037\, Re^{0.8}\, Pr^{1/3} \text{ and } \delta/x = 0.381\, Re^{-0.2}$$
Therefore, $h = 0.028 \times 0.037 \times (5.348 \times 10^6)^{0.8} \times (0.7)^{0.333}/2$
$$= 110.99 \text{ W/m}^2\text{K and } \delta = 9.02\text{cm}$$
Rate of heat transfer, $Q = 110.99 \times (2 \times 1) \times (95 - 25) = 15.54$ kW
Assuming that laminar boundary layer exists upto $Re = 5 \times 10^5$,

$$h = (0.028/2) \times (0.037 \times (5.348 \times 10^6)^{0.8} - 850) \times (0.7)^{0.333}$$
$$= 100.427 \text{ W/m}^2\text{K}; \ \delta/x = 0.381 Re_x^{-0.2} - 10256/Re \Rightarrow \delta = 7.102 \text{ cm}$$

Rate of heat transfer, $\dot{Q} = 100.427 \times (2 \times 1) \times (95 - 25) = 14.06$ kW
the value decreases by 9.5%.
Drag force = Skin friction coefficient $\times \rho\, U^2/2 \times$ Surface area
$$= 0.074/ (5.348 \times 10^6)^{0.2} \times 1.1 \times 2500/2 \times (2 \times 1) = 9.18 \text{ N}$$
Drag force when the boundary layer is partly laminar and partly turbulent,

$$= [0.074 (5.348 \times 10^6)^{-0.2} - 1700/5.34 \times 10^6] \times 1.1 \times 1250 \times 2$$
$$= 8.3\text{N}$$
i.e., the drag force decreases by a small value and the boundary layer thickness decreases by about 22 percent.

8. Turbulent Flow Heat Transfer with Constant Heat Flux

Experimental results show that for constant wall heat flux in turbulent flow, the local Nusselt number is only about 4 percent higher than for isothermal surfaces. With the current interest in solar energy and flat plate solar collectors, experiments have been conducted on inclined plates of finite width and inclined at various angles of attack to air flow. The proposed correlation is accurate within + 10%.

$$Nu_{av} = 0.86\, Re^{1/2}\, Pr^{1/3}, \quad 20000 < Re < 90000, \ 90° < \alpha < 25° \qquad (7.21)$$

where the Reynolds number is based on the characteristic length,

$$L_c = 4 \text{ (plate surface area) / (plate circumference).}$$

9. Turbulent Flow in Tubes — Velocity Profile

The fully developed velocity profile for turbulent flow in a tube is shown in Fig 7.4. The profile can be divided into three distinct zones. The flow is essentially laminar in a very thin region close to the wall and is termed the 'viscous sublayer'. In a region adjacent to the axis of the tube, there is a fully turbulent region which is called the 'turbulent core'. The third region, called the 'buffer layer' separates the viscous sublayer and the turbulent core.

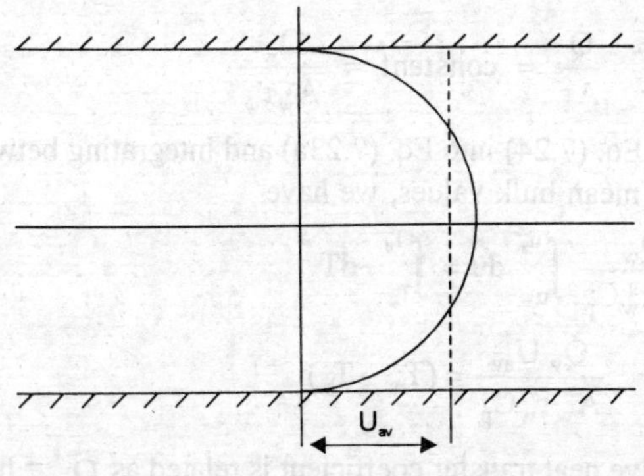

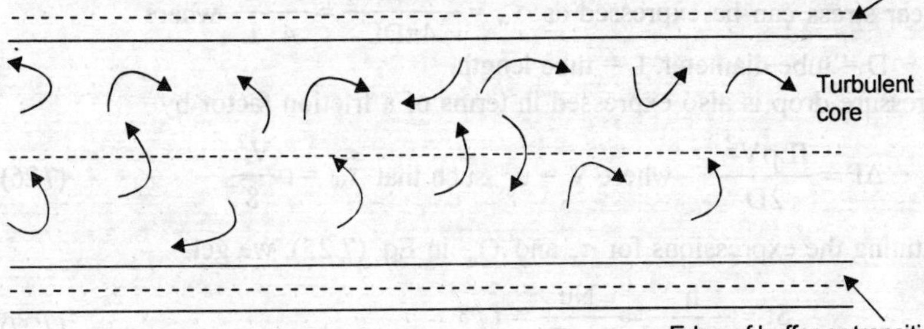

**Fig. 7.4 Fully developed velocity profile and the flow pattern
for the turbulent flow in a pipe**

10. Turbulent Flow through Tubes—Reynolds Analogy, Evaluation of 'h'

In order to compute the convective heat transfer coefficient in turbulent flow
through the tube, it is assumed that the transport of heat energy can be represented
by

$$\frac{\dot{Q}}{\rho C_p A} = -(\alpha + \varepsilon_H)\frac{dT}{dy} \quad \text{where } \varepsilon_H \text{ is eddy thermal diffusivity.} \quad (7.22)$$

The above equation represents the total heat conducted as a sum of the
molecular conduction and the macroscopic eddy conduction. Similarly, the shear
stress in turbulent flow is expressed as

$$\frac{\tau}{\rho} = (v + \varepsilon)\frac{du}{dy}, \text{ where } \varepsilon \text{ is eddy momentum viscosity.} \quad (7.23)$$

Further it is assumed that the momentum and heat energy are transported at
the same rate, i.e., $\varepsilon = \varepsilon_H$ and $v = \alpha$, or $Pr = 1$.
Dividing Eq. (7.22) by Eq. (7.23), we get

$$\frac{Q}{AC_p\tau}.du = -dT \quad (7.23a)$$

Since it is assumed that the heat and momentum are transported at the same
rate, it can also be assumed that the ratio of the heat transfer per unit area to the
shear stress is constant across the flow field,

Or,
$$\frac{\dot{Q}}{A\tau} = \text{constant} = \frac{\dot{Q}_w}{A_w \tau_w} \tag{7.24}$$

Substituting Eq. (7.24) into Eq. (7.23a) and integrating between the conditions at the wall and mean bulk values, we have

$$\frac{\dot{Q}_w}{A_w \tau_w C_p} \int_o^{u_{av}} du = \int_{T_w}^{T_b} -dT$$

or
$$\frac{\dot{Q}_w U_{av}}{A_w \tau_w C_p} = (T_w - T_b) \tag{7.25}$$

The convective heat transfer coefficient is related as $\dot{Q}_w = h\,A_w\,(T_w - T_b)$, and

the shear stress can be expressed as $\tau_w = \dfrac{\Delta P(\pi D^2)}{4\pi DL} = \dfrac{\Delta P}{4}\dfrac{D}{L}$ where

$\qquad\qquad$ D = tube diameter, L = tube length.
The pressure drop is also expressed in terms of a friction factor by

$$\Delta P = \frac{fL\rho V^2}{2D} \quad \text{where } V = u_{av} \text{ such that } \tau_w = \rho\frac{V^2}{8} \tag{7.26}$$

Substituting the expressions for τ_w and $\dot{Q}_w$ in Eq. (7.25), we get

$$St = \frac{h}{\rho C_p V} = \frac{Nu}{Re.Pr} = f/8 \tag{7.27}$$

which is the Reynolds analogy for turbulent flow in pipes.

Blasius has recommended that the turbulent friction factor upto Reynolds numbers of about 2×10^5 for flow in smooth tubes is given by

$$f = 0.316/(Re)^{\frac{1}{4}} \tag{7.28a}$$

and that gives $\quad Nu/(Re.Pr) = 0.0395/(Re)^{-\frac{1}{4}} \tag{7.28b}$

or, $\qquad\qquad\qquad Nu = 0.0395\,Re^{4/5}$, for Pr = 1.

For flow over a flat plate, it has been indicated that fluid-friction-heat-transfer analogy shows a Prandtl number dependence of $Pr^{2/3}$ and it has been observed that this dependence works very well for turbulent flow in tubes also. Thus,

$$Nu = 0.0395\,Re^{0.8}\,Pr^{1/3} \tag{7.29}$$

Dittus and Boelter have recommended the following correlations for fully developed turbulent flow inside tubes

$$Nu = 0.023\,Re^{0.8}\,Pr^{0.4} \text{ for heating} \tag{7.30}$$

$$Nu = 0.023\,Re^{0.8}\,Pr^{0.3} \text{ for cooling} \tag{7.31}$$

Another correlation suggested by Colburn is

$$Nu = 0.023\,Re^{0.8}\,Pr^{1/3} \tag{7.32}$$

When large temperature differences are present in the flow, Seider and Tate recommend the following relation

$$Nu = 0.027\,Re^{0.8}\,Pr^{1/3}\,(\mu/\mu_w)^{0.14} \tag{7.33}$$

Experiments reveal that turbulent velocity profiles require 40 to 50 diameters or more of pipe length before they are fully developed. But the distance required for the friction factor to become constant is usually less. The dimensionless distance required for the friction factor to become constant has been suggested as

$$L/D = 0.623 \ (Re)^{1/4} \tag{7.34}$$

and Nusselt has recommended that the following relation should be used in the entrance region

$$Nu = 0.036 \ Re^{0.8} \ Pr^{1/3} \ (D/L)^{0.055} \ \text{for} \ 10 < L/D < 400 \tag{7.35}$$

Most pipes in engineering structure cannot be regarded as being hydraulically smooth, atleast at higher Reynolds number and therefore, it is imperative that the effects of surface roughness, ε , on the velocity profile and resistance to flow is considered. The calculation of the friction factor is required in calculating the pumping power and in determining the convective heat transfer coefficient using Reynolds analogy.

Nikuradse conducted a very systematic, extensive and careful measurements on rough pipes and suggested that if the roughness parameter ε/R lies between 0.00197 and 0.0327, the the power law exponent n in velocity profile empirical equation, given by $\bar{u}/U_\infty = (y/R)^{1/n}$ lies between 4 and 5. Schlichting, after comparing the results obtained by Nikuradse has suggested that the resistance formula for completely rough pipe can be approximated by

$$f = [2 \ \log_{10} \ (R/\varepsilon) + 1.74]^{-2.0}$$

$$\tag{7.36}$$

11. Effect of Surface Irregularities

If the surface irregularities do not disturb the laminar pattern of fluid flow in the sublayer, the fluid flows over them without separation or eddy formation. Then, there is no difference between a rough tube and smooth tube. If the surface roughness is greater than the thickness of the laminar sublayer, the fluid flow takes on a separated eddy pattern. Since the thermal resistance to heat transfer is concentrated in the sublayer, any change in flow pattern will cause an increase in heat transfer coefficient and pressure drop.

Example 7.6   Calculate the heat transfer coefficient for water flowing through a 25 mm inner diameter tube at the rate of 2 kg/s, when the mean bulk temperature is 40°C.

Solution: The physical properties of water at 40°C are:

$\rho = 1000 \ \text{kg/m}^3, \quad \mu = 6.5 \times 10^{-4} \ \text{Pa-s}, \ Pr = 4.3, \ k = 0.632 \ \text{W/mK}$

Volume rate of flow = 2.0 / 1000 = 0.002 m^3/s

Average velocity, $V = 0.002 \times 4/(3.142 \times 0.025^2) = 4.074$ m/s

Reynolds number, $Re = 10^3 \times 4.074 \times 0.025/6.5 \times 10^{-4}$

$$= 1.567 \times 10^5, \text{a turbulent flow}$$

From Eq. (7.32), $Nu = 0.023 \, (Re)^{0.8} \, (Pr)^{0.333}$

$$= 0.023 \, (1.567 \times 10^5)^{0.8} \, (4.3)^{0.333} = 535.46$$

The convective heat transfer coefficient, $h = 535.46 \times 0.632/0.025$

$$= 13.53 \text{ kW/m}^2 \text{ K.}$$

Example 7.7 Air is heated by passing it through a 25 mm inner diameter copper tube which is maintained at 300°C. The air enters at 25°C and comes out at 270°C at a mean velocity of 30 m/s. Estimate the length of the tube and the pumping power required.

Solution: The mean temperature $= \frac{1}{2}(T_w + T_b) = \frac{1}{2}\left(300 + \frac{270 + 25}{2}\right) = 223.75°C$

The properties of air are:

$\rho = 0.7 \text{ kg/m}^3$, $\quad \mu = 2.7 \times 10^{-5}$ Pa-s, $\quad Pr = 0.684$,

$k = 0.04$ W/mK, $\quad C_p = 1030$ J/kgK

Reynolds number, $\quad Re = 0.7 \times 30 \times 0.025/2.7 \times 10^{-5}$

$$= 1.94 \times 10^4, \text{ a turbulent flow}$$

From Eq. (7.28a), the friction factor, $\quad f = 0.316/Re^{\frac{1}{4}}$

$$= 0.316/(1.94 \times 10^4)^{\frac{1}{4}}$$

$$= 0.02677$$

From Eq. (7.27), $\quad h = (\rho \, C_p \, V) \cdot f/8 = 0.7 \times 1030 \times 30 \times 0.02677/8$

$$= 72.38 \text{ W/mK}$$

Mass flow of air $= (3.142/4) \times (0.025)^2 \times 30 \times 0.7 = 0.0103$ kg/s

Heat received by air $= \dot{m} \, C_p \, (\Delta T) = 0.0103 \times 1030 \times (270 - 25) = 2599.2$ W

$\dot{Q} = h \, A \, (LMTD)$; or, $2599.2 = 72.38 \times A \times [(300 - 25) - (300 - 270)]/\ln(275/30)$

Therefore, $A = 0.325$ m^2, and the tube length, $L = 0.325/(3.142 \times 0.025)$

$$= 4.137 \text{ m}$$

P/L, pumping power per unit length $= \tau_\omega \pi D V$

Heat flow per unit area, $\qquad \dot{Q} = \tau_\omega C_p \, (\Delta T)/V$, and therefore,

$\dot{Q}$/L, heat flow per unit length, $\qquad = \tau_\omega C_p \, (\Delta T) \, \pi \, D/V$; where $\Delta T = LMTD$

$$= 110.58$$

By taking the ratio of the two expressions,

Pumping power, $P = \dot{Q} \times V^2/[C_p(\Delta T)] = 2599.2 \times 900/(1030 \times 110.58)$

$$= 20.54 \text{ W.}$$

Example 7.8 Water at 25°C is flowing through a rough pipe, 80 cm in diameter, at the rate of 2 m^3/s. Calculate the power loss in overcoming friction per km length of the pipe if the height of the roughness element is 0.5 mm.

Solution: From Eq. (7.36), the friction factor is given as

$$f = [2 \log_{10}(R/\varepsilon) + 1.74]^{-2.0} = [2 \log_{10}(40/0.05) + 1.74]^{-2.0} = 0.0176$$

The average velocity, V = volume/area = $2/(3.142 \times 0.64/4) = 3.98$ m/s

Head lost in friction, $h_f = f L V^2/(2 g D)$

$$= 0.0176 \times 1000 \times (3.98)^2 / (2 \times 9.81 \times 0.8) = 17.76 \text{ m}$$

$$\text{Power loss} = \rho Q g h_f = 10^3 \times 2 \times 9.81 \times 17.76 = 348.89 \text{ kW}$$

Convective heat transfer coefficient, $h = \rho C_p V \times f/8$

$$h = (10^3 \times 4182 \times 3.98) \times 0.0176/8 = 36.61 \text{ kW/m}^2\text{K}.$$

Example 7.9   Air at 1 bar and 200°C is heated when it flows through a long tube 25 mm inner diameter at a velocity of 20 m/s. Calculate the rate of heat transfer per unit length of the tube. The tube receives heat energy at a constant rate such that the wall temperature of the tube is 20°C higher than the air temperature all along the length. Estimate the temperature of air at outlet if the length of the tube is 2 m.

Solution: Assuming that the temperature of air as 200°C, the properties are:

$$\rho = 10^5/(287 \times 473) = 0.7366 \text{ kg/m}^3, \ \ Pr = 0.68, \ \ \mu = 2.56 \times 10^{-5} \text{ Pa-s},$$

$$k = 0.0387 \text{ W/mK}, \ \ C_p = 1025 \text{ J/kgK}$$

Reynolds number, $Re = 0.7366 \times 20 \times 0.025/2.56 \times 10^{-5}$

$$= 14387, \text{ a turbulent flow.}$$

Since the air is being heated, we use the Eq. (7.30),

$$Nu = 0.023 \, Re^{0.8} \, Pr^{0.4}$$

$$= 0.023 \times (14387)^{0.8} \times (0.68)^{0.4} = 41.793$$

Convective heat transfer coefficient, $h = 41.793 \times 0.0387/0.025$

$$= 64.69 \text{ W/m}^2\text{K}$$

Rate of heat transfer per unit length, $\dot{Q}/L = 64.69 \times 3.142 \times 0.025 \times 20$

$$= 101.628 \text{ W/m}$$

In order to calculate the air exit temperature, we make an energy balance

$$\dot{Q} = \dot{m} C_p (T_0 - 200), \text{ where,}$$

$$\dot{m} = 0.7366 \times 3.142 \times (0.025/2)^2 \times 20 = 7.2325 \times 10^{-3} \text{ kg/s}$$

or, $101.628 \times 2 = 7.2325 \times 10^{-3} \times 1025 \times (T_0 - 200)$ which gives $T_0 = 213.7°C$.

Example 7.10   Water at 25°C flows through a pipe 10cm in diameter and 5m long. Estimate the water temperature at exit when the tube wall temperature is maintained at 75°C, average velocity of water 2m/s.

Solution: The ratio $L/D = 5/0.1 = 50$ which is less than 400.

Therefore, we use Eq. (7.35)

The mean temperature can be assumed as $(75 + 25)/2 = 50°C$.

The properties of water is: $\rho = 971$ kg/m^3, Pr = 3.42, $\mu = 5.28 \times 10^{-4}$ Pa-s

$$k = 0.645 \text{ W/mK}, \qquad C_p = 4182 \text{ J/kgK}$$

Reynolds number, Re = $971 \times 2 \times 0.1/5.28 \times 10^{-4} = 3.678 \times 10^5$, turbulent flow

$$\text{Nu} = 0.036 \text{ Re}^{0.8} \text{ Pr}^{1/3} (D/L)^{0.055}$$

$$= 0.036 (3.678 \times 10^5)^{0.8} (3.42)^{0.333} (0.1/5)^{0.055} = 1239.3$$

and $\qquad$ h = $1239.3 \times 0.645/0.1 = 7.99$ kW/m^{2}K

Water flow rate, $\dot{m} = \rho A V = 971 \times 3.142 (0.05)^2 \times 2 = 15.25$ kg/s

By making an energy balance,

$$\dot{m} C_p (T_O - 25) = h \times \pi DL \times [T_w - (T_o + 25)/2]$$

Solving for T_o, we get $T_o = 33.96°$C.

Example 7.11 Air enters at 25°C into a long annulus (inner diameter 4 cm outer diameter 6cm) with a velocity of 30 m/s. The temperature of the outer surface of the inner tube is maintained at 75°C and air comes out of the annulus at 50°C. Estimate the convective heat transfer coefficient.

Solution: The mean temperature is $(75 + (25 + 50)/2)/2 = 56.25°$C.

The properties of air are: $\rho = 1.07$ kg/m^3, k = 0.028 W/mK, Pr = 0.7

$$\mu = 19.8 \times 10^{-6} \text{ Pa–s.}, C_p = 1005 \text{ J/kgK}$$

The hydraulic diameter, $D_h = 4A/p = D_o - D_i = 2$ cm.

Reynolds number, Re = $1.07 \times 30 \times 0.02/19.8 \times 10^{-6} = 3.24 \times 10^4$.

Since the air is being heated, we use the relation

$$\text{Nu} = 0.023 \text{ Re}^{0.8} \text{ Pr}^{0.4}$$

$$= 0.023 (3.24 \times 10^4)^{0.8} \times (0.7)^{0.4} = 80.95$$

Convective heat transfer coefficient, h = $80.95 \times 0.028/0.02$

$$= 113.33 \text{ W/m}^2\text{K}.$$

Mass flow of air = $3.142 \times (6^2 - 4^2) \times 10^{-4} \times 1.07 \times 30 = 5.043 \times 10^{-2}$ kg/s

Heat received by air = $\dot{m} C_p (\Delta T) = 5.043 \times 10^{-2} \times 1005 \times (50 - 25) = 1267$ W

Since the wall temperature is constant, the log mean temperature difference,

$$\text{LMTD} = ((75 - 25) - (75 - 50))/\ln 2 = 36.06°\text{C}$$

By making an energy balance, h A (LMTD) = 1267

Therefore Area, A = $1267/(113.33 \times 36.06) = 0.31$ m^2.

and this would require a length, L = $0.31/(0.02 \times 3.142) = 4.93$ m.

12. Heat Transfer with Liquid Metals—Merits and Demerits

Liquid metals are best suited for use as convective fluids when large heat transfer rates are to be obtained in relatively small spaces, such as in nuclear reactors. The advantages of using liquid metals are:

1. high thermal capacity (product of specific heat and density),
2. low viscosity,
3. they remain as liquid at high temperatures.

The disadvantages are:
1. solidify at room temperature,
2. relatively expensive, and
3. some react violently with air or water and some are toxic.

The commonly used liquid metals are mercury, sodium and lead-bismuth alloys. The heat transfer mechanism in cas e of flowing liquid metal is significantly different than that from other fluids because they have high thermal conductivities and very low Prandtl number and this leads to a very thick thermal boundary layer.

13. Liquid Metals Flowing over Flat Plate—Peclet Number

Since the thermal conductivity of liquid metals is very high, the primary mode of heat transfer in the boundary layer would be by conduction, in both laminar and turbulent flows. Further, it is assumed that the velocity profile inside the thermal boundary would be nearly uniform (thickness of the momentum boundary layer is very small in comparison with the thickness of the thermal boundary layer). Fig. 7.5. Thus, the energy equation (6.11) can be written as

$$U_\infty \partial T/\partial x = \alpha \partial^2 T/\partial y^2 \tag{7.37}$$

and the solution of this equation for a constant surface temperature is

$$\dot{Q}/A = k(T_w - T_\infty)(U_\infty/\pi\alpha x)^{1/2} \tag{7.38}$$

Since $\dot{Q}/A = h_x(T_w - T_\infty)$, we write

$$\frac{h_x x}{k} = \left(\frac{U_\infty k}{\pi\alpha}\right)^{\frac{1}{2}} = \left(\frac{1}{\pi}\right)^{\frac{1}{2}}\left(\frac{U_\infty x}{v}\right)^{\frac{1}{2}}\left(\frac{v}{\alpha}\right)^{\frac{1}{2}}$$

or,

$$Nu_x = \frac{\sqrt{Re_x \, Pr}}{\sqrt{\pi}} = 0.564(Pe_x)^{0.5} \tag{7.39}$$

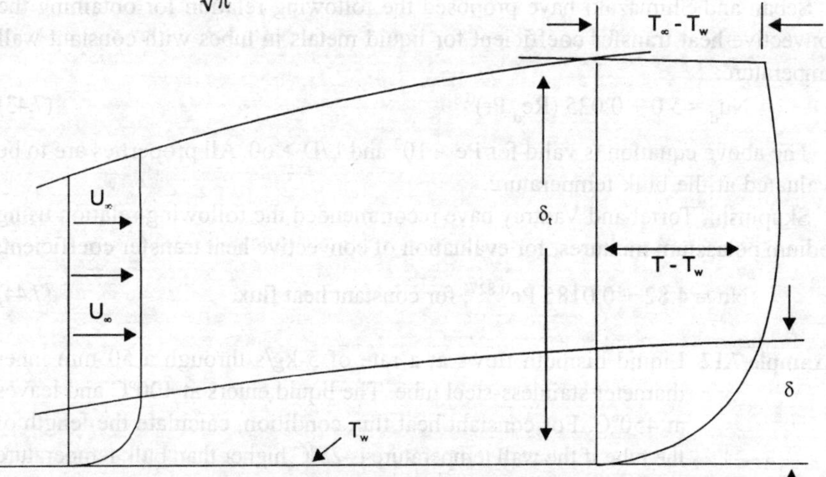

Fig. 7.5 Boundary-layer regimes for analysis of liquid-metal heat transfer

where the local Peclet number Pe_x is the product of the local Reynolds number and the fluid Prandtl number. And, the Peclet number is the ratio of the energy transport by convection to energy transport by conduction.

When the integral energy equation is considered and written as

$$\frac{d}{dx}\left[\int_0^{\delta_t}(T_\infty - T)u\,dy\right] = \alpha \left.\frac{dT}{dy}\right|_{y=0} \tag{7.40}$$

and it is assumed that the temperature profile is given by a cubic parabola

$$\frac{\theta}{\theta_w} = \frac{T - T_w}{T_\infty - T_w} = \frac{3}{2}\left(\frac{y}{\delta_t}\right) - \frac{1}{2}\left(\frac{y}{\delta_t}\right)^3$$

The convective heat transfer coefficient is evaluated from the relation

$$Nu_x = \frac{h_x x}{K} = 0.530(Re_x\, Pr)^{1/2} = 0.530(Pe_x)^{0.5} \tag{7.41}$$

The value of h given by equation (7.39) is about 7 percent higher than that given

by Eq (7.41). The thickness of the thermal boundary layer can be computed from

the relation: $\dfrac{\delta}{\delta_t} = 1.64(Pr)^{\frac{1}{2}}$

14. Liquid Metals flowing inside Tubes—Empirical Relations

Lubarsky and Kaufman have recommended the following relation for calculation of heat transfer coefficient in fully developed turbulent flow of liquid metals in smooth tubes with constant heat flux at the wall:

$$Nu_d = h\, D/k = 0.625\, (Re_d\, Pr)^{0.4} \tag{7.42}$$

The above equation is valid for $10^2 < Pe < 10^4$ and for $L/D > 60$. All properties are to be evaluated at the bulk temperature.

Seban and Shimazaki have proposed the following relation for obtaining the convective heat transfer coefficient for liquid metals in tubes with constant wall temperature:

$$Nu_d = 5.0 + 0.025\, (Re_d\, Pr)^{0.8} \tag{7.43}$$

The above equation is valid for $Pe > 10^2$ and $L/D > 60$. All properties are to be evaluated at the bulk temperature.

Skupinshi, Tortel and Vautrey have recommended the following relation using sodium potassium mixtures, for evaluation of convective heat transfer coefficient:

$$Nu = 4.82 + 0.0185\, Pe^{0.827}, \text{ for constant heat flux.} \tag{7.44}$$

Example 7.12 Liquid bismuth flows at a rate of 5 kg/s through a 50 mm inner diameter stainless-steel tube. The liquid enters at 400°C and leaves at 450°C. For constant heat flux condition, calculate the length of the tube if the wall temperature is 25°C higher than bulk temperature of bismuth all along the tube. What would be the length of the tube if wall of the tube is maintained at a constant temperature of 475°C?

Solution: The properties of bismuth at bulk temperature are:

$\mu = 1.34 \times 10^{-3}$ Pa–s, $C_p = 149$ J/kgK, $k = 15.7$ W/mK, Pr = 0.013

Cross-sectional area of the tube = $3.142 \times (0.025)^2 = 1.9637 \times 10^{-3}$ m^2

Density × Velocity = Mass flow rate/Cross-sectional area

$$= 5/1.9637 \times 10^{-3} = 2.546 \times 10^3$$

$$\text{Re} = 2.546 \times 10^3 \times 0.05/1.34 \times 10^{-3} = 95000$$

$$\text{Pe} = \text{Re} \times \text{Pr} = 95000 \times 0.013 = 1235$$

Since the Eq. (7.44) is valid for $3.6 \times 10^3 < \text{Re} < 9.05 \times 10^5$ and $10^2 < \text{Pe} < 10^4$, we can use this equation

or,   Nu = $4.82 + 0.0185 \, (1235)^{0.827} = 11.49$ and h $= 11.49 \times 15.7/0.05$
$$= 3607.8 \text{ W/m}^2\text{K}$$

The quantity of energy supplied = $\dot{m} \, C_p \, (T_o - T_i) = 5 \times 149 \times 50 = 37250$ W

The required surface area of the tube can be computed from

$$37250 = \text{h} \, (\,\pi\text{D L}) \, (\Delta\,\text{T}) = 3607.8 \times 3.142 \times 0.05 \times \text{L} \times 25$$

or, the length,   L = 2.31 m.

When the tube wall temperature is maintained as constant, we use Eq. (7.43)

$$\text{Nu} = 5.0 + 0.025 \, (\text{Pe})^{0.8} = 5.0 + 0.025(1235)^{0.8}$$
$$= 12.435, \text{ and h} = 12.435 \times 15.7/0.05 = 3878 \text{ W/m}^2\text{K}$$

The log mean temperature difference, LMTD = $[(475 - 400) - (475 - 450)]/\ln(3)$
$$= 45.5°\text{C}$$

By making an energy balance,

$$37250 = \text{h} \, (\,\pi\text{D L}) \, (\text{LMTD})$$

Therefore,   L = $37250/(3878 \times 3.142 \times 0.05 \times 45.5) = 1.344$ m

and   L/D = $1.344/0.05 = 26.88 < 60$, therefore it becomes an entry region problem.

(Thus there are many open questions concerning liquid-metal heat transfer and they require attention.)

15. Combined Free and Forced Convection Heat Transfer—Mechanism

A number of practical situations involve convection heat transfer which is neither forced nor free in nature. Free convection takes place whenever there is an unstable temperature gradient. Forced convection is significant when a fluid is flowing over a solid surface or through a tube. The two significant dimensionless parameter: Grashof and Reynolds number are used to identify the convection flow regimes. It has been observed that when

$Gr/Re^2 \gg 1$, we have free convection (forced convection negligible)

$Gr/Re^2 \approx 1$, we have free and forced convection of the same order

and, $Gr/Re^2 \ll 1$, we have forced convection (free convection negligible).

Thus, when a fluid is flowing over a heated flat plate or with a very low velocity through a horizontal or vertical tube with a radial temperature distribution, we get a natural convection superimposed on the forced convection. In a horizontal tube with hot walls, natural convection occurs as shown in Fig. 7.6. This motion when superimposed on axial flow, distorts the axial velocity distribution of Fig. 6.7.

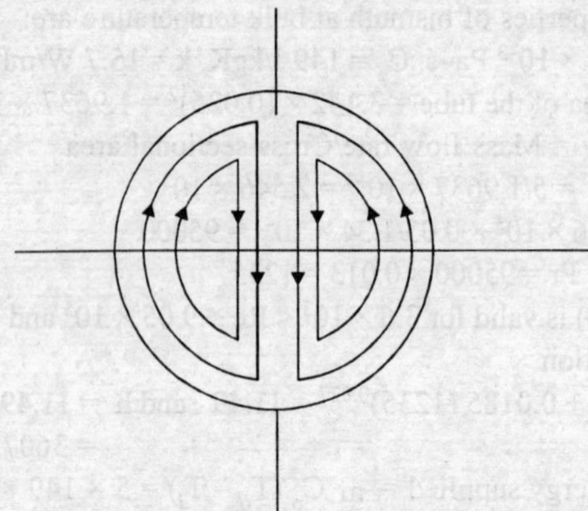

Fig. 7.6 Natural convection in horizontal hot tubes

Brown and Gauvin have developed an empirical correlation for the mixed convection in laminar flow region:

$$Nu = 1.75(\mu_b / u_w)^{0.14}[Gz + 0.012(Gz.Gr^{1/3})^{4/3}]^{1/3} \qquad (7.45)$$

where μ_b is evaluated at the bulk temperature.

Eubank and Proctor have surveyed the available data for laminar flow of petroleum oils in horizontal tubes and the approximate empirical relation suggested by them is:

$$Nu = 1.75 [\pi/4 \, (Gz) + 0.04 \, (Gr \, Pr \, D/L)^{0.75}]^{1/3} \qquad (7.46)$$

where $Gz = Re \, Pr \, D/L$ and Gr is based on the tube diameter and ΔT_a. The product $Gr.Pr$ ranged from 3.3×10^5 to 8.6×10^8 and the relation is based on data on petroleum oils with Prandtl number varying from 140 to 15200. The properties are to be evaluated at the mean temperature and ΔT_a is the arithmetic mean temperature difference between entrance and exit.

Example 7.13 Air at 1 atm and 300 K is forced through a horizontal 25 mm diameter tube at an average velocity of 30 cm/s. The tube wall is maintained at 140°C. Calculate the heat transfer coefficient for this situation when the tube is 0.5 m long.

Solution: The film temperature, $T = (140 + 27)/2 = 83.5°C = 356.5$ K

And the properties are: $\rho = p/RT = 1.0132 \times 10^5/(287 \times 355.5) = 0.99$ kg/m³

$\beta = 1/T = 2.805 \times 10^{-3}$ K^{-1}, $\mu_w = 2.337 \times 10^{-5}$ Pa–s

$\mu_f = 2.102 \times 10^{-5}$ Pa-s. k = 0.0305 W/mK, Pr = 0.695

Assuming the bulk temperature of air at 27°C, $\mu_b = 1.846 \times 10^{-5}$ Pa–s

$$Re = \rho V D/\mu = 0.99 \times 0.3 \times 0.025/2.102 \times 10^{-5} = 353$$

$$Gr = \rho^2 g\beta \, (\Delta T) \, D^3/\mu^2 = \frac{(0.99)^2 \times 9.81 \times 2.805 \times 10^{-3} \times (140 - 27)(0.025)^3}{(2.102 \times 10^{-5})^2}$$

$$= 1.007 \times 10^5$$

$$Gr/Re^2 = 1.007 \times 10^5/(353)^2 = 0.807 \approx 1, \text{ therefore, we use Eq. (7.45)}$$
$$Gz = Re \ Pr \ D/L = 353 \times 0.695 \times 0.025/0.5 = 12.27.$$
From Eq. (7.45): $Nu = 1.75 (1.846 \times 10^{-5}/2.337 \times 10^{-5})^{0.14} \times$
$$[12.27 + 0.012 \{12.27 (1.007 \times 10^5)^{1/3}\}^{4/3}]^{1/3}$$
$$= 6.945$$
And the average heat transfer coefficient
$$h = 6.945 \times k/D = 6.945 \times 0.0305/0.025 = 8.47 \text{ W/m}^2\text{K}.$$

Example 7.14 A rectangular plate 0.6 m long and 0.5 m wide is maintained at 140°C. Air at 1 atm pressure and 27°C flows over the plate with a velocity of 15 cm/s. Estimate the rate of heat transfer.

Solution: The mean or film temperature, $T = (140 + 27)/2 = 83.5°C = 356.5$ K
$$\rho = p/RT = 0.99 \text{ kg/m}^3 \quad \beta = 1/T = 2.805 \times 10^{-3} \text{ K}^{-1}$$
$$\nu = 2.123 \times 10^5 \text{ m}^2/\text{s. } k = 0.0305 \text{ W/mK, } Pr = 0.695$$
Characteristic length for calculating Grashof number = Area / Perimeter
$$(0.6 \times 0.5)/(2(0.6 + 0.5) = 0.136$$
$$Gr = 9.81 \times 2.805 \times 10^{-3} (140 - 27) (0.136)^3 / (2.123 \times 10^{-5})^2$$
$$= 1.735 \times 10^7$$
Reynolds number, $Re = V \ L/\nu = 0.15 \times 0.6 / (2.123 \times 10^{-5}) = 4239$
$$Gr/Re^2 = 1.735 \times 10^7/(4239)^2 = 0.965. \text{ We have a free and forced convection}$$
$$Gr \ Pr = 1.735 \times 10^7 \times 0.695 = 1.2 \times 10^7$$
For free convection, $Nu = 0.15 (1.2 \times 10^7)^{1/3} = 34.34$ (from Table 5.1).
$$h = 34.34 \times 0.0305/0.136 = 7.7 \text{ W/m}^2\text{K}$$
For forced convection and laminar flow,
$$Nu = 0.664 \ Re_L^{0.5} \ Pr^{1/3} = 0.664 (4239)^{0.5} (0.695)^{1/3} = 38.29$$
$$h = 38.29 \times 0.0305/0.6 = 1.95 \text{ W/m}^2\text{K}$$
Rate of heat transfer $= h \ A \ (\Delta T) = (1.95 + 7.7) (0.6 \times 0.5) (140 - 27)$
$$= 327 \text{ W.}$$

Example 7.15 A liquid is flowing through a tube maintained at a constant wall temperature. For turbulent flow conditions, compare the heat transfer coefficients when the (a) velocity of flow is doubled, keeping the diameter constant, and (b) diameter is reduced by half, keeping the velocity constant.

Solution: For a turbulent flow through a tube, maintained at constant wall temperature
$$Nu = 0.023 (Re)^{0.8} (Pr)^{1/3}$$
or
$$\frac{hD}{k} = 0.023 \left(\frac{VD}{\nu} \right)^{0.8} (Pr)^{1/3}$$

and $h = \text{constant} \times V^{0.8}/D^{0.2}$ when all other parameters remain the same.

(a) when V is doubled, diameter remaining the same

$$\frac{h_2}{h_1} = \left(\frac{V_2}{V_1}\right)^{0.8} = (2)^{0.8} = 1.741$$

i.e. heat transfer coefficient will increase by 1.741 times.

(b) when the diameter is reduced by half, velocity remaining the same

$$\frac{h_2}{h_1} = \left(\frac{D_1}{D_2}\right)^{0.2} = \left(\frac{D_1}{D_1/2}\right)^{0.2} = (2)^{0.2} = 1.149$$

i.e. heat transfer coefficient will increase by 1.149 times.

SUMMARY

1. The motion of fluid particles in a turbulent boundary layer is highly irregular and is characterized by velocity fluctuations. These fluctuations increase the transfer of momentum and energy and consequently, the surface friction and convection heat transfer increase.

2. A turbulent boundary layer can be divided into three distinct regions:
   Laminar sublayer – where the momentum transport is dominated by diffusion and the velocity profile is nearly linear,
   Buffer layer – where diffusion and turbulent mixing are comparable, and
   Turbulent zone – where the transport is dominated by turbulent mixing.

3. For flow over a flat plate, the value of Re_x for which transition begins varies from 10^5 to 3×10^6 depending upon surface roughness and the turbulence level of the free stream. A representative value of $Re_{x,c} = 5 \times 10^5$ is usually assumed for boundary layer calculations.

4. The thickness of the turbulent boundary layer grows with 4/5 power of x, and the boundary layer grows faster and is thicker than a laminar boundary layer at a given distance from the leading edge.

5. A turbulent Prandtl number is defined as the ratio of eddy viscosity and eddy diffusivity. Both increase in the same proportion as compared with their molecular values and therefore, the two Prandtl numbers are assumed to be equal.

6. The local Nusselt number for constant wall heat flux is about 4 percent higher than for isothermal surfaces.

7. Reynolds analogy for turbulent flow in pipes (Re greater than 2300) suggest that $St = f/8$ where the shear stress at the wall $\tau_\omega = \Delta pD/4L = \rho V^2/8$. and $f = 0.316/Re^{1/4}$.

8. In flow through rough pipes, if the surface roughness is greater than the laminar sublayer, the flow takes on a separated eddy pattern and this causes an increase in heat transfer coefficient and pressure drop.

9. Peclet number is the product of Reynolds number and fluid Prandtl number.

10. Liquid metals are best suited for use as convective fluids when large heat transfer rates are to be obtained in relatively small spaces, such as in nuclear reactors.

11. Since the thermal conductivity of liquid metals is very high, the primary mode of heat transfer in the boundary layer would be by conduction in both laminar and turbulent flows.
12. We have combined free and forced convection when $Gr/Re^2 \approx 1$.

MULTIPLE CHOICE QUESTIONS

1. Assertion (A): The analogy between heat and momentum transfer is a useful tool for analysing turbulent heat transfer processes.
    Reasoning (R): Because mathematical equations describing the temperature distribution is based on a simplified model.
    Code: (a) Both A and R are true   (b) A is true, R is false
            (c) A is false, R is true   (d) Both A and R are false   **(a)**
2. The eddy viscosity is
    (a) a physical property of the fluid (b) analogous to the kinematic viscosity
    (c) a property of flow   (d) only (b) and (c)   **(d)**
3. The convective heat transfer coefficient for laminar and turbulent flow over a flat plate varies respectively as:
    (a) $x^{\frac{1}{2}}$ and $x^{0.2}$ (b) $x^{\frac{1}{2}}$ and $x^{0.2}$
    (c) $x^{\frac{1}{2}}$ and $x^{-0.2}$ (d) $x^{-\frac{1}{2}}$ and $x^{-0.2}$ **(d)**
4. Assertion (A): In flow over a flat plate, the turbulent heat transfer coefficient is much larger than the laminar-heat-transfer coefficient at a given value of the Reynolds number
    Reasoning (R): Because in turbulent flow the conduction mechanism is aided by innumerable eddies which carry lumps of fluid across the streamlines.
    Code: (a) A is false, R is true   (b) Both A and R are false
            (c) A is true, R is false   (d) Both A and R are true   **(d)**
5. For flow over inclined plates of finite width, the Reynolds number is calculated on the basis of a characteristic length given by
    (a) Plate surface area/plate circumference
    (b) 4 × plate surface area/plate circumference
    (c) L sin θ , where θ is the angle of attack   **(b)**
    (d) 4 × plate circumference/plate surface area
6. In liquid metal heat transfer, the thickness of the thermal boundary layer in comparison with the thickness of the momentum boundary is
    (a) very small   (b) very large   (c) of the same order
    (d) dependent on the thermal conductivity of the surface   **(b)**
7. Consider the following conditions for heat transfer:
    1. $\delta_t = \delta$ if $Pr = 1$;   2. $\delta_t \gg \delta$ if $Pr \ll 1$;   3. $\delta_t \ll \delta$ if $Pr \gg 1$.
    Which of these conditions apply for convective heat transfer
    (a) 1 and 2   (b) 2 and 3   (c) 1 and 3   (d) 1, 2 and 3   **(d)**
8. Peclet number is defined as
    (a) $Nu \times Pr$ (b) $Gr \times Pr$ (c) $Re \times Pr$ (d) $St \times Pr$ **(c)**
9. In liquid metal heat transfer, Nusselt number is a function of
    (a) Reynolds number only   (b) Prandtl number only

    (c) Peclet number only   (d) Reynolds and Peclet number.   **(c)**

10. In flow through rough tubes, if the height of the surface roughness is greater than the thickness of the laminar sublayer, we should expect
    (a) increased heat transfer rate   (b) increased pressure drop
    (c) both (a) and (b)
    (d) reduced heat transfer rate but increased pressure drop.   **(c)**

11. Assertion (A): In flow through a tube, if the roughness is created in an irrational way, the heat transfer coefficient may even be lower than for a plain tube.
    Reasoning (R): Because high surface irregularities may lead to the formation of a stagnation zone between surface irregularity and the wall surface.
    Code: (a) Both A and R are false   (b) A is true and R is false
           (c) A is false and R is true   (d) Both A and R are true   **(d)**

12. When a fluid is flowing through a horizontal hot tube with a radial temperature distribution, the mechanism of heat transfer is expected to be
    (a) natural convection   (b) forced convection
    (c) both (a) and (b)   (d) natural convection when flow is laminar   **(c)**

13. Assertion (A): In a turbulent boundary layer over a flat plate, the wall shear stress cannot be calculated by the velocity profile given by one-seventh power law.
    Reasoning (R): Because the laminar sub-layer is very thin.
    Code: (a) Both A and R are true   (b) A is true and R is false
           (c) A is false, R is true
           (d) R is not the correct explanation of A   **(d)**

14. Local Nusselt number for turbulent flow over a flat plate with constant heat flux, in comparison with constant wall temperature, is
    (a) the same   (b) higher by about 4 percent
    (c) lower by about 5 percent   (d) much higher   **(b)**

15. The average turbulent shear stress is given by
    (a) $-\rho \overline{u'v'}$ (b) $+\rho \overline{u'v'}$

    (c) $-\overline{\rho u'} \cdot v'$ (d) $-\overline{\rho u'v'}$ **(d)**

16. When we have free and forced convection simultaneously, Gr/Re^2 is
    (a) greater than 1   (b) almost equal to 1
    (c) less than 1   (d) equal to Peclet number   **(b)**

17. The thickenss of a turbulent boundary layer over a flat plate at a distnce x from the leading edge is 5mm. When the velocity of flow is doubled, the thickness at that point would be (when all other parameters remain same)
    (a) 10mm   (b) 5.74 mm
    (c) 2.5 mm   (d) 1.0 mm   **(b)**

18. The local coeffecient of friction in turbulent bounday layer over a flat plate at a distance x from the edge is f_1. When the velocity of flow is doubled, the new coeffecient of friction would be
    (a) $2f_1$ (b) f_1 (c) $0.84f_1$ (d) $0.5f_1$ **(c)**

NUMERICALS

1. A rectangular plate 60 cm in length and 30 cm wide is maintained at 100°C. Air at 20°C flows over the plate with a velocity of 20 m/s along its length. If the critical Reynolds number is 5×10^5, estimate the rate of heat transfer. For air: $\rho = 1.06$ kg/m^3, k = 0.03 W/mK, Pr = 0.696, $\mu = 2.1 \times 10^{-5}$ Pa-s

(455 W)

2. The wing of an aeroplane flying with a speed of 300 km/h, is 150 cm wide. Considering the wing as a flat plate, estimate the rate of heat transfer per metre length of the wing. The wing is being maintained at 20°C and the surroundings are at 2°C.
   If the flow is assumed to be completely turbulent, calculae the drag force per metre length of the wing.
   Take $\rho = 1.05$ kg/m^3, $v = 18 \times 10^{-6}$ m^2/s, k = 0.025 W/mK, C = 1005 J/kgK

(4033.6 W, 17.33 N/m)

3. Air flows over a 2 m long flat plate with a velocity of 5 m/s. Estimate the thickness of the boundary layer at the trailing edge and the drag force per unit width of the plate.
   $\rho = 1.1$ kg/m^3, $v = 8.6 \times 10^{-6}$ m^2/s

(29 mm. 0.1246 N/m)

4. A flat plate 1.5 m long and 1.0 m wide is maintained at 90°C. Air at 25°C is flowing over the plate along its length. Estimate the air velocity if the rate of heat loss from the plate is 4 kW.
   Take $v = 19 \times 10^{-6}$ m^2/s, Pr = 0.7, k = 0.028 W/mK

(19.7 m/s)

5. Water at 20°C flows over a flat plate 2 m long with a velocity of 3.5 m/s. Estimate the thickness of the boundary layers, convective heat transfer coefficient and the drag force when the plate is maintained at 60°C.
   For water: $\mu = 5.8 \times 10^{-4}$ Pa-s,   Pr = 3.77,   k = 0.64 W/mK,
   μ at 60°C = 4.2×10^{-4} Pa-s, μ at 20°C = 8.6×10^{-4} Pa-s

(28.4 mm, 18.93 kW/m^{2}K, 23.61 N/m)

6. Air at 15°C and 1 bar is heated to 285°C while flowing at 34.2 m^3/h through a 25 mm diameter tube which is maintained at 455°C. Assuming simple Reynolds analogy is valid, taking $f = 0.0791/Re^{0.25}$ estimate the length of the tube required, when
   $\rho = 0.73$ kg/m^3, $v = 3.6 \times 10^{-5}$ m^2/s,   k = 0.039 W/mK,
   Pr = 0.68,   C = 1027 J/kgK

(1.34 m)

7. Air flows through a 20 mm diameter tube 2 m long with a mean velocity of 40 m/s. The tube wall temperature is 150°C and the air temperature increases from 15 to 100°C. Using simple Reynolds analogy estimate the pressure loss in mm of water in the tube due to friction and the pumping power required.
   $\rho = 0.98$ kg/m^3, $v = 2.17 \times 10^{-6}$ m^2/s.

(182 mm of water, 22.4 W)

8. A gas ($C_p = 1.13$ kJ/kgK) is flowing through a 25mm inner diameter tube with an average velocity of 24 m/s. The pressure drop per metre length is 0.0002 bar. Using Reynolds analogy, estimate the heat transfer coefficient.

    (5.58 W/m²K)

9. Calculate the heat transfer coefficient for water flowing through a 25 mm diameter tube at the rate of 1.5 kg/s when the mean bulk temperature is 40°C. Use the relation
    $$Nu = 0.0243\ Re^{0.8}\ Pr^{0.4}$$
    Take $\mu = 6.51 \times 10^{-4}$ Pa-s,   Pr = 4.3,   C = 4182 J/kgK

    (12.5 kW/m²K)

10. Water at 20°C flows through a 8 cm inner diameter pipe with a velocity of 2 m/s. The tube wall temperature is maintained at 80°C. Estimate the water temperature at exit if the length of the pipe is 4 m.
    For water: $\rho = 970$ kg/m³,   Pr = 3.42, $\mu = 5.3 \times 10^{-4}$ Pa-s.
    $k = 0.645$ W/mK, $C_p = 4182$ J/kgK

    (30°C approx.)

11. Air at 1 bar and 150°C is heated when it flows through a long tube 25 mm inner diameter at a velocity of 18 m/s. The tube receives heat energy at a constant rate such that the wall temperature is 25°C higher than the air temperature all along the length. Estimate (i) the length of the tube where the air temperature is 155°C and (ii) the rate of heat transfer per unit length of the tube.
    for Air: $\nu = 3.0 \times 10^{-5}$ m²/s, $k = 0.036$ W/mK, $C_p = 1005$ J/kgK

    (0.3 m 122.6 W/m)

12. In an oil cooler, the oil enters a 10 mm diameter tube at 160°C and is cooled to 40°C; the mean velocity of oil being 1.5 m/s. Calculate the heat transfer coefficient. Use the relation:
    For turbulent flow: $Nu = 0.0265\ Re^{0.8}\ Pr^{0.3}$
    For laminar flow: $Nu = 3.65$ **(50 W/m²K)**

    The properties of oil are:

T°C	ρ, kg/m³	$\nu \times 10^6$ m²/s	k (W/mK)	C(kJ/kgK)
40	878	251.0	0.144	1.96
100	839	20.4	0.137	2.22
160	806	5.7	0.131	2.48

13. Liquid sodium is to be heated from 115°C to 154°C at a rate of 2.5 kg/s in a 25 mm diameter tube. Calculate the length of the tube if the wall temperature is 25°C higher than the bulk temperature of sodium all along the tube. What would be the length of the tube if the wall temperature is maintained constant at 200°C.
    For sodium: $\rho = 915$ kg/m³, Pr = 0.0087, $\nu = 0.595 \times 10^{-6}$ m²/s,
    $k = 85$ W/mK

    (1.5m, 0.485 m)

CHAPTER 8

Forced Convection over Exterior Surfaces

Heat transfer from or to bodies subjected to forced convective external flow is encountered in many practical applications. Many heat-exchanger designs depend upon external convective heat transfer from cylinders subjected to cross flow, and heat transfer in packed beds and fluidized systems involves convective heat transfer from spheres.

1. Flow Considerations in Cross-flow over a Single Cylinder

In a manner similar to the analysis of the flow over a flat plate, the flow past a curved surface can be divided into two regions: a boundary layer near the surface and a nonviscous (inviscid) region away from the surface. Let us consider a long thin cylinder of diameter D, placed transversely in a fluid stream as shown in Fig. 8.1(a). Since the cylinder diameter is fixed, the Reynolds number will vary with the velocity of fluid stream. The velocity of fluid particles reduces to zero at point P, known as the 'forward stagnation point', and the velocity head is completely converted into pressure head under ideal conditions. The fluid particles tend to accelerate as they flow around the forward portion of the cylinder and then decelarate when they flow around the rear portion. This results in a decreasing pressure on the forward portion and an increasing pressure on the rear portion. The relatively slow moving fluid particles situated adjacent to the cylinder wall are hard pressed to continue their forward motion in the face of an increasing pressure gradient on the rear portion. At some point, where $\partial u/\partial y = 0$ at $y = 0$, the fluid particles separate away from the surface (point of separation, S, shown in Fig. 8.1 (b) and begin to flow backward. The variation in the pressure gradient also causes changes in the velocity profile in the boundary layer, Fig. 8.1 (b).

2. Effect of Reynolds Number on Local Nusselt Number Variation

Since the flow separation process is complicated, it is not possible to calculate analytically the average heat transfer coefficient in cross-flow. The heat transfer characteristics of a circular cylinder placed in cross-flow have been studied

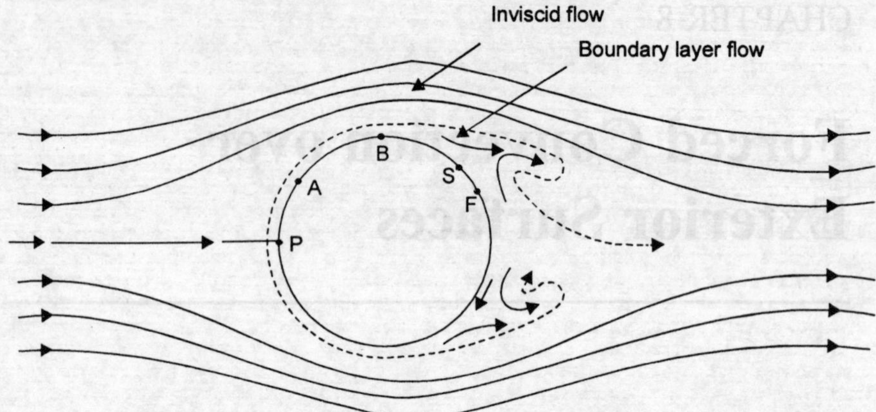

Fig. 8.1(a) Flow over a cylinder placed in a uniform stream; P–Stagnation Point, S–Separation Point

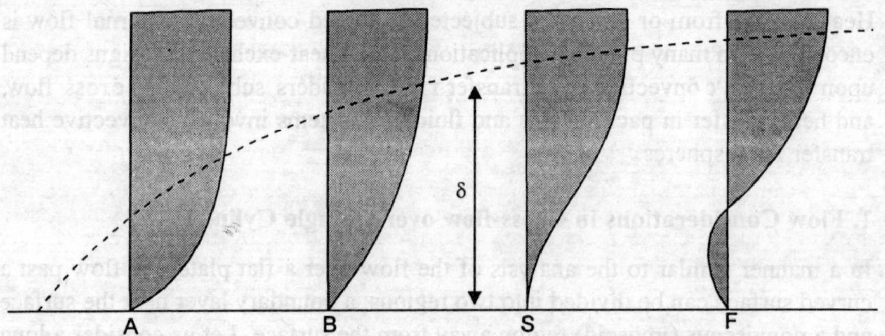

Fig. 8.1(b) Velocity profiles at different points on the cylinder

extensively by Geidt. His results, shown in Fig. 8.2, show that there is a considerable circumferential variation in the local value of the Nusselt number[*]. When $Re \leq 10^5$. Nu_θ decreases with θ as a result of laminar boundary layer development and at $\theta \approx 80°$, where separation occurs, Nu_θ is minimum. Nu_θ increases with θ because of mixing associated with vortex formation in the wake. For $Re \geq 10^5$, there are two minima for Nu_θ. Since the boundary layer is laminar up to $\theta \approx 80°$, the Nu_θ first decreases from the value at the stagnation point and, for $80° < \theta < 100°$, Nu_θ increases because of boundary layer transition to turbulence. During the development of turbulent boundary layer, Nu_θ again begins to decline till the separation point ($\theta \approx 140°$), and Nu_θ increases with θ as a result of mixing in the wake region. With increasing Re, Nu_θ increases because of reduced boundary layer thickness.

---

[*] Over the forward portion of the cylinder ($0° < \theta < 80°$), the empirical equation for convective heat transfer coeffecient is :

$$Nu(\theta) = \frac{h_c(\theta)}{k} = 1.14 \left(\frac{\rho U_\infty D}{\mu} \right)^{0.5} Pr^{0.4} \left[1 - \left(\frac{\theta}{90} \right)^3 \right]$$

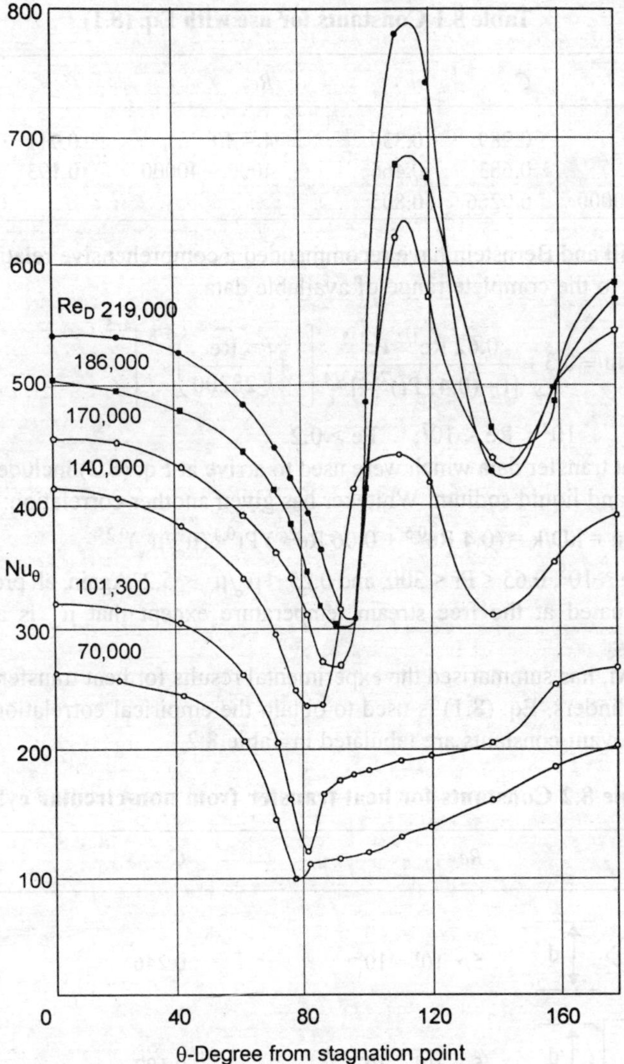

Fig. 8.2 Circumferential variation of the convective heat transfer coefficient with Re for a circular cylinder in cross-flow

3. Empirical Correlations for 'h' in Cross-flow over Circular and Non-Circular Cylinders

Experimental correlations of data for gases by Hilpert. R, and for liquids by Knudsen and Katz indicate that the average heat-transfer coefficients may be calculated by

$$Nu = hD/k = C(Re)^n Pr^{1/3} \tag{8.1}$$

where the constants C and n are tabulated in Table 8.1 and the fluid properties are to be evaluated at the mean temperature.

Table 8.1 Constants for use with Eq. (8.1)

Re	C	n	Re	C	n
0.4 – 4	0.989	0.330	4 – 40	0.911	0.385
40 – 4000	0.683	0.466	4000 – 40000	0.193	0.618
40000 – 400000	0.0266	0.805			

Churchill and Bernstein have recommended a comprehensive relation that can be applied to the complete range of available data:

$$Nu = 0.3 + \frac{0.62\,Re^{1/2}\,Pr^{1/3}}{[1+(0.4/Pr)^{2/3}]^{3/4}}\left[1+\left(\frac{Re}{28200}\right)^{5/8}\right]^{4/5} \tag{8.2}$$

for $\quad 10^2 < Re < 10^7; \quad Pe > 0.2.$

The heat transfer data which were used to arrive at Eq. (8.2) include fluids like air, water, and liquid sodium. Whiteker has given another correlation:

$$Nu = hD/k = (0.4\,Re^{0.5} + 0.06\,Re^{2/3})\,Pr^{0.4}\,(\mu_\infty/\mu_w)^{0.25} \tag{8.3}$$

for $40 < Re < 10^5$, $0.65 < Pr < 300$, and $0.25 < \mu_\infty/\mu_w < 5.2$. Again, all properties are to be evaluated at the free stream temperature except that μ_w is at the wall temperature.

Jakob. M, has summarised the experimental results for heat transfer from non-circular cylinders. Eq. (8.1) is used to obtain the empirical correlation for gases and the relevant constants are tabulated in Table 8.2.

Table 8.2 Constants for heat transfer from non-circular cylinders

Geometry	Re	C	n
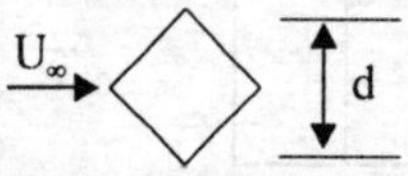 $U_\infty \rightarrow$ ◇ (diamond, d)	$5 \times 10^3 – 10^5$	0.246	0.588
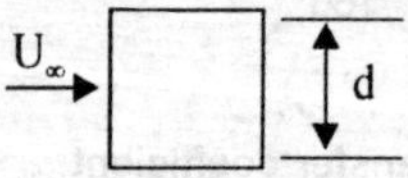 $U_\infty \rightarrow$ □ (square, d)	$5 \times 10^3 – 10^5$	0.102	0.675
$U_\infty \rightarrow$ ⬡ (hexagon, d)	$5 \times 10^3 – 1.95 \times 10^4$	0.160	0.638
	$1.95 \times 10^4 – 10^5$	0.0385	0.782
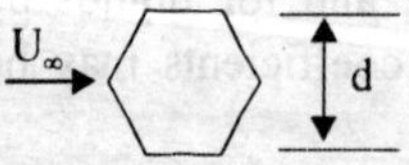 $U_\infty \rightarrow$ ⬡ (hexagon, d)	$5 \times 10^3 – 10^5$	0.153	0.638
 $U_\infty \rightarrow$ │ (vertical plate, d)	$4 \times 10^3 – 1.5 \times 10^4$	0.228	0.731

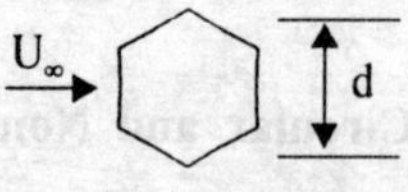

Example 8.1   Determine the rate of heat transfer per metre length to a 15 mm circular refrigerant line at –10°C subjected to a cross flow of air at 25°C flowing with a speed of 20 m/s.

Solution: The fluid properties of air are to be evaluated at mean temperature

$\quad$ T = (25 – 10)/2 = 7.5°C. The properties are:

$\quad \rho$ = 1.26 kg/m^3, ν = 14.14 × 10^{-6} m^2/s, k = 22.7 × 10^{-3} W/mK, Pr = 0.712

Eq. (8.1) is applicable to either heating or cooling of air flowing across a single cylinder. The Reynolds number, Re = 20 × 0.015/14.14 × 10^{-6}

$$= 2.12 \times 10^4$$

From Table 8.1, the value for C = 0.193 and n = 0.618

$\quad$ Nu = 0.193 × (21200)$^{0.618}$ × (0.712)$^{1/3}$ = 81.3

$\quad$ h = 81.3 × 0.227/0.015 = 123.03 W/m^{2}K

Rate of heat transfer to the refrigerant line = 123.03 × 3.142 × 0.015 × 35

$$= 202.95 \text{ W/m.}$$

Example 8.2   Air at 1 bar and 35°C flows over a cylinder, 50 mm in diameter with a velocity of 55 m/s. Calculate the quantity of heat energy required to maintain the cylinder surface at 155°C.

Solution: The mean temperature is (35 + 155)/2 = 95°C

The properties are:$\quad \rho$ = 10^5/(287 × 368) = 0.947 kg/m^3, μ = 2.16 × 10^{-5} Pa-s,

$\quad$ k = 0.0312 W/mK, Pr = 0.695

Re, Reynolds number = 0.947 × 55 × 0.05/2.16 × 10^{-5} = 1.273 × 10^5.

From Table 8.1:$\quad$ C = 0.0266 and n = 0.805

From Eq. (8.1),$\quad$ Nu = 0.0266 × (1.273 × 10^5)$^{0.805}$ (0.695)$^{1/3}$ = 303.11

$\quad$ h = 303.11 × 0.0312/0.05 = 189.14 W/m^{2}K

Energy required,$\quad \dot{Q}$ = 189.14 × 3.142 × 0.05 × 1 × (155 – 35)

$$= 3565.67 \text{ W/m.}$$

If we use Eq. (8.2),

$$Nu = 0.3 + \frac{0.62(1.273\times10^5)^{1/2}(0.695)^{1/3}}{[(1+0.4/0.695)^{2/3}]^{1/4}}\left[1+\left(\frac{27300}{28200}\right)^{1/2}\right] = 287.58$$

$\quad$ h = 287.58 × 0.0312/0.05 = 179.45 W/m^{2}K

and$\quad \dot{Q}$ = 179.45 × 3.142 × 0.05 × 1 × 120 = 3383 W/m (5.12% lower value)

$\quad$ [Eq. (8.1) is the easiest to use from a computational standpoint. Eq. (8.2) is more comprehensive and is preferred for computer set ups. But for air either relations give satisfactory results.]

Example 8.3   A copper wire 5mm in diameter is placed in an air-stream at 1 bar and 25°C having a speed of 20 m/s perpendicular to the wire. Calculate the rate of heat loss per metre length of the wire and the current carrying capacity if the surface temperature of the wire is maintained at 65°C and the resistivity of copper is 0.018 Ω mm^2/m.

Solution: The properties of air at free stream temperature is

$$\rho = 1.162 \text{ kg/m}^3, \quad \nu = 15.89 \times 10^{-6} \text{ m}^2/\text{s}, \quad Pr = 0.707, \quad k = 0.0263 \text{ W/mK}$$

Reynolds number, $Re = 20 \times 0.005/15.89 \times 10^{-6} = 6293 < 10^5$;

Since Prandtl number lies between 0.65 and 300

and $\left(\dfrac{\mu_\infty}{\mu_w}\right) = 184/204 = 0.902$, lies between 0.25 & 5.2,

we can use Eq (8.3)

or, $Nu = [0.4 \times (6293)^{0.5} + 0.06 \, (6293)^{2/3}] \, (0.707)^{0.4}(0.902)^{0.25}$

$\qquad = 44.319$

$\qquad h = 44.319 \times 0.0263/0.005 = 233.118 \text{ W/m}^2\text{K}$

We can calculate the current carrying capacity by making an energy balance,

$$\dot{Q} = h \, (\pi DL)\,(\Delta T) = 1^2 R, \text{ where the resistance } R = 4\rho L/\pi D^2$$

or, $\dot{Q} = 233.118 \times 3.142 \times 0.005 \times 1 \times (65 - 25) = 146.49 \text{ W/m}$

and $\quad R = 4 \times 0.018 \times 1/(3.142 \times 25) = 0.00092 \ \Omega/\text{m}$

Current, $\quad I = (146.49/0.00092)^{\frac{1}{2}} = 403.8 \text{ Amp.}$

Example 8.4 Air at 27°C flows across a tube 25 mm in diameter, with a velocity of 30 m/s. Compare the coefficient of heat transfer with a square tube having the same surface area when (i) the vertical face of the square, and (ii) one of the diagonals of the square, stand perpendicular to the air stream. The tube wall temperature is maintained at 77°C in all the three cases.

Solution: The mean temperature is 52°C and the properties are:

$$\nu = 17.4 \times 10^{-6} \text{ m}^2/\text{s}; \quad Pr = 0.703; \quad k = 0.028 \text{ W/mK}$$

Case I: Characteristic dimension for the circular tube = diameter = 0.025 m

$Re = 30 \times 0.025/17.4 \times 10^{-6} = 4.3 \times 10^4$.

From Table 8.1, $C = 0.0266$, and $n = 0.805$

$\qquad Nu = 0.0266 \, (4.3 \times 10^4)^{0.805} \, (0.703)^{1/3} = 127.01$

$\qquad h = 127.01 \times 0.028/0.025 = 142.258 \text{ W/m}^2\text{K}$

Case II: Characteristic dimension of the square, having the same surface area and one side stands normal to the flow, $D_1 = 3.142 \times 0.025/4 = 0.0196375$ m

$$\text{Reynolds number } Re = \dfrac{30 \times 0.0196375}{17.4 \times 10^6}$$

$$= 3.3858 \times 10^4$$

From Table 8.2, $\quad C = 0.102, \ n = 0.675$

$\qquad Nu = 0.102 \, (33858)^{0.675} \, (0.703)^{1/3} = 103.55$

and, $\qquad h = 103.55 \times 0.028/0.0196375 = 147.65 \text{ W/m}^2\text{K}$

Case III: Characteristic length is equal to the diagonal of the square:

or, $\qquad D_2 = 1.414 \times 0.0196375 = 0.027767$ m

Reynolds number, $Re = 30 \times 0.027767/17.4 \times 10^{-6} = 4.7 \times 10^4$

From Table 8.2, $\quad C = 0.246$ and $n = 0.588$

$\qquad Nu = 0.246 \, (47800)^{0.588} \, (0.703)^{1/3} = 123.445$

$\qquad h = 123.445 \times 0.028/0.027767 = 124.48 \text{ W/m}^2\text{K}.$

(We find that although the surface area of the tube in all the three cases are the same and the temperature difference causing the heat transfer is also the same, the orientation of the tube makes the difference in convective heat transfer coefficients.)

4. Empirical Correlation for the Evaluation of 'h' in Flow over a Sphere

The phenomenon of flow over a sphere is similar to that over a cylinder in cross-flow where the boundary layer usually experiences separation. Consequently the calculations for the heat transfer coefficient for this configuration are also based upon empirical correlations.

For the flow of air over a single sphere, McAdams has recommended

$$\mathrm{Nu} = hD/k = 0.37\,(\mathrm{Re})^{0.6}, \text{ for } 17 < \mathrm{Re} < 70{,}000. \tag{8.4}$$

and the properties are to be evaluated at the mean temperature. This same expression can be used for other gases since the Prandtl number does not usually differ substantially from that of air.

For liquids, flowing over a single sphere, Vliet and Leppert have recommended.

$$\mathrm{Nu} = [1.2 + 0.53(\mathrm{Re})^{0.54}]\,\mathrm{Pr}^{0.3}\,(\mu_\infty/\mu_w)^{0.25} \tag{8.5}$$

For $1 < \mathrm{Re} < 2000{,}000$; and all properties are to be evaluated at the free stream temperature except for μ_w. For very slow flow over a sphere, Johnston et al, have shown theoretically that $\mathrm{Nu} \to 2$.

Whitaker has developed a single equation for gases and liquids flowing past spheres:

$$\mathrm{Nu} = 2 + (0.4\,\mathrm{Re}^{0.5} + 0.06\,\mathrm{Re}^{2/3})\,\mathrm{Pr}^{0.4} \tag{8.6}$$

For $3.5 < \mathrm{Re} < 8\times10^4$ and $0.7 < \mathrm{Pr} < 380$, properties to be evaluated at free stream temperature.

Example 8.5 Air at 1 bar and 27°C flows across a 15 mm diameter sphere at a free stream velocity of 5 m/s. Estimate the heat energy lost by the sphere if the surface temperature is maintained at 77°C.

Solution: The mean temperature is $(27 + 77)/2 = 52°C$. The properties of air are:
$$\nu = 18.23 \times 10^{-6}\ \mathrm{m^2/s},\ \ k = 0.028\ \mathrm{W/mK},$$
Reynolds number, $\mathrm{Re} = (5 \times 0.015)/(18.23 \times 10^{-6}) = 4114$

Using Eq. (8.4),
$$\mathrm{Nu} = 0.37\,(4114)^{0.6} = 54.545$$
$$h = 54.545 \times 0.028/0.015 = 101.817\ \mathrm{W/m^2K}$$
Heat lost to the surroundings, $\dot{Q} = 101.817 \times 4 \times 3.142 \times (0.0075)^2 \times 50 = 3.6\ \mathrm{W}$

Example 8.6 An air stream at 27°C flows across an electric bulb approximated as a sphere, 75 mm in diameter, with a speed of 0.5 m/s. If 7 percent of the power (200 W electric bulb) is lost by convection, estimate the surface temperature of the bulb.

Solution: The properties of air at free stream temperature, 27°C are:
$$\nu = 15.89 \times 10^{-6}\ \mathrm{m^2/s},\ \mu = 1.84 \times 10^{-5}\ \mathrm{Pa\text{-}s},\ k = 0.0263\ \mathrm{W/mK},\ \mathrm{Pr} = 0.707$$

$$Re = 0.5 \times 0.075/15.89 \times 10^{-6} = 2360;$$
$$\mu_w = 2.35 \times 10^{-5} \text{ Pa-s at } 107°C$$

From Eq. (8.6), $Nu = 2 + [0.4 \, (2360)^{0.5} + 0.06 \, (2360)^{0.667}] \, (0.707)^{0.4} \left(\dfrac{1.84}{2.35}\right)^{1/4}$

$$= 26.64 \quad \text{and} \quad h = 26.64 \times 0.063/0.075 = 9.34 \text{ W/m}^2\text{K}$$

Heat lost, $\quad \dot{Q} = 0.07 \times 200 = 14W = 9.34 \times 4 \times 3.142 \times (0.0375)^2 \, (T - 27)$

and $T = 111.8°C$, a second trial will not change the result significantly.

(When the particles have an irregular shape, the data for spheres will yield satisfactory results if the sphere diameter is replaced by an equivalent diameter, i.e., the diameter of the spherical particle be so chosen that the surface area of the irregular particle and the surface area of the chosen sphere are equal.)

5. Heat Transfer from Tube Bundles in Cross-flow—Its Significance

The computation of the convective heat transfer coefficient between a bank of tubes and a fluid flowing at right angles to the tubes is an important step in the design and performance analysis of many types of commercial heat exchangers. Fig. 8.3 depicts several arrangements of tubular air preheaters in which the products of combustion, after they leave the boiler, economiser, or superheater, are used to preheat the air going to the furnace of the steam generating unit.

The heat transfer in flow over tube bundles depends mostly on the flow pattern and the degree of turbulence, which in turn are functions of the velocity of fluid and the size and arrangements of the tubes. Two general geometric arrangements are widely used in heat-exchanger designs. One consists of rows (parallel with the direction of the free stream) of in-line tubes, Fig. 8.4, and the other has staggered tube rows, Fig. 8.5.

Grimson (Trans.ASME, 59, 1937) has evaluated the results of several researchers and has suggested that the convective heat transfer coefficient can be obtained by using the relation

$$Nu = hD/k = C_1 \, (V \, D/v \,)^n \, Pr^{1/3} \tag{8.7}$$

where V is the maximum velocity in the bank. The values of constants C_1 and n are given in Table 8.3. The velocity would be maximum at the minimum flow area, which would be either between two adjacent tubes in a row or between two diagonally o pposed tubes (staggered tube b anks only).

The area of flow is: $(S_n - D)$ or $[(S_n^2 + S_p^2)^{1/2} - D]$ per unit depth, and the velocity, $V = U_\infty \, S_n$/smaller of the two area of flow. The free stream velocity, U is obtained by assuming no tubes in the shell container, i.e., the free-passage flow.

The Nusselt number given by Eq. (8.7) with appropriate constants from Table 8.3 are satisfactory for tube banks, which are 10 or more tubes deep in the direction of U_∞. When the number of tubes is less, the average Nusselt number is affected less by the lower turbulence in the flow over the first few tubes. Kays and Lo have obtained correction coefficients for less than 10 tubes, given in Table 8.4.

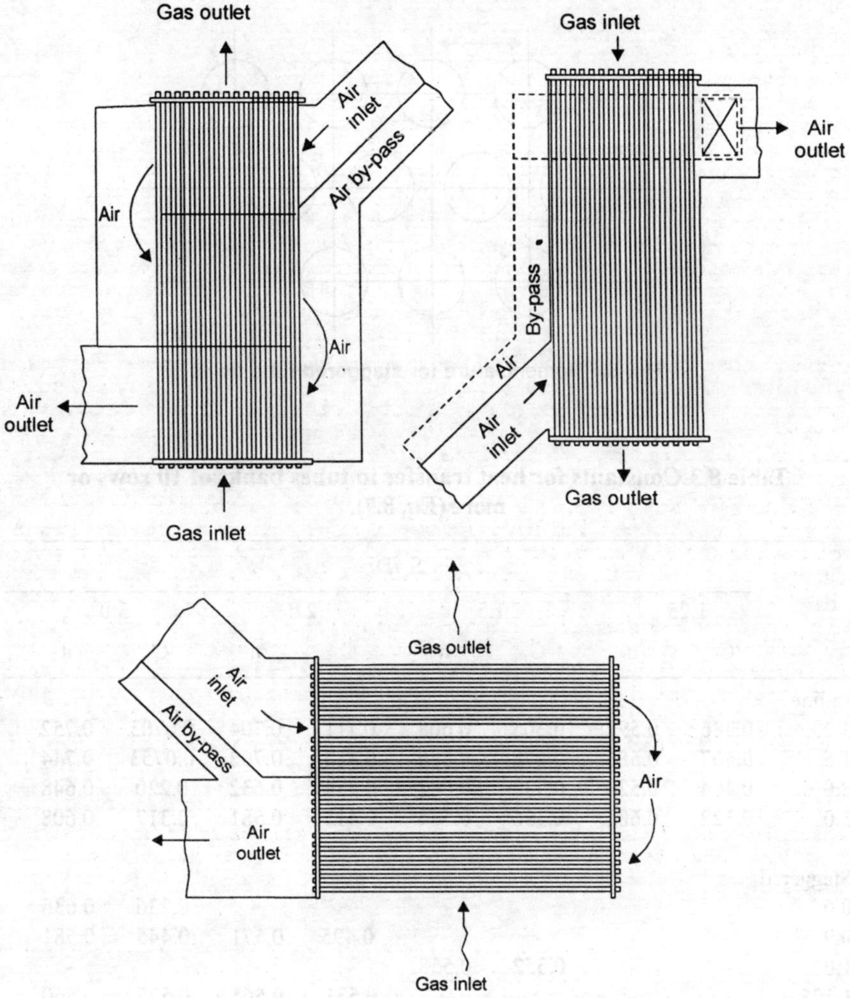

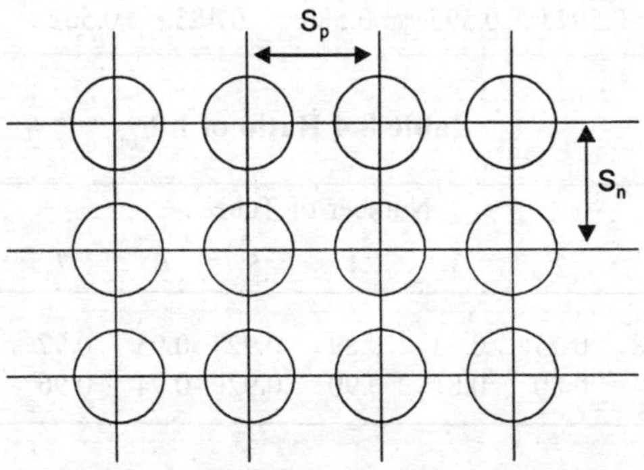

Fig. 8.3 Some arrangements of tubular air heaters to suit various directions of gas and air flow

Fig. 8.4 Nomenclature for in-line tube flows

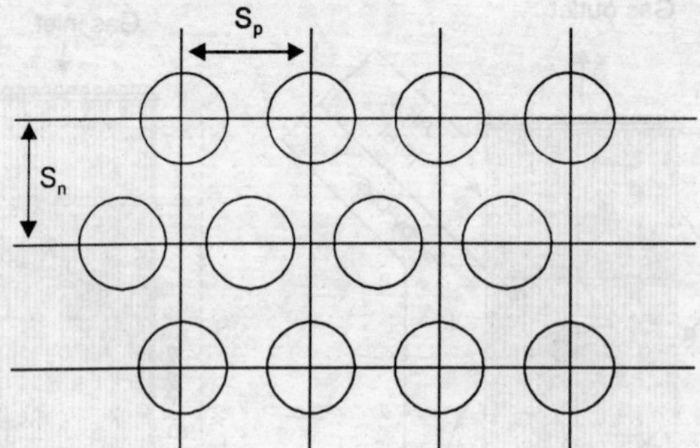

Fig. 8.5 Nomenclature for staggered tube flows

Table 8.3 Constants for heat transfer in tubes banks of 10 rows or more (Eq. 8.7).

$\frac{S_p}{d}$	S_n/D							
	1.25		1.5		2.0		3.0	
	C	n	C	n	C	n	C	n
In line								
1.25	0.386	0.592	0.305	0.608	0.111	0.704	0.0703	0.752
1.5	0.407	0.586	0.278	0.620	0.112	0.702	0.0753	0.744
2.0	0.464	0.520	0.332	0.602	0.254	0.632	0.220	0.648
3.0	0.322	0.601	0.396	0.584	0.415	0.581	0.317	0.608
Staggered								
0.6							0.236	0.636
0.9	-	-	-	-	0.495	0.571	0.445	0.581
1.0	-	-	0.552	0.558	-	-	-	-
1.125	-	-	-	-	0.531	0.565	0.575	0.560
1.25	0.575	0.556	0.561	0.554	0.576	0.556	0.579	0.562
1.5	0.501	0.568	0.511	0.562	0.502	0.568	0.542	0.568
2.0	0.448	0.572	0.462	0.568	0.535	0.556	0.498	0.570
3.0	0.344	0.592	0.395	0.580	0.488	0.562	0.467	0.574

Table 8.4 Ratio of h/h_{10}

	Number of Tubes									
	1	2	3	4	5	6	7	8	9	10
Staggered	0.68	0.75	0.83	0.89	0.92	0.95	0.97	0.98	0.99	1.00
In line	0.64	0.80	0.87	0.90	0.92	0.94	0.96	0.98	0.99	1.00

Pressure drop for flow of gases over a bank of tubes may be obtained with the following relation

$$\Delta p = \frac{2f' \, G_{max}^2 \, N}{\rho} \left(\frac{\mu_w}{\mu_b}\right)^{0.14} \; N/m^2 \tag{8.8}$$

where G_{max} = mass velocity at minimum flow area, $kg/m^2.s$
  ρ = density evaluated at free stream conditions, kg/m^3
  N = number of transverse rows
  μ_b = average free stream viscosity

Jakob has suggested that the empirical friction factor f' can be obtained from:

$$f' = \left[0.044 + \frac{0.08 S_p / D}{[(S_n - D)/D]^{0.43+1.13D/S_p}} \right] Re^{-0.15} \tag{8.9}$$

for in-line arrangements, and from

$$f' = Re^{-0.16} \left[0.25 + \frac{0.118}{[(S_n - D)/D]^{1.08}} \right] \tag{8.10}$$

for staggered tube arrangements.

For wide ranges of Reynolds numbers and property variations, Zukauskas has recommended the following correlation:

$$Nu = \frac{hD}{k} = CRe^n Pr^{0.36} (Pr/Pr_w)^{1/4} \tag{8.11}$$

where all properties except Pr_w are evaluated at free stream temperature and values of constants are tabulated in Table 8.5. The equation is valid for $0.7 < Pr < 500$ and $10 < Re < 10^6$. Table 8.6 gives the correction factor to be applied when the number of rows in the direction of flow is less than 20.

Table 8.5 Constants for use in Eq. 8.11 (after Zukauskas)

Goemetry	Re_d, max	C	n
In-line	$10 - 100$	0.8	0.4
	$100 - 10^3$	treat as individual tubes	
	$10^3 - 2 \times 10^5$	0.27	0.63
	$> 2 \times 10^5$	0.21	0.84
Staggered	$10 - 100$	0.9	0.4
	$100 - 10^3$	treat as individual tubes	
	$10^3 - 2 \times 10^5$	$0.35(S_n/S_L)^{0.2}$ for $S_n/S_L = 2$	0.6
	$10^3 - 2 \times 10^5$	0.4 for $(S_n/S_L) = 2$	0.6
	$> 2 \times 10^5$	0.022	0.84

Table 8.6: Ratio of h for N rows deep to that for 20 rows deep (h/h_{20})

N	2	3	4	5	6	8	10	16	20
Staggered	0.77	0.84	0.89	0.92	0.94	0.97	0.98	0.99	1.00
In-line	0.70	0.80	0.90	0.92	0.94	0.97	0.98	0.99	1.00

Example 8.7   Air at 27°C flows across a bundle of tubes 15 rows in the direction of flow and 8 rows in the direction transverse to the flow. The free stream velocity is 7.5 m/s and the tube wall temperature is maintained at 75°C. The tubes are 5 cm in diameter, arranged in an in-line manner so that the spacing in parallel and normal directions are 6.25 cm and 7.5 cm respectively. Calculate the total heat transfer rate and the exit air temperature if the length of the tubes is 1.5 m.

Solution: $S_p/D = 6.25/5.0 = 1.25$; $S_n/D = 7.5/5.0 = 1.5$

From Table 8.3, $C = 0.305$, $n = 0.608$,   The mean temperature = 51°C.

The properties are: $\rho = 10^5/(287 \times 324) = 1.075$ kg/m^3, $k = 0.028$ W/mK,

$$\nu = 18.4 \times 10^{-6} \text{m}^2/\text{s}, \quad Pr = 0.703, \quad C_p = 1008 \text{ J/kgK}$$

To calculate the maximum velocity, we should calculate the minimum flow area. The two flow areas are: $(S_n - D) = 7.5 - 5.0 = 2.5$ cm^2 per unit depth, $(S_n^2 + S_p^2)^{\frac{1}{2}} - D = 4.92$ cm^2 per unit depth. therefore, the maximum velocity,

$$V = U_\infty \, S_n/\text{minimum area}$$
$$= 7.5 \times 7.5/2.5 = 22.5 \text{ m/s}$$

Reynolds number Re $= 22.5 \times 0.05/18.4 \times 10^{-6} = 6.1 \times 10^4$

From Eq. (8.7)   Nu $= 0.305 \, (6.1 \times 10^4)^{0.608} \, (0.703)^{1/3} = 220.2$

$$h = 220.2 \times 0.028/0.05 = 123.31 \text{ W/m}^2\text{K}$$

Since there are 15 rows in a direction parallel to the stream, no correction is required. The total surface area for heat transfer is —

$$A = \text{No. of tubes} \times \text{Surface area} = 120 \times 3.142 \times 0.05 \times 1.5 = 28.278 \text{ m}^2$$

Rate of heat transfer, $\dot{Q} = (123.31 \times 28.278) \times [75 - (27 + T_o)/2]$ W

Mass flow rate of air at entrance to 8 tubes, $\dot{m} = 1.075 \times 7.5 \times 8 \times 7.5 \times 10^{-2}$
$$= 4.8375 \text{ kg/s}$$

By making an energy balance

$$4.8375 \times 1008 \times (T_o - 27) = (123.31 \times 28.278) \times [75 - (27 + T_o)/2]$$

Solving for the exit temperature, $T_o = 52.3$°C; and $\dot{Q} = 123.37$ kW

If we interchange the tube arrangement, keeping the total number of tubes the same, we would have 8 rows in the parallel direction and the correction factor from Table 8.4 is 0.98.

The rate of heat transfer would then be 0.98 h A (Δ T)

Mass flow rate of air $\dot{m} = 1075 \times 7.5 \times 15 \times 7.5 \times 10^{-2} = 9.07$ kg/s

By making the energy balance,

$$9.07 \times 1008 \times (T_o - 27) = (0.98 \times 123.31 \times 28.278) \times [75 - (T_o + 27)/2]$$

And solving for T_o, we get $T_o = 42.12$°C;   and $\dot{Q} = 138.23$ kW

an increase in rate of heat transfer.

Example 8.8   Air at 27°C flows over a bundle of tubes, diameter 5 cm, number of tubes 10 in the direction of flow, arranged in a staggered number, spacing in the parallel and transverse direction are equal to 7.5 cm. Calculate the heat transfer rate per metre length of the tube, wall temperature 75°C and the pressure drop across the tube bundle, when the free stream velocity is 8m/s.

Solution: For staggered arrangements, the minimum flow area will also be $S_n - D$ unless S_p/S_n is so small that $(S_p^2 + S_n^2)^{1/2} < (S_n + D)/2$ between diagonally opposed tubes. Therefore in this case, the minimum flow area is : $S_n - D = 2.5$ cm per unit depth of tube.

The maximum velocity, $V = U_\infty S_n/(S_n - D) = 8 \times 7.5/2.5 = 24$ m/s

The properties of air at free stream temperature are:

$$\rho = 1.61 \text{ kg/m}^3, \quad \nu = 15.89 \times 10^{-6} \text{ m}^2/\text{s}, \quad k = 0.0263 \text{ W/mK}, \quad Pr = 0.707$$

Reynolds number, $Re = 24 \times 0.05/15.89 \times 10^{-6} = 7.5 \times 10^4$

We make use of Eq. (8.11). The constants are to be evaluated from Table 8.5.

For, $S_n/S_p = 1$, $C = 0.35$ and $n = 0.6$ and $Pr_w = 0.700$

or, $Nu = 0.35 \,(75000)^{0.6} (0.707)^{0.36} (0.707/0.700) = 260.6$

The correction factor for 10 tubes is 0.98 and

$$h = 0.98 \times 260.6 \times 0.0263/0.5 = 134.33$$

Air flow rate, $m = 1.61 \times 8 \,(0.025 \times 10) = 3.22$ kg/s per metre length

If T_o is the temperature of air at exit, by making an energy balance

$$\dot{Q} = 3.22 \times 1005 \times (T_o - 27)$$

$$= 134.33 \times 3.142 \times 10 \times 0.05 \times [75 - (T_o + 27)/2]$$

Solving for T_o, we get $T_o = 30.03$ per metre length

and $\dot{Q} = 3.22 \times 1005 \times 3.03 = 9.8$ kW per metre length.

In order to compute the pressure drop, we calculate the friction f' from Eq. (8.10) or, $f' = (75000)^{-0.16} [0.25 + 0.118/(0.5)^{1.08}] = 3.00$

$$G_{max} = 3.22/0.025 = 128.8 \text{ kg/m}^2.\text{s per metre length}$$

$$\Delta P = 2 \times 3 \times (128.8)^2 \times 10 \,(2.08/1.98)^{0.14}/1.61 = 6.2 \text{ bar.}$$

6. Heat Transfer in High Speed Flow—Aerodynamic Heating

When a fluid is flowing over a solid surface with a very high velocity, the surface temperature increases even without a heat transfer. The reasons are: (i) when the fluid is brought to rest at the solid surface, its kinetic energy is converted into internal thermal energy, and (ii) at very high velocities, viscous stresses within the boundary layer do shearing work on the fluid elements and raise its temperature appreciably. This phenomenon of aerodynamic heating or thermal barrier is becoming increasingly important because the aeroplanes, missiles and satellites are moving with supersonic speed and therefore the problems associated with the properties of aircraft materials have to be solved, the limitations on personnel and instruments on the aircraft have to be studied and dissipation of heat energy generated in the boundary layer at high speeds is to be analysed.

7. Adiabatic Wall Temperature and Recovery Factor—Defined

When a gas flowing with a free stream velocity U_∞ is brought to rest adiabatically, the process can be described by the steady flow energy equation for an adiabatic process:

$$h_o = h_\infty + U_\infty^2/2$$

where h_∞ is the free stream enthalpy and h_0 is called stagnation enthalpy. We can express the above equation in terms of temperature, or,

$$C_p(T_0 - T_\infty) = U_\infty^2/2$$

where T_0 is the stagnation temperature and T_∞ is the free stream temperature.

Using the definition of the Mach number of the flowing gas M_∞, and the specific heat ratio $\gamma = C_p/C_v$, we can write

$$T_0/T_\infty = 1 + (\gamma - 1) M_\infty^2 /2 \tag{8.11a}$$

where $M_\infty = U_\infty/C_\infty$ and C_∞ is the local velocity of sound, equal to $(\gamma RT_\infty)^{1/2}$

When the gas flows past an insulated or adiabatic surface, the temperature at the surface will be more than the local temperature of the gas but may not be equal to the total or stagnation temperature of the gas. The actual temperature at an adiabatic surface is called the 'adiabatic wall temperature' T_a. In practice, it has been found convenient to relate T_a and T_0 by the 'recovery factor' R, which is a measure of the fraction of the free stream temperature rise recovered at the wall. Or,

$$R = (T_a - T_\infty) / (T_0 - T_\infty) \tag{8.12}$$

The recovery factor may be determined experimentally, or, for some flow systems, analytical calculations may be made.

The influence of heat transfer to or from the plate on the temperature distribution is shown in Fig. 8.6. We can see that heat can flow to the surface even when the surface temperature is higher than the free stream temperature. This rather unexpected phenomenon is a result of aerodynamic heating. And, in high speed flows, the heat transfer rates should be calculated from the relation

$$\dot{Q} = h\,A\,(T_w - T_a) \tag{8.13}$$

For gases with Prandtl number near unity, the recovery factor may be obtained from:

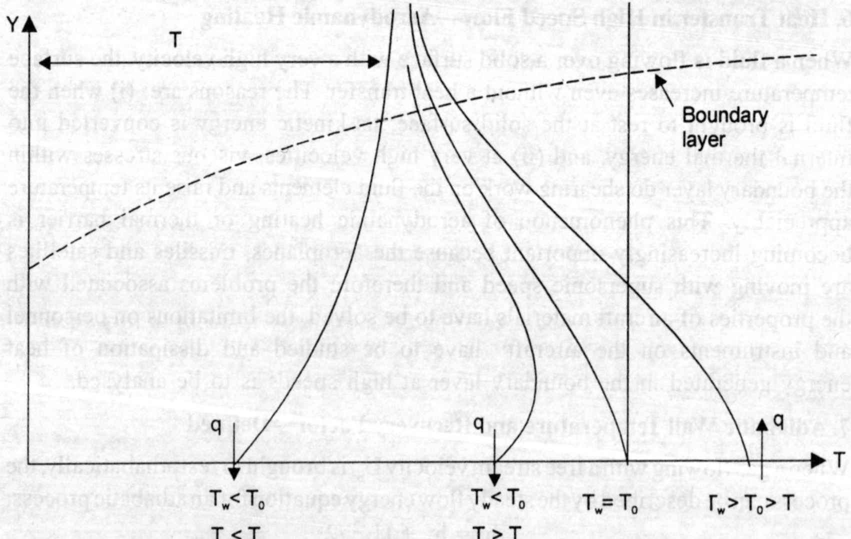

Fig. 8.6 Temperature distributions in a boundary layer with viscous dissipation

Laminar flow: $R = Pr^{1/2}$
Turbulent flow: $R = Pr^{1/3}$ (8.14)

Since in high velocity boundary layer flows substantial temperature gradients are likely to occur, Eckert has recommended that the fluid properties be evaluated at a reference temperature T*, defined as:

$$T^* = T_\infty + 0.50\,(T_w - T_\infty) + 0.22\,(T_a - T_\infty)$$ (8.15)

8. Evaluation of 'h' in High Speed Flow – Reynolds Analogy

The convective heat transfer coefficient can be evaluated by using the Reynolds analogy provided the friction coefficient is known. Thus, the following relations have been recommended for computing the heat transfer coefficients:

Laminar boundary layer ($Re_x^* < 5 \times 10^5$)

$$St_x^*\,Pr^{*2/3} = 0.332\,Re_x^{*-1/2}$$ (8.16)

Turbulent boundary layer ($5 \times 10^5 < Re_x^* < 10^7$)

$$St_x^*\,Pr^{*2/3} = 0.0288\,Re_x^{*-1/5}$$ (8.17)

Turbulent boundary layer ($10^7 < Re_x^* < 10^9$)

$$St_x^*\,Pr^{*2/3} = 2.46\,(\log\,Re_x^*)^{-2.584} = C_{f_x}/2$$ (8.18)

In the above expressions, all properties are to be evaluated at T*. In order to obtain an average heat-transfer coefficient, the above expressions are to be integrated over the length of the plate. Since the reference temperature for laminar and turbulent boundary layers are different, care should be taken while integrating Eq. (8.17).

Example 8.9 Air at 20°C flows over a plate 25 mm long at a velocity of 200 m/s. If the plate is maintained at 35°C. Calculate the adiabatic wall temperature.

Solution: The properties of air at 20°C are: $\nu = 15.8 \times 10^{-6}$ m²/s,

$$Pr = 0.71, \qquad C_p = 1.007 \text{ kJ/KgK}.$$

Reynolds number, $Re = 200 \times 0.025/15.8 \times 10^{-6} = 3.16 \times 10^5$,

A laminar flow recovery factor, $R = Pr^{1/2} = 0.71^{1/2} = 0.8426$

Stagnation temperature, $T_o = 20 + (200)^2/(2 \times 1007)$
$$= 39.86°C$$

Adiabatic wall temperature, $T_a = 20 + 0.8426 \times (39.86 - 20)$
$$= 36.72°C$$

Since the adiabatic wall temperature is a little higher than the plate temperature, heat energy will flow to the plate.

Example 8.10 A flat plate 70 mm long and 75 mm wide is placed in a wind tunnel where the flow conditions are M = 2.5, p = 1/20 bar, and temperature = – 40°C. Calculate the required cooling rate of the plate if the plate temperature is maintained at 40°C.

Solution: The free stream velocity of sound,
$$C = [1.4 \times 287 \times (273 - 40)]^{1/2} = 306 \text{ m/s}$$
The free stream velocity, $U_\infty = M \times C = 2.5 \times 306 = 765$ m/s
The density of air in the wind tunnel, $\rho = p/RT = 10^5/(20 \times 287 \times 233)$
$$= 0.0748 \text{ kg/m}^3$$
Viscosity of air at $-40°C = 1.43 \times 10^{-5}$ Pa-s $C_p = 1006.5$ J/KgK
Reynolds number at the trailing edge,
$$Re = (0.0748 \times 765 \times 0.07)/(1.43 \times 10^{-5}) = 2.79 \times 10^5, \text{ a laminar}$$
boundary layer over the entire plate.
Assuming a Prandtl number of 0.7,
The recovery factor for laminar flow, $R = (0.7)^{1/2} = 0.837$

The adiabatic wall temperature $T_a = T_\infty + R(T_0 - T_\infty)$

$$= T_\infty + R \cdot U_\infty^2 / 2C_p$$
$$= 233 + 0.837 \times (765)^2/(2 \times 1006.5)$$
$$= 476.335 \text{ K} = 203.335°C$$

Reference temperature, $T^* = 233 + 0.5 (40 + 40) + 0.22 (476.335 - 233)$
$$= 326.5 \text{ K}$$
The properties of air at this reference temperature are
$$Pr = 0.703, \quad \rho = 10^5/(20 \times 287 \times 326.5) = 0.0533 \text{ kg/m}^3;$$
$$\mu = 1.96 \times 10^{-5} \text{ Pa-s,}$$
$$k = 0.028 \text{ W/mK, } C_p = 1008 \text{ J/kgK}$$
$$Re = 0.0533 \times 765 \times 0.07/1.96 \times 10^{-5} = 1.456 \times 10^5.$$
From Eq. (8.16), Nu $= 0.332$ Re$^{1/2}$ Pr$^{1/3} = 0.332(1.456 \times 10^5)^{1/2} (0.703)^{1/3}$
$$= 112.656 \text{ and h} = 112.656 \times 0.028/0.07 = 45.06 \text{ W/m}^2\text{K}$$
For laminar flow, the average heat transfer coefficient, h = 2h = 90.12 W/m^{2}K
Since the adiabatic wall temperature is higher than the plate temperature, heat energy will flow to the plate and therefore the cooling rate required is

$$\dot{Q} = 90.12 \times (0.07 \times 0.075) \times (203.335 - 40) = 77.28 \text{ W.}$$

Example 8.11 A missile travels at M = 3.25 at an altitude of 3000 m where the temperature is –6°C and the density is 0.9 kg/m^3. Find the local heat transfer coefficient at a point 30 cm behind the leading edge, where the temperature is 150°C.

Solution: From Eq. (8.11a), The stagnation temperature,
$$T_0 = 267 \times [1 + 0.2(3.25)^2] = 831 \text{ K}$$
Assuming a turbulent flow and Pr = 0.68, the recovery factor, $R = 0.88 = (Pr)^{1/3}$
Adiabatic wall temperature, $T_a = 267 + 0.88 \times (831 - 267) = 763.3$ K
Reference temperature $T^* = 267 + 0.5 (423 - 267) + 0.22 (763.3 - 267) = 454.2$K
The properties are to be evaluated at 454.2K. The properties are:
$$\text{Density} = 0.53 \text{ kg/m}^3, \quad \text{Viscosity} = 2.5 \times 10^{-5} \text{ Pa-s, } k = 0.0375 \text{ W/mK,}$$
$$Pr = 0.686 \text{ (initial assumption was reasonable), } C_p = 1021 \text{ J/kgK}$$

Free stream velocity of sound, $C = (1.4 \times 287 \times 267)^{1/2} = 327.54$ m/s
Free stream velocity, $U_\infty = 327.54 \times 3.25 = 1064.5$ m/s
Reynolds number, $\quad Re = 0.53 \times 1064.5 \times 0.3/2.5 \times 10^{-5} = 6.77 \times 10^6 < 10^7$
From Eq. (8.17) $\quad Nu = 0.0296 \, (Re)^{0.8} \, (Pr)^{1/3}$
$$= 0.0296 \, (6.77 \times 10^6)^{0.8} \, (0.686)^{0.333} = 760.78$$
$$h = 760.78 \times 0.0375/0.30 = 95.1 \text{ W/m}^2\text{K}.$$

(If it is assumed that the entire flow is turbulent, the average Nusselt number can be calculated by the relation:
$$Nu = (0.036 \, Re^{0.8} \, Pr^{1/3})$$
At extremely high velocities, the internal energy of the gas may rise to a point that the gas may partly dissociate or even ionize the molecules. And, the specific heat of the gas in the boundary layer will vary significantly. Eckert has suggested that the convective heat transfer coefficient be defined in terms of enthalpy and not in terms of temperature.

or, $\quad h = \dot{Q}/A \, (h_w - h_a)$, where h_w and h_a are the enthalpy at T_w and T_a.
The Nusselt number is defined as: $Nu = C_p \, h \, x/k$, and

the Stanton number $\quad St = h/(\rho U_w C_p) = C_f/2$

Similarly the recovery factor is defined as: $R = (h_a - h_\infty)/ \, (U^2_\infty/2)$,

and the properties are to be evaluated at the reference enthalpy
$$h^* = h_\infty + 0.5 \, (h_w - h_\infty) + 0.22 \, (h_a - h_\infty).$$

SUMMARY

1. Evaluation of heat transfer coefficients in external flows over cylinders and spheres are required in designing many types of heat exchangers and packed beds respectively.

2. In cross-flow over cylinders, the local heat transfer coefficient at the surface of the cylinder varies with Reynolds number and θ, the angle measured from the forward stagnation point.

3. The convective heat transfer coefficients in cross-flow over circular and non-circular cylinders and spheres are evaluated only by empirical relations because of the complicated flow separation processes.

4. In high speed flow over solid surfaces, evaluation of 'h' is important for aircrafts and missiles. In such flows, the flow field is described by Reynolds and Mach numbers.

5. The high temperature at the surface of the solids is the combined effect of the heating due to viscous dissipation and the temperature rise due to the deceleration of fluid particles within the boundary layer.

6. Since the fluid particles cannot be brought to rest reversibly, because of the viscous shearing process (an irreversible process), a recovery factor R is defined as

$$R = (T_a - T_\infty)/(T_0 - T_\infty)$$

Missiles entering the earth's atmosphere at a very high speed, require cooling of its surface within a very short time. A special type of cooling called 'ablation' is employed. The heat energy generated is utilized to melt a part of its body instead of being conducted.

where T_a is the adiabatic wall temperature, T_o is the stagnation temperature and T_∞ is the free stream temperature.

7. The recovery factor will always be less than 1 and for gases, it can be obtained from:

$$R = (Pr)^{1/2} \text{ for laminar flow,}$$
$$= (Pr)^{1/3} \text{ for turbulent flow.}$$

8. If the surface temperature is equal to the adiabatic wall temperature, there will be no flow of heat energy even if the free stream temperature is less than the surface temperature.

MULTIPLE CHOICE QUESTIONS

1. Assertion (A) : In cross-flow over a circular cylinder, it is not possible to calculate analytically the value of the average heat transfer coefficient.
   Reasoning (R): Because the flow separation process is complicated.
   Code: (a) Both A and R are true, R is not the only reason for A
         (b) Both A and R are true and R is the correct reason for A
         (c) A is true, R is false
         (d) A is false, R is true   **(b)**

2. The local Nusselt number along the circumference of a circular cylinder in cross-flow varies with
   (a) Prandtl number   (b) Peclet number
   (c) Reynolds number   (d) Rayleigh number   **(c)**

3. The heat transfer in flow over tube bundles depends mostly on
   (a) size and arrangement of the tubes
   (b) velocity of the fluid and degree of turbulence
   (c) flow pattern
   (d) only (a) and (b)   **(d)**
   (e) all of the above

4. In cross-flow over a bundle of tubes, the friction factor
   (a) increases with Reynolds number
   (b) is independent of Reynolds number
   (c) decreases with Reynolds number
   (d) first increases and then decreases with Reynolds number   **(c)**

5. Assertion (A): In high speed flow, the temperature of the solid surface increases without a heat transfer
   Reasoning (R): Because viscous stresses within the boundary layer do the shearing work on the fluid elements
   Code: (a) Both A and R are true and R is the only explanation of A
         (b) Both A and R are true and R is not the only explanation of A.
         (c) A is true, R is false
         (d) A is false, R is true   **(b)**

6. Recovery factor is defined as:
   (a) $(T_\infty - T_a)/(T_o - T_\infty)$ (b) $(T_o - T_\infty)/(T_\infty - T_a)$
   (c) $(T_a - T_\infty)/(T_o - T_\infty)$ (d) $(T_a - T_\infty)/(T_\infty - T_o)$ **(c)**

where T_∞ is the free stream temperature, T_0 is the stagnation temperature and T_a is the adiabatic wall temperature.

7. The Recovery factor in flow of gases is a function of
   (a) Reynolds number   (b) Peclet number
   (c) Prandtl number   (d) Rayleigh number   **(c)**

8. In high speed flow over a plane wall, heat energy will flow to the wall even if the temperature of the free stream is less than the wall temperature. In that case
   (a) $T_w > T_0$ (b) $T_w = T_0$ (c) $T_w < T_0$ (d) $T_a = T_\infty$ **(c)**

9. Adiabatic wall temperature is
   (a) equal to the stagnation temperature
   (b) greater than the stagnation temperature
   (c) lower than the stagnation temperature
   (d) the temperature of an adiabatic surface   **(c)**

10. In high speed flows, the heat transfer rates are calculated by the relation:
    (a) $h A (T_w - T_\infty)$ (b) $h A (T_a - T_0)$
    (c) $h A (T_w - T_0)$ (d) $h A (T_w - T_a)$ **(d)**

11. The local convective heat transfer coefficient at any point on the circumference of a cylinder in cross-flow increases with increasing Reynolds number because
    (a) the separation is delayed (b) the wake is thinner
    (c) both (a) and (b)   (d) boundary layer thickness decreases   **(d)**

12. The local Nusselt number along the circumference of a cylinder in cross-flow for $Re \le 10^5$, is minimum at
    (a) the stagnation point   (b) the point of separation
    (c) at about $90°$ (d) at about $140°$ **(b)**

13. Assertion (A): The local Nusselt number along the circumference of a cylinder in cross-flow for Reynolds number greater than 10^5 has two minimum values.
    Reasoning (R): Because heat transfer coefficient is minimum at the point of separation.
    Code: (a) Both A and R are true   (b) A is true but R is false
               (c) Both A and R are false   (d) A is false and R is true   **(a)**

14. The heat transfer by conduction from a spherical surface to a stationary infinite medium around the surface can be obtained by putting Nu equal to
    (a) 2.0   (b) 1.8   (c) 1.6   (d) 1.0   **(a)**

15. For turbulent flow, the Prandtl number of a fluid is 0.85. The recovery factor can be approximated as:
    (a) 1.12   (b) 0.95   (c) 0.90   (d) 0.64   **(b)**

NUMERICALS

1. Air at 1 atm and 27°C flows across a 5 cm outside diameter pipe with a velocity of 5 m/s. Calculate the rate of heat transfer from the pipe per unit length if the temperature of the pipe surface is maintained at 127°C.
   For air: $k = 0.03$ W/mK, $\nu = 18 \times 10^{-6}$ m²/s, $Pr = 0.7$.
   (580.96 W)

2. An object can be approximately represented by a vertical cylinder 30cm in diameter and 1.7m in length. If the object faces a wind of velocity 36km/h at 10°C, when its surface temperature is 30°C, estimate the rate of heat transfer. For air: $k = 0.0259$ W/mK, Pr $= 0.707$, $v = 15 \times 10^{-6}$m²/s. **(723.5 W)**

3. Air at 30°C is flowing across a 100W street light bulb with a velocity of 2m/s. Assuming that the bulb can be represented by a 60mm diameter sphere and its surface temperature is 125°C, calculate the rate of heat transfer and the percentage of power lost due to convection.
   For air: $v = 20 \times 10^{-6}$ m²/s, $k = 0.03$ W/mK,   Pr $= 0.70$

4. A copper wire 10 mm in diameter is placed in an air stream at 1 bar and 27°C. The wind velocity is 15 m/s perpendicular to the wire. Calculate the current carrying capacity of the wire if the surface temperature does not exceed 75°C and the resistivity of wire is 0.018Ω mm²/m.
   Take: $v = 16 \times 10^{-6}$ m²/s, Pr $= 0.707$, $k = 0.026$ W/mK.   **(928.8 Amp)**

5. Air at 27°C flows across a tube with a velocity of 28 m/s. Calculate the convective heat transfer coefficient when the walls of the tube are at 77°C. What would be the value of the heat transfer coefficient when a hexagonal tube whose diagonal equals the diameter of the tube is placed across the air stream. For the properties of air, use the values from Q. No. 4.

6. A fine wire (diameter 0.03 mm) is placed in air stream at 25°C having a velocity of 60 m/s perpendicular to the wire. An electric current is passed through the wire such that its temperature is raised to 65°C. Calculate the loss of heat energy per unit length of the wire.

7. A 20 mm sphere ($\rho = 8940$ kg/m³, $k = 400$ W/mK, C $= 0.387$ kJ/kgK) has its surface temperature at 75°C and is placed in air stream at 1 bar and 25°C having a velocity of 10 m/s. Calculate the time required to cool the sphere to 30°C. For air: $\mu = 190 \times 10^{-7}$ Pa.S,   Pr 0.7, $k = 0.0258$ W/mK. **(3.84 min)**

8. Air at 427°C flows across a bundle of tubes having 10 rows in the direction of flow and 50 tubes in each row. The free stream velocity is 5 m/s and the tube (diameter 10 mm) wall temperature is maintained at 27°C. The tubes are arranged in an in-line manner so that the spacing in parallel and transverse direction = 20 mm. Calculate the rate of heat transfer per unit length of the tube in the bank.

9. A flat plate is placed in a supersonic wind tunnel with air flowing over it at M = 2.0, a pressure = 25 kN/m², and an ambient temperature of – 15°C. If the plate is 30cm long in the direction of flow, calculate the rate of cooling per unit area of the plate of the surface temperature does not exceed 120°C.

10. A satellite re-enters the earth's atmosphere at a velocity of 2700 m/s. Estimate the maximum temperature the heat shield would reach. The temperature of the upper layer of atmosphere is – 50°C. Neglect the effects of radiation.
    (3577°C)

CHAPTER 9

Thermal Radiation

1. Essentials of Heat Transfer by Radiation

The transfer of heat energy by conduction and convection requires the presence of a material medium. But when a hot body is placed in vacuum, it has been observed that it also loses heat energy and it is said that the energy is lost by radiation. Energy from the sun passes through the space and then through the atmosphere before it reaches and warms the earth's surface. This energy transfer by radiation does not make itself evident unless it falls on matter. Therefore, the other important characteristic of energy transfer by radiation is that it travels from a hot system through a cold non-absorbing or partially absorbing medium to reach another warmer system. Whereas, in conduction and convection heat transfer, heat energy is always transferred from a region of higher temperature region to a region of lower temperature.

All objects emit radiant energy continuously. Thermal radiation is a process of propagation of the internal energy of an emitting body by means electromagnetic disturbances. These disturbances originating from the radiating body travel in space with the velocity of light (= 3×10^{10} cm/s). And, when absorbed by another system, are converted back into thermal-motion energy of molecules.

2. Physical Mechanism of Thermal Radiation

The exact mechanism of radiation heat transfer is yet to be established. However, two theories have been proposed to explain the phenomenon:

(i) *The Quantum Theory* – Energy emitted by an object is not continuous. The objects emit energy through individual discrete particles, called photons. The emitted photon is a particle of matter possessing energy, momentum and electromagnetic mass. Thermal radiation can, therefore, be considered as a photon gas. The energy, E, is given by

$$E = h\nu \tag{9.1}$$

where h is the Planck's constant and is equal to 6.625×10^{-34} J-s and ν is the frequency. Using the relation, $E = m c^2$,

$$m = h\nu/c^2; \quad \text{and}$$

$$\text{Momentum} = \frac{h\nu}{c^2} \times c = \frac{h\nu}{c}$$

(ii) *The Wave Theory* – All space is believed to be filled up by a hypothetical medium called 'ether'. Since each object is always in contact with the ether, the vibrating molecules of the object set up waves in the ether. These waves are called electromagnetic waves and they travel with the velocity of light.

Thus, we can think of radiation which possesses both the continuity properties of electromagnetic waves, and the properties of discreteness of typical photons and energy propagates in the form of waves. Thus, radiation is characterized by wavelength (λ) or oscillation frequency $(\nu = c/\lambda)$. These electromagnetic waves have been classified as:

Table 9.1 Classification of electromagnetic radiation by wavelength

Kind of radiation	Wavelength, m	Kind of radiation	Wavelength, m
Cosmic rays	upto 4×10^{-13}	Visible light	3.9×10^{-7} to 7.8×10^{7}
Gamma rays	4×10^{-13} to 1.4×10^{-10}	Infrared rays	7.8×10^{-7} to 1×10^{-3}
X-rays	1×10^{-11} to 2×10^{-8}	Heat rays	1×10^{-7} to 1×10^{-4}
Ultraviolet rays	1×10^{-8} to 3.9×10^{-7}	Radio waves	0.2×10^{-3}

3. Basic Terminology—Definitions

Emissive Power, E: It denotes the total emitted thermal radiation leaving a surface, per unit time, per unit area of the emitting surface. It depends upon the temperature of the emitting surface, the material of which the surface is composed and the nature of the surface structure.

Radiosity, J: It indicates the total radiant energy leaving a surface, per unit surface area, per unit time. It includes the reflected energy as well as the original emission.

Irradiation, G: It denotes the total incident radiation upon a surface, per unit time, per unit area of irradiated surface. It is the result of emissions and reflections from other surfaces.

When radiation falls on a surface, a part of it may be absorbed by the body, a part may be reflected away from the surface and a part may be transmitted through the body.

Absorptivity, α: the fraction of the total incident radiation absorbed by the body is called absorptivity, i.e., $\alpha = G_\alpha/G$. Fig. 9.1

Reflectivity, ρ: it is the fraction of the total incident radiation reflected by the body; or, $\rho = G_\rho/G$.

Transmissivity, τ: it is the fraction of the total incident radiation transmitted through the body; or, $\tau = G_\tau/G$.

And, from the conservation of energy principle:

$$G_\alpha/G + G_\rho/G + G_\tau/G = 1; \quad \text{or,} \quad \alpha + \rho + \tau = 1 \tag{9.2}$$

Specular and Diffuse Reflection: Two types of reflection phenomena may be observed when radiation strikes a surface. If the angle of incidence is equal to the angle of reflection, the reflection is called specular. On the other hand, when an incident beam is distributed uniformly in all directions after the reflection, the reflection is called diffuse.

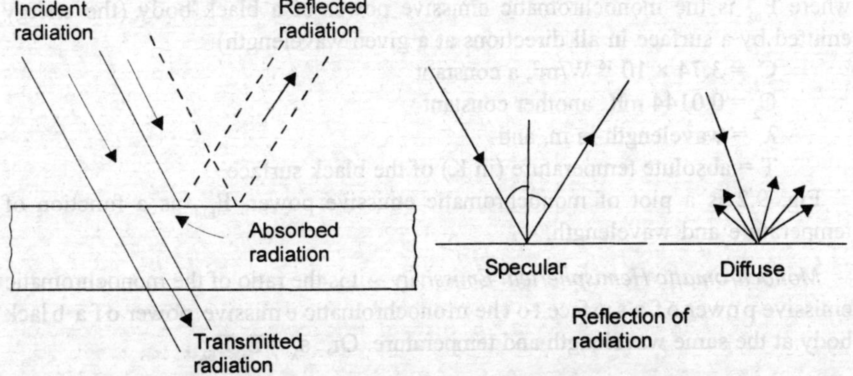

Fig. 9.1 Reflection, absorption and transmission of radiation

Opaque, Black and Gray Body — Emissivity

If the transmissivity, $\tau = 0$, the body is called an opaque body and then $\alpha + \rho = 1$. Since almost all engineering materials (solids and liquids) are quite thick (thickness > 10^{-6} m), they may be considered opaque. Glass is an exception. It is transparent to shorter wavelength radiation waves and is opaque for longer wavelength radiation waves.

When $\rho = \tau = 0$, $\alpha = 1$ i.e., the total incident radiation will be absorbed by the body and the body is called a black body. If $\rho = 1$, ($\alpha = \tau = 0$), the body is an ideal reflector.

Emissivity, ε is the ratio of the emissive power of a body to the emissive power of a black body at the same temperature. Or, $\varepsilon = E/E_b$. When the absorptivity of a surface does not vary with temperature and wavelength of the incident radiation, the body is called a 'Gray Body'. (In general, unoxidised surfaces of good electrical conductors have low value of α and surfaces of poor conductors and oxidised surfaces have high value of α.)

Further, the radiosity and irradiation are related as:

$$J = E + \rho G = E + (1 - \alpha) G = \varepsilon E_b + (1 - \alpha)G \tag{9.3}$$

4. Monochromatic Emissive Power of a Black Body — Planck's Distribution Law

The radiant energy emitted by a black body at a given wavelength increases as the temperature increases. And, at a given temperature, the amount of radiation emitted by a black body varies with the wavelength. Thus, the amount of radiation emitted by a black body is a function of the wavelength of the radiation waves and the temperature of the body.

Max Planck, in 1900, using the concept of quantum theory, derived an expression:

$$E_{b\lambda} = C_1 \, \lambda^{-5}/(\exp(C_2/\lambda T) - 1) \tag{9.4}$$

Rayleigh-Jean had earlier suggested that $E_{b\lambda} = C_1 T/C_2 \lambda^4$ but it does not agree with experimental results over the entire wavelength range.

where $E_{b\lambda}$ is the monochromatic emissive power of a black body (the energy emitted by a surface in all directions at a given wavelength)

$C_1 = 3.74 \times 10^{-16}$ W/m², a constant

$C_2 = 0.0144$ mK, another constant

λ = wavelength in m, and

T = absolute temperature (in K) of the black surface.

Fig. 9.2 is a plot of monochromatic emissive power, $E_{b\lambda}$, as a function of temperature and wavelength.

Monochromatic Hemispherical Emissivity – it is the ratio of the monochromatic emissive power of a surface to the monochromatic emissive power of a black body at the same wavelength and temperature. Or, $\varepsilon_\lambda = E_\lambda / E_{b\lambda}$.

5. Wien's Displacement Law

It can be observed from Fig. 9.2, that as the temperature of the emitting surface decreases, there is an increase in the wavelength, λ_{max}, at which the maximum energy is emitted. That is, for a given temperature, we can determine, λ_{max} by differentiating Eq. (9.4) and setting it equal to zero and solve for λ. This leads to the relation known as 'Wien's Displacement Law'.

or, $\qquad \lambda_{max} T = 2897.6 \times 10^{-6}$ mK $= 0.2897$ cmK $\qquad$ (9.5)

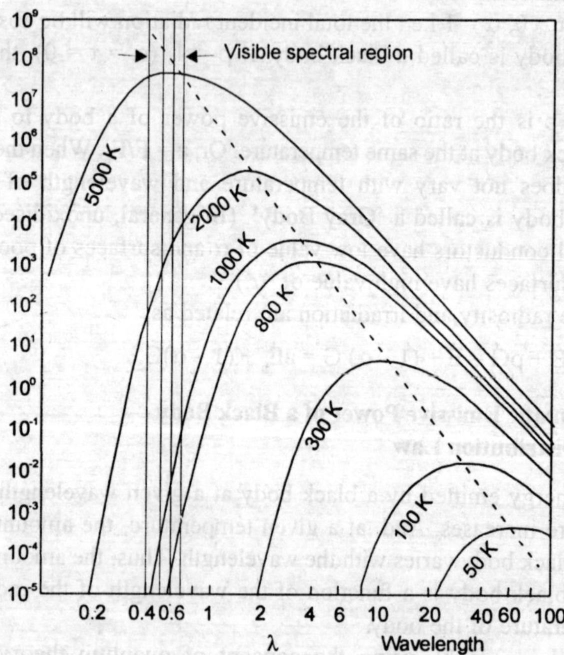

Fig. 9.2 Monochromatic emissive power of a black body as a function of wavelength

The hemispherical monochromatic absorptivity of a surface is defined as the fraction of the total irradiation at wavelength λ that is absorbed by the surface

$$\alpha_\lambda = G_{\lambda \text{ absorbed}}(T)/G_\lambda(T)$$

Example 9.1 A black body emits radiation at 2000 K. Calculate (i) the monochromatic emissive power at 1μm wavelength, (ii) wavelength at which the emission is maximum, and (iii) the maximum emissive power.

Solution: (i) From Planck's law:

$$E_{b\lambda} = C_1 \lambda^{-5} / [\exp (C_2/\lambda T) - 1]$$
$$= 3.74 \times 10^{-16} \times (1 \times 10^{-6})^{-5} / [\exp (0.0144/10^{-6} \times 2000) - 1]$$
$$= 2.79 \times 10^{11} \text{ W/m}^3$$

(ii) From Wien's law: $\lambda_{max} = 2897 \times 10^{-6}/2000 = 1.4485 \times 10^{-6}$ m.

(iii) The maximum emissive power from a black body at a given temperature would occur at the wavelength, $\lambda = 2897.6 \times 10^{-6}/T$

$$\therefore \quad C_1 (\lambda_{max})^{-5} = 3.74 \times 10^{-16} \times (2897.6 \times 10^{-6}/T)^{-5}$$
$$= 1.83 \times 10^{-3} \times T^5$$

and, $\exp (C_2/\lambda_m T) - 1 = \exp (0.0144 / 2897.6 \times 10^{-6}) - 1 = 142.9736$

$$\therefore \quad E_{b\lambda} = (1.83 \times 10^{-3} \text{ } T^5)/(142.9736) = 1.28 \times 10^{-5} \text{ } T^5 \text{ W/m}^2/\text{m}$$

and, when $\quad T = 2000$ K, the maximum emissive power,

$$E_{b\lambda} = 4.096 \times 10^{11} \text{ W/m}^2/\text{m}.$$

Example 9.2 An optical pyrometer using a filter of mean wavelength of 0.65 μm records the temperature of a body as 1600 K. If the emissivity of the body at that wavelength is 0.7, estimate its true temperature.

Solution: The working of an optical pyrometer is based on the principle that at a given wavelength λ, the radiant intensity (brightness) of a black body varies with temperature T and it obeys Planck's Law. The pyrometer is calibrated in such a way that it measures the temperature which a black body would have if it were emitting radiation equal to that of the body under investigation. That is, the brightness of the target image matches with that of the filament of the instrument. Thus, if T is the temperature of the body under investigation at wavelength λ, and emissivity ε_λ , we have

$$C_1 \lambda^{-5} / [\exp (C_2/\lambda T_b) - 1] = \varepsilon_\lambda C_1 \lambda^{-5} / [\exp (C_2/\lambda T) - 1]$$

For temperature less than 4000°C, the term $\exp (C_2/\lambda T)$ is much greater than unity. Hence we can write, after taking logarithm of both sides,

$$\exp (-C_2/\lambda T_b) = \varepsilon_\lambda \exp (- C_2/\lambda T)$$

or, $\quad\quad 1/T = 1/T_b + (\lambda/C_2) \ln (\varepsilon_\lambda)$.

$$1/T = 1/1600 + (0.65 \times 10^{-6} / 0.0144) \ln(0.7)$$

which gives $\quad\quad T = 1642$ K.

Example 9.3 An optical pyrometer using two filters having mean wavelengths of 0.65 microns and 0.54 microns record the temperature of a body as 1650 K and 1680 K respectively. Estimate the true temperature and the emissivity of the body assuming it to be a gray body.

Solution: When a body is far away and is inaccessible, it may be difficult to measure its emittance. And, an inaccurate value of emittance results in error in the temperature measurement by all types of radiation pyrometers. However, the use of two wavelengths to measure temperature has found some support. It is assumed that the investigating body is a gray body and the two wavelengths are taken fairly close so that the emittance may be assumed constant.

Thus, using the relation developed in Ex. 9.2, we write

$$1/T = 1/T_{b1} + (\lambda_1/C_2) \ln (\varepsilon_\lambda)$$

and $\quad 1/T = 1/T_{b2} + (\lambda_2/C_2) \ln (\varepsilon_\lambda)$

or, $\quad 1/T = 1/1650 + (0.65 \times 10^{-6}/0.0144) \ln (\varepsilon_\lambda)$; and

$\quad\quad 1/T = 1/1680 + (0.54 \times 10^{-6}/0.0144) \ln (\varepsilon_\lambda)$

or, $\quad\quad 1/1650 - 1/1680 = \ln (\varepsilon_\lambda) (- 7.639 \times 10^{-6})$

$\therefore \quad\quad\quad\quad \ln(\varepsilon_\lambda) = - 1.4167;$ and $\varepsilon_\lambda = 0.2425$

And, temperature, $\quad T = 1844.64$ K.

Example 9.4   Estimate the surface temperature of the sun and its emissive power, assuming it to be a black body and it emits maximum radiation at $\lambda = 0.5$ microns. Also estimate (i) the energy received by the surface of the earth and (ii) the energy received by a 2m × 2m solar collector whose normal is inclined at 60° to the sun. Take the diameter of the sun as 1.4×10^9 m, diameter of the earth as 13×10^6 m and the distance of earth from the sun as 15×10^{10} m.

Solution: From Wien's displacement law: $T = 2897 \times 10^{-6}/ 0.5 \times 10^{-6} = 5794$ K. The emissive power of the sun can be estimated by using Stefan-Boltzmann law:

$$E_b = 5.667 \times 10^{-8} (5794)^4 = 63.865 \text{ MW/m}^2.$$

The radiation reaching the earth's atmosphere would be:

= Emissive power of the sun × (Radius of the sun/Distance from the earth)^2

$= 63.865 \times [(0.7 \times 10^9)/(15 \times 10^{10})]^2$

$= 1.39 \text{ kW/m}^2.$

Surface area of the solar collector in the direction normal to solar radiation

$$= A \cos \theta = 4 \cos 60° = 2\text{m}^2$$

Energy received $= 2.78$ kW.

6. Derivation of Stefan-Boltzmann Law from Planck's Law of Radiation

In order to derive the Stefan-Boltzman law from Planck's law, we write:

$$E_b = \int_0^\infty E_{b_\lambda} \times d\lambda = \int_0^\infty C_1 \lambda^{-5}/ [\exp(C_2/\lambda T) - 1]d\lambda$$

Let $\quad x = C_2/\lambda T, \quad$ or $\quad \lambda = C_2/xT; \quad d\lambda = (-C_2/x^2 T)dx$

and $\quad E_b = -C_1 \int_0^\infty \dfrac{(xT)^5}{C_2^5} \dfrac{1}{[\exp(x)-1]} \dfrac{C_2}{x^2 T} \cdot dx$

$$= (C_1/C_2^4)T^4 \int_0^\infty (x^3/[\exp(x)-1]) \cdot dx$$

$$= \dfrac{C_1 T^4}{C_2^4}\left(\dfrac{6\pi^4}{90}\right) = 5.67 \times 10^{-8} T^4 = \sigma \cdot T^4.$$

7. Statement and Proof of Kirchhoff's Law of Radiation

Kirchhoff's law (1882) states that, at thermal equilibrium, the ratio of the emissive power of a surface to its absorptivity is the same for all bodies.

Let us consider a body 1 having surface area A_1, absorptivity α_1, is placed in a large evacuated enclosure, Fig. 9.3, perfectly insulated from its surroundings. Radiation energy exchange will take place between the body and the inner walls of the enclosure until thermal equilibrium is attained and the body and the walls reach the same temperature. If G is the irradiation, the energy absorbed by the body will be $\alpha_1 GA_1$. Under thermal equilibrium, the energy emitted by the body, E_1A_1, must be equal to $\alpha_1 GA_1$.

Or, $\qquad E_1 = \alpha_1 G$ and $E_1/\alpha_1 = G$ $\hfill (9.6)$

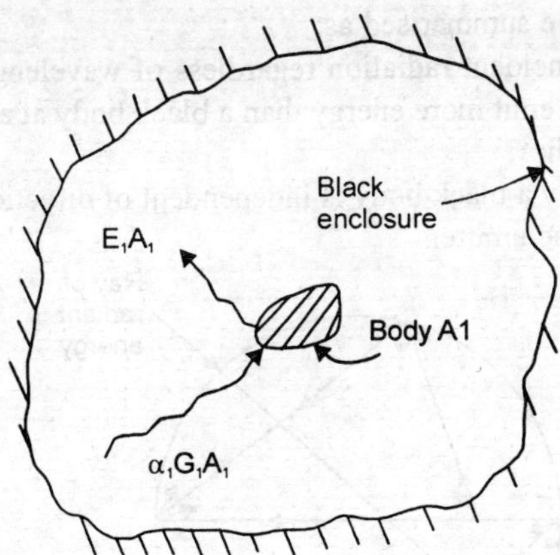

Fig 9.3 Kirchhoff's law model

Similarly, for a second body having the surface area A_2, absorptivity α_2, emissive power E_2, we get: $E_2/\alpha_2 = G$

Thus, $\qquad\qquad E_1/\alpha_1 = E_2/\alpha_2 = \ldots = E/\alpha = G$ $\hfill (9.7)$

If we have a black body of surface area A_b, emissive power E_b, we have $E_b = G$, because the absorptivity of the black body is equal to 1

or, $\qquad\qquad \dfrac{E_1}{\alpha_1} = \dfrac{E_2}{\alpha_2} = E_b ;$ and $\dfrac{E_1}{E_b} = \alpha_1 ;$ $\dfrac{E_2}{E_b} = \alpha_2$

or, $\qquad\qquad \varepsilon_1 = \alpha_1 ;$ $\varepsilon_2 = \alpha_2$

That is, at thermal equilibrium, the absorptivity and the emissivity of a body are equal. For a black body, both are equal to 1.

Kirchhoff's law is also valid for monochromatic radiation. In other words, at a given wavelength and temperature, the ratio of the monochromatic emissive power E_λ to the monochromatic absorptivity α_λ is also a constant for all bodies when they are in thermal equilibrium with their surroundings. That is,

$$E_{\lambda 1}/\alpha_{\lambda 1} = E_{\lambda 2}/\alpha_{\lambda 2} = E_{b\lambda} = \text{constant.}$$

From the above, it follows that $\varepsilon_\lambda = \alpha_\lambda$, i.e., if a body is in thermal equilibrium with its surroundings, its monochromatic absorptivity is equal to its monochromatic emissivity ε_λ.

8. The Concept of a Black Body

A black body is a theoretical concept and can only be approximated in practice. Let us consider a hollow sphere, shown in Fig. 9.4, whose interior surface is maintained at a uniform temperature. Any radiation entering through a small hole provided in the wall of the sphere will be partly absorbed and partly reflected diffusely at the interior surface. Each time, the reflected radiation strikes the wall and a part of it is absorbed and when the original radiation beam, if at all, reaches the hole and escapes, it is so weak that the energy leaving the cavity is negligibly small. This would be true regardless of the surface and the composition of the walls of the cavity. Thus, a small hole in a spherical enclosure behaves like a black body because almost all incident radiation is absorbed by it. The properties of a black body can be summarised as:

a. It absorbs all incident radiation regardless of wavelength and direction;
b. No surface can emit more energy than a black body at a given temperature and wavelength;
c. The radiation by a black body is independent of direction. That is, a black body is a diffuse emitter.

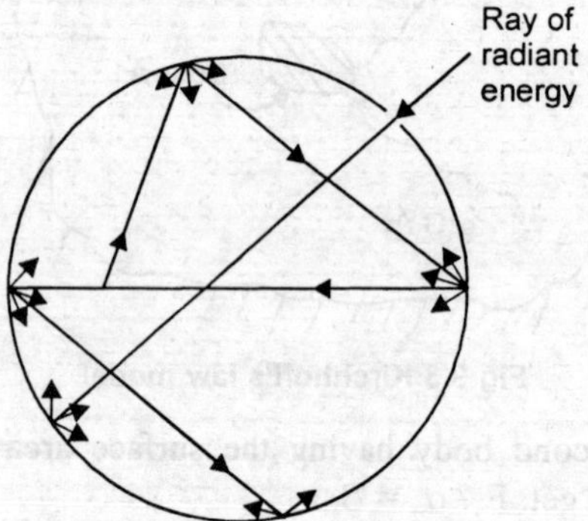

Fig 9.4 Reflection of radiation in a cavity – a black body enclosure

9. Intensity of Radiation

Radiation emitted by a surface propagates in all possible directions. It has been observed that most surfaces emit a greater part of their energy in a direction normal to the surface. The 'Intensity of Radiation' I is the radiant energy leaving the surface per unit area through unit solid angle along a normal to the surface, per unit time. For black surfaces, the intensity is uniform in all directions. The same is true for any surface which is a diffuse emitter. The intensity of radiation in any other direction at any angle ϕ to the normal is given by 'Lambert's Cosine Law' which states:

> 'The intensity of radiation in a direction ϕ from the normal to a black emitter is proportional to the cosine of angle ϕ.'

That is, if I_n represents the normal intensity and I_ϕ represents the intensity at an angle ϕ from the normal, then

$$I_\phi = I_n \cos \phi = I \cos \phi. \tag{9.8}$$

10. An Expression for Obtaining I, the Intensity of Radiation

Let us consider a small area dA at O with OP as normal to it, (Fig. 9.5). We choose another small element on the surface of a hemisphere such that OA makes an angle ϕ with the normal OP and $\angle AOB = d\phi$. The width of the element is the length of the arc, subtending an angle $d\phi$ and radius r. Thus, AB = r dϕ.

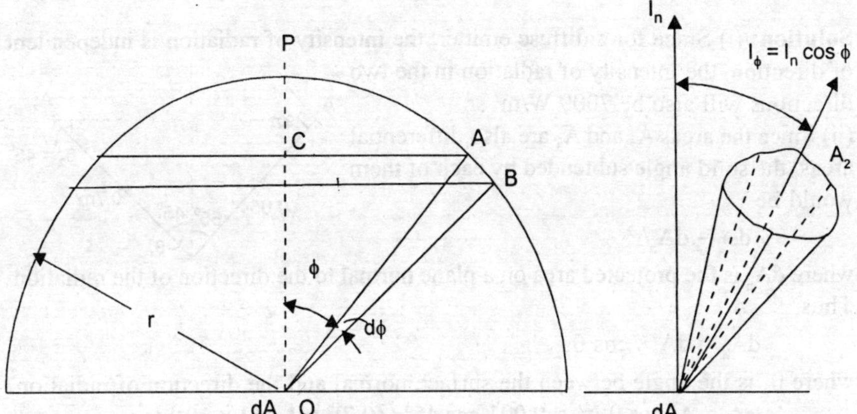

(a) Radiation from a small element to a hemisphere **(b) Solid angle subtended by an area A2 at a differential area**

Fig. 9.5

The radius of the element, CA = r sin ϕ and therefore, the surface area of the element on the sphere = width × circumference = r dϕ · 2πr sin ϕ. Since a surface subtends a solid angle at a point at a distance r from all points on the surface, equal to the surface area divided by r^2, the solid angle subtended at dA

$$d\omega = \frac{rd\phi \cdot 2\pi r \sin \phi}{r^2} = 2\pi \sin \phi \, d\phi$$

The total energy emitted by dA is given by

$$EdA = \int_0^{\pi/2} I_\phi \, d\omega \, dA$$

$$= \int_0^{\pi/2} dA \, I_\phi \, 2\pi \sin \phi \, d\phi$$

$$= 2\pi dAI \int_0^{\pi/2} \sin \phi \cos \phi \, d\phi$$

where E is the emissive power of the element dA.

$$\therefore \qquad E = 2\pi I \int_0^{\pi/2} \frac{\sin 2\phi}{2} \, d\phi = \pi I$$

or $\qquad\qquad I = E/\pi$ $\qquad\qquad\qquad\qquad\qquad\qquad$ (9.9)

and for a black body, $I_b = E_b / \pi$.

Example 9.5 A small surface of area A_1 = 0.0012 m² emits diffusely and the intensity of radiation in the normal direction is 7000 W/m².sr. Radiation emitted from this surface is intercepted by two other surfaces of area A_2 and A_3 equal to 0.001 m² and 0.0015 m² respectively, as shown in the figure. Calculate (i) intensity associated with emission in both directions, (ii) the solid angles subtended by the two surfaces when viewed from the emitting surface, and (iii) the rate at which radiation emitted by A_1 is intercepted by the two surfaces.

Solution: (i) Since for a diffuse emitter, the intensity of radiation is independent of direction, the intensity of radiation in the two directions will also be 7000 W/m² sr.

(ii) Since the areas A_2 and A_3 are also differential areas, the solid angle subtended by each of them would be

$$d\omega = dA_n/r^2$$

where dA_n is the projected area on a plane normal to the direction of the radiation. Thus,

$$dA_n = dA \times \cos\theta_2$$

where θ_2 is the angle between the surface normal and the direction of radiation.

$$\omega_{2\text{-}1} = A_2 \cos\theta_2/r^2 = 0.001 \cos 45°/(0.7)^2 = 1.443 \times 10^{-3} \text{ sr.}$$
$$\omega_{3\text{-}1} = A_3 \cos 0°/r^2 = 0.0015 \times 1/(0.65)^2 = 3.55 \times 10^{-3} \text{ sr.}$$

(iii) The rate at which radiation is intercepted by the two surfaces is estimated by the relation

$$q_{1\text{-}2} = I\,A \cos\theta_1 \times \text{solid angle subtended by the area}$$
$$\therefore \quad q_{1\text{-}2} = 7000 \times 0.001 \cos 45° \times 1.443 \times 10^{-3} = 7.1425 \times 10^{-3} \text{ W.}$$
$$\text{and,} \quad q_{1\text{-}3} = 7000 \times 0.0015 \cos 60° \times 3.55 \times 10^{-3} = 18.638 \times 10^{-3} \text{ W.}$$

Example 9.6 A black body, surface area 0.9 m² has its temperature 1000 K. Calculate (i) total rate of energy emission, (ii) the intensity of radiation, and (iii) intensity of radiation in a direction 45° to the normal.

Solution: From Stefan Boltzmann law, $E_b = \sigma T^4$

Total rate of energy emission $= E_b \times$ area $= 5.67 \times 10^{-8}(1000)^4\,0.9$
$$= 51.03 \text{ kW}$$

Intensity of radiation, $I = E_b/\pi = 5.67 \times 10^{-8}(1000)^4/3.142$
$$= 18.046 \text{ kW/m}^2\text{.sr}$$

Since a black body is a diffuse emitter, the intensity of radiation in any other direction would be the same.

11. Wavelength Dependent Characteristics—Solar Radiation

A surface emits radiation in all directions and at all wavelengths. And, the spectral distribution of thermal radiation varies with the nature and temperature of the surface. A surface emits more energy at increasing temperature but the amount of emitted radiation is strongly influenced by the wavelength. The shift in the peak point of the radiation curve, Fig. 9.2, explains the change in the colour of a surface when it is heated. As the temperature of the surface increases, the maximum intensity shifts to the shorter wavelengths and the surface appears as dark-red. As the temperature increases more and more, the colour changes to bright red, then to bright yellow and finally white.

The concept of a black body is an idealization, i.e., a perfect black body does not exist. But this concept helps in solving many practical problems. Max Planck has derived an equation, Eq. (9.5), for the emissive power of a black body at a particular wavelength and at a given temperature. The solar radiation is the primary source of energy for our life on the earth. It is a form of thermal radiation having a particular wavelength distribution. Its intensity strongly depends upon the atmospheric conditions, time of the year and the angle of incidence of the sun's rays on the surface of the earth.

If we desire to calculate the amount of radiant energy, emitted by a surface, in a specified wavelength band, say between λ_1 and λ_2, Fig 9.6, we write

$$E_b(\lambda_1 \rightarrow \lambda_2) = \int_0^{\lambda_2} E_{b_\lambda}\, d\lambda - \int_0^{\lambda_1} E_{b_\lambda}\, d\lambda = E_b(0 - \lambda_2 T) - E_b(0 - \lambda_1 T)$$

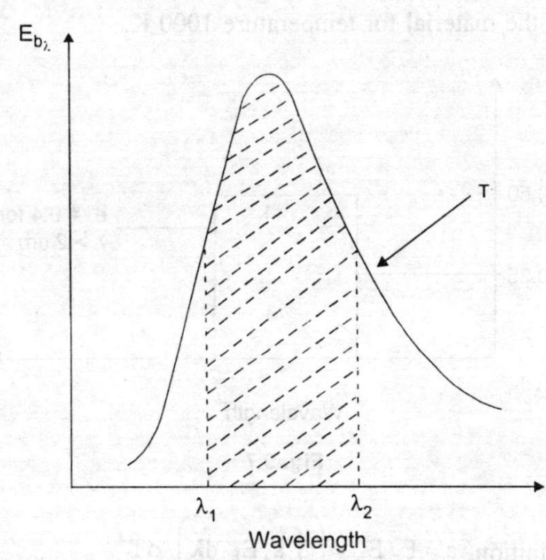

Fig. 9.6

The values of E_b $(0 - \lambda)$ depends upon λ and on the temperature of the surface. Therefore, we make use of 'Radiation Functions' tabulated in Table 9.3 where the values of E_b $(0 - \lambda T) / \sigma T^4$ are available as a function of λT.

Example 9.7 Estimate the fraction of energy emitted by the sun in the visible spectrum of wavelength $\lambda_1 = 0.35$ μm and $\lambda_2 = 0.7$ μm.

Solution: Assuming that the temperature of the sun be 5800 K

$\lambda_1 T = 0.35 \times 10^{-6} \times 5800 = 2030$ μK; $\lambda_2 T = 0.7 \times 10^{-6} \times 5800 = 4060$ μK

From Table 9.3, by interpolation:

$\lambda_1 T = 2030$; $E_b(0 - \lambda_1 T)/E_b) = 0.07165$;

$\lambda_2 T = 4060$; $E_b(0 - \lambda_2 T)/E_b = 0.49156$

Therefore, $E_b (\lambda_1 - \lambda_2)/E_b = 0.49156 - 0.07165 = 0.41991$

Or, 4 1.99% of the sun's energy is available in the visible spectrum and its absolute value is $[0.4199 \times 5.67 \times 10^{-8} (5800)^4] = 26.94$ MW/m².

Example 9.8 A black body is maintained at 1000 K. Calculate the wavelength such that the emission from 0 to λ is 40 percent of the emissive power from λ to ∞.

Solution: $E_b(0 - \lambda T) = 0.4 \, E_b(\lambda T - \infty)$

$$= (0.4 / 1.4) \, E_b(0 - \infty) = 0.2857$$

By interpolation from Table 9.3, we get:

$$E_b(0 - \lambda T) / \sigma T^4 = 0.2857, \quad \text{when} \quad \lambda T = 3055$$

$$\therefore \qquad \lambda = 3.055 \, \mu m.$$

Example 9.9 The monochromatic emissivity of a material varies with the wavelength as shown in Fig. 9.7. Estimate the average emissivity of the material for temperature 1000 K.

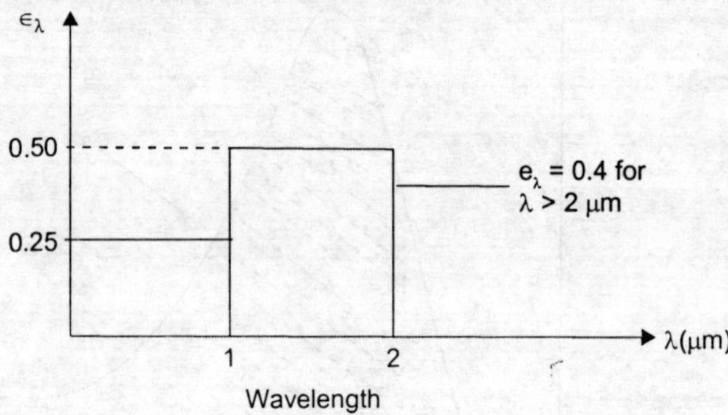

Fig. 9.7

Solution: By definition, $\varepsilon = E / E_b = \left(\int_0^\infty \varepsilon_\lambda E_{b_\lambda} \, d\lambda \right) / \sigma T^4$

For the values given above in Fig 9.7

$$\varepsilon = \left[\int_0^1 \varepsilon_\lambda E_{b_\lambda} \, d\lambda + \int_1^2 \varepsilon_\lambda E_{b_\lambda} \, d\lambda + \int_2^\infty \varepsilon_\lambda E_{b_\lambda} \, d\lambda \right] / \sigma T^4$$

We can show this calculation in the tabular form:

$\lambda_1 - \lambda_2$	$\varepsilon_{\lambda 1} - \varepsilon_{\lambda 2}$	T	$\lambda_1 T$	$\lambda_2 T$	$\dfrac{E_b(0 - \lambda_1 T)}{\sigma T^4}$	$\dfrac{E_b(0 - \lambda_2 T)}{\sigma T^4}$	$\dfrac{(\varepsilon_{\lambda_1} - \varepsilon_{\lambda_2})[E_b(\lambda_1 T - \lambda_2 T)]}{\sigma T^4}$
0 – 1	0.25	1000	0	1000	0	0.0003	0.000075
1 – 2	0.50	1000	1000	2000	0.0003	0.0667	0.0332
2 – ∞	0.40	1000	2000		0.0667	1.000	0.3733

$$\therefore \qquad \varepsilon_{av} = 0.000075 + 0.0332 + 0.3733 = 0.406$$

$$E = \varepsilon \sigma T^4 = 0.406 \times 5.67 \times 10^{-8} \times (1000)^4 = 23.02 \, kW/m^2.$$

12. Greenhouse Effect

Solar radiation presents an unusual phenomenon. This is because the radiation is concentrated at short wavelengths and materials like ordinary glass are transparent to short wavelength radiation ($\lambda < 2\mu$) but are opaque to long wavelength radiations ($\lambda > 3\mu$). Thus a glass house will allow much more radiation to come in than can e scape and t his results i n producing the f amiliar heating e ffect. This phenomenon is called 'Greenhouse Effect'.

Example 9.10 The surface of a body has been covered with aluminium paint. The hemispherical emissivity of the surface is 0.4 at wavelengths below 3μm and 0.8 at longer wavelengths. Calculate the total emissivity of the s urface at 27°C and 800 K. Also calculate the effective absorptivity of the surface if the body receives solar radiation. Assume the sun as a black body at 5800 K.

Solution: (i) Temperature = 27°C = 300 K; $\lambda_1 T = 900$ μm;
From Table 9.1: for $\lambda_1 T = 900$ μm, $E_b(0 - \lambda_1 T) / \sigma T^4 = 0.0001$
Therefore: $E_b (\lambda_1 T - \infty) / \sigma T^4 = 0.9999$. And,

$$\varepsilon = \left[\int_0^3 \varepsilon_\lambda E_{b_\lambda} \, d\lambda + \int_3^\infty \varepsilon_\lambda \varepsilon \, E_{b_\lambda} \, d\lambda \right] = (0.4 \times 0.0001) + (0.8 \times 0.9999) \equiv 0.8$$

(ii) Temperature 800K, $\lambda_1 T = 2400$ μm, from Table 9.1 for $\lambda_1 T = 2400$ μm,
$$E_b (0 - \lambda_1 T) / \sigma T^4 = 0.14 \text{ and } E_b (\lambda_1 T -) / \sigma T^4 = 1.0 - 0.14 = 0.86$$

And, $$\varepsilon = \left[\int_0^3 \varepsilon_\lambda E_{b_\lambda} \, d\lambda + \int_3^\infty \varepsilon_\lambda E_{b_\lambda} \, d\lambda \right] = 0.4 \times 0.14 + 0.8 \times 0.86 = 0.744$$

Since at lower temperatures, practically all the radiation is emitted at longer wavelengths. The value of emissivity is higher at 300 K.
(iii) Temperature = 5800K. $\lambda_1 T = 3 \times 5800$ μm $= 17.4 \times 10^{-3}$ mK
and from Table 9.1: $E_b(0 - \lambda_1 T) / \sigma T^4 = 0.98$
This shows t hat 98% of t he solar radiation falls o n the surface a t wavelength below 3 μm . Further, $\varepsilon_\lambda = \alpha_\lambda$
$$\alpha(\lambda, T) = 0.4 \times 0.98 + 0.8 \times 0.02 = 0.408.$$

Example 9.11 The material of Ex. 9.9 receives solar radiation. Assuming the temperature of the sun as 5800 K, Calculate the percentage of solar radiation absorbed by the surface.

Solution: The emissivity of the surface is given as:
$$\lambda_1 \leq 1, \varepsilon_1 = 0.25; \qquad 1 \leq \lambda_2 \leq 2: \qquad \varepsilon_2 = 0.5; \qquad \lambda_3 > 2, \varepsilon_3 = 0.4$$
$$\lambda_1 T = 1 \times 5800 = 5800 \text{ μK}; \qquad \lambda_2 T = 2 \times 5800 = 11.6 \times 10^{-3} \text{ mK}.$$
From Table 9.1: $E_b(0 - \lambda_1 T) / \sigma T^4 = 0.7202;$ $E_b(0 - \lambda_2 T) / \sigma T^4 = 0.941$
$$\varepsilon = \varepsilon_1 E_b (0 - \lambda_1 T) / \sigma T^4 + \varepsilon_2 \left[E_b (0 - \lambda_2 T) - E_b (0 - \lambda_1 T] \right. / \sigma T^4$$
$$+ \varepsilon_3 \left[E_b (0 - \infty) - E_b (0 - \lambda_2 T) \right] / \sigma T^4$$
$\therefore$ $\alpha = [0.25 + 0.7202 \times 0.5 (0.941 - 0.7202) + 0.4 (1.0 - 0.941)]$
$= 0.314$ i.e. 31.4% of the solar radiation would be absorbed by the surface.

Example 9.12 A glass plate has a transmissivity of 0.92 for the wavelengths between 0.4 µm and 2.5 µm and is opaque to all other wavelengths. Calculate the fraction of the incident solar energy transmitted through the glass.

Solution: The sun is considered as a black body having its surface temperature equal to 5800 K.

$$\therefore \qquad \lambda_1 T = 0.4 \times 5800 \ \mu mK = 2320 \ \mu mK$$
$$\lambda_2 T = 2.5 \times 5800 \ \mu mK = 14500 \ \mu mk$$

From Table 9.3, for $\lambda_1 T = 2320 \ \mu mK$, $E_b (0 - \lambda_1 T)/\sigma \ T^4 = 0.1242$
$$\lambda_2 T = 14500 \ \mu mK, \qquad E_b (0 - \lambda_2 T) / \sigma \ T^4 = 0.9657$$
$$\therefore \qquad E_b(\lambda_2 T - \lambda_1 T) / E_b = 0.9657 - 0.1242 = 0.8415$$

Thus, 84.15% of the incident solar energy is in the wavelength band from 0.4 µm and 2.5 µm . Since the transmissivity is 0.92, the energy transmitted would be $0.92 \times 0.8415 = 0.7742$, or 77.42% of the incident energy.

Example 9.13 Calculate the radiation equilibrium temperature for a plate painted white (α for solar radiation 0.12, α for low temperature radiation 0.9) when the plate is exposed to a solar flux of of 800 W/m² in a surrounding at 300 K.

Solution: For thermal equilibrium, the energy absorbed must be equal to the exchange of energy with the surroundings.

Energy absorbed = $\alpha \times$ incident energy = $0.12 \times 800 = 96$ W/m²
Energy exchange = $\varepsilon\sigma \ (T^4 - 300^4)$; because the surroundings are always treated as a black body.

$$\therefore \qquad (T/100)^4 = 96/(0.9 \times 5.67) + 3^4 = 99.81; \quad \text{and} \quad T = 316.08 \ = 43.08°C.$$

Example 9.14 A surface emits radiation as a black body at 1880 K. Calculate the emission from the surface in direction corresponding to $0 \leq \theta \leq 60°$ and in the wavelength interval 1.5 µm $\leq \lambda \leq$ 4.5µm.

Solution: The emission from a surface is given by

$$E = \int_{\lambda_1}^{\lambda_2} \int_{\phi=0}^{\phi=2\pi} \int_{\theta=0}^{\pi/3} I_{\lambda_b} \cos\theta \sin\theta \ d\theta \ d\phi \ d\lambda \ = $$
$$\int_{1.5}^{4.5} I_{\lambda_b} \left(\int_0^{2\pi} \int_0^{\pi/3} \cos\theta \sin\theta \ d\theta \ d\phi \ d\lambda \right)$$

Since a black body is a diffuse emitter,

$$E = \int_{1.5}^{4.5} I_{\lambda b} (2\pi \sin^2 \theta / 2)_0^{\pi/3} \ d\lambda = 0.75 \int_{1.5}^{4.5} \pi I_{b\lambda} d\lambda = 0.75 E_b \int_{1.5}^{4.5} \frac{E_{\lambda b}}{E_b} d\lambda$$

From Table 9.3, $\lambda_1 T = 1.5 \times 1880 = 2820 \ \mu mK$; $E_b (0 - \lambda_1 T)/\sigma \ T^4 = 0.2280$
$$\lambda_2 T = 4.5 \times 1880 = 12960 \ \mu mK; \quad E_b(0 - \lambda_2 T)/\sigma \ T_4 = 0.9550$$
$$E = 0.75 \ E_b(0.9550 - 0.2280)$$
$$= 0.75 \times 5.67 \times 10^{-8} \times (1880)^4 \times 0.727$$
$$= 0.386 \ MW/m^2.$$

[The total hemispherical emissive power is reduced by 25% and 27.3% due to the directional and spectral restrictions respectively.]

Example 9.15 Estimate the surface area of the filament of a 100 W light bulb when the filament temperature is 2500 K and its efficiency is 15 percent.

Solution: The wavelength band of visible light lies between 0.35 μm and 0.75 μm. Therefore,

$$\lambda_1 T = 0.35 \times 2500 = 875 \;\mu mK$$
$$\lambda_2 T = 0.75 \times 2500 = 1875 \;\mu mK$$

From Table 9.3

$$E_b(0 - \lambda_1 T) / \sigma T^4 = 0.6589 \times 10^{-4}$$
$$E_b(0 - \lambda_2 T) / \sigma T^4 = 0.04886$$
$$\therefore \qquad E_b(\lambda_1 - \lambda_2) / E_b = 0.04886 - 0.00006589 = 0.04879$$

By energy balance: $0.04879 \times E_b \times$ required surface area $= 0.15 \times 100W$

$$\therefore \qquad \text{Area } A = 15/[0.04879 \times 5.67 \times (25)^4]$$
$$= 1.338 \times 10^{-4} \; m^2 = 133.8 \; mm^2.$$

Example 9.16 A furnace with black interior walls maintained at 1400 K, has an opening in the side covered with a glass window having the following properties:

$$0 < \lambda < 3\mu m, \quad \tau = 0.8, \quad \varepsilon = 0.2, \quad \rho = 0$$
$$3\mu m < \lambda < \infty, \quad \tau = 0.0, \quad \varepsilon = 0.8, \quad \rho = 0.2$$

Calculate the radiation loss through the window to a large room at 25°C.

Solution: Temperature of the furnace wall, T = 1400 K

$$\lambda_1 T = 3\mu m \times 1400 = 4200 \;\mu mK$$

From Table 9.3, $\qquad E_b (0 - \lambda_1 T) / E_b = 0.5159$

and $\qquad E_b (\lambda_1 T - \infty) / E_b = 1.0 - 0.5159 = 0.4841$

Total incident radiation transmitted between the wavelength band

$$0 < \lambda < 3 \;\mu m = 0.5159 \times 5.67 \times 14^4 \times 0.8 = 89.9 \; kW/m^2$$

Radiation absorbed $= 0.2 \times 0.5159 \times 5.67 \times 14^4 = 22.475 \; kW/m^2 \; (0 < \lambda < 3\mu m)$

$$= 0.8 \times (1 - 0.5159) \times 5.67 \times 14^4 = 84.357 \; kW/m^2 \; (3\mu m < \lambda < \infty)$$

Radiation transmitted through the opening $= 89.9 \; kW/m^2$

Radiation absorbed/emitted through the opening $= (22.475 + 84.357) \; kW/m^2$

$$= 106.832$$

Energy received by the glass window from the environment $= 0.8 \times \sigma T^4$

$$= 0.8 \times 5.67 \times 2.98^4 = 0.357 \; kW/m^2$$

$$\therefore \qquad \text{Total loss} = 89.9 + 106.832 - 0.357 = 196.375 \; kW/m^2.$$

13. Characteristics of Real Surfaces

The radiation from real surfaces differs from black or gray body radiation in several aspects. A black body is an ideal emitter and no surface can emit more radiation than a black body at the same temperature. It is therefore convenient to choose the black body as a reference in describing emission from any other surface. Gray surfaces radiate a constant fraction, $\in_g$. of the monochromatic emissive power of a black surface at the same temperature over the entire spectrum. Real surfaces also radiate a fraction, $\in_\lambda$ at any wavelength and this fraction is not constant and varies with wavelength, Fig. 9.8.

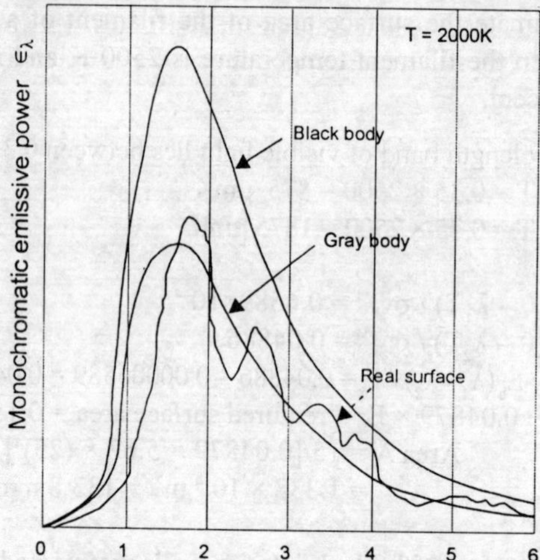

Fig. 9.8 Comparison of hemispherical monochromatic emission for a black, gray (ϵ_g = 0.6) and a real surface

Solids which are good conductors of electricity (such as polished aluminium and copper) generally have low values of emissivity or absorptivity and are therefore good reflectors. Solids which are non-conductors of electricity (concrete, cement etc) generally have high values of emissivity, Fig. 9.9.

Fig. 9.10 shows the effect of temperature on hemispherical total emissivity of several materials, because we require an average value of emissivity or absorptivity for the wavelength band in which the bulk of the radiation is emitted or absorbed Siber has evaluated the total absorptivity of the surface of several materials as a function of source temperature, with the receiving surfaces at room temperature and the emitter a black body, Fig. 9.11.

The emissivity of many surfaces varies also with direction. Such surfaces do not obey the Lambert's cosine law as a black surface does. In Fig. 9.12, the directional emissivities of several surfaces have been plotted in polar diagrams. It may be noted that a surface obeying the cosine law would have a constant emissivity and the emissivity curve on a polar diagram would be a semi-circle. It is evident that for non-conductors like glass, wood paper, etc. ϵ_θ is nearly constant up to $\theta = 75°$ and beyond this value of θ, the value drops off sharply. For electrical conductors, emissivity increases with angle θ and reaches a maximum value very close to $\theta = 90°$.

14. Radiation Energy Exchange between Two Black Bodies Placed in a Non-absorbing Medium — the Shape Factor

Let us consider two black surfaces of areas A_1 and A_2, at temperatures T_1 and T_2 and separated by a distance R, Fig. 9.13. The line joining the centres makes angles ϕ_1 and ϕ_2 and to their normals. Let the elemental surface dA_2 subtend a solid angle $d\omega_1$ at the centre of the surface dA_1. Thus, the rate of energy emitted by dA_1 and reaching $dA_2 = I_{b1} \cos \phi_1 \, d\omega_1 \, dA_1$.

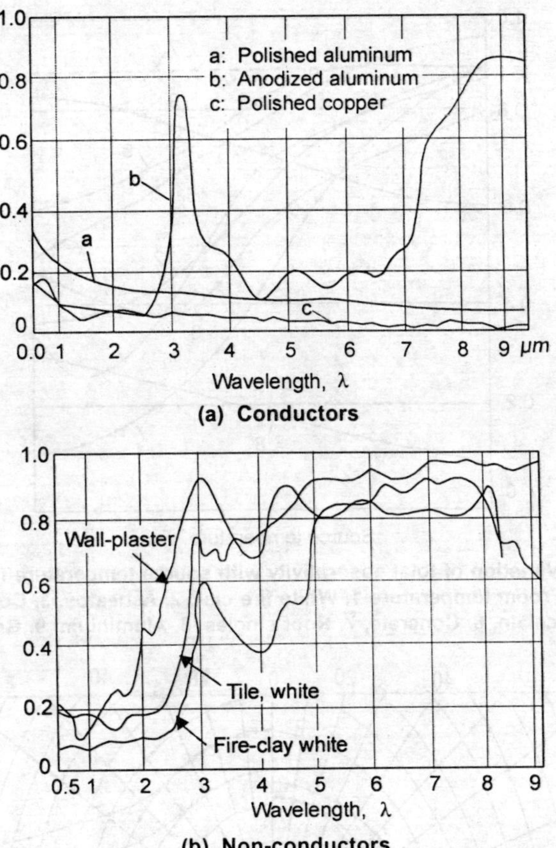

(a) Conductors

(b) Non-conductors

Fig. 9.9 Variation of monochromatic absorptivity (or emissivity) with wavelength for different materials.

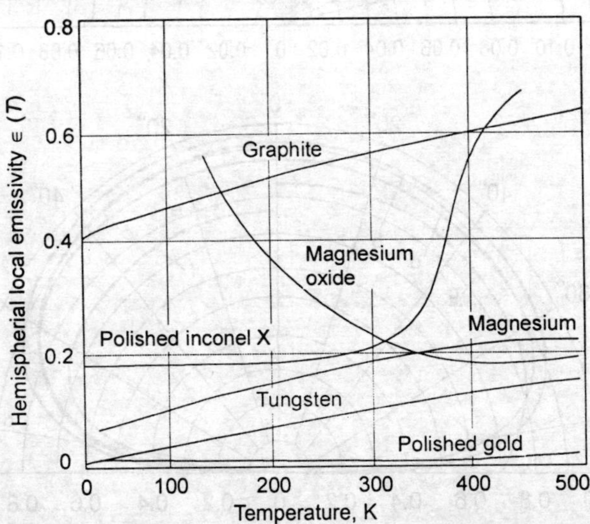

Fig. 9.10 Effect of temperature on hemispherical total emissivity of several metals and one dielectric

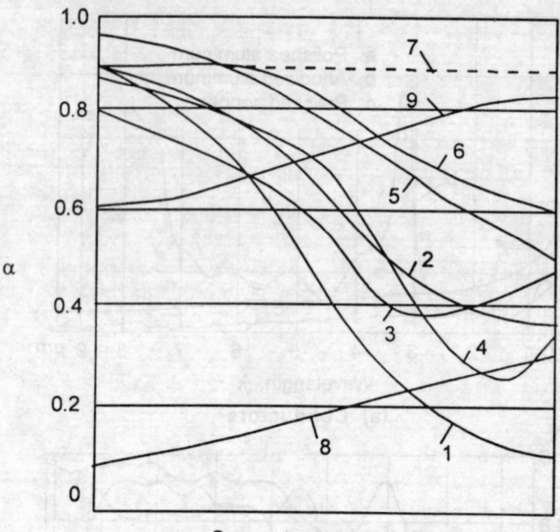

Fig. 9.11 Variation of total absorptivity with source temperature for several materials at room temperature 1. White fire clay, 2. Asbestos, 3. Cork, 4. Wood, 5. Porcelain, 6. Concrete, 7. Roof singles, 8. Aluminium, 9. Graphite

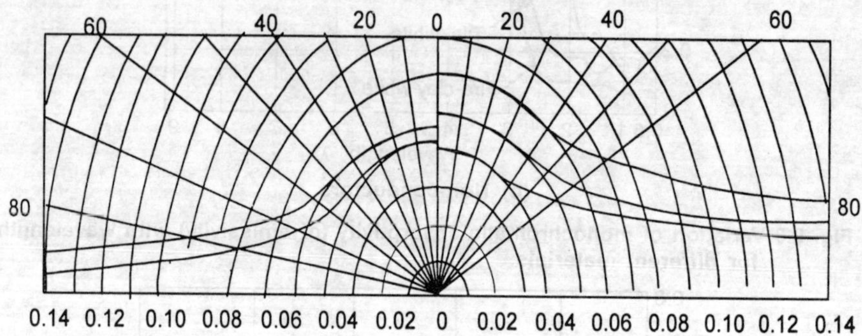

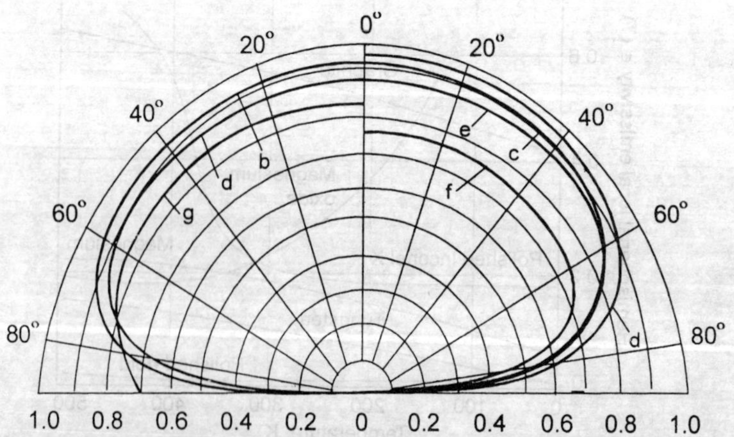

Fig. 9.12 Variation of directional emissivity with θ for (a) several metals and (b) non-conductors a. Wet ice, b. Wood, c. Glass, d. Paper, e. Clay, f. Copper oxide, g. Aluminium oxide

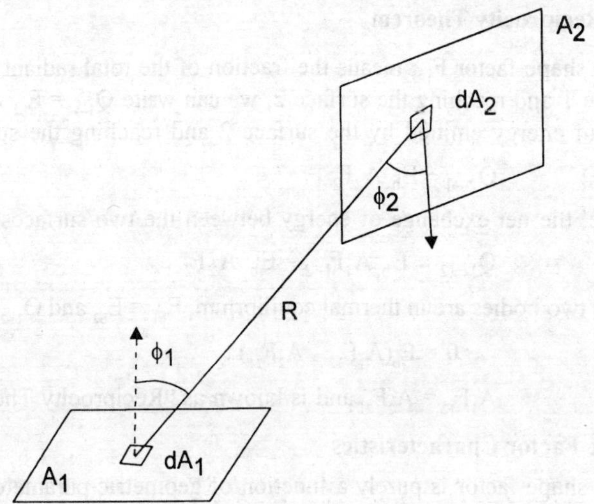

Fig. 9.13 Two black surfaces exchanging radiation heat energy

$$\therefore \qquad dQ_{1\to 2} = I_{b_1} \cos \phi_1 \, \frac{dA_2 \cos \phi_2}{R^2} \, dA_1 \qquad \left(\because d\omega_1 = \frac{dA_2 \cos \phi_2}{R^2} \right)$$

Similarly, $dQ_{2\to 1} = I_{b2} \cos \phi_2 \cos \phi_1 \, dA_1 \, dA_2 / R^2$

and the total net heat transfer:

$$\dot{Q}_{1\Leftrightarrow 2} = (I_{b_1} - I_{b_2}) \int_{A_1} \int_{A_2} \frac{\cos \phi_1 \cos \phi_2 dA_1 dA_2}{R^2}$$

$$= \frac{E_{b_1} - E_{b_2}}{\pi} \int_{A_1} \int_{A_2} \frac{\cos \phi_1 \cos \phi_2 \, dA_1 \, dA_2}{R^2}$$

The total energy emitted by $A_1 = E_{b_1} A_1$

The energy emitted by A_1 and reaching A_2 is given by

$$E_{b_1} \int_{A_1} \int_{A_2} \cos \phi_1 \cos \phi_2 \, dA_1 \, dA_2 / (\pi R^2)$$

The fraction of energy leaving A_1 and reaching A_2 is given by,

$$F_{1-2} = \frac{E_{b_1} \int_{A_1} \int_{A_2} (\cos \phi_1 \cos \phi_2 dA_1 dA_2) / \pi R^2}{E_{b_1} A_1}$$

$$= \frac{1}{A_1} \int_{A_1} \int_{A_2} \frac{\cos \phi_1 \cos \phi_2 \, dA_1 \, dA_2}{\pi R^2} \qquad (9.10a)$$

where F_{1-2} is called the shape factor, or view factor or configuration factor.

Similarly, $$F_{2-1} = \frac{1}{A_2} \int_{A_1} \int_{A_2} (\cos \phi_1 \cos \phi_2 \, dA_1 \, dA_2) / \pi R^2 \qquad (9.10b)$$

The geometric factors given by Eqs. (9.10a) and (9.10b) can be obtained analytically or graphically. Charts are available for some of the more common configurations. Figures 9.14, 9.15 and 9.16.

15. The Reciprocity Theorem

Since the shape factor F_{1-2} means the fraction of the total radiant energy emitted by surface 1 and reaching the surface 2, we can write $Q_{1\rightarrow 2} = E_{b1} A_1 F_{1-2}$ and the quantity of energy emitted by the surface 2 and reaching the surface 1 can be written as: $\qquad Q_{2\rightarrow 1} = E_{b_2} A_2 F_{2-1}$

Therefore, the net exchange of energy between the two surfaces is given by

$$Q_{1\Leftrightarrow 2} = E_{b_1} A_1 F_{1-2} - E_{b_2} A_2 F_{2-1}. \qquad (9.11)$$

When the two bodies are in thermal equilibrium, $E_{b1} = E_{b2}$ and $Q_{1=2} = 0$, , and thus,

$$0 = E_b (A_1 F_{12} - A_2 F_{21}) .$$

$$\therefore \qquad A_1 F_{12} = A_2 F_{21} \text{ and is known as "Reciprocity Theorem."}$$

16. Shape Factor Characteristics

a. The shape factor is purely a function of geometric parameters only.

b. If there are N surfaces exchanging heat energy amongst themselves then

$$F_{1-1} + F_{1-2} + \ldots + F_{1-N} = 1.0, \qquad \text{or,} \quad \sum_{J=1}^{N} F_{1-j} = 1.0$$

and $\qquad F_{2-1} + F_{2-2} + \ldots + F_{2-N} = 1.0, \qquad \text{or,} \quad \sum_{J=1}^{N} F_{2-j} = 1.0 \text{ ; and so on.}$

c. If a radiating surface is a flat or convex, $F_{1-1} = 0$, but for a concave surface, F_{1-1} is not zero.

d. If a receiving surface 2 is divided into two parts, say 3 and 4, then,

$$A_1 F_{1-2} = A_1 F_{1-3} + A_1 F_{1-4}; \qquad \text{or } F_{1-2} = F_{1-3} + F_{1-4}.$$

Example 9.17 Calculate the shape factor from a small area dA_1 to a parallel finite circular disc, radius R, and separated by a distance L, as shown in Fig. 9.17.

Solution: We select an annulus, on the circular disc, of radius r and thickness dr. Thus: $\quad dA_2 = 2\pi r \, dr$. Since $\phi_1 = \phi_2$ from Eq. (9.10a), we get

$$dA_1 F_{dA_1 - dA_2} = dA_1 \int_{A_2} \cos^2 \phi_1 \, 2\pi r \, dr \Big/ \pi x^2$$

Since $\quad x^2 = L^2 + r^2, \qquad$ and $\; \cos \phi_1 = L/(L^2 + r^2)^{1/2},$

$$dA_1 F_{dA_1 - dA_2} = dA_1 \int_0^R (2L^2 r \, dr)/(L^2 + r^2)^2$$

$$= dA_1 \int_{L^2}^{L^2+R^2} \frac{L^2 dz}{z^2}$$

where $\quad z = L^2 + r^2 \quad$ and $\quad dz = 2r \, dr$

$$\therefore \qquad F_{dA_1 - dA_2} = L^2 \left[\frac{1}{L^2} - \frac{1}{L^2 + R^2} \right] = \frac{R^2}{(L^2 + R^2)} \quad \left(\text{where } \sin \alpha = \frac{R}{(L^2 + R^2)^{\frac{1}{2}}} \right)$$

$$= \sin^2 \alpha$$

The values of shape factor, F_{1-2} are given for some common configurations in Table 9.2.

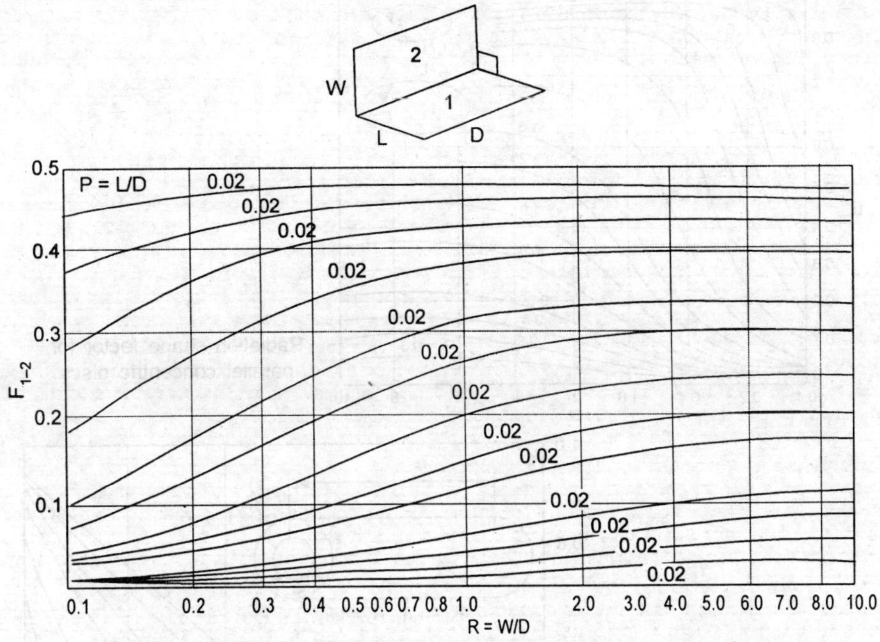

Radiation shape factor for perpendicular rectangles with a common edge

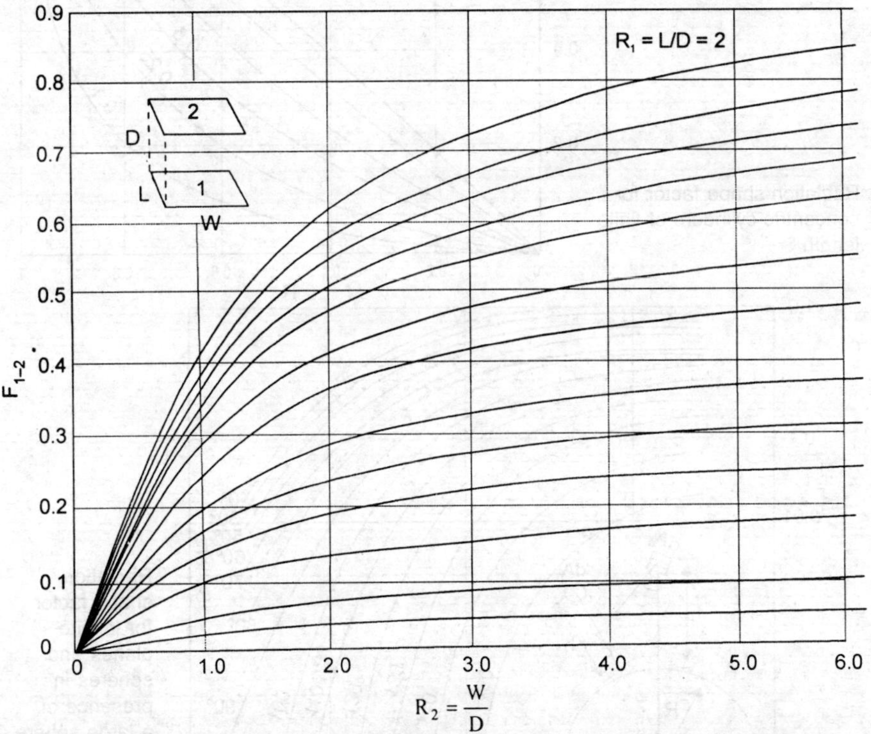

Radiation shape factor for parallel directly opposed rectangles of equal areas

Fig. 9.14

R_2 = 8
6
5
4
3
2
1.5
1.25
1.0
0.8
0.6
0.5
0.4
R_2 = 0.3
F_{1-2}
R_1 = r_1 / L
R_2 = r_2 / L
1/R_1
Radiation shape factor for
parallel concentric discs
L
A_2
R_1 = r_1/r_2, R_2 = L/r_2
A_1
r_2
F_{2-1}
R_2 = ∞
2
1
0.5
0.25
0.1
R_1
Radiation shape factor for
concentric cylinders of finite
length
F_{dA_1-2}
dA_1
h
h
A_2
R
0°
30°
50°
60°
70°
Sphere
80°
150°
140°
130°
120°
110°
100°
90°
Radiation
shape factor
for infinite
planes and
spheres in
presence of
a large sphere

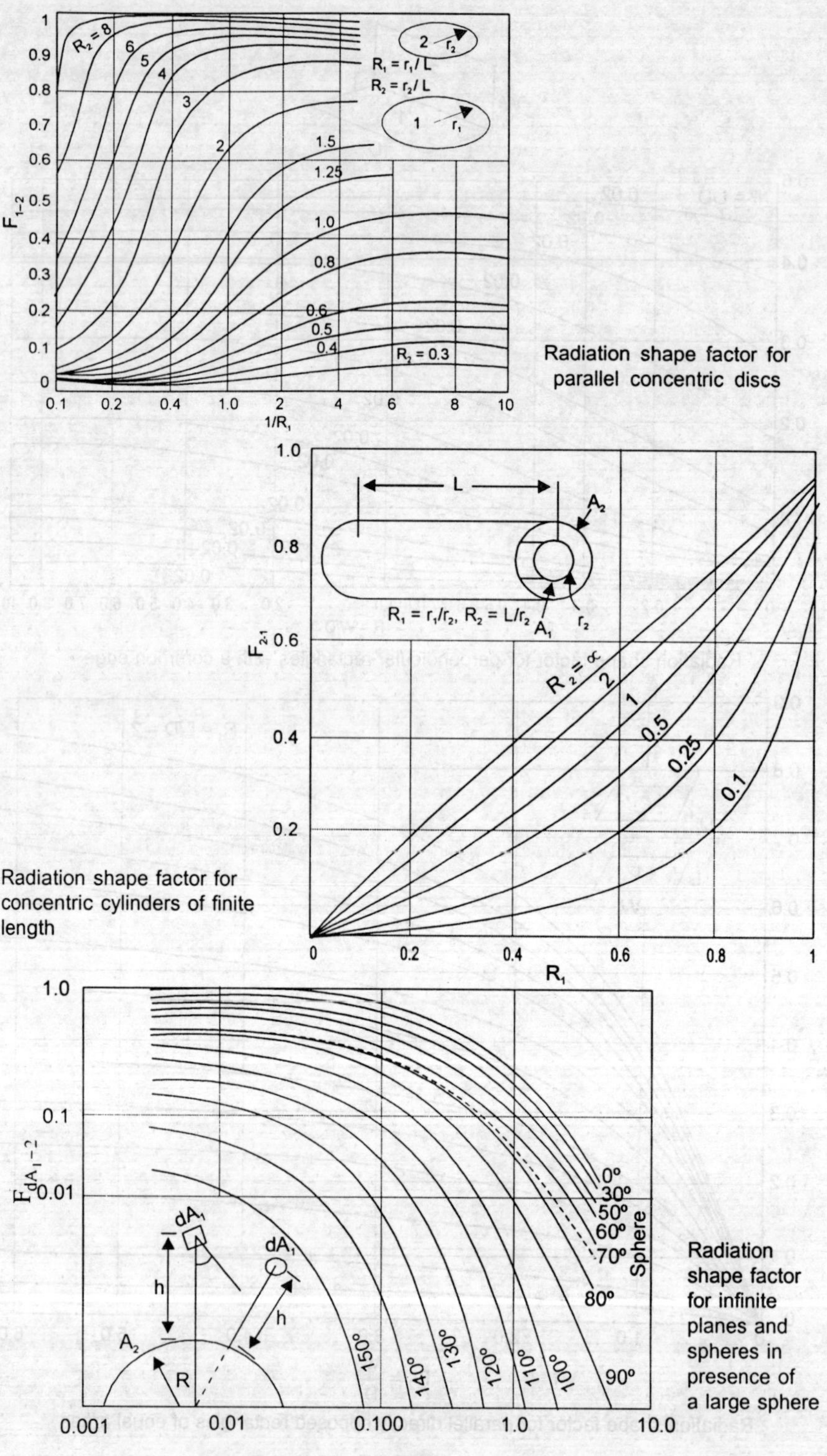

Fig. 9B

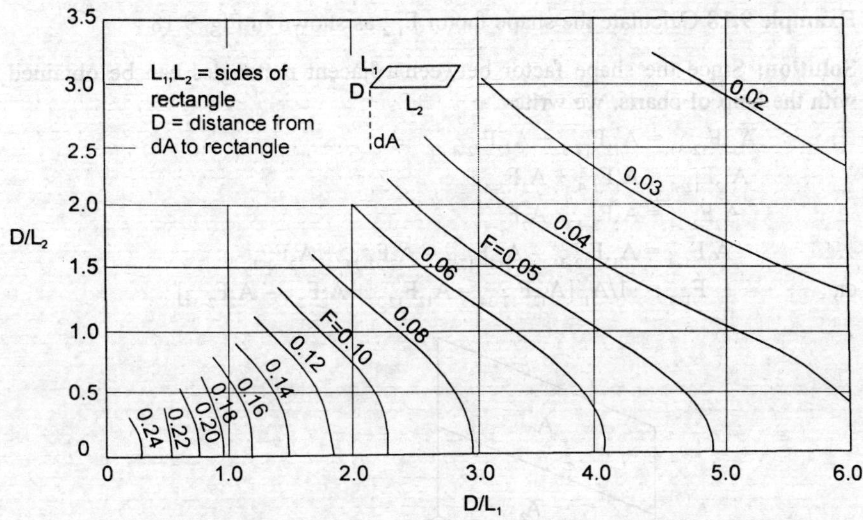

Fig. 9.16 Configuration factor for a surface element and parallel rectangle

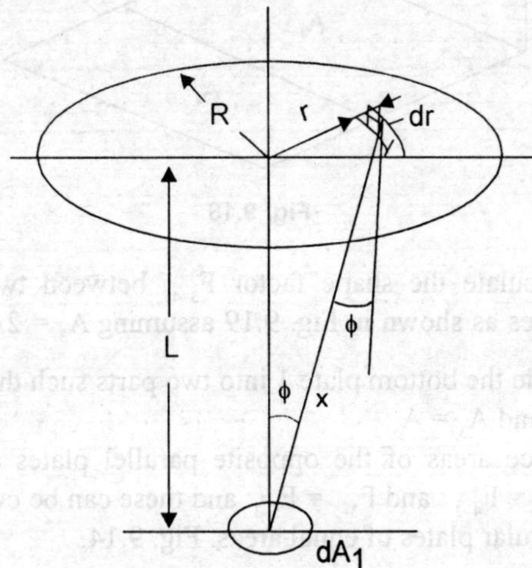

Fig. 9.17 Parallel circular discs

Table 9.2 Some shape factors

Configuration	Geometric factor/Shape factor, F_{1-2}
(a) Body 1 completely enclosed by body 2	1
(b) Parallel Circular Discs radii r_1 and r_2, distance x apart on a common axis	$\dfrac{(x^2 + r_1^2 + r_2^2) - \sqrt{[(x^2 + r_1^2 + r_2^2)^2 - 4r_1^2 r_2^2]}}{2r_1^2}$
(c) Small Sphere opposite a circular plate of radius R at a perpendicular distance L	$\dfrac{1}{2}\left[1 - \dfrac{L}{\sqrt{(L^2 + R^2)}}\right]$

Example 9.18 Calculate the shape factor F_{1-4} as shown in Fig. 9.18

Solution: Since the shape factor between adjacent rectangles can be obtained with the help of charts, we write

$$A_{12}F_{12-34} = A_{12}F_{12-3} + A_{12}F_{12-4}$$
$$A_{12}F_{12-4} = A_1F_{1-4} + A_2F_{2-4}$$
$$A_2F_{2-34} = A_2F_{2-3} + A_2F_{2-4}$$
$$\therefore \qquad A_1F_{1-4} = A_{12}F_{12-34} - A_{12}F_{12-3} - A_2F_{2-34} + A_2F_{2-3}$$
$$\text{or,} \qquad F_{1-4} = 1/A_1 \left[A_{12}F_{12-34} - A_{12}F_{12-3} + A_2F_{2-3} - A_2F_{2-34} \right]$$

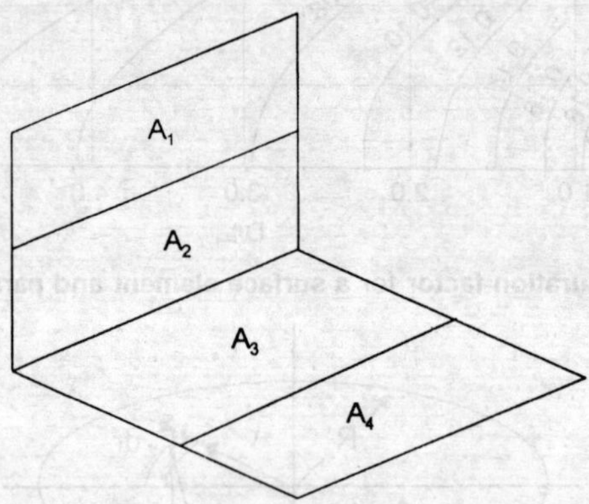

Fig. 9.18

Example 9.19 Calculate the shape factor F_{3-45} between two opposed parallel plates as shown in Fig. 9.19 assuming $A_3 = 2A_2$.

Solution: We divide the bottom plate 1 into two parts such that $A_5 = 2A_4$ and that the areas $A_2 = A_4$ and $A_3 = A_5$.

Since the surface areas of the opposite parallel plates are equal, we have $F_{23-45} = F_{45-23}$; $F_{2-4} = F_{4-2}$ and $F_{3-5} = F_{5-3}$ and these can be evaluated from charts for parallel rectangular plates of equal areas. Fig. 9.14.

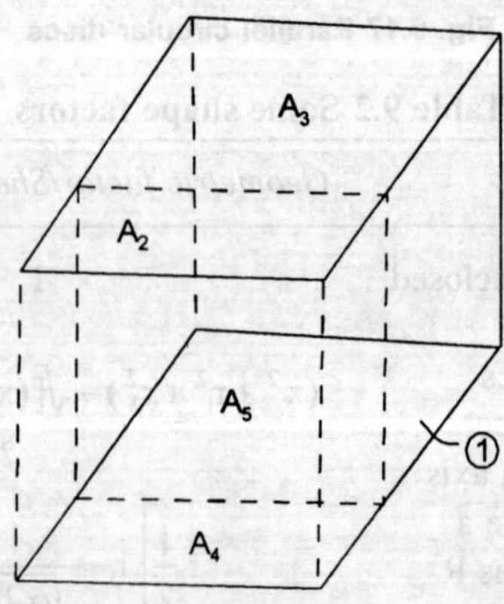

Fig. 9.19

$$A_{23}\,F_{23-45} = A_2 F_{2-4} + A_2 F_{2-5} + A_3 F_{3-4} + A_3 F_{3-5}$$

$$\therefore \quad A_2 F_{2-5} + A_3 F_{3-4} = A_{23} F_{23-45} - A_2 F_{2-4} - A_3 F_{3-5}$$

By symmetry,

$$A_2 F_{2-5} = A_3 F_{3-4}$$

$$\therefore \quad 2A_3 F_{3-4} = A_{23} F_{23-45} - A_2 F_{2-4} - A_3 F_{3-5}$$

$$A_3 F_{3-45} = A_3 F_{3-4} + A_3 F_{3-5}$$

$$= \tfrac{1}{2}\left[A_{23}\,F_{23-45} - A_2\,F_{2-4} - A_3 F_{3-5}\right] + A_3 F_{3-5}$$

and

$$F_{3-45} = \frac{1}{2A_3}\left[A_{23} F_{23-45} - A_2 F_{2-4}\right] + \tfrac{1}{2} F_{3-5}$$

Example 9.20 A hemispherical cavity, radius 0.75 m, is covered with a plate having a hole of 0.25 m diameter drilled at its center. The inner surface of the plate is maintained at 550 K by a heater embedded in the surface. Assuming the surfaces to be black and the hemisphere to be well insulated, calculate

(i) the temperature of the surface of the hemisphere, and

(ii) the power input to the heater.

Solution: Let the inner surface of the plate be 1, the surface of the hemisphere be 2 and the projected surface of the hole be 3, Fig. 9.20.

Since the surface 1 is completely surrounded, we have

$$F_{1-1} + F_{1-2} + F_{1-3} = 1$$

But, the surface 1 can neither see itself nor surface 3, therefore,

$$F_{1-1} = F_{1-3} = 0; \quad \text{and} \quad F_{1-2} = 1$$

By reciprocity theorem, $A_1 F_{1-2} = A_2 F_{2-1}$

$$\therefore \quad F_{2-1} = \frac{A_1}{A_2} = [\pi(0.75^2 - 0.125^2)]/(2\pi \times 0.75^2) = 0.4861$$

Again for surface 3:

$$F_{3-3} + F_{3-2} + F_{3-1} = 1; \quad \text{or } F_{3-2} = 1 \qquad (\because F_{3-3} = F_{3-1} = 0)$$

$$\therefore \quad F_{2-3} = A_3 / A_2 = \pi\,(0.125)^2 / 2\pi\,(0.75)^2 = 0.0139$$

Assuming that the rate of energy entering through the hole from outside is negligible because the surroundings are very large and are at normal temperature, the rate of energy incident on surface of the hemisphere

$$= A_1 F_{1-2}\,\sigma\,T_1^4 = A_1 \sigma\,T_1^4$$

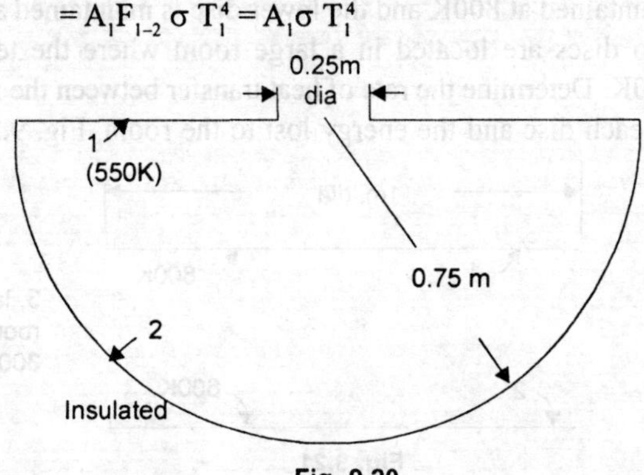

Fig. 9.20

Rate of energy emitted by surface 2 would be:

$$A_2 F_{2-1}\,\sigma\,T_2^4 + A_2 F_{2-3}\,\sigma\,T_2^4 = A_2\,\sigma\,T_2^4(0.4861 + 0.0139)$$
$$= 0.5\,A_2\,\sigma\,T_2^4$$

Under steady state conditions,

$$A_1\sigma\,T_1^4 = 0.5\,A_2\,\sigma\,T_2^4$$

$$\therefore \qquad \left(T_2/T_1\right)^4 = (A_1/A_2) \times \frac{1}{0.5} = 0.9722$$

and $\qquad T_2 = (0.9722)^{0.25} \times 550 = 546.14 K$

Heat input to the heater:

$$\dot{Q} = A_1 F_{1-2}\,\sigma\,(T_1^4 - T_2^4)$$
$$= \pi[0.75^2 - 0.125^2] \times 1 \times 5.67(5.5^4 - 5.4^4) = 256.53 \text{ W.}$$

Example 9.21 A spherical ball 6 cm in diameter and at 310 K is placed inside a large spherical furnace at 600K. Estimate the diameter of the spherical furnace such that 1/5 of the energy emitted by the furnace reaches the spherical ball. Assume surfaces to be black.

Solution: When an object is completely enclosed by another object, the number of surfaces exchanging energy by radiation is 2. Since the small spherical ball is a convex surface body, $F_{1-1} = 0$, and therefore $F_{1-2} = 1$. By reciprocity theorem: $A_1 F_{1-2} = A_2 F_{2-1}$ Energy emitted by the inside surface of the furnace is $A_2 E_2$ and the fraction reaching the spherical ball would be:

$$A_2 F_{2-1} E_2 = 0.2 A_2 E_2; \qquad\qquad \therefore\ \ F_{2-1} = 0.2$$

and, $$A_2 = \frac{A_1 F_{1-2}}{F_{2-1}} = \frac{A_1}{F_{2-1}} = 4\pi r_1^2 / 0.2$$

$$\therefore \qquad r_2 = (r_1^2 / 0.2)^{1/2}$$

$$= 3/\sqrt{0.2} = 6.71 \quad \text{and} \quad \text{Diameter} = 13.42 cm$$

The net exchange of heat energy between the two surfaces:

$$= A_1 F_{1-2}\,\sigma\,(T_2^4 - T_1^4)$$
$$= 4\pi\,(0.03)^2 \times 1 \times 5.67(6^4 - 3.1^4) = 77.195 \text{ W.}$$

Example 9.22 Two parallel discs 1m in diameter are spaced 25 cm apart such that one disc is located directly above the other. The upper disc is maintained at 800K and the lower disc is maintained at 600 K. The two discs are located in a large room where the temperature is 300K. Determine the rate of heat transfer between the inside surface of each disc and the energy lost to the room, Fig. 9.21.

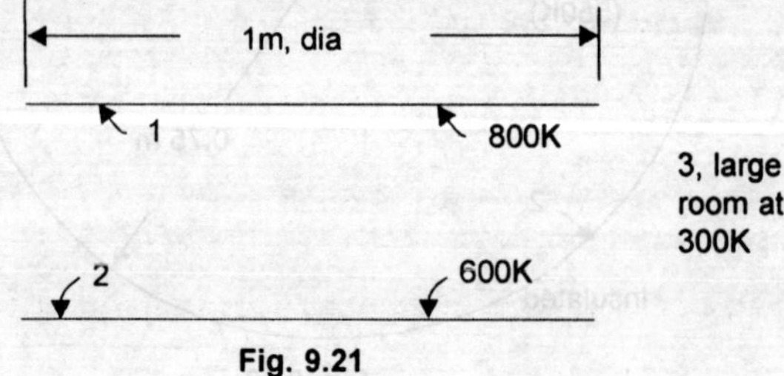

Fig. 9.21

Solution: From charts for parallel discs, $F_{1-2} = F_{2-1} = 0.6$;
Also, $F_{1-1} + F_{1-2} + F_{1-3} = 1$.   Since $F_{1-1} = 0$, $F_{1-3} = F_{2-3} = 1-0.6 = 0.4$
Rate of heat transfer between the inside surfaces of each disc

$$Q_{1-2} = A_1 F_{1-2} (E_{b1} - E_{b2}) = \tfrac{\pi}{4} d^2 \times 0.6 \times \sigma(T_1^4 - T_2^4)$$

$$= \tfrac{\pi}{4} \times 1 \times 0.6 \times 5.67(8^4 - 6^4) = 7.48 \text{kW}$$

Rate of heat transfer to the walls of the room = Rate of heat transfer from both discs.

Or, $A_1 F_{1-3} (E_{b1} - E_{b3}) + A_2 F_{2-3} (E_{b2} - E_{b3}) = \tfrac{\pi}{4}(1)^2 \times 0.4[E_{b_1} + E_{b_2} - 2E_{b_3}]$

$$= \tfrac{\pi}{4} \times 0.4 \times 5.67 \times [8^4 + 6^4 = 2 \times 3^4] = 9.46 \text{kW}.$$

Example 9.23 A circular rod 10 cm in diameter is placed along the geometric centreline of a circular half cylinder, diameter 60 cm, as shown in Fig. 9.22. The system is surrounded by a large enclosure. Calculate F_{1-1}, F_{1-2} and F_{1-3}.

Solution: From symmetry, we have:
$F_{2-1} = F_{2-3} = 0.5$, Also $F_{1-1} + F_{1-2} + F_{1-3} = 1.0$
We choose an arbitrary flat plate surface
shown by dotted line.
For this surface $F_{4-4} = 0$ and $F_{4-1} = 1.0$
Since all radiation leaving surface 1 will
either go to 2 or 3 or will arrive at the
imaginary surface 4, we have

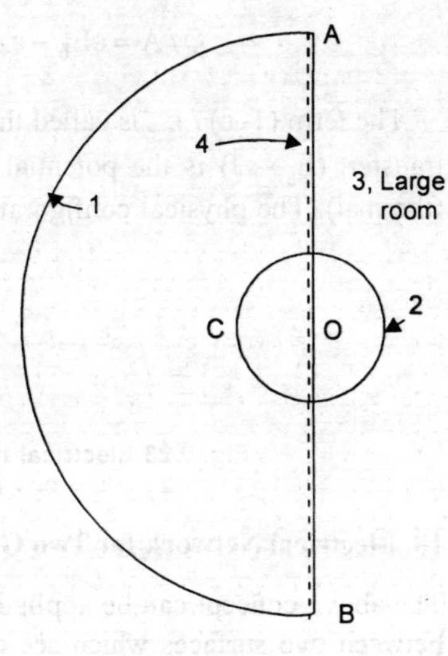

$$F_{1-2} + F_{1-3} = F_{1-4}$$
From reciprocity theorem,
$$A_1 F_{1-4} = A_4 F_{4-1}$$
For unit length, the areas are:
$$A_1 = 2\pi r/2 = \pi \times 0.3 = 0.9426 \text{ m}^2;$$
$$A_2 = 2\pi r = 0.3142 \text{ m}^2;$$
$$A_4 = 0.60$$

$\therefore$ $F_{1-4} = \dfrac{A_4}{A_1} = \dfrac{0.6}{0.9426} = 0.6365$

Fig. 9.22

and, $A_1 F_{1-2} = A_2 F_{2-1}$

$\therefore$ $F_{1-2} = \dfrac{A_2 \times 0.5}{A_1} = \dfrac{0.3142 \times 0.5}{0.9426} = 0.1667$

$F_{1-3} = F_{14} - F_{1-2} = 0.6365 - 0.1667 = 0.4698$

and $F_{1-1} = 1.0 - 0.1667 - 0.4698 = 0.3635.$

17. Heat Exchange between Gray Bodies—Radiation Network

From the preceeding examples it is clear that with the help of geometrical considerations, we can easily calculate the shape factors and radiant heat exchange between black bodies. With gray bodies, the situation is a little more complex because when radiant energy strikes a surface, all of it will not be absorbed. A part will be reflected back to another heat transfer surface and a part may be reflected out of the system entirely. Thus, such problems require careful handling to get meaningful results.

The analysis of energy exchange between gray bodies is based on the following assumptions:

(i) all surfaces are gray and opaque,
(ii) all reflections and emissions are diffuse,
(iii) the temperature of each surface is uniform, and
(iv) the enclosure is filled with a transparent gas.

From Eq. (9.3) and Kirchhoff's law:

$$J = \varepsilon E_b + (1 - \varepsilon)G \tag{9.12}$$

The net energy leaving the surface is the difference between the radiosity and irradiation, i.e.,

$$\dot{Q}/A = J - G = \varepsilon E_b + G - \varepsilon G - G = \varepsilon E_b - \varepsilon G$$

From Eq. (9.12), $G = (J - \varepsilon E_b)/(1 - \varepsilon)$

$$\therefore \quad \dot{Q}/A = \varepsilon E_b - \varepsilon \cdot \frac{J - \varepsilon E_b}{1 - \varepsilon} = \frac{E_b - J}{(1-\varepsilon)/\varepsilon} \tag{9.13}$$

The term $(1-\varepsilon)/\varepsilon$ is called the surface resistance (thermal) to radiation heat transfer, $(E_b - J)$ is the potential difference (thermal) and $\dot{Q}/A$ is the current (thermal). The physical configuration can be represented as Fig. 9.15.

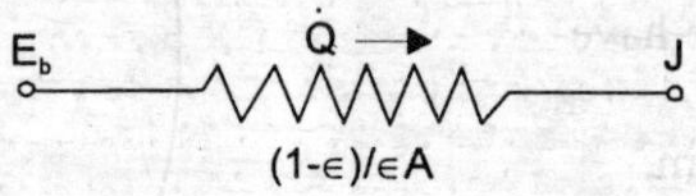

$$E_b \quad\xrightarrow{\dot{Q}}\quad J$$
$$(1-\epsilon)/\epsilon A$$

Fig. 9.23 Electrical network to radiation heat transfer

18. Electrical Network for Two Gray Bodies

The above concept can be applied to calculate the radiant heat energy exchange between two surfaces which see each other only. Let the two surfaces 1 and 2 have surface areas A_1 and A_2, emissivity ε_1 and ε_2. The fraction of total energy leaving 1 and reaching 2 will be $J_1 A_1 F_{1-2}$. Similarly, the fraction of total energy leaving 2 and reaching 1 will be $J_2 A_2 F_{2-1}$. Therefore, the net energy exchange,

$$\dot{Q}_{12} = (J_1 - J_2) A_1 F_{1-2} = A_2 F_{2-1} (J_1 - J_2)$$

or $\quad \dot{Q}_{12} = (J_1 - J_2)/ 1/A_1 F_{1-2}$ where $1/A_1 F_{1-2}$ is called the space resistance,

$$= (E_{b_1} - E_{b_2}) / \left[\frac{1-\varepsilon_1}{\varepsilon_1 A_1} + \frac{1}{A_1 F_{1-2}} + \frac{1-\varepsilon_2}{\varepsilon_2 A_2} \right] \tag{9.14}$$

and its equivalent network would be as shown in Fig. 9.24

Eq. (9.14) can be simplified for simple geometrical arrangements: Table 9.2 A

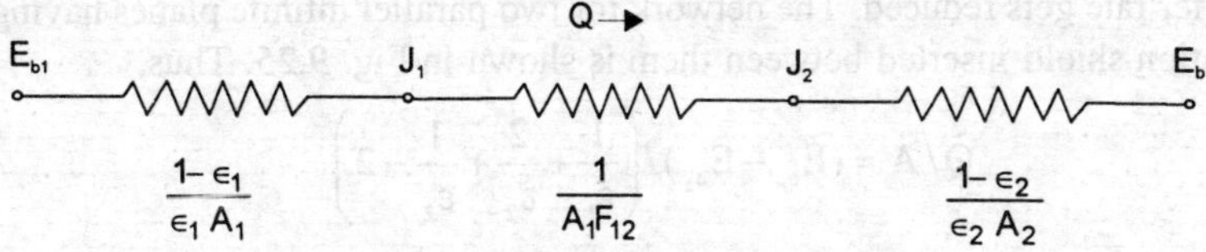

Fig 9.24 Electrical network for two surfaces which see each other and nothing else

(a) *Infinite Parallel Planes* — When we have two infinite parallel planes, their areas A_1 and A_2 are taken equal and the radiation shape factor, $F_{1-2} = 1.0$, because all the radiation leaving one plane reaches the other. The electrical network is the same as in Fig. 9.24 and the rate of heat transfer per unit area is obtained by Eq. (9.14) in which $A_1 = A_2 = 1$ and $F_{1-2} = 1.0$. Thus,

$$\dot{Q}/A = (E_{b_1} - E_{b_2})/\left[\frac{1}{\varepsilon_1} + \frac{1}{\varepsilon_2} - 1\right] \tag{9.15}$$

(b) *Concentric Cylinders or Spheres* — Since the inner cylinder or sphere has convex surfaces, $F_{1-2} = 1.0$ and therefore,

$$\dot{Q} = (E_{b_1} - E_{b_2})/\left[\frac{1-\varepsilon_1}{A_1\varepsilon_1} + \frac{1}{A_1} + \frac{1-\varepsilon_2}{A_2\varepsilon_2}\right]$$

$$= A_1(E_{b_1} - E_{b_2})/\left(\frac{1}{\varepsilon_1} + \frac{A_1}{A_2}(1/\varepsilon_2 - 1)\right) \tag{9.16}$$

(c) *Small Body in a Large Enclosure* — We come across cases like a pipe carrying steam in a large room or a thermocouple bead located inside a duct for measuring temperature. In such cases, $F_{1-2} = 1.0$ and $A_2 \gg A_1$.

Therefore, $\dot{Q} = \varepsilon_1 A_1 (E_{b1} - E_{b2})$ $\hfill$ (9.17)

Eq. (9.17) would be applicable even if the body 2 is not a black body because a very large enclosure is always conceived as a black body.

Example 9.24 It is desired to reduce the rate of radiant heat transfer by inserting an aluminium foil ($\varepsilon_3 = 0.05$) midway between two very large parallel planes maintained at 1000K ($\varepsilon_1 = 0.8$) and 600 K ($\varepsilon_2 = 0.9$). Calculate the percentage reduction in heat loss by radiation using the aluminium foil.

Solution: When the aluminium foil is not inserted between the two infinite parallel planes, the rate of heat transfer is given by Eq. (9.15)

$$\dot{Q}/A = \sigma(T_1^4 - T_2^4)/\left(\frac{1}{\varepsilon_1} + \frac{1}{\varepsilon_2} - 1\right)$$

$$= 5.67(10^4 - 6^4)/\left(\frac{1}{0.8} + \frac{1}{0.9} - 1\right) = 36.26 \text{kW}/\text{m}^2$$

Highly reflective materials (low emissivity materials) are used as 'radiation shields'. These shields neither deliver nor remove any heat from the overall system. They only add a resistance in the heat flow path so that the overall heat transfer rate gets reduced. The network for two parallel infinite planes having one radiation shield inserted between them is shown in Fig. 9.25. Thus,

$$\dot{Q}/A = (E_{b_1} - E_{b_2})/\left(\frac{1}{\varepsilon_1} + \frac{2}{\varepsilon_2} + \frac{1}{\varepsilon_2} - 2\right)$$

$$= 5.67(10^4 - 6^4)/\left(\frac{1}{0.8} + \frac{2}{0.05} + \frac{1}{0.9} - 2\right) = 1.2\,\text{kW}/\text{m}^2$$

% reduction = $(32.26 - 1.2)/32.26 = 96.6\%$

We can also compute the temperature of the radiation shield (aluminium foil)

$$\dot{Q}/A = (E_{b_1} - E_{b_3})/\left(\frac{1}{\varepsilon_1} + \frac{1}{\varepsilon_3} - 1\right) = 5.67\,[10^4 - (T_3/100)^4]/\left[\frac{1}{0.8} + \frac{1}{0.05} - 1\right]$$

Solving for T_3, we get $T_3 = 866.43$ K

(when $\varepsilon_1 = \varepsilon_2 = \varepsilon$, we have $\dot{Q}/A = \dfrac{E_{b_1} - E_{b_2}}{(2/\varepsilon - 1)}$ and when one shield having the

same emissivity is placed between the planes, we get,

$$\dot{Q}/A = (E_{b_1} - E_{b_2})/\left(\frac{4}{\varepsilon} - 2\right) = (E_{b_1} - E_{b_2})/2\left(\frac{2}{\varepsilon} - 1\right)$$

Thus, when all emissivities are equal,

$$\left(\frac{\dot{Q}}{A}\right)_{\text{withshield}} = \frac{1}{2}\left(\frac{\dot{Q}}{A}\right)_{\text{withoutshield}}$$

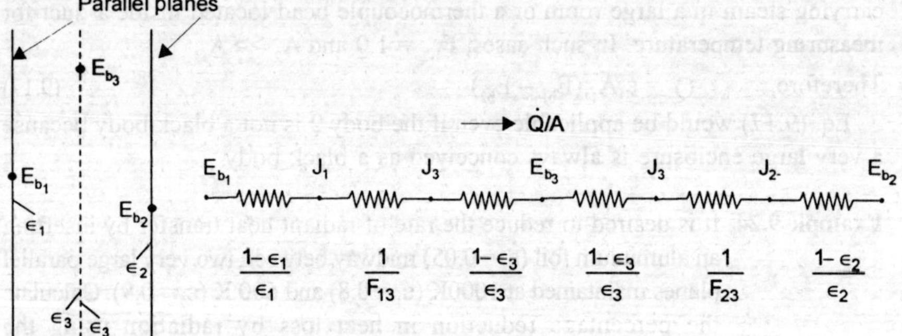

Fig. 9.25 Electrical network for two parallel infinite planes with one radiation shield inserted between them

With this approach, we can analyse a system having n shields inserted between two infinite parallel planes. Assuming that the emissivities of all the shields are same, the value of each space resistance is unity and the surface resistances would be the same. For n shields, we would have (n + 1) spaces and a total of (n + 2) parallel planes. Therefore, the number of surface resistances would be (2n + 2) and space resistances would be n+1. And, the total resistance is then

$$(2n + 2)\,(1/\varepsilon - 1) + n + 1 = (n + 1)\,(2/\varepsilon - 1)$$

And, $\left(\dfrac{\dot{Q}}{A}\right)_{\text{with n shields}} = \dfrac{1}{n+1}\left(\dfrac{\dot{Q}}{A}\right)_{\text{without shield}}$ (9.18)

Example 9.25 Calculate the rate of heat transfer by radiation between two concentric spheres (diameter 25 cm and 40 cm, emissivities 0.04) when the outer surface of the inner sphere is maintained at 100K and the inner surface of the outer sphere is maintained at 320K.

Solution: The rate of heat transfer between concentric spheres is given by Eq. (9.16):

$A_1 = 4\pi r_1^2 = \pi D_1^2 = \pi(0.25)^2 = 0.1964$ m^2.

$\varepsilon_1 = \varepsilon_2 = 0.04;$ $A_1/A_2 = (0.25/0.40)^2 = 0.39$

$$\dot{Q} = A_1(E_{b_1} - E_{b_2}) / \left[\dfrac{1}{\varepsilon_1} + \dfrac{A_1}{A_2}\left(\dfrac{1}{\varepsilon_2} - 1\right)\right]$$

$$= 0.1964 \times 5.67 \times (1^4 - 3.2^4) / \left[\dfrac{1}{0.04} + 0.39\left(\dfrac{1}{0.04} - 1\right)\right] = -3.364 \text{ W}.$$

Example 9.26 A circular heating element (0.3 m in diameter, $\in = 0.8$), temperature 450K is embedded in the ceiling of a room 3 m high. Calculate the radiant heat exchange between the heating element and a circular plate (diameter 0.2 m, $\in = 0.6$) temperature 300K, placed on the floor directly below the element.

Solution: The heating element and the circular plate can be considered as small gray bodies placed in a large enclosure. The quantity of radiant energy emitted by the heating element is $\in_1 A_1 \sigma T_1^4$. The amount of this radiant energy reaching the plate at the floor is $\in_1 A_1 F_{1-2} \sigma T_1^4$ and the plate will absorb a part of this energy

$$\alpha_2 \in_1 A_1 F_{1-2} \sigma T_1^4 = \in_2 \in_1 A_1 F_{1-2} \sigma T_1^4$$

Similarly, the amount of energy emitted by the bottom plate and absorbed by the heating element is: $\sigma \in_1 \in_2 A_2 F_{2-1} T_2^4$

And the net heat transfer: $\sigma \in_1 \in_2 A_1 F_{1-2} (T_1^4 - T_2^4)$

From curves of shape factor for parallel circular discs, $F_{1-2} = 0.014$

∴ $Q_{1-2} = 5.67 \times 10^{-8} \times 0.8 \times 0.6 \times \pi (0.15)^2 \times 0.014 (450^4 - 300^4)$

   $= 0.886$ W.

Example 9.27 Two circular discs of diameter 25 cm each are placed 2 m apart. Calculate the radiant heat exchange between the two discs when they are maintained at 1000 K and 500 K, and the respective emissivities are 0.4 and 0.65.

Solution: The shape factor for parallel circular discs of equal radius r and separated by a distance x on a common axis is given by

$$F_{1-2} = [(x^2 + 2\,r^2)] - \sqrt{[(x^2 + 2r^2)^2 - 4r^4)]} / 2r^2$$
$$= 0.004; \quad \text{for } x = 2m \quad \text{and } r = 0.125 \text{ m}$$

The disc 1 (temperature 1000 K, $\epsilon_1 = 0.4$) will emit radiation $= \epsilon_1 A_1 \sigma T_1^4$

The radiation emitted by 1 and reaching the disc 2 $= \epsilon_1 A_1 F_{1-2} \sigma T_1^4$

The radiation absorbed by the disc 2 will be $= \epsilon_1 \alpha_2 A_1 F_{1-2} \sigma T_1^4$
$$= \epsilon_1 \epsilon_2 A_1 F_{1-2} \sigma T_1^4$$

Similarly, the radiation emitted by disc 2 and absorbed by disc 1
$$= \epsilon_1 \epsilon_2 A_2 F_{2-1} E_{b2}$$

Therefore the net exchange of energy between the two
$$= \epsilon_1 \epsilon_2 A_1 F_{1-2} \sigma (T_1^4 - T_2^4)$$
$$= 2.174 \text{ W.}$$

Example 9.28 A small disc artificial satellite, (1m in diameter, $\epsilon_1 = 0.35$, temperature 260 K) circles the earth (radius 6250 km, $\epsilon_2 = 0.75$, temperature 300 K) at a distance of 6550 km from the surface of the earth. Estimate the net rate at which the energy is leaving the satellite.

Solution: Since the satellite is very small in size relative to the earth's surface, the shape factor can be approximated by the relation (Ex 9.17)
$$F_{1-2} = R^2/(R^2 + L^2); \text{ where R, the radius of the earth} = 6250 \text{ km}$$
and $\quad L = 6550$ km,
$$\therefore \quad F_{12} = 0.4766$$

The radiation emitted by the earth and absorbed by the satellite is
$$= \epsilon_1 \epsilon_2 A_2 F_{1-2} \sigma T_2^4 = \epsilon_1 \epsilon_2 A_1 F_{1-2} \sigma T_2^4$$
$$= 45.133 \text{ W}$$

Total energy emitted by the both surfaces of the satellite disc
$$= 2 \times 0.35 \times 3.142 \times (0.5)^2 \times 5.67 \times (2.6)^4 = 142.47 \text{ W}$$

Net rate at which the energy leaves the satellite $= 142.47 - 45.133 = 97.38$ W.

Example 9.29 Calculate the shape factor of a cylindrical cavity, radius R and height H, with respect to itself.

Solution: Let the cylindrical cavity be denoted by 1. We cover the cavity by a flat surface, 2, such that the two surfaces form an enclosure. Therefore, $F_{11} + F_{12} = 1$ and $F_{22} + F_{21} = 1$.
But, $F_{22} = 0$ and $F_{21} = 1.0$
Also, $A_2 F_{21} = A_1 F_{12}$ $\quad \therefore \quad F_{12} = A_2/A_1 = \pi R^2 / (\pi R^2 + 2\pi R H)$
$$\therefore \quad F_{11} = 1 - R/(R + 2H).$$

19. Electrical Network for Three Gray Bodies

When three gray surfaces exchange radiant heat energy with one another, there are three surface resistances (one for each surface present) and three spatial resistance. The electrical network for such a configuration can be drawn as shown in Fig. 9.26.

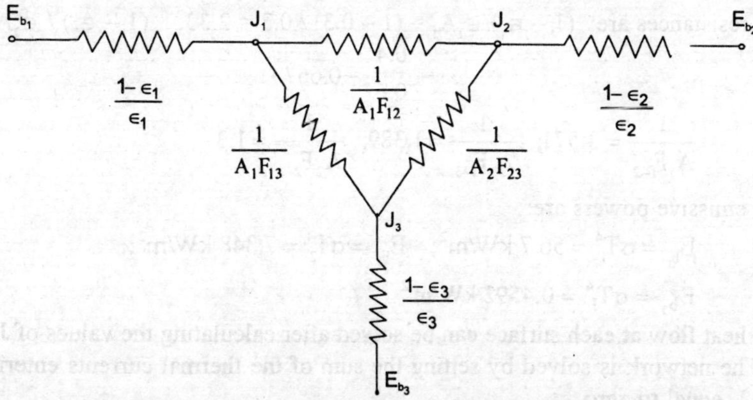

Fig. 9.26 Radiation network for three surfaces which see each other and nothing else

The shape factors F_{1-2}, F_{1-3}, and F_{2-3} have to be determined from the given geometry.

Example 9.30 Two parallel plates 1m × 1m are spaced 0.75m apart. One plate has temperature $T_1 = 1000$ K, $\epsilon_1 = 0.3$ and the other plate has temperature $T_2 = 600$ K, $\epsilon_2 = 0.6$. The plates are located in a very large room where the temperature is 300 K. Find the net exchange of heat energy between the plates and the room.

Solution: It is assumed that plate surfaces facing each other exchange heat energy only between them and the room. The geometrical configuration is shown in Fig. 9.27. The given data are:

$$T_1 = 1000 \text{ K}, \quad A_1 = 1\text{m}^2, \quad \epsilon_1 = 0.3, \quad T_2 = 600 \text{ K}, \quad A_2 = 1\text{m}^2,$$
$$\epsilon_2 = 0.6; \quad T_3 = 300 \text{ K}$$

Since the area A_3 is very large, the surface resistance $(1-\epsilon_3)\,/\,A_3\epsilon_3 = 0$, and $E_{b3} = J_3$ in Fig 9.26. The shape factor $F_{1-2} = 0.28$ (from chart) $= F_{2-1}$ and

$$F_{1-3} = F_{2-3} = 1 - 0.28 = 0.72.$$

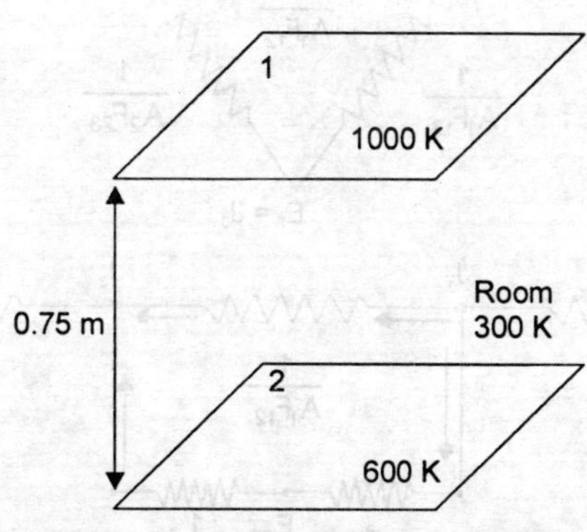

Fig. 9.27

The resistances are: $(1 - \epsilon_1)/\epsilon_1 A_1 = (1 - 0.3)/0.3 = 2.33; \quad (1 - \epsilon_2)/\epsilon_2 A_2$

$$= \frac{0.4}{0.6} = 0.667$$

$$\frac{1}{A_1 F_{1-2}} = 3.571; \quad \frac{1}{A_1 F_{1-3}} = 1.389; \quad \frac{1}{A_2 F_{2-3}} = 1.3$$

The emissive powers are:

$$E_{b_1} = \sigma T_1^4 = 56.7 \text{ kW/m}^2; \quad E_{b_2} = \sigma T_2^4 = 7.348 \text{ kW/m}^2;$$

$$E_{b_3} = \sigma T_1^4 = 0.4592 \text{ kW/m}^2$$

The heat flow at each surface can be solved after calculating the values of J_1 and J_2. The network is solved by setting the sum of the thermal currents entering J_1 and J_2 equal to zero.

Node J_1: $(E_{b1} - J_1)/2.33 + (J_2 - J_1)/3.571 + (E_{b3} - J_1)/1.389 = 0$ (a)

Node J_2: $(E_{b2} - J_2)/0.667 + (J_1 - J_2)/3.571 + (E_{b3} - J_2)/1.389 = 0$ (b)

Solving for J_1 and J_2, we get: $J_1 = 18.536$ kW/m²; $J_2 = 6.615$ kW/m².

Total heat energy lost by plate 1: $(E_{b_1} - J_1)/\left(\dfrac{1-\epsilon_1}{A_1 \epsilon_1}\right) = \dfrac{56.7 - 18.536}{2.33} = 16.38$ kW

Total energy lost by plate 2: $(E_{b_2} - J_2)/\left(\dfrac{1-\epsilon_2}{\epsilon_2 A_2}\right) = \dfrac{7.348 - 6.615}{0.667} = 1.099$ kW

Total heat received by the room:

$$(J_1 - J_3)/\frac{1}{A_1 F_{13}} + \frac{J_2 - J_3}{1/A_2 F_{23}} = \frac{18.536 - 0.4592}{1.389} + \frac{6.615 - 0.4592}{1.389}$$

$$= 17.44 \text{ kW}.$$

(b) If the two parallel plates of Ex. 9.30 are placed in a re-radiating environment, the equivalent network reduces to Fig. 9.28, and then the total resistance is

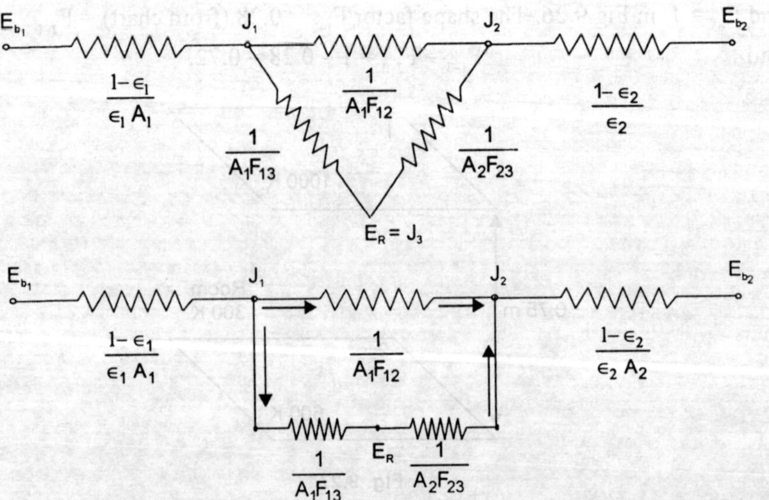

Fig 9.28 Equivalent network for two gray bodies placed in a re-radiating enclosure

$$\frac{1-\varepsilon_1}{\varepsilon_1 A_1} + \frac{1-\varepsilon_2}{\varepsilon_2 A_2} + \frac{1}{A_1 F_{1-2} + \dfrac{1}{1/A_1 F_{13} + 1/A_2 F_{23}}} \tag{9.19}$$

$$= 2.33 + 0.667 + 1/[0.28 + 1/(1.389 + 1.389)] = 4.56$$

The rate of heat flow from the hotter plate to colder plate is

$$\dot{Q} = (E_{b_1} - E_{b_2})/R = (56.7 - 7.348)/4.56 = 10.82 \text{ kW}$$

We can also compute the temperature of the reradiating enclosure:

$$E_{b_1} - J_1 = \dot{Q} \times \frac{1-\varepsilon_1}{\varepsilon_1 A_1} = 10.82 \times 2.33 = 25.21 \text{ kW/m}^2$$

$$\therefore \qquad J_1 = 31.49 \text{ kW/m}^2$$

$$J_2 - E_{b_2} = \dot{Q} \times \frac{1-\varepsilon_2}{\varepsilon_2 A_2} = 10.82 \times 0.667 = 7.217 \text{ kW/m}^2$$

$$\therefore \qquad J_2 = 14.565 \text{ kW/m}^2$$

Rate of heat flow through the re-radiating surface = $10.82 - 4.74 = 6.08$ kW/m²

and $\qquad J_1 - J_3 = 8.445$

$\therefore \qquad J_3 = 23.045$ kW/m²   and $T_3 = 798$ K

Example 9.31 In a muffle furnace, the floor 4.5m × 4.5m is constructed of refractory material ($\epsilon = 0.7$). Two rows of oxidised steel tubes are placed 3m above and parallel to the floor, but for the purpose of analysis these can be replaced by a 4.5m × 4.5m plane having an effective emissivity of 0.9. The average temperatures for the floor and tubes are 900 and 270°C respectively. Taking the shape factor for radiation from floor to tubes as 0.32, calculate (i) the net rate of heat transfer to the tubes, and (ii) the mean temperature of the refractory walls of the furnace, assuming these are well insulated.

Solution: The emissivity of oxidized steel is 0.8. Since some of the radiation reflected from the first row of tubes will be absorbed by the second row, the effective emissivity of the tubes will be higher than 0.8. As such the tubes are supposed to be replaced by an imaginary parallel surface of area equal to the floor and having an effective emissivity 0.9. Further, the walls of the furnace are insulated and therefore, the heat transfer at the surface will be zero and they can be considered as re-radiating enclosure. The equivalent network will be as shown in Fig 9.28, where the floor is surface 1, the tubes are surface 2, and the walls are surface 3. Thus:

$$E_{b_1} = \sigma T_1^4 = 5.67 (11.73)^4 = 107.34 \text{ kW/m}^2;$$

$$\frac{(1-\varepsilon_1)}{A_1 \varepsilon_1} = \frac{1-0.7}{0.7 \times (4.5)^2} = 0.021$$

$$E_{b_2} = \sigma T_2^4 = 5.67 (5.43)^4 = 4.93 \text{ kW/m}^2;$$

$$\frac{1-\varepsilon_2}{A_2 \varepsilon_2} = \frac{1-0.9}{0.9 \times (4.5)^2} = 0.00549$$

$$F_{1-2} = 0.32; \qquad \therefore \quad F_{13} = 1 - F_{12} = 0.68 = F_{23}$$

$$\frac{1}{A_1 F_{12}} = \frac{1}{(4.5)^2 \times 0.32} = 0.154$$

$$\frac{1}{A_1 F_{13}} = 0.0726 = \frac{1}{A_2 F_{23}}$$

The network is a series-parallel system, and as such the total resistance,

$$R = \frac{1-\varepsilon_1}{A_1 \varepsilon_1} + \frac{1-\varepsilon_2}{A_2 \varepsilon_2} + \cfrac{1}{A_1 F_{12} + \cfrac{1}{1/A_1 F_{13} + 1/A_2 F_{23}}}$$

$$= 0.021 + 0.00549 + 0.0748 = 0.1013$$

$$\dot{Q} = (E_{b_1} - E_{b_2})/R = (107.94 - 4.93)/0.1013 = 1010 \text{ kW}$$

$$J_1 = E_{b_1} - \dot{Q}\left(\frac{1-\varepsilon_1}{\varepsilon_1 A_1}\right) = 86.13 \text{ kW/m}^2$$

$$J_2 = E_{b_2} - \dot{Q}\left(\frac{1-\varepsilon_2}{\varepsilon_2 A_2}\right) = 4.93 + 1010 \times 0.00549 = 10.475 \text{ kW/m}^2$$

$$\dot{Q}_1 = \frac{J_1 - J_2}{1/A_1 F_{12}} = \frac{86.13 - 10.475}{0.154} = 491.26 \text{ kW}$$

$$\dot{Q}_2 = 1010 - 491.26 = 518.74 \text{ kW}$$

$$J_3 = J_1 - \dot{Q}_2 \times \frac{1}{A_1 F_{13}} = 86.13 - 518.74 \times 0.0726 = 48.47 \text{ kW/m}^2$$

$$= \sigma T_3^4 = 48.47$$

$$\therefore \qquad T_3 = 961.55 \text{ K} \equiv 688.55°C.$$

Example 9.32 A vertical cylindrical container 2 m in diameter has its base at 1500 K with $\epsilon = 0.3$. The walls of the container are insulated and 1 m high. The top is open to the atmosphere at 300 K. Draw the radiation network and estimate the temperature of the wall and the rate of heat loss to the surroundings.

Solution: We represent the base as surface 1, walls as surface 2 and the atmosphere as surface 3. Therefore, the data given are:

$$T_1 = 1500\text{K}, \qquad \epsilon_1 = 0.3, \qquad A_1 = \pi r^2 = 3.142 \text{ m}^2$$
$$A_2 = \pi D L = \pi \cdot 2.1 = 6.284 \text{ m}^2, \qquad T_2 = ?$$

A_3 is very large, $T_3 = 300$ K

The radiation network would be as shown in Fig. 9.29

$$E_{b_1} = \sigma T_1^4 = 287 \text{ kW/m}^2; \qquad E_{b_3} = \sigma T_3^4 = 0.459 \text{ kW/m}^2$$

$$\frac{1-\varepsilon_1}{A_1 \varepsilon_1} = 0.7426; \quad \text{from curves,} \quad F_{13} = 0.38; \quad F_{11} = 0; \qquad \therefore F_{12} = 0.62$$

$$A_1 F_{12} = A_2 F_{21}; \quad F_{21} = \frac{A_1 F_{12}}{A_2} = 0.31 \quad \text{By symmetry,} \quad F_{23} = 0.31$$

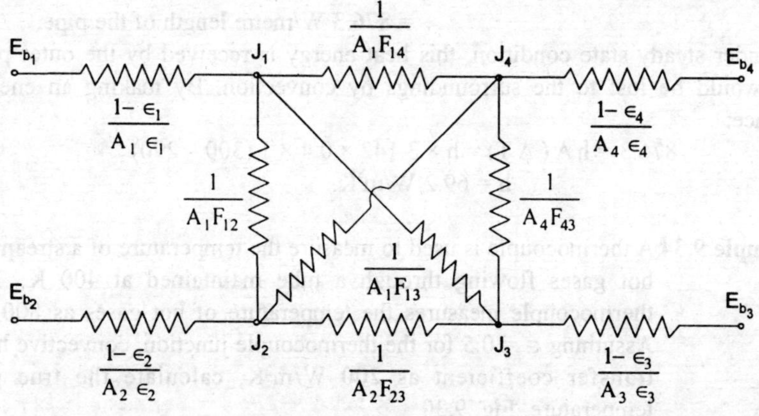

Fig 9.29 Electrical network

$$A_1 F_{13} = A_3 F_{31}; \quad F_{31} = 0.38; \quad F_{33} = 0, \text{ and } \quad F_{32} = 0.62$$

$$\frac{1}{A_1 F_{12}} = \frac{1}{3.142 \times 0.62} = 0.513; \quad \frac{1}{A_1 F_{13}} = 0.8375; \quad \frac{1}{A_2 F_{23}} = 0.513$$

The network has series and parallel resistances, therefore the total resistance,

$$R = \frac{1-\varepsilon_1}{A_1\varepsilon_1} + \cfrac{1}{A_1 F_{13} + \cfrac{1}{1/A_1 F_{12} + 1/A_2 F_{23}}}$$

$$= 0.7426 + 1/[1.194 + 1/(0.513 + 0.513)] = 1.2$$

$$\dot{Q} = (E_{b_1} - E_{b_3})/R = (287 - 0.459)/1.2 = 238 \text{ kW/m}^2$$

$$J_1 = E_{b_1} - \dot{Q}\left(\frac{1-\varepsilon_1}{A_1\varepsilon_1}\right) = 287 - 238 \times 0.7426 = 110.26 \text{ kW/m}^2$$

$$\dot{Q}_1 = (J_1 - J_3)/(1/A_1 F_{1-3}) = (110.26 - 0.459)/0.8375 = 131.1 \text{ kW}$$

$$\dot{Q}_2 = 238 - 131.1 = 106.9 \text{ kW}$$

$$E_{b_2} = J_2 = J_1 - \dot{Q}_2/(1/A_1 F_{1-2}) = 110.26 - 106.9 \times 0.513 = 55.42$$

$$= \sigma T_2^4; \quad \therefore T_2 = 994.3, \text{ or, } 721.3°C.$$

20. Electrical Network for a System Consisting of Four Gray Surfaces

The electrical circuit is shown in the accompanying figure. The net rate of heat flow from the four surfaces can be computed as follows:

$$\dot{Q}_1 = \frac{E_{b_1} - J_1}{(1 - \varepsilon_1)/A_1\varepsilon_1} = (J_1 - J_2)A_2F_{21} + (J_1 - J_3)A_1F_{13} + (J_1 - J_4)A_1F_{14}$$

$$\dot{Q}_2 = (E_{b2} - J_2)/(1 - \epsilon_2)/(A_2\epsilon_2) = (J_2 - J_1)A_2F_{21} + (J_2 - J_3)A_2F_{23} + (J_2 - J_4)A_2F_{24}$$

$$\dot{Q}_3 = (E_{b3} - J_3)/(1 - \epsilon_3)/(A_3\epsilon_3) = (J_3 - J_1)A_1F_{13} + (J_3 - J_2)A_3F_{32} + (J_3 - J_4)A_3F_{34}$$

$$\dot{Q}_4 = (E_{b4} - J_4)/(1 - \epsilon_4)/(A_4\epsilon_4) = (J_4 - J_1)A_4F_{41} + (J_4 - J_3)A_4F_{43} + (J_4 - J_2)A_4F_{42}$$

where $A_1F_{14} = A_4F_{41};\ A_1F_{12} = A_2F_{21};\ A_1F_{13} = A_3F_{31}$ and so on.

21. Radiation—Convection System

There are many practical examples where simultaneous exchange of heat energy by convection and radiation takes place and this would require a simultaneous solution of radiation and convection equation.

Example 9.33 A 20 cm outer diameter steam pipe, surface temperature 450K $\epsilon = 0.8$, is placed co-axially in another pipe 40 cm inner diameter pipe, $\epsilon = 0.85$ a nd inside surface t emperature 300 K. The two pipes pass through a large room, temperature 290 K. Calculate the radiation heat loss from the steam pipe and the surface heat transfer coefficient for the outer pipe.

Solution: The steam pipe cannot see itself and is completely enclosed in another pipe. The electrical network for the radiation heat transfer would be the same as Fig. 9.24, where A_1 is the surface area of the inner pipe.

$A_1 = \pi \times D \times L = 3.142 \times 0.20 \times 1 = 0.6284$ m^2 per metre length

and $A_2 = \pi \times D \times L = 3.142 \times 0.40 \times 1 = 1.2568$ m^2 per metre length

$F_{12} = 1.0$; $(1 - \epsilon_1)A_1\epsilon_1 = (1 - 0.8)/(0.8 \times 0.6284) = 0.3978$

$(1 - \epsilon_2)/A_2\epsilon_2 = (1 - 0.85)/(0.85 \times 1.2584) = 0.14$

$$\left.\begin{array}{l}\text{Rate of heat flow from}\\ \text{steam pipe by radiation}\end{array}\right| \frac{E_{b_1} - E_{b_2}}{R} = \frac{(5.67)(4.5^4 - 3^4)}{[0.3978 + (1/0.6284) + 0.14]}$$

$$= 876.3 \text{ W/metre length of the pipe.}$$

Under steady state condition, this heat energy is received by the outer pipe and would be lost to the surroundings by convection. By making an energy balance:

$$876.3 = h A (\Delta T) = h \times 3.142 \times 0.4 \times 1 (300 - 290)$$

or, $\qquad\qquad\qquad\qquad$ h $= 69.2$ W/m^{2}K.

Example 9.34 A thermocouple is used to measure the temperature of a stream of hot gases flowing through a tube maintained at 400 K. The thermocouple measures the temperature of hot gases as 800 K. Assuming $\epsilon = 0.5$ for the thermocouple junction, convective heat transfer coefficient as 200 W/m^{2}K, calculate the true gas temperature, Fig. 9.30.

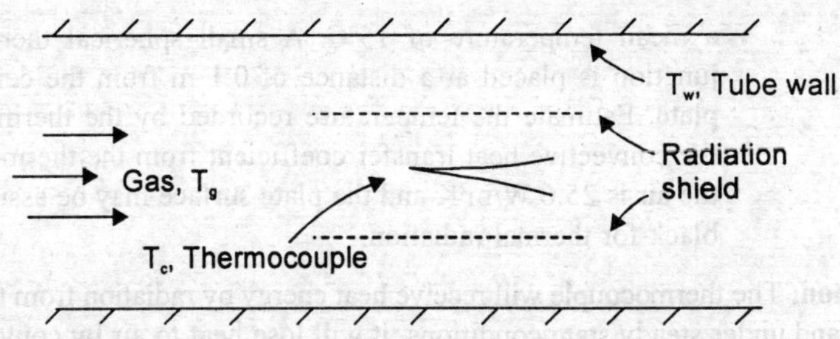

Fig. 9.30 Thermocouple in a gas stream with radiation shield

Solution: The thermocouple placed in the gas stream, which is transparent to thermal radiation, will receive heat energy by convection and under steady state condition, this gain in heat energy will be lost to the cold walls of the tube by radiation.

or, $$\dot{Q}_{conv} = \dot{Q}_{rad}$$

If T_g is the temperature of the gas and T_c is the temperature of the thermocouple,

If $$\dot{Q}_c = h\,A\,(T_g - T_c). \text{ If } \epsilon \text{ is the emissivity of the thermocouple,}$$

$$\dot{Q}_r = \epsilon A\sigma(T_c^4 - T_w^4)$$

Therefore, $$T_g = T_c + \epsilon\sigma\,(T_c^4 - T_w^4)/h$$
$$= 800 + 0.5 \times 5.67 \times (8^4 - 4^4)/200 = 854.43 \text{ K,}$$

The error in temperature ($= 54.43°C$) measurement can be considerably reduced by employing highly reflective (low emissivity) radiation shields is shown in Fig. 9.30.

After the shield is employed, it is assumed that: (i) the shield completely surrounds the thermocouple, (ii) convection occurs on both sides of the shield, (iii) radiant energy exchange takes place between the thermocouple bead and the inner surface of the shield. Under these assumptions, by making an energy balance for the shield and the tube wall:

$$2hA_s(T_g - T_s) = \epsilon_s A_s\sigma(T_s^4 - T_w^4)$$

or, $$2h\,(T_g - T_s) = \epsilon_s\sigma(T_s^4 - T_w^4) \qquad\qquad (a)$$

where T_s is the shield temperature, A_s shield surface area and ϵ_s is the shield emissivity.

For exchange of energy between the thermocouple bead and the shield, we have

$$hA_c\,(T_g - T_c) = \epsilon_c A_c\sigma(T_c^4 - T_s^4) \qquad\qquad (b)$$

Assuming that the shield has an emissivity of 0.3, we get from (a)

$$2 \times 200\,(854.43 - T_s) = 0.3 \times 5.67 \times 10^{-8}\,[T_s^4 - 400^4]$$

The above equation can only be solved by trial and error, which gives $T_s = 835$ K.

Again from (b): $$h\,(T_g - T_c) = \epsilon_c\sigma(T_c^4 - T_s^4)$$

or, $$200\,(854.43 - T_c) = 0.5 \times 5.67 \times 10^{-8}\,[T_c^4 - 835^4]$$

And we obtain, $$T_c = 849.6 \text{ K, an error of about } 5°C.$$

Example 9.35 A circular plate of radius 0.1 m is at a temperature of 500°C in a large room, the walls of which are at 10°C. The air in the room is at a mean temperature of 15°C. A small spherical thermo-couple junction is placed at a distance of 0.1 m from the centre of the plate. Estimate the temperature recorded by the thermocouple if the convective heat transfer coefficient from the thermocouple to the air is 25.6 W/m²K and the plate surface may be assumed to be black for thermal radiation.

Solution: The thermocouple will receive heat energy by radiation from the circular plate and under steady state conditions, it will lose heat to air by convection and to the walls of the room by radiation. The shape factor for the plate and the thermocouple is given by the relation:

$$F_{12} = \frac{1}{2}\left[1 - \frac{L}{(L^2 + R^2)^{1/2}}\right] \qquad \text{from Table 9.2(c)}$$

$$= \frac{1}{2}[1 - 0.1/(0.1^2 + 0.1^2)^{1/2}] = 0.146$$

Let the temperature of the thermocouple be T_c,

$$\therefore \qquad \sigma\,[(500 + 273)^4 - T_c^{\,4}] \times 0.146 = h\,(T_c - 288) + \sigma\,[T_c^{\,4} - (273 + 10)^4]$$

Solving by trial and error, $\qquad\qquad T_c = 97°C.$

22. Heat Transfer Coefficient for Radiation

We can always define a heat transfer coefficient for radiation analogous to surface heat transfer coefficient. Let us consider two black bodies A_1 and A_2 at temperature T_1 and T_2 respectively. They exchange energy by thermal radiation. Therefore, we write:

$$\dot{Q}_r = A_1 F_{12}\sigma(T_1^{\,4} - T_2^{\,4}) = \dot{Q}_c = h_r A_1\,(T_1 - T_2)$$

Therefore, the heat transfer coefficient for radiation, $\quad h_r = \dfrac{F_{12}\sigma(T_1^4 - T_2^4)}{(T_1 - T_2)}.$

Example 9.36 A cylinder, surface area 4 m², $\epsilon = 0.7$ temperature 550 K is placed in a large room at 300 K. Calculate the heat transfer coefficient for radiation and the total loss of heat energy by convection and radiation if the convective heat transfer coefficient is 10 W/m²K.

Solution: Since the cylinder is placed in a very large room, $F_{12} = 1.0$, and the rate of heat transfer by radiation is given by

$$\dot{Q}_r = \epsilon A F_{12}\sigma(T_1^{\,4} - T_2^{\,4}) = 0.7 \times 4 \times 1 \times 5.67\,(5.5^4 - 3.0^4)$$

$$= 13.24 \text{ kW}$$

$$h_r = 13.24 \times 1000/(4 \times 250) = 13.24 \text{ W/m}^2\text{K}.$$

The rate of heat transfer by convection $= h_c\,A(\Delta T)$

$$= 10 \times 4\,(550 - 300) = 10 \text{ kW}$$

Total loss of energy = 23.24 kW and the overall heat transfer coefficient

$$= h_c + h_r = (10 + 13.24) = 23.24 \text{ W/m}^2\text{K}.$$

23. Gas Radiation

In the preceding sections, it has been assumed that the space between the bodies exchanging heat energy by radiation is filled with gases which are transparent to thermal radiation waves. Inert gases like argon and gases with symmetric diatomic molecules (i.e, oxygen, nitrogen) neither absorb nor emit radiation waves. But gases like carbon dioxide, carbon monoxide, sulphur dioxide, nitrous oxide, water vapour (i.e., gases with asymmetric molecular structures) are not transparent and radiation is absorbed from and emitted to surrounding surfaces. The radiation absorbed is dependent on the frequency of the radiation waves striking the molecules because of their rotational and vibrational motions. Thus, the absorption and emission of radiation in gases is selective, occuring in only certain bands of wavelength (Table 9.4) whereas radiation from solids cover all wavelengths. Moreover, radiation entering a gas volume is not absorbed within a small distance from the surface as is the case in normal solids. The intensity of incident radiation decreases slowly as the radiation passes through the gas. And, therefore, the absorptivity of a gas volume depends upon its pressure, temperature, shape and size.

24. Absorptivity of Gases

Let us assume that a beam of monochromatic radiation of intensity I_{λ_0} strikes the body of the gases at x = 0. Since the quantity of energy would be absorbed in a layer of gas, we assume that the intensity of radiation impinging on the gas layer of thickness dx is $I_{\lambda x}$, Fig. 9.31. The decrease in intensity due to the absorption in the layer is assumed to be proportional to the thickness of the layer and the intensity of radiation at that point. Thus

$$dI_\lambda = - K_\lambda I_\lambda dx \qquad (9.20)$$

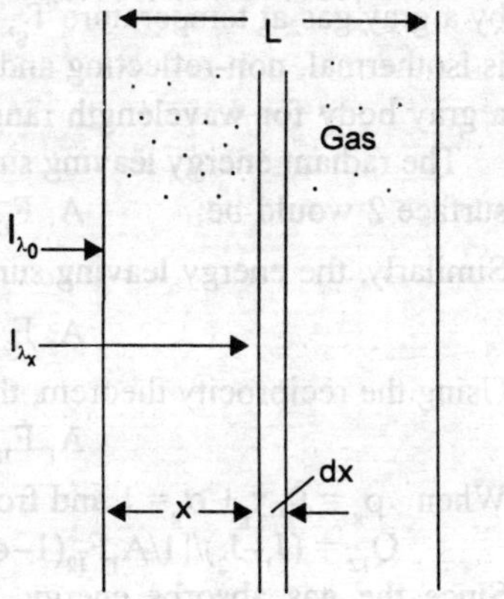

Fig. 9.31 Absorption in a gas layer

where K_λ is the monochromatic absorption or extinction coefficient. Integrating Eq. (9.20), we get

$$\int_{I_{\lambda_0}}^{I_{\lambda_L}} \frac{dI_\lambda}{I_\lambda} = \int_0^L -K_\lambda dx; \qquad \text{or,} \qquad \frac{I_{\lambda_L}}{I_{\lambda_0}} = e^{-K_\lambda L} \qquad (9.21)$$

The Eq. (9.21) is called Beer's Law.

The monochromatic absorptivity, of the gas is given by

$$\alpha_\lambda = (I_{\lambda_0} - I_{\lambda_L})/I_{\lambda_0} = 1 - I_{\lambda_L}/I_{\lambda_0} = 1 - e^{-K_\lambda L} \qquad (9.22)$$

and it depends upon the characteristic coefficient K_λ of the gas and on the length x of the gas layer. If the gas is non-reflecting, then

$$\alpha_\lambda + \tau_\lambda = 1$$

and transmissivity is given as: $\tau_\lambda = e^{-K_\lambda L}$ $\qquad (9.23)$

According to the Kirchoff's Law, the monochromatic emissivity, $\epsilon_\lambda = \alpha_\lambda$. It may be noted that when L is very large, $\alpha_\lambda = \epsilon_\lambda = 1$ i.e., gas radiation will approach a black body radiation within the wavelength band for very thick layers.

Example 9.37 Calculate the absorption coefficient for radiation passing through a gas layer 100 mm thick if the spectral intensity of radiation is reduced by 92 percent.

Solution: From Eq. (9.21), $I_{\lambda_L}/I_{\lambda_0} = e^{-K_\lambda L}$ we get the absorption coefficient,
$$K_\lambda = -(1/L)\ \ln(I_{\lambda_L}/I_{\lambda_0})$$
Since 92% of the intensity of radiation is absorbed in the gas of thickness 100 mm. $I_{\lambda_L} = 0.08\,I_{\lambda_0}$
Therefore, $K_\lambda = -(1/0.1)\ \ln(0.08) = 25.25$ per metre.

25. Radiant Heat Exchange between Two Infinite Parallel Planes Separated by a Gray Gas

Let us consider two infinite parallel planes at temperatures T_1 and T_2 separated by a gray gas at temperature T_g, as shown in Fig. 9.32. It is assumed that the gas is isothermal, non-reflecting and the nature of the gas is such that it behaves like a gray body for wavelength ranges relevant to temperature T_g.

The radiant energy leaving surface 1, transmitted through the gas and reaching surface 2 would be; $A_1\,F_{12}\,J_1\,\tau_g$.

Similarly, the energy leaving surface 2 and reaching surface 1 would be
$$A_2\,F_{21}\,J_2\,\tau_g.$$
Using the reciprocity theorem, the net energy exchange by direct transmission is:
$$A_1\,F_{12}\,\tau_g\,(J_1 - J_2)$$
When $\rho_g = 0$, $\tau_g + \alpha_g = 1$ and from Kirchhoff's law, $\alpha_g = \epsilon_g$ and $\tau_g = 1 - \epsilon_g$,
$$Q_{12} = (J_1 - J_2)/[1/A_1 F_{12}(1-\epsilon_g)]$$ and the network is shown in Fig. 9.33.
Since the gas absorbs energy, it must emit energy and therefore, the energy emitted by gas: $J_g = \epsilon_g E_{bg}$
The energy leaving the gas and reaching surface 1 and 2 would be given by:

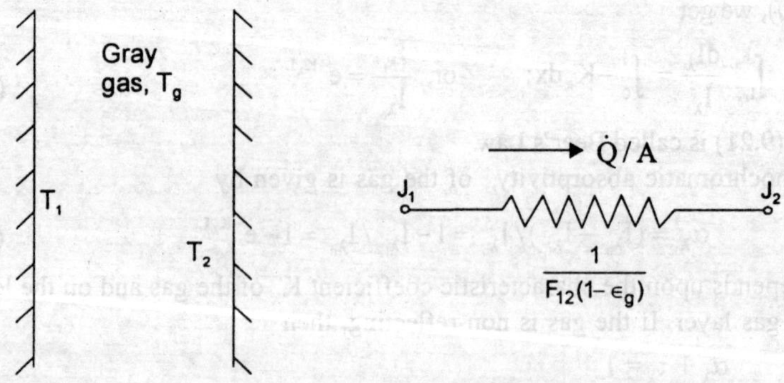

Fig. 9.32 Gray gas between parallel planes

Fig. 9.33 Space resistance for infinite parallel planes separated by gray gas

$A_g F_{g1} \in_g E_{bg}$ and $A_g F_{g2} \in_g E_{bg}$ respectively.

Thus the net energy loss by the gas would be:

To surface 1:

$$A_g F_{g-1} \in_g E_{bg} - A_1 F_{1-g} J_1 \in_g = \frac{(E_{bg} - J_1)}{1/A_1 F_{1g} \varepsilon_g}$$

To surface 2:

$$A_g F_{g2} \in_g E_{bg} - A_2 F_{2g} J_2 \in_g = \frac{(E_{bg} - J_2)}{1/A_2^* F_{2g} \varepsilon_g}$$

and the analogous electrical network would be as shown in Fig. 9.34.

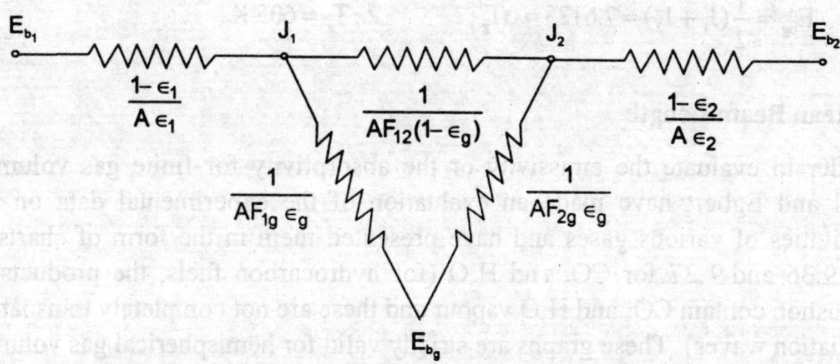

Fig. 9.34 Network for radiant energy exchange between infinite parallel planes with gas between them

Example 9.38 Two very large parallel planes are at: $T_1 = 850$ K, $\in_1 = 0.25$, and $T_2 = 350$ K, $\in_2 = 0.65$ are separated by a gray gas having $\in_g = 0.2$, $\tau_g = 0.8$. Calculate the heat transfer rate between the planes with and without the presence of gas.

Solution: The analogous network is shown in Fig. 9.34.

With
$$F_{12} = F_{1g} = F_{2g} = 1.0$$

We have,
$$\frac{1-\varepsilon_1}{\varepsilon_1} = \frac{0.75}{0.25} = 3, \qquad \frac{1}{F_{12}(1-\varepsilon_g)} = \frac{1}{0.8} = 1.25$$

$$\frac{1-\varepsilon_2}{\varepsilon_2} = \frac{0.35}{0.65} = 0.538, \qquad \frac{1}{F_{1g}\varepsilon_g} = \frac{1}{F_{2g}\varepsilon_g} = \frac{1}{0.2} = 5.0;$$

$$E_{b_1} = \sigma T_1^4 = 29.6 \text{ kW/m}^2; \qquad E_{b_2} = \sigma T_2^4 = 0.85 \text{ kW/m}^2$$

The total resistance, $R = 3 + 0.538 + \dfrac{1}{1/1.25 + 1/(5+5)} = 4.649$

$$\frac{Q}{A} = \frac{29.6 - 0.85}{4.649} = 6.184 \text{ kW/m}^2$$

When there is no gas $\dfrac{\dot{Q}}{A} = \dfrac{29.6 - 0.85}{1/\varepsilon_1 + 1/\varepsilon_2 - 1} = \dfrac{28.75}{4 + 1.538 - 1} = 6.33 \text{ kW/m}^2$

Thus, the introduction of a participating gas increases the resistance between nodes of potentials J_1 and J_2 and the gas actually behaves like a radiation shield. In order to determine the temperature of the gas, we calculate J_1 and J_2:

$$J_1 = E_{b_1} - \frac{1-\varepsilon_1}{\varepsilon_1} \times \frac{\dot{Q}}{A} = 29.6 - 3 \times 6.184 = 11.048 \text{ kW/m}^2$$

$$J_2 = E_{b_2} + \frac{1-\varepsilon_2}{\varepsilon_2} \times \frac{\dot{Q}}{A} = 0.85 + 0.538 \times 6.184 = 4.177 \text{ kW/m}^2$$

and, $\quad E_{bg} = \dfrac{1}{2}(J_1 + J_2) = 7.6125 = \sigma T_g^4; \qquad \therefore T_g = 605 \text{ K.}$

26. Mean Beam Length

In order to evaluate the emissivity or the absorptivity for finite gas volumes, Hottel and Egbert have made an evaluation of the experimental data on the emissivities of various gases and have presented them in the form of charts in Figs. 9.36 and 9.37 for CO_2 and H_2O (for hydrocarbon fuels, the products of combustion contain CO_2 and H_2O vapour and these are not completely transparent to radiation waves). These graphs are strictly valid for hemispherical gas volumes of radius L radiating to an elemental surface at the centre of the gas of the hemisphere shown in Fig. 9.35. However for other gas shapes, an effective or mean beam length, L, is approximated by

$$L = 3.6 \times (\text{Volume V}) / (\text{Surface area, A}) \tag{9.24}$$

where V is the total volume and A is the area of the bounding surface of the gas. Table 9.5 lists the values of L for commonly encountered geometries. The curves in Fig 9.36 and 9.37 give the effective emissivity at different temperatures for various values of the product pL where p is the partial pressure in atmosphere and L is the mean beam length in metres.

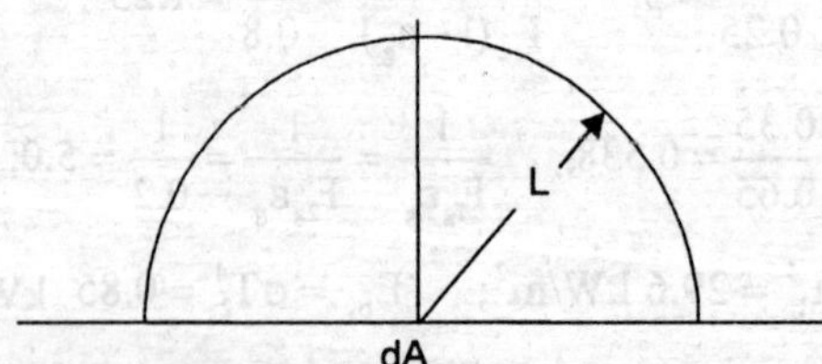

Fig. 9.35 Gas volume for Hottel's graph

Example 9.39 A gas mixture contains 12% CO_2 and 10% H_2O by volume. The total pressure is 1 atm and the gas temperature is 1000°C. If the mean beam length is 0,15 m, calculate the emissivity of the gas.

Solution: By Dalton's law of partial pressure, the partial pressure of CO_2 = 0.12 atm and the partial pressure of H_2O = 0.1 atm.

$$P_{H_2O}L = 0.1 \times 0.15 = 0.015$$

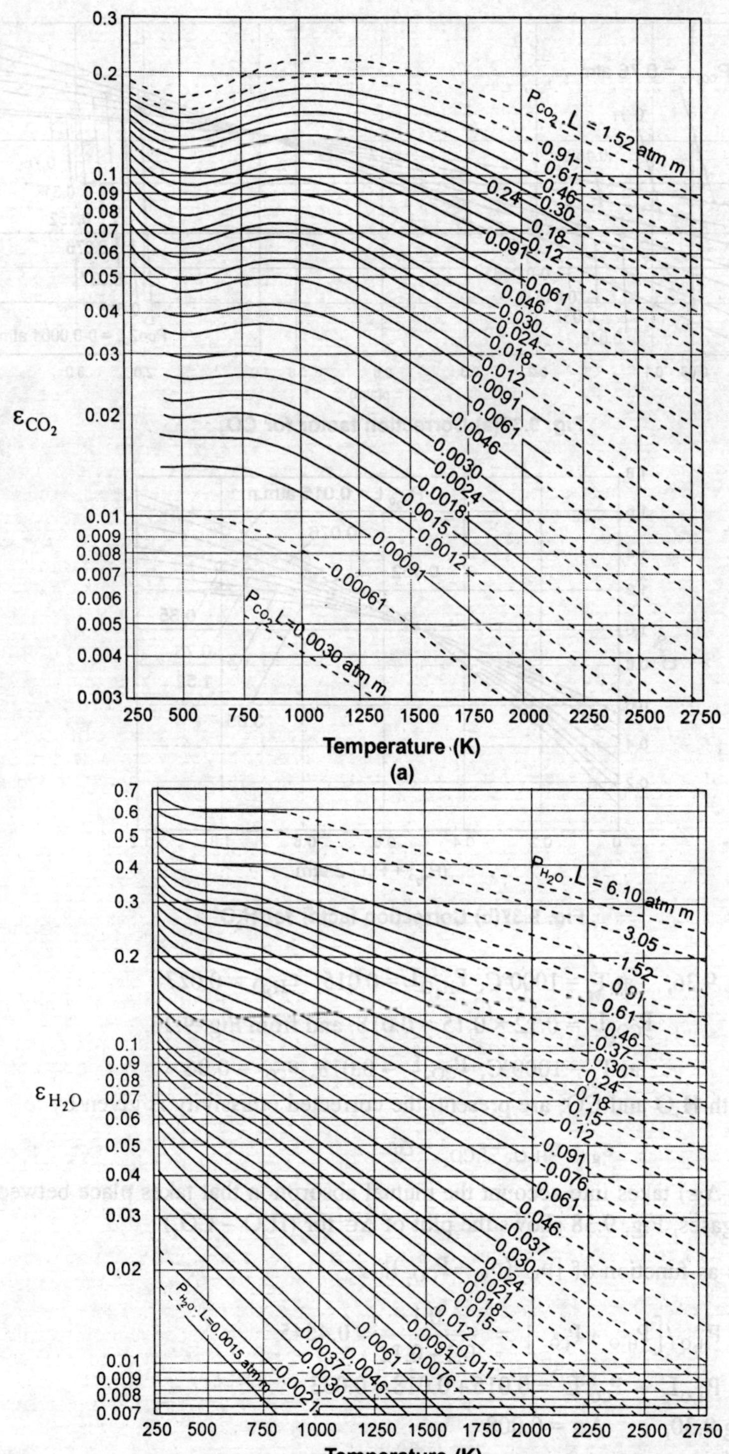

Fig. 9.36 Hottel's graph for emissivities of CO_2 and H_2O

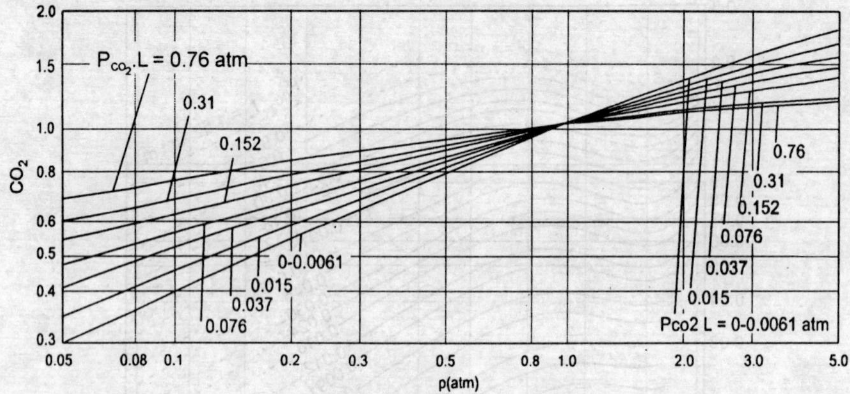

Fig. 9.37(a) Correction factor for CO₂

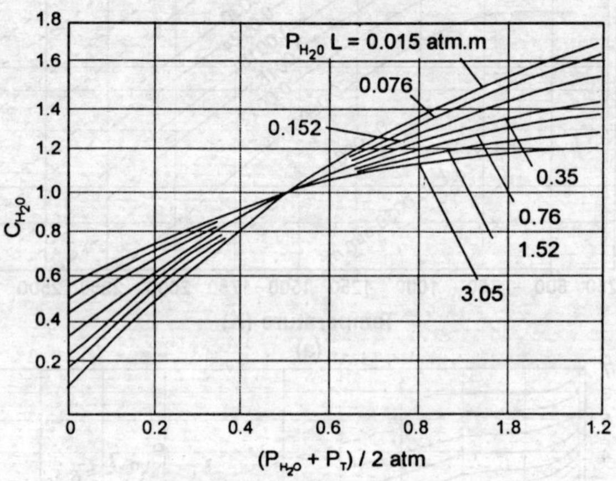

Fig. 9.37(b) Correction factor for H₂O

From Fig. 9.36, at $T_g = 1000°C$, $P_{H_2O}L = 0.015$, $\varepsilon_{H_2O} = 0.027$

Again, $P_{CO_2}L = 0.12 \times 0.15 = 0.018$, and from Fig. 9.36,

at $T_g = 1000°C$, $P_{CO_2}L = 0.018$, $\varepsilon_{CO_2} = 0.06$

Since both H_2O and CO_2 are present, the corrected emissivity is given by

$$\varepsilon_g = \varepsilon_{H_2O} + \varepsilon_{CO_2} - \Delta\varepsilon$$

Where $(-\Delta\epsilon)$ takes into account the mutual absorption that takes place between the two gases. Fig. 9.38 shows the plot of $\Delta\epsilon$ for $(H_2O - CO_2)$

Mixtures as function of $[p_w /(p_w + P_{CO_2})]$

$$P_{H_2O} / \left[P_{H_2O} + P_{CO_2} \right] = \frac{0.1}{0.1 + 0.12} = 0.4545$$

and, $P_{H_2O}L + P_{CO_2}L = 0.015 + 0.018 = 0.033$

From Fig 9.30, $\Delta\epsilon = 0.008$

$\therefore \qquad \varepsilon_g = \varepsilon_{H_2O} + \varepsilon_{CO_2} - \Delta\epsilon = 0.027 + 0.06 - 0.008 = 0.079$

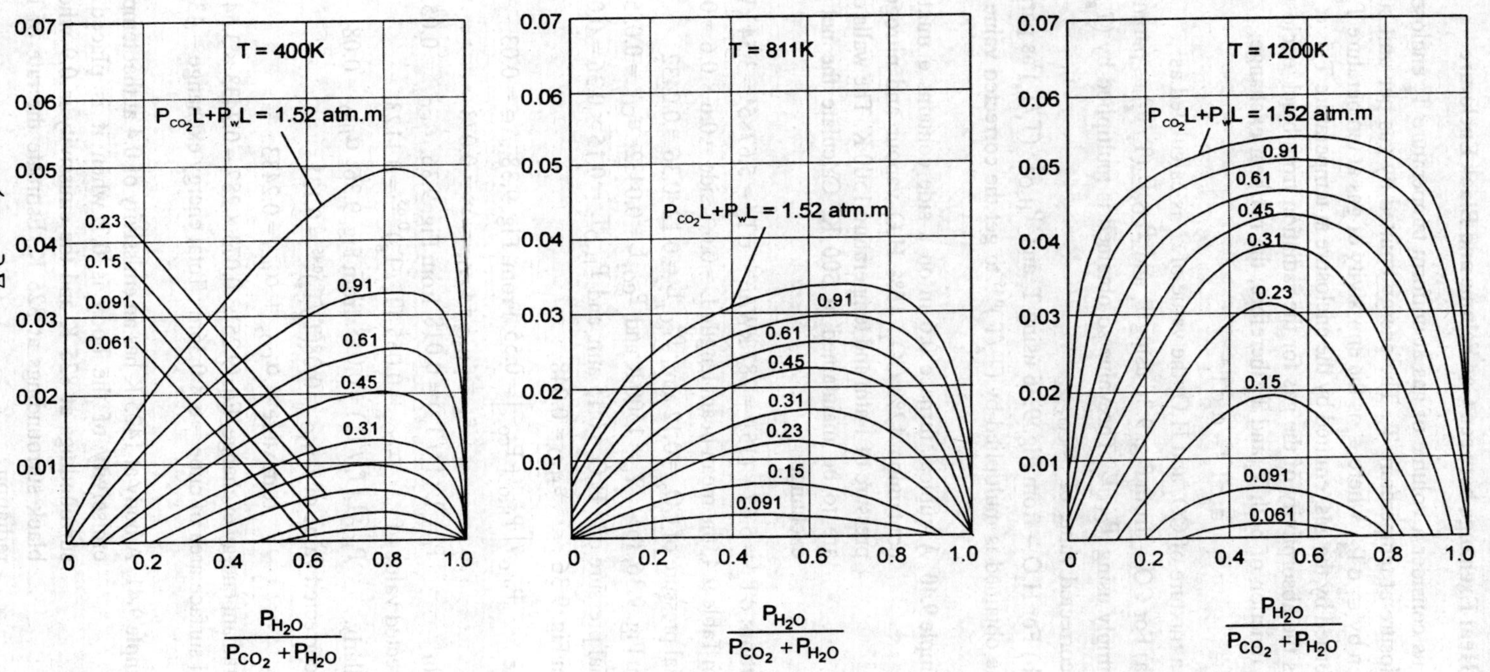

Fig. 9.38 Correction factor for mixture of CO_2 and H_2O

27. Heat Exchange between Gas Volume and Black Enclosure

Let us consider a volume of gas at uniform temperature T_g enclosed by a black enclosure at temperature T_s. The energy emitted by gas per unit surface area is given by $\epsilon_g \sigma T_g^4$, where ϵ_g is the emissivity of gas at temperature T_g. The energy absorbed by the gas, emitted by the enclosure at temperature T_s is $\alpha_g \sigma T_s^4$, where α_g is the absorptivity of the gas for the radiation from black enclosure at T_s and is a function of both T_s and T_g therefore, the net heat exchange:

$$\epsilon_g \sigma T_g^4 - \alpha_g \sigma T_s^4 \tag{9.25}$$

For a mixture of CO_2 and H_2O, the value of α_g is calculated as:

(a) For CO_2 – from Fig. 9.36 using T_s and $P_{CO_2} L$ (T_s/T_g) as parameters instead of simply using $P_{CO_2} L$. The value so obtained is multiplied by $(T_g/T_s)^{0.65}$ to get the corrected value of α_{CO_2}

(b) For H_2O – from Fig. 9.36 using T_s and $P_{H_2O}L$ (T_s/T_g) as parameters. The value obtained is multiplied by $(T_g/T_s)^{0.45}$ to get the corrected value of $P_{H_2O}L$.

Example 9.40 A cubical furnace 0.6 m on a side contains a mixture of gases containing 12% CO_2, 15% H_2O vapour and nitrogen at a total pressure of 1 atm and temperature 1500 K. The walls of the furnace are to be maintained at 500 K. Calculate the net radiant heat exchange.

Solution: $\sigma T_g^4 = 5.67 \times (15)^4 = 287 \text{ kW/m}^2$; $\sigma T_s^4 = 5.67 \times 5^4 = 3.54 \text{ kW/m}^2$

From Table 9.4, the mean beam length, $L = 0.6 \times$ side $= 0.6 \times 0.6 = 0.36$ m

Partial pressure of $CO_2 = 0.12$ atm; $P_{CO_2} L = 0.12 \times 0.36 = 0.0432$

From Fig. 9.36, for $T_g = 1500$ K and $P_{CO_2} L = 0.0432$; $\varepsilon_{CO_2} = 0.055$

Partial pressure of $H_2O = 0.15$ atm, and $P_{H_2O}L = 0.15 \times 0.36 = 0.054$

From Fig. 9.36 $\varepsilon_{H_2O} = 0.048$

Since $P_{H_2O} / \left[P_{H_2O} + P_{CO_2} \right] = 0.55$ From Fig. 9.38, $\Delta \epsilon = 0.03$

and $\varepsilon_g = \varepsilon_{H_2O} + \varepsilon_{CO_2} - \Delta \varepsilon = 0.073$

Again $P_{CO_2} L (T_s / T_g) = 0.018$ from Fig. 9.36, $\alpha_{CO_2} = 0.055$

Corrected value of $\alpha_{CO_2} = 0.055 \, (T_s / T_g)^{0.65} = 0.1123$

Similarly, $P_{H_2O}L(T_s / T_g) = 0.018$ from Fig. 9.36, $\alpha_{H_2O} = 0.08$

and corrected value of $\alpha_{H_2O} = 0.08 (T_g/T_s)^{0.45} = 0.131$

$\therefore$ α mixture $= \alpha_{CO_2} + \alpha_{H_2O} = 0.2433.$

Net radiant energy exchange, Eq. (9.25): $0.073 \times 287 - 0.2433 \times 3.54 = 20 \text{ kW/m}^2$

Total surface area of cube $= 6 \times 0.36 \text{m}^2$; Total energy exchange $= 43.47$ kW.

Example 9.41 A body at 1425 K has an emissivity of 0.4 at that temperature. The emissivity of the body is 0.7 when it is placed in a black surroundings at 825 K and the emissivity is 0.9 when placed in black surroundings at 325 K. Estimate the rate of heat loss by radiation :

(a) when the body is assumed to be gray with emissivity 0.4

(b) when the body is not gray.

Solution: When a gray body is placed in a black surroundings or in a very large enclosure, the rate of heat loss by radiation is given by

$$\dot{Q} = \varepsilon \, (E_{b_1} - E_{b_2}) = 0.4 \times 5.67 \times 10^{-8} \, (T_1^4 - T_2^4)$$

When the surroundings are at 825 K,

$$\dot{Q} = 0.4 \times 5.67 \, (14.25^4 - 8.25^4) = 83 \text{ kW/m}^2$$

When the surroundings are at 325 K,

$$\dot{Q} = 0.4 \times 5.67 \, (14.25^4 - 3.25^4) = 93.266 \text{ kW/m}^2$$

When the body does not behave like a gray body, its absorptivity when the surroundings are at T_2 will be equal to the emissivity when the body is at temperature T_2.

Therefore, the rate of energy emission = $\in \sigma T_1^4 = 0.4 \times 5.67 \times 14.25^4$

and the rate of energy absorption = $\alpha \sigma T_2^4 = 0.7 \times 5.67 \times 8.25^4$

$$\dot{Q} = 0.4 \times 5.67 \times 14.25^4 - 0.7 \times 5.67 \times 8.25^4 = 75.163 \text{ kW/m}^2$$

When the surroundings are at 325 K. the rate of heat loss by radiation is

$$\dot{Q} = 0.4 \times 5.67 \times 14.25^4 - 0.9 \times 5.67 \times 3.25^4 = 92.95 \text{ kW/m}^2$$

Thus, when the surroundings are at 825 K and we assume that body is gray, the error is = $(83 - 75.163)/75.163 \equiv 10.4\%$; and when the surroundings are at 325 K, the gray body assumption would give an error of

$$= (93.266 - 92.95)/92.95 \equiv 0.34\%$$

(It is evident from the above example that the gray body assumption gives a very good assumption when one of the temperatures is very small in comparison with the other. Further, the gray body assumption will also give a fairly accurate result when both temperatures are small.)

Example 9.42 A long cylindrical (0.5 m diameter) gas turbine combustion chamber has walls at temperature 505°C. The products of combustion are at 1 atm and 1005°C and contain 15% by volume of both CO_2 and H_2O. Calculate the net radiant energy exchange between the gases and the combustion chamber walls.

Solution: Partial pressure of CO_2 and $H_2O = 0.15$ atm.

Beam length for a long cylinder = diameter = 0.5 m

$\therefore \qquad P_{H_2O}L = 0.075$ and from Fig. 9.36 (b), $\in = 0.07$

and $\qquad P_{CO_2}L = 0.075$ and from Fig. 9.36 (a) $\in = 0.09$

$$P_{CO_2}L + P_{H_2O}L = 0.15 \qquad \text{and} \qquad \frac{P_{H_2O}}{P_{H_2O} + P_{CO_2}} = 0.5 \, ; \quad \Delta\varepsilon = 0.02$$

$$\varepsilon_g = \varepsilon_{H_2O} + \varepsilon_{CO_2} - \Delta\varepsilon = 0.07 + 0.09 - 0.02 = 0.14$$

$$P_{CO_2} L (T_s / T_g) = 0.075 (778/1278) = 0.0456$$

From Fig. 9.28, $\alpha_{CO_2} = 0.03$

and corrected value of $\alpha_{CO_2} = 0.03 (1278/778)^{0.65}$

$$= 0.0414$$

Similarly, $P_{H_2O} L(T_s / T_g) = 0.0456$

and $\alpha_{H_2O} = 0.095$

corrected value of $\alpha_{H_2O} = 0.095 (1278/778)^{0.45}$

$$= 0.1187$$

$$\alpha_{mixture} = \alpha_{CO_2} + \alpha_{H_2O}$$
$$= 0.1601$$

Net radiant energy exchange $= \epsilon_g \sigma T_g^4 - \alpha \sigma T_s^4$

$$= 0.14 \times 5.67 \times 12.78^4 - 0.1601 \times 5.67 \times 7.78^4 = 17.85 \text{ kW/m}^2.$$

Example 9.43 A gas mixture at a total pressure of 5 atm and 1000 C is contained in a cubical chamber having one side equal to 100 cm. The partial pressure of CO_2 in the mixture is 0.35 atm. The other gases present in the mixture may be considered transparent. Estimate the emissivity of the gas volume.

Solution: The mean beam length, L = 0.67 × one side = 0.67 × 1 = 0.67 m.

$$P_{CO_2} L = 0.35 \times 0.67 = 02345 \text{ m.atm.}$$

From Fig. 9.36, for $P_{CO_2} L = 0.2345$ and T = 1000°C, $\epsilon = 0.15$

From Fig. 9.37, at 5 atm pressure, the correction factor for

$$P_{CO_2} L = 0.2345 \text{ is } 1.2.$$

Therefore, the emissivity of the gas volume

$$= 1.2 \times 0.15 = 0.18.$$

Table 9.3 A F_{12} and $(F_g)_{12}$ for various configurations

Configuration	Shape factor F_{12}	$(Fg)_{12}$
1. Infinite parallel planes	1	$1/(1/\epsilon_1 + 1/\epsilon_2 - 1)$
2. Concentric spheres or Infinitely long concentric cylinders	1	$1/(1/\epsilon_1 + (1/\epsilon_2 - 1)A_1/A_2)$
3. A small body 1 completely enclosed by body 2	1	ϵ_1
4. A large body completely enclosed by body 2	1	$1/(1/\epsilon_1 + (1/\epsilon_2 - 1)A_1/A_2)$
5. Two rectangles with a common side at right angles to each other	1	$\epsilon_1 \cdot \epsilon_2$

* $(Fg)_{12}$ is known as shape factor for gray body.

Table 9.4 Radiation functions

λT (µK)	$F_{0-\lambda T} = \dfrac{E_{b_0 - \lambda T}}{\sigma T^4}$	λT (µK)	$F_{0-\lambda T} = \dfrac{E_{b_0 - \lambda T}}{\sigma T^4}$
555.6	0	4000.0	0.4809
666.7	0	4222.2	0.5199
777.7	0	4444.4	0.5558
888.9	0.0001	4666.7	0.5890
1000.0	0.0003	5000.0	0.6337
1111.1	0.0009	6000.0	0.7378
1222.2	0.0025	7000.0	0.8081
1333.4	0.0053	8000.0	0.8563
1444.4	0.0098	8888.9	0.8868
1155.6	0.0164	9444.4	0.9017
1666.6	0.0254	10000.0	0.9142
1777.7	0.0368	10555.6	0.9247
1888.9	0.0506	11111.1	0.9335
2000.0	0.0667	11666.7	0.9411
2111.1	0.0850	12222.2	0.9475
2222.2	0.1051	12777.8	0.9531
2333.3	0.1267	13333.3	0.9589
2444.4	0.1496	13888.9	0.9621
2555.6	0.1734	14444.4	0.9657
2666.7	0.1979	15000.0	0.9689
2777.7	0.2229	15555.6	0.9718
2888.9	0.2481	16111.1	0.9742
3000.0	0.2733	16666.7	0.9765
3111.1	0.2983	22222.7	0.9881
3222.2	0.3230	27777.7	0.9941
3333.3	0.3474	33333.3	0.9963
3444.4	0.3712	38898.9	0.9981
3555.6	0.3945	44444.4	0.9987
3666.7	0.4171	50000.0	0.9990
3777.8	0.4391	55555.5	0.9992
3888.9	0.4604	∞	1.000

Table 9.5 Emission bands of CO_2 and H_2O

	CO_2		H_2O	
Band number	Wavelength range µ	Band width $\Delta\mu$	Wavelength range µ	Band width $\Delta\mu$
First	2.64-2.84	0.2	2.24-3.37	1.13
Second	4.01-4.80	0.71	4.8-8.5	3.7
Third	12.5-16.5	4.0	12-25	13.0

Table 9.6 Mean beam length for various gas volumes

No.	Gas volume	Characteristic dimension, D	Beam length, L
1	Sphere	Diameter	0.67 D
2	Infinite cylinder radiating to walls	Diameter	D
3.	Cube	Side	0.67 D
4.	Volume between infinite parallel planes	Separation distance, D	1.8 D
5.	Cylinder of height equal to diameter radiating to whole surface	Diameter	0.67 D
6.	Infinite cylinder radiating to elemental surface on centre of base	Diameter	D
7.	Cylinder of height equal to diameter radiating to elemental surface at centre of base	Diameter	0.77 D
8.	Space outside infinite bank of tubes with centres on equilateral triangles, tube diameter being equal to clearance	Clearance	3.4 D
9.	Same as (8) except that tube diameter equals ½ clearance	Clearance	4.5 D
10.	Same as (8) except that tube centres are on squares	Clearance	4.1 D
11.	1×2×6 rectangular parallelopiped radiating to	Shortest edge	
	2 × 6 face		1.18 D
	1 × 6 face		1.24 D
	1 × 2 face		1.18 D
	all face		1.02 D
12.	Any other shape	Hydraulic mean radius	4 D

MULTIPLE CHOICE QUESTIONS

1. Match List I with List II and select the correct answer using the code give below:

List I	List II
A. Infinite parallel planes	1. ε_1
B. A small body completely enclosed in a large enclosure	2. $\varepsilon_1\varepsilon_2$
C. Two rectangles with a common edge	3. $1/[1/\varepsilon_1 + 1/\varepsilon_2 - 1]$
D. Concentric cylinders	4. $1/\left[\frac{1}{\varepsilon_1} + \frac{A_1}{A_2}\left(\frac{1}{\varepsilon_2} - 1\right) \right]$

Code:	A	B	C	D
(a)	1	2	4	3
(b)	3	1	4	2
(c)	2	1	3	4
(d)	3	1	2	4

(d)

2. What would be the net radiant heat exchange per square metre for two very large parallel planes at temperature 800 K and 500 K when the emissivity of the hot and cold plates are 0.8 and 0.6 respectively.
    (a) 1.026 kW/m^2 (b) 10.26 kW/m^2
    (c) 102.6 kW/m^2 (d) 1026 kW/m^2 **(b)**

3. The spectral emissive power for a diffusely emitting surface is
    $E_\lambda = 0$ for $\lambda < 3\,\mu m$; $E_\lambda = 150$ W/m$^2\mu$m
    for $3 < \lambda < 12\ \mu m$; $E_\lambda = 300$ W/m$^2\mu$m
    for $12 < \lambda < 25$ mm; $E_\lambda = 0$ for $\lambda > 25\,\mu m$
    The total emissive power of the surface over the entire spectrum is
    (a) 1250 W/m^2 (b) 2500 W/m^2
    (c) 4000 W/m^2 (d) 5250 W/m^2 **(d)**

4. Consider the following statements:
    1. For metals, the value of absorptivity is high
    2. For non-conducting materials, reflectivity is low
    3. For polished surfaces, reflectivity is high
    4. For gases, reflectivity is low
    Of these statements:
    (a) 2, 3 and 4 are correct   (b) 3 and 4 are correct
    (c) 1, 2 and 4 are correct   (d) 1 and 2 are correct   **(a)**

5. On a summer day, a scooter rider feels more comfortable while on the move than while at stop light because
    (a) an object in motion captures less radiation
    (b) air is transparent to radiation and hence it is cooler than the body
    (c) more heat is lost by convection and radiation while in motion
    (d) Air has low specific heat and hence it is cooler.   **(c)**

6. If the temperature of a solid surface changes from 27°C to 627°C, its emissive power will increase in the ratio of
    (a) 3   (b) 9
    (c) 27   (d) 81   **(d)**

7. A spherical aluminium shell of inside diameter 2 m is evacuated and used as a radiation test chamber. If the inner surface is coated with carbon black and maintained at 600K, the irradiation on a small test surface placed inside the chamber is
    (a) 1000 W/m^2 (b) 3400 W/m^2 (c) 5680 W/m^2 (d) 7348 W/m^2 **(d)**

8. Match List I with List II and select the correct answer from the code given below

List I	List II
A. Stefan-Boltzmann Law	1. $E_{b\lambda} = C_1\lambda^{-5}/\exp(C_2/\lambda T)-1$
B. Kirchhoff's Law	2. $E_{b\lambda} = C_1 T/C_2\lambda^4$
C. Planck's Law	3. $E = \varepsilon\, E_b$
D. Rayleigh-Jeans Law	4. $\dot{Q} = A\varepsilon\sigma(T_1^4 - T_2^4)$

Code

	A	B	C	D
(a)	4	1	3	2
(b)	4	3	1	2
(c)	2	1	3	4
(d)	2	3	1	4

(b)

9. A radiation shield of emissivity $\in$ (on both sides) is placed between two infinite parallel planes of emissivity $\in$, and temperatures T_1 and T_2. The ratio of radiant energy flux with shield and without shield would be

(a) 0.25   (b) 0.5   (c) 0.75   (d) 1.0   **(b)**

10. It is desired to reduce the radiation energy exchange between two infinite parallel planes   by inserting radiation shields of the same emissivity. The number of shields required for 80% reduction would be

(a) two   (b) three   (c) four   (d) five   **(c)**

11. The mean beam length in gas radiation is defined as

(a) volume V/ surface area A   (b) volume V/3.6 (surface area A)

(c) 3.6 V/A   (d) 4 V/A   **(c)**

12 Choose the correct statement

(a) $A_1F_{1-3,4} = A_1F_{1-3} + A_3F_{3-4}$ (b) $\sum_{j=1}^{n-1} F_{ij} = 1.0$

(c) $A_{12}F_{12-34} = A_1F_{1-34} + A_2F_{2-34}$ (d) $A_1F_{1-3} = A_2F_{3-1}$ **(c)**

13. The intensity of radiation is obtained by multiplying the emissive power by a factor

(a) π (b) $\dfrac{1}{\sqrt{2}}\pi$

(c) $\dfrac{1}{\pi}$ (d) $\dfrac{\sqrt{2}}{\pi}$ **(c)**

14. Assertion (A): Gas radiation approaches a black body radiation within the wavelength band for very thick layers of gas.

Reasoning (R): Because the monochromatic absorptivity of the gas depends upon the extinction coefficient and the length of the gas layer.

Code: (a) Both A and R are false   (b) A is false, R is true

(c) Both A and R are true   (d) A is true and R is false   **(c)**

15. Beer's Law is defined as: $I_{\lambda_L} / I_{\lambda_0} =$

(a) $1 - e^{-K_\lambda L}$ (b) $1 + e^{-K_\lambda L}$

(c) $e^{-K_\lambda L}$ (d) $e^{K_\lambda L}$ **(a)**

16. Assertion (A): Absorption and emission of radiation in gases occur only in certain bands of wavelength.

Reasoning (R): Because radiation entering a gas volume is not absorbed within a small distance from the surface.

Code: (a) Both A and R are true and R is the correct reason for A

(b) A is true, R is false

(c) A is false, R is true

(d) Both A and R are true and R is not the explanation of A.   **(d)**

17. The absorptivity of a gas volume depends upon
    - (a) pressure only
    - (b) temperature only
    - (c) shape and size
    - (d) all of the above
    - (e) both (a) and (b)   **(d)**

18. Choose the correct statement
    For a given shape, partial pressure and temperature, the absorptivity of
    - (a) oxygen is more than nitrogen
    - (b) nitrogen is more than carbon monoxide
    - (c) carbon dioxide is more than oxygen
    - (d) water vapour is less than nitrogen   **(c)**

19. For an efficient greenhouse effect, glass should be
    - (a) transparent to long wavelength radiation
    - (b) opaque to short wavelength radiation
    - (c) transparent to short wavelength radiation
    - (d) transparent to all wavelengths   **(c)**

20. The mechanism of heat transfer in a flat plate solar collector is
    - (a) radiation only
    - (b) convection only
    - (c) conduction only
    - (d) all of the above
    - (e) both (a) and (b)   **(e)**

21. Two large parallel gray plates with a small gap, exchange radiation at the rate of 1000 W/m^2 when their emissivities are 0.5 each. By coating one plate, its emissivity is reduced to 0.25, the temperature remains unchanged. The new rate of heat exchange will be
    (a) 500 W/^{2} (b) 600 W/m^2 (c) 700 W/m^2 (d) 800 W/m^2 **(b)**

22. Two long parallel plates of the same emissivity 0.5 are maintained at different temperatures a nd h ave radiation h eat e xchange b etween them. A r adiation shield of emissivity 0.25 placed in the middle will reduce the radiation heat exchange to
    (a) 1/2 (b) 1/4 (c) 3/10 (d) 3/5   **(c)**

23. Match List I (type of radiation) with List II (characteristic) and select the correct answer using the codes given below:

List I	List II
A. Black body	1. Emissivity is independent of wavelength
B. Gray body	2. Mirror like reflection
C. Specular	3. Zero reflectivity
D. Diffuse	4. Intensity same in all direction

Code:	A	B	C	D
(a)	2	1	3	4
(b)	3	4	2	1
(c)	2	4	3	1
(d)	3	1	2	4

    (d)

24. The value of the shape factor for two inclined flat plates having a common edge of equal width and with an angle of 20 degrees is
    (a) 0.83   (b) 1.17   (c) 0.66   (d) 1.34   **(a)**

25. Assertion (A): A gray surface may be defined as one for which $\in$ and α are independent of λ over the spectral regions of the irradiation and the surface emission.

    Reasoning (R): Because, the irradiation and surface emission are concentrated in a region for which the spectral properties of the surface are approximately constant.

    Code: (a) Both A and R are false   (b) Both A and R are true
               (c) A is true, R is false   (d) A is false, R is true.   **(b)**

NUMERICALS

1. A small surface of area $A_1 = 10^{-4}$ m^2 emits diffusely with a total hemispherical emissive power of $E_1 = 5 \times 10^4$ W/m^2. Calculate the rate of which the radiation emitted by A_1 is intercepted by a small surface of area $A_2 = 5 \times 0^{-4}$ m^2; oriented as shown. What is the irradiation on A_2?

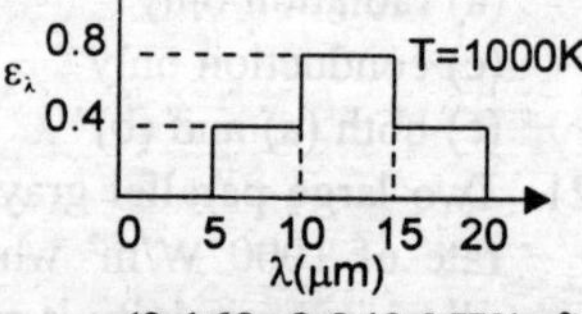

$$(1.378 \times 10^{-3} \text{W}, \ 13.78 \text{W/m}^2)$$

2. The spectral distribution of the radiation emitted by a diffuse surface may be approximated as follows:

    Calculate the following:

    (a) total hemispherical emissivity

    (b) total emissive power

    $$(0.163, \ 9.242 \text{ kW/m}^2)$$

3. A certain plastic has a transmissivity of 0.85 for wavelength between 0.5 μm and 3.5 μm and has a transmissivity of zero for all other wavelengths. Estimate the fraction of incident solar energy that will be transmitted through the plastic.   **(64.5%)**

4. A cylindrical storage tank, 1 m in diameter and 1.2 m long, has an outside temperature at the surface as 60°C and an emissivity of 0.9. Calculate the rate of heat loss by radiation when the tank is placed in a large room, the walls of which are of 15°C. Determine the reduction in the rate of heat loss if the tank is painted with an aluminium paint of emissivity 0.4.

    (1475 W, 820 W)

5. A gas turbine combustion chamber can be approximated as a cylinder of 0.3 m diameter. The temperature of the steel surface (emissivity 0.79) is 773 K and the surroundings temperature is 298 K. Estimate the percentage heat loss by radiation when the combustion chamber is enclosed in another cylindrical shell 0.6 m in diameter, the inside and outside surfaces of which are painted with aluminium paint of emissivity 0.4.   **(61.3%)**

6. A hot water radiator has overall dimensions 2 m × 1 m × 0.2 m, is used to heat a room at 18°C. The surface temperature of the radiator is 60°C and its surface can be treated as black. Estimate the radiation heat transfer coefficient. If the effective heat transfer surface area for convection is 2.5 times of the overall dimensions and the convective heat transfer coefficient is 4.55 W/m^2 K, calculate the total loss from the radiator surface.   **(6.92 W/m^{2}K, 4 kW)**

7. The outside wall of a house consists of two brick walls separated by an air gap. Calculate the heat energy exchange by radiation across the gap when the inner surface temperatures are 286 K and 280 K. The emissivity of both surfaces is 0.9.

If a thin aluminium sheet (emissivity 0.1) is placed in the air gap, estimate the radiation heat transfer across the gap assuming that the surface temperature of the walls remain the same. **(25.2 W/m² 1.525 W/m²)**

8. Two rectangles 75 cm × 75 cm are placed perpendicularly with a common edge. One surface has T_1 = 100 K, ϵ = 0.5 while the other surface is insulated and in thermal equilibrium with a large surrounding room at 300 K. Determine the temperature of the insulated surface and the heat lost by the surface 1.

(878 K, 15.489 kW)

9. Two parallel planes 1.2 by 1.2 m are separated by a distance of 1.2 m. The emissivities of the planes are 0.4 and 0.6 and the temperatures are 760 and 300 C, respectively. A 1.2 by 1.2 m radiation shield having an emissivity of 0.05 on both sides is located equidistant between the two planes. Calculate (a) the heat transfer rate from each of the two planes with and without the shield, and (b) the temperature of the shield. The plates are placed in a large room maintained at 300 K.

10. Two long concentric cylinders have diameters of 30 and 10 cm respectively. The inside cylinder is at 1000 K and the outer cylinder is at 300 K. Calculate the percentage reduction in the heat transfer if a cylindrical shield having a diameter of 20 cm is placed between the two cylinders. Take the emissivity of each surface as 0.05. **(41.65%)**

11. A thermocouple is used to measure the temperature of a hot gas flowing in a tube maintained at 100°C. The thermocouple indicates a temperature of 500°C. If the emissivity of the thermocouple junction is 0.5 and the convective heat transfer coefficient is 250 W/m²K. Calculate the true temperature of the gas. **(538⁰C)**

12. A thermocouple indicates a temperature of 800°C when placed a tube where the hot gases at 870°C are flowing. If the convective heat transfer coefficient between the thermocouple and the gas is 60 W/m²K, estimate the temperature of the pipe. If the pipe wall temperature were 700°C. Calculate the value of the convective heat transfer coefficient.

13. A gas turbine gas chamber 50 cm in diameter and its wall are maintained at 800°C. The products of combustion are at 1400°C, a pressure of 1 atmosphere, and contain 10 percent by volume of CO_2 and 20 percent by volume of H_2O. Assuming the combustion chamber to be a very long cylinder determine the net radiant energy exchange between the gases and the combustion chamber walls.

14. A gas mixture in a cubical volume of 0.25 m each side contains 12% CO_2 and 12% H_2O by volume at a total pressure of 1 atm and a temperature of 900°C. Calculate the emissivity of the mixture. If the gas mixture is confined in a black enclosure whose walls are at 500°C. Find the net radiant heat exchange per square metre of wall area. **(3.9 kW)**

CHAPTER 10

Heat Exchangers

1. Heat Exchangers: Regenerators and Recuperators

A heat exchanger is an equipment where heat energy is transferred from a hot fluid to a colder fluid. The transfer of heat energy between the two fluids could be carried out (i) either by direct mixing of the two fluids and the mixed fluids leave at an intermediate temperature determined from the principles of conservation of energy, (ii) or by transmission through a wall separating the two fluids. The former types are called direct contact heat exchangers such as water cooling towers and jet condensers. The latter types are called regenerators, recuperators or surface exchangers.

In a regenerator, hot and cold fluids alternately flow over a surface which provides alternately a sink and source for heat flow. Fig. 10.1 (a) shows a cylinder containing a matrix that rotates in such a way that it passes alternately

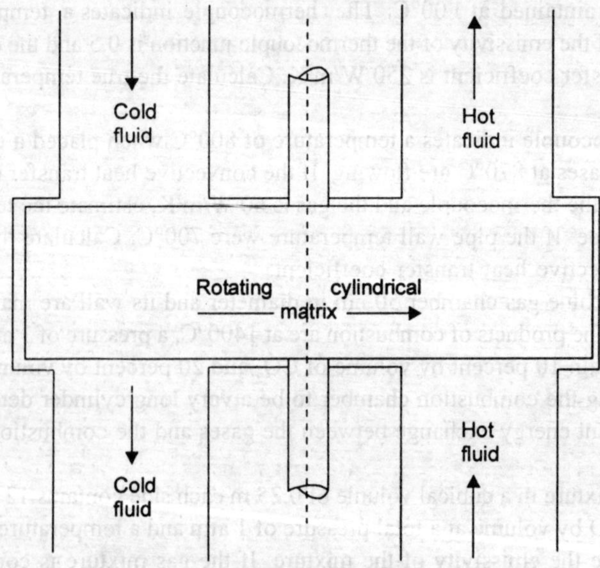

Fig. 10.1 (a) Rotating matrix regenerator

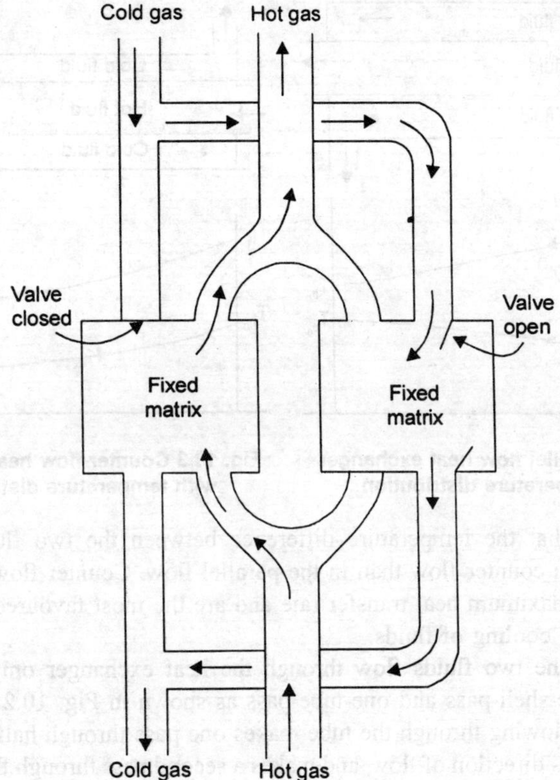

Fig. 10.1 (b) Stationary matrix regenerator

through cold and hot gas streams which are sealed from each other. Fig. 10.1 (b) shows a stationary matrix regenerator in which hot and cold gases flow through them alternately.

In a recuperator, hot and cold fluids flow continuously following the same path. The heat transfer process consists of convection between the fluid and the separating wall, conduction through the wall and convection between the wall and the other fluid. Most common heat exchangers are of recuperative type having a wide variety of geometries:

2. Classification of Heat Exchangers

Heat exchangers are generally classified according to the relative directions of hot and cold fluids:

(a) *Parallel Flow* – the hot and cold fluids flow in the same direction. Fig. 10.2 depicts such a heat exchanger where one fluid (say hot) flows through the pipe and the other fluid (cold) flows through the annulus.

(b) *Counter Flow* – the two fluids flow through the pipe but in opposite directions. A common type of such a heat exchanger is shown in Fig. 10.3. By comparing the temperature distribution of the two types of heat exchanger,

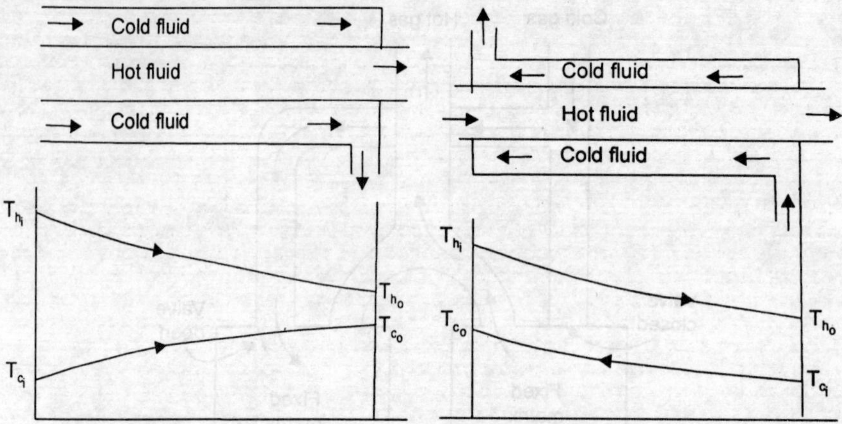

Fig. 10.2 Parallel flow heat exchanger with temperature distribution

Fig. 10.3 Counter-flow heat exchanger with temperature distribution

we find that the temperature difference between the two fluids is more uniform in counter flow than in the parallel flow. Counter flow exchangers give the maximum heat transfer rate and are the most favoured devices for heating or cooling of fluids.

When the two fluids flow through the heat exchanger only once, it is called one-shell-pass and one-tube-pass as shown in Fig. 10.2 and 10.3. If the fluid flowing through the tube makes one pass through half of the tube, reverses its direction of flow, and makes a second pass through the remaining half of the tube, it is called 'one-shell-pass, two-tube-pass' heat exchanger, Fig. 10.4. Many other possible flow arrangements exist and are being used. Fig. 10.5 depicts a 'two-shell-pass, four-tube-pass' exchanger.

(c) *Cross-flow* – A cross-flow heat exchanger has the two fluid streams flowing at right angles to each other. Fig. 10.6 illustrates such an arrangement. An automobile radiator is a good example of cross-flow exchanger. These exchangers are 'mixed' or 'unmixed' depending upon the mixing or not mixing of either fluid in the direction transverse to the direction of the flow stream and the analysis of this type of heat exchanger is extremely complex because of the variation in the temperature of the fluid in and normal to the direction of flow.

(d) *Condenser and Evaporator* – In a condenser, the condensing fluid temperature remains almost constant throughout the exchanger and the temperature of the colder fluid gradually increases from the inlet to the exit, Fig. 10.7 (a). In an evaporator, the temperature of the hot fluid gradually decreases from the inlet to the outlet whereas the temperature of the colder fluid remains the same during the evaporation process, Fig. 10.7 (b). Since the temperature of one of the fluids can be treated as constant, it is immaterial whether the exchanger is parallel flow or counter flow.

(e) *Compact Heat Exchangers* – these devices have close arrays of finned tubes or plates and are typically used when atleast one of the fluids is a gas. The tubes are either flat or circular as shown in Fig. 10.8 and the fins may

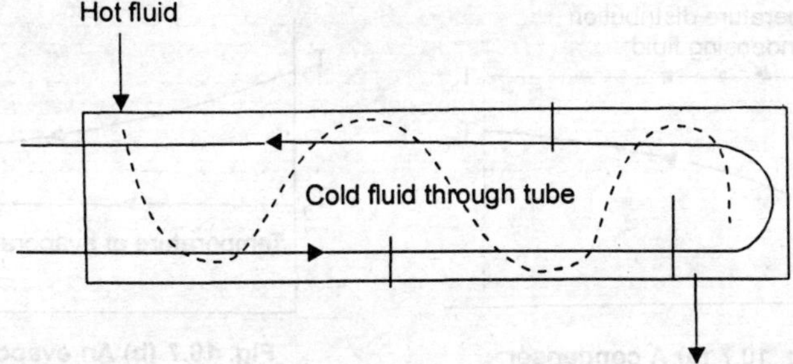

Fig. 10.4 Multi-pass exchanger one-shell pass, two-tube passes

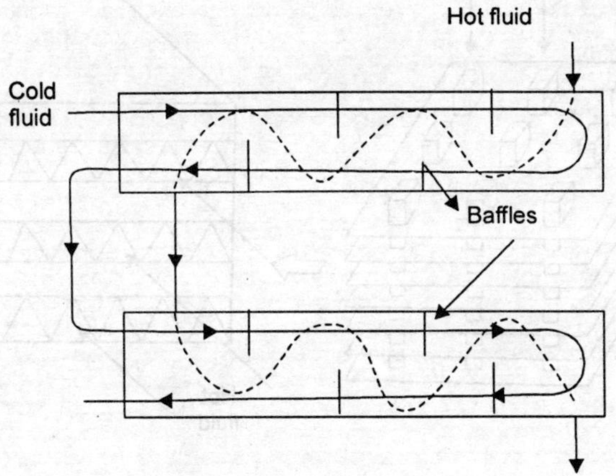

Fig. 10.5 Two-shell passes, four-tube passes heat exchanger (baffles increase the convection coefficient of the shell side fluid by inducing turbulence and a cross-flow velocity component)

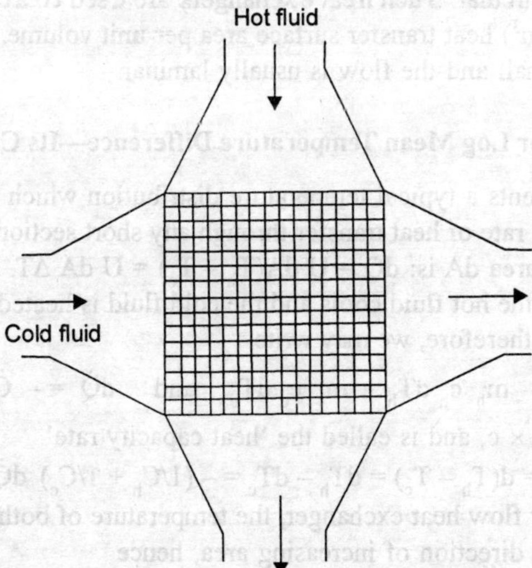

Fig. 10.6 A cross-flow exchanger

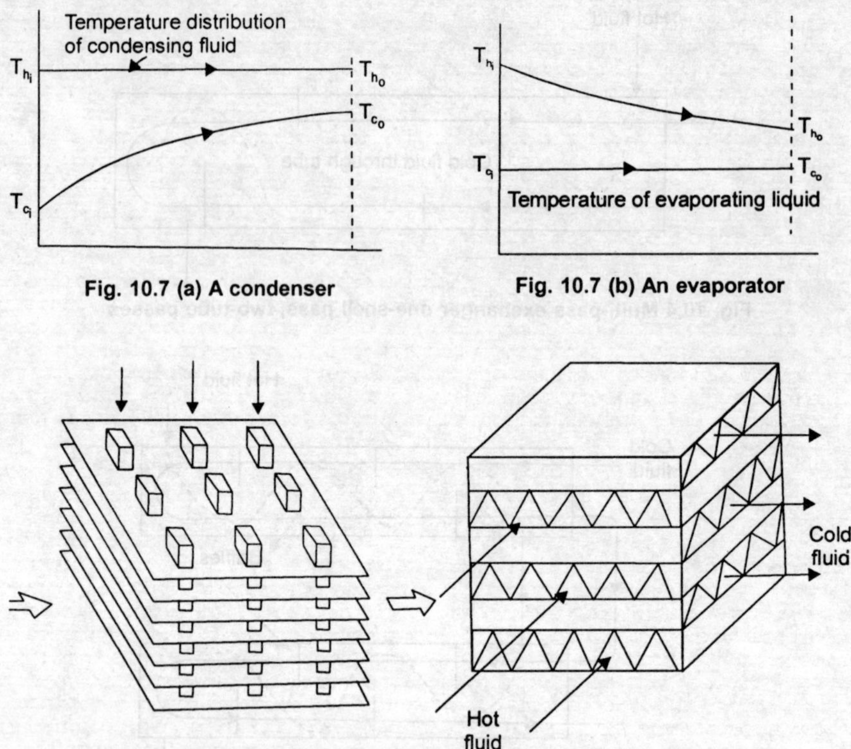

Fig. 10.7 (a) A condenser   Fig. 10.7 (b) An evaporator

Fig. 10.8 Compact heat exchangers: (a) flat tubes, continuous plate fins,
(b) plate fin (single pass)

be flat or circular. Such heat exchangers are used to achieve a very large (≥ 700 m²/m³) heat transfer surface area per unit volume. Flow passages are typically small and the flow is usually laminar.

3. Expression for Log Mean Temperature Difference—Its Characteristics

Fig. 10.9 represents a typical temperature distribution which is obtained in heat exchangers. The rate of heat transfer through any short section of heat exchanger tube of surface area dA is: $dQ = U\,dA(T_h - T_c) = U\,dA\,\Delta T$. For a parallel flow heat exchanger, the hot fluid cools and the cold fluid is heated in the direction of increasing area, therefore, we may write

$$d\dot{Q} = -\,\dot{m}_h\,c_h\,dT_h = \dot{m}_c\,c_c\,dT_c \quad \text{and} \quad d\dot{Q} = -\,\dot{C}_h\,dT_h = +\,\dot{C}_c\,dT_c$$

where $\dot{C} = \dot{m} \times c$, and is called the 'heat capacity rate'.

Thus, $d(\Delta T) = d(T_h - T_c) = dT_h - dT_c = -(1/C_h + 1/C_c)\,d\dot{Q}$ \hfill (10.1)

For a counter flow heat exchanger, the temperature of both hot and cold fluid decreases in the direction of increasing area, hence

$$d\dot{Q} = -\,\dot{m}_h\,c_h\,dT_h = -\,\dot{m}_c\,c_c\,dT_c, \quad \text{and} \quad d\dot{Q} = -\,C_h\,dT_h = -\,C_c\,dT_c$$

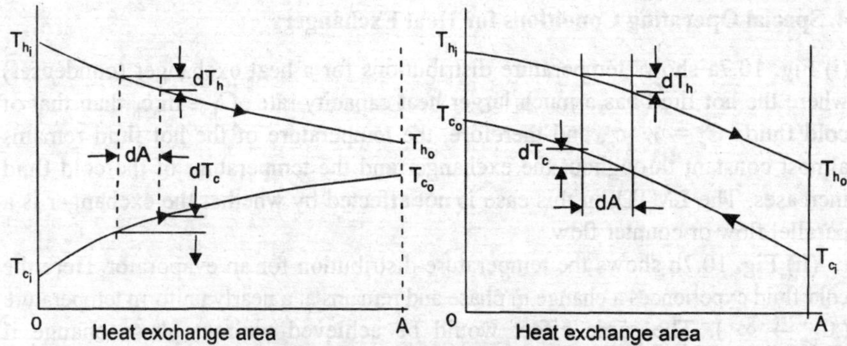

Fig. 10.9 Parallel flow and Counter flow heat exchangers and the temperature distribution with length

or, $\qquad d(\Delta T) = dT_h - dT_c = - (1/C_h - 1/C_c)\, d\dot{Q}$ (10.2)

Integrating equations (10.1) and (10.2) between the inlet and outlet. and assuming that the specific heats are constant, we get

$$- (1/C_h \pm 1/C_c)\, \dot{Q} = \Delta T_o - \Delta T_i$$ (10.3)

The positive sign refers to parallel flow exchanger, and the negative sign to the counter flow type. Also, substituting for dQ in equations (10.1) and (10.2) we get

$$- (1/C_h \pm 1/C_c)\, U\, dA = d(\Delta T) / \Delta T$$ (10.3a)

Upon integration between inlet i and outlet o and assuming U as a constant,

We have $\qquad - (1/C_h \pm 1/C_c) U\, A = 1n\, (\Delta T_o / \Delta T_i)$ (10.4)

By dividing (10.3) by (10.4), we get

$$\dot{Q} = U\, A\, [(\Delta T_o - \Delta T_i) / \ln(\Delta T_o / \Delta T_i)]$$ (10.5)

Thus the mean temperature difference is written as

Log Mean Temperature Difference, $\text{LMTD} = (\Delta T_o - \Delta T_i) / \ln(\Delta T_o / \Delta T_i)$ (10.6)

(The assumption that U is constant along the heat exchanger is never strictly true but it may be a good approximation if at least one of the fluids is a gas. For a gas, the physical properties do not vary appreciably over moderate range of temperature and the resistance of the gas film is considerably higher than that of the metal wall or the liquid film, and the value of the gas film resistance effectively determines the value of the overall heat transfer coefficient U.)

It is evident from Fig.10.9 that for parallel flow exchangers, the final temperature of fluids lies between the initial values of each fluid whereas in counter flow exchanger, the temperature of the colder fluid at exit is higher than the temperature of the hot fluid at exit. Therefore, a counter flow exchanger provides a greater temperature range, and the LMTD for a counter flow exchanger will be higher than for a given rate of mass flow of the two fluids and for given temperature changes, a counter flow exchanger will require less surface area.

4. Special Operating Conditions for Heat Exchangers

(i) Fig. 10.7a shows temperature distributions for a heat exchanger (condenser) where the hot fluid has a much larger heat capacity rate, $\dot{C}_h = \dot{m}_h c_h$ than that of cold fluid, $\dot{C}_c = \dot{m}_c c_c$, and therefore, the temperature of the hot fluid remains almost constant throughout the exchanger and the temperature of the cold fluid increases. The LMTD, in this case is not affected by whether the exchanger is a parallel flow or counter flow.

(ii) Fig. 10.7b shows the temperature distribution for an evaporator. Here the cold fluid experiences a change in phase and remains at a nearly uniform temperature ($\dot{C}_c \rightarrow \infty$). The same effect would be achieved without phase change if $\dot{C}_c >> \dot{C}_h$, and the LMTD will remain the same for both parallel flow and counter flow exchangers.

(iii) In a counter flow exchanger, when the heat capacity rate of both the fluids are equal, $\dot{C}_c = \dot{C}_h$, the temperature difference is the same all along the length of the tube. And in that case, LMTD should be replaced by $\Delta T = \Delta T_a = \Delta T_b$, and the temperature profiles of the two fluids along its length would be parallel straight lines.

(Since $\quad d\dot{Q} = -\dot{C}_c\, dT_c = -\dot{C}_h\, dT_h;\quad dT_c = -\,d\dot{Q}/\dot{C}_c,\quad$ and $dT_h = -\,d\dot{Q}/\dot{C}_h.$

and, $\quad dT_c - dT_h = d\theta = -\,dQ\,(1/\dot{C}_c - 1/\dot{C}_h) = 0,\quad$ (because $\dot{C}_c = \dot{C}_h$)

Or, $\quad d\theta = 0$, gives $\theta =$ constant and the temperature profiles of the two fluids along its length would be parallel straight lines.)

5. LMTD for Cross-flow Heat Exchangers

LMTD given by Eq (10.6) is strictly applicable to either parallel flow or counter flow exchangers. When we have multipass parallel flow or counter flow or cross flow exchangers, LMTD is first calculated for single pass counter flow exchanger and the mean temperature difference is obtained by multiplying the LMTD with a correction factor F which takes care of the actual flow arrangement of the exchanger. Or,

$$\dot{Q} = U\,A\,F\,(LMTD) \tag{10.7}$$

The correction factor F for different flow arrangements are obtained from charts given in Fig. 10.10 (a, b, c, d).

6. Fouling Factors in Heat Exchangers

Heat exchanger walls are usually made of single materials. Sometimes the walls are bimetallic (steel with aluminium cladding) or coated with a plastic as a protection against corrosion, because, during normal operation surfaces are subjected to fouling by fluid impurities, rust formation, or other reactions between the fluid and the wall material. The deposition of a film or scale on the surface greatly increases the resistance to heat transfer between the hot and cold fluids. And, a scale coefficient of heat transfer h_s is defined as:

$$R_s = 1/h_s A, \text{ °C/W or K/W}$$

where A is the area of the surface before scaling began and $1/h_s$ is called 'Fouling

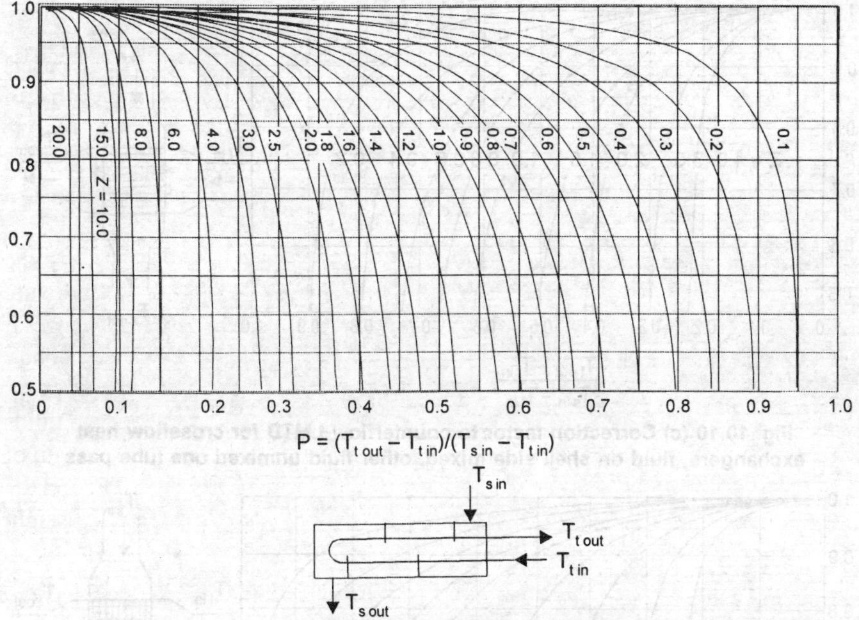

Fig. 10.10(a) Correction factor to counter flow LMTD for heat exchanger with one shell pass and two, or a multiple of two, tube passes

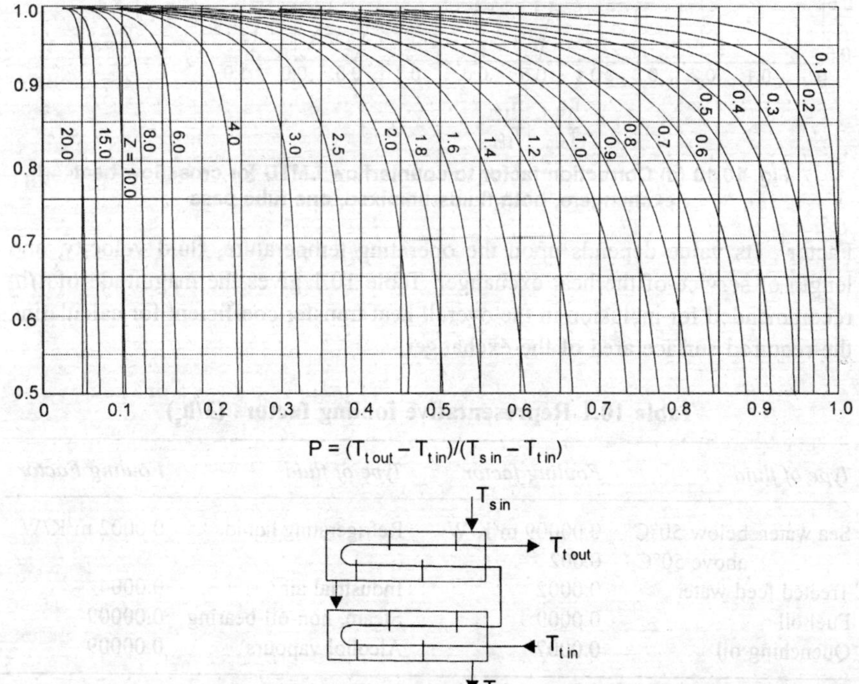

Fig. 10.10 (b) Correction factor to counter flow LMTD for heat exchanger with two shell passes and a multiple of two tube passes

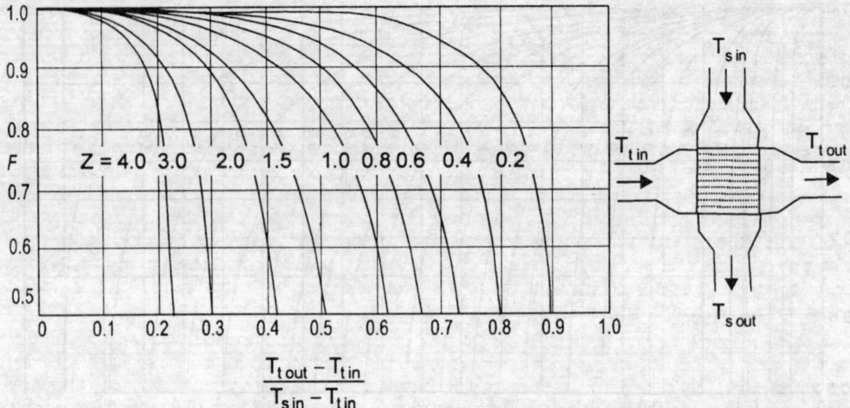

Fig. 10.10 (c) Correction factor to counterflow LMTD for crossflow heat exchangers, fluid on shell side mixed, other fluid unmixed one tube pass

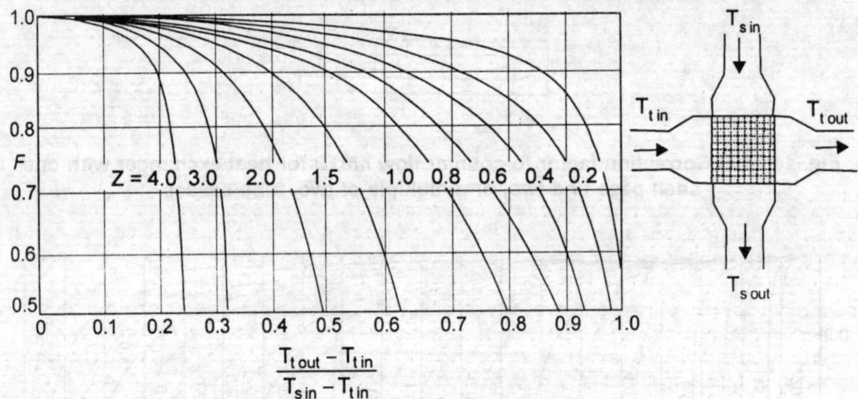

Fig. 10.10 (d) Correction factor to counterflow LMTD for crossflow heat exchangers, both fluids unmixed, one tube pass

Factor'. Its value depends upon the operating temperature, fluid velocity, and length of service of the heat exchanger. Table 10.1 gives the magnitude of $1/h_s$ recommended for inclusion in the overall heat transfer coefficient for calculating the required surface area of the exchanger.

Table 10.1 Representative fouling factors ($1/h_s$)

Type of fluid	Fouling factor	Type of fluid	Fouling Factor
Sea water below 50°C	0.00009 m²K/W	Refrigerating liquid	0.0002 m²K/W
above 50°C	0.002		
Treated feed water	0.0002	Industrial air	0.0004
Fuel oil	0.0009	Steam, non-oil-bearing	0.00009
Quenching oil	0.0007	Alcohol vapours	0.00009

However, fouling factors must be obtained experimentally by determining the values of U for both clean and dirty conditions in the heat exchanger.

7. The Overall Heat Transfer Coefficient

The determination of the overall heat transfer coefficient is an essential, and often the most uncertain, part of any heat exchanger analysis. We have seen that if the two fluids are separated by a plane composite wall the overall heat transfer coefficient is given by:

$$1/U = (1/h_i) + (L_1/k_1) + (L_2/k_2) + (1/h_o) \qquad (10.8)$$

If the two fluids are separated by a cylindrical tube (inner radius r_i, outer radius r_o), the overall heat transfer coefficient is obtained as:

$$1/U_i = (1/h_i) + (r_i/k)\ln(r_o/r_i) + (r_i/r_o)(1/h_o) \qquad (10.9)$$

where h_i and h_o are the convective heat transfer coefficients at the inside and outside surfaces and U_i is the overall heat transfer coefficient based on the inside surface area. Similarly, for the outer surface area, we have:

$$1/U_o = (1/h_o) + (r_o/k)\ln(r_o/r_i) + (r_o/r_i)(1/h_i) \qquad (10.10)$$

and $U_i A_i$ will be equal to $U_o A_o$; or, $U_i r_i = U_o r_o$.

The effect of scale formation on the inside and outside surfaces of the tubes of a heat exchanger would be to introduce two additional thermal resistances to the heat flow path. If h_{si} and h_{so} are the two heat transfer coefficients due to scale formation on the inside and outside surface of the inner pipe, the rate of heat transfer is given by

$$Q = (T_i - T_o)/[(1/h_iA_i) + 1/h_{si}A_i + \ln(r_o/r_i)/2\pi Lk + 1/h_{so}A_o + (1/h_oA_o)] \quad (10.11)$$

where T_i and T_o are the temperature of the fluid at the inside and outside of the tube. Thus, the overall heat transfer coefficient based on the inside and outside surface area of the tube would be:

$$1/U_i = 1/h_i + 1/h_{si} + (r_i/k)\ln(r_o/r_i) + (r_i/r_o)(1/h_{so}) + (r_i/r_o)(1/h_o); \quad (10.12)$$

and

$$1/U_o = (r_o/r_i)(1/h_i) + (r_o/r_i)(1/h_{si}) + \ln(r_o/r_i)(r_o/k) + 1/h_{so} + 1/h_o.$$

Example 10.1 In a parallel flow heat exchanger water flows through the inner pipe and is heated from 25°C to 75°C. Oil flowing through the annulus is cooled from 210°C to 110°C. It is desired to cool the oil to a lower temperature by increasing the length of the tube. Estimate the minimum temperature to which the oil can be cooled.

Solution: By making an energy balance, heat received by water must be equal to the heat given out by oil.

$$\dot{m}_w c_w (75 - 25) = \dot{m}_o c_o (210 - 110); \quad \dot{C}_w / \dot{C}_o = 100/50 = 2.0$$

In a parallel flow heat exchanger, the minimum temperature to which oil can be cooled will be equal to the maximum temperature to which water can be heated, Fig. 10.2: $(T_{ho} = T_{co})$

therefore, $C_w (T - 25) = C_o (210 - T)$;

$$(T - 25)/(210 - T) = 1/2 = 0.5; \qquad \text{or,} \quad T = 260/3 = 86.67°C.$$

For the same capacity rates the oil can be cooled to 25°C (equal to the water inlet temperature) in a counter-flow arrangement.

Example 10.2 Water at the rate of 1.5 kg/s is heated from 30°C to 70°C by an oil (specific heat 1.95 kJ/kg C). Oil enters the exchanger at 120°C and leaves the exchanger at 80°C. If the overall heat transfer coefficient remains constant at 350 W/m²°C, calculate the heat exchange area for (i) parallel-flow, (ii) counter-flow, and (iii) cross-flow arrangement.

Solution: Energy absorbed by water, $\dot{Q} = \dot{m}_w c_w (\Delta T) = 1.5 \times 4.182 \times 40 = 250.92$ kW

(i) Parallel flow: Fig. 10.9; $\Delta T_a = 120 - 30 = 90$; $\Delta T_b = 80 - 70 = 10$

$$LMTD = (90 - 10)/\ln(90/10) = 36.4;$$
$$Area = Q/U \, (LMTD) = 250920/(350 \times 36.4) = 19.69 \text{ m}^2.$$

(ii) Counter flow: Fig 10.9; $\Delta T_a = 120 - 70 = 50$, $\Delta T_b = 80 - 30 = 50$

Since $\Delta T_a = \Delta T_b$, LMTD should be replaced by $\Delta T = 50$
$$Area \, A = \dot{Q}/U \, (\Delta T) = 250920/(350 \times 50) = 14.33 \text{ m}^2.$$

(iii) Cross flow: assuming both fluids unmixed – Fig. 10.10d

using the nomenclature of the figure and assuming that water flows through the tubes and oil flows through the shell,

$$P = (T_{to} - T_{ti})/(T_{si} - T_{ti}) = (70 - 30)/(120 - 30) = 0.444$$
$$Z = (T_{si} - T_{so})/(T_{to} - T_{ti}) = (120 - 80)/(70 - 30) = 1.0$$

and the correction factor, $F = 0.93$
$$\dot{Q} = U A F (\Delta T); \quad \text{or} \quad Area \, A = 250920/(350 \times 0.93 \times 50) = 15.41 \text{ m}^2.$$

Example 10.3 0.5 kg/s of exhaust gases flowing through a heat exchanger are cooled from 400°C to 120°C by water initially at 25°C. The specific heat capacities of exhaust gases and water are 1.15 and 4.19 kJ/kgK respectively, and the overall heat transfer coefficient from gases to water is 150 W/m²K. If the cooling water flow rate is 0.7 kg/s, calculate the surface area when (i) parallel-flow (ii) cross-flow with exhaust gases flowing through tubes and water is mixed in the shell.

Solution: The heat given out by the exhaust gases is equal to the heat gained by water.

or, $0.5 \times 1.15 \times (400 - 120) = 0.7 \times 4.19 \times (T - 25)$

Therefore, the temperature of water at exit, $T = 79.89$°C

For parallel-flow: $\Delta T_a = 400 - 25 = 375$; $\Delta T_b = 120 - 79.89 = 40.11$
$$LMTD = (375 - 40.11)/\ln(375/40.11) = 149.82$$
$$\dot{Q} = 0.5 \times 1.15 \times 280 = 161000 \text{ W};$$

Therefore Area $A = 161000/(150 \times 149.82) = 7.164 \text{ m}^2.$

For cross-flow: $\dot{Q} = U A F (LMTD);$

and LMTD is calculated for counter-flow system.

$$\Delta T_a = (400 - 79.89) = 320.11; \quad \Delta T_b = 120 - 25 = 95$$
$$LMTD = (320.11 - 95)/\ln(320.11/95) = 185.3.$$

Using the nomenclature of Fig 10.10c,

$$P = (120 - 400)/(25 - 400) = 0.747$$
$$Z = (25 - 79.89)/(120 - 400) = 0.196 \qquad \therefore \quad F = 0.92$$

and the area $A = 161000/(150 \times 0.92 \times 185.3) = 6.296 \text{ m}^2.$

Example 10.4 In a certain double pipe heat exchanger hot water flows at a rate of 5000 kg/h and gets cooled from 95°C to 65°C. At the same time 5000 kg/h of cooling water enters the heat exchanger. The overall heat transfer coefficient is 2270 W/m²K. Calculate the heat transfer area and the efficiency assuming two streams are in (i) parallel flow (ii) counter flow. Take C_p for water as 4.2 kJ/kgK, cooling water inlet temperature 30°C

Solution: By making an energy balance:

Heat lost by hot water $= 5000 \times 4.2 \times (95 - 65)$

$\qquad\qquad$ = heat gained by cold water

$\qquad\qquad = 5000 \times 4.2 \times (T - 30)$

$\qquad\qquad\therefore\qquad\quad T = 60°C$

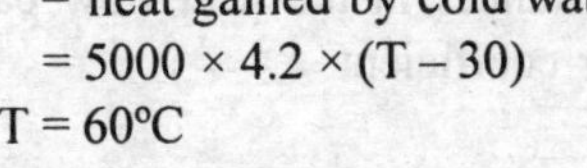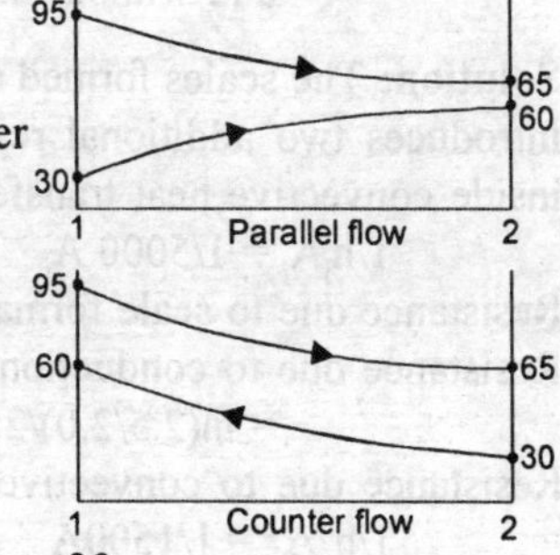

(i) Parallel flow

$\quad \theta_1 = (95 - 30) = 65$

$\quad \theta_2 = (65 - 60) = 5$

$\quad$ LMTD $= (65 - 5)/\ln(65/5) = 23.4$

Area, $\quad A = \dot{Q}/(U \times \text{LMTD}) = \dfrac{5000 \times 4.2 \times 10^3 \times 30}{3600 \times 2270 \times 23.4} = 3.295 \text{ m}^2$

(ii) Counter flow: $\qquad \theta_1 = (95 - 60) = 35$

$\qquad\qquad\qquad\qquad\quad \theta_2 = (65 - 30) = 35$

$\qquad\qquad\qquad\quad$ LMTD $= \Delta T = 35$

Area $\qquad\qquad A = 500 \times 4200 \times 30/(3600 \times 2270 \times 35) = 2.2 \text{ m}^2.$

ϵ, Efficiency = Actual heat transferred/Maximum heat that could be transferred.
Therefore, for parallel flow, $\epsilon = (95 - 65)/(95 - 60) = 0.857$
For counter flow, $\qquad\qquad \epsilon = (95 - 65)/(95 - 30) = 0.461.$

Example 10.5 The flow rates of hot and cold water streams running through a double pipe heat exchanger (inside and outside diameter of the tube 80 mm and 100 mm) are 2 kg/s and 4 kg/s. The hot fluid enters at 75°C and comes out at 45°C. The cold fluid enters at 20°C. If the convective heat transfer at the inside and outside surface of the tube is 150 and 180 W/m²K, thermal conductivity of the tube material 40 W/mK, calculate the area of the heat exchanger assuming counter flow.

Solution: Let T is the temperature of the cold water at outlet.
By making an energy balance, $\dot{Q} = m_h\, c_h\, (T_{h1} - T_{h2}) = \dot{m}_c\, c_c\, (T_{c2} - T_{c1})$

since $\quad c_h = c_c, = 4.2 \text{ kJ/kgK}; 2 \times (75 - 45) = 4 \times (T - 20); \qquad T = 35°C,$

and $\qquad \dot{Q} = 252 \text{ kW}$

for counter flow: $\theta_1 = (75 - 35) = 40; \quad \theta_2 = (45 - 20) = 25$

$\qquad\qquad$ LMTD $= (40 - 25)/\ln(40/25) = 31.91$

overall heat transfer coefficient based in the inside surface of tube

$\qquad 1/U = (1/h_i) + (r_i/k)\ln(r_o/r_i) + (r_o/r_i)\,(1/h_o)$

$\qquad\qquad = 1/150 + (0.04/40)\ln(50/40) + (50/40)(1/180) = 0.0138$

and $\qquad U = 72.28$

area $\qquad A = \dot{Q}/(U \times \text{LMTD}) = 252 \times 10^3/(72.28 \times 31.91) = 109.26 \text{ m}^2.$

Example 10.6 Water flows through a copper tube (k = 350 W/mK, inner and outer diameter 2.0 cm and 2.5 cm respectively) of a double pipe heat exchanger. Oil flows through the annulus between this pipe and steel pipe. The convective heat transfer coefficient on the inside and outside of the copper tube are 5000 and 1500 W/m^{2}K. The fouling factors on the water and oil sides are 0.0022 and 0.00092 K/W. Calculate the overall heat transfer coefficient with and without the fouling factor.

Solution: The scales formed on the inside and outside surface of the copper tube introduces two additional resistances in the heat flow path. Resistance due to inside convective heat transfer coefficient

$$1/h_i A_i = 1/5000\ A_i$$

Resistance due to scale formation on the inside = $1/h_s A_i$ = 0.0022

Resistance due to conduction through the tube wall = $\ln(r_o/r_i)/2\pi Lk$

$$= \ln(2.5/2.0)/2\pi \times L \times 350 = 1.014 \times 10^{-4}/L$$

Resistance due to convective heat transfer on the outside

$$1/h_o A_o = 1/1500 A_o$$

Resistance due to scale formation on the outside = $1/h_s A_o$ = 0.00092

Since $\qquad Q = \Delta T/\Sigma R = U_i\ A_i\ (\Delta T) = \Delta T/(1/U_i A_i)$; we have

(a) With fouling factor:-

Overall heat transfer coefficient based on the inside pipe surface

$\quad U_i = 1/(1/5000 + \pi \times 0.02(0.0022 + 0.00092) + 0.02\pi \times 1.014 \times 10^{-4} + 8.33 \times 10^{-4})$

$\quad\quad = 809.47$ W/m^{2}K per metre length of pipe

(b) Without fouling factor

$\quad\quad U_i = 1/(1/5000 + 0.02\pi \times 1.014 \times 10^{-4} + 8.33 \times 10^{-4})$

$\quad\quad\quad = 962.12$ W/m^{2}K per m of pipe length

The heat transfer rate will reduce by $(962.12 - 809.47)/962.12 = 15.9$ percent when fouling factor is considered.

Example 10.7 In a surface condenser, dry and saturated steam at 50°C enters at the rate of 1kg/s. The circulating water enters the tube, (25 mm inside diameter, 28 mm outside diameter, k = 300 W/mK) at a velocity of 2 m/s. If the convective heat transfer coefficient on the outside surface of the tube is 5500 W/m^{2}K, the inlet and outlet temperatures of water are 25°C and 35°C respectively, calculate the required surface area.

Solution: For calculating the convective heat transfer coefficient on the inside surface of the tube, we calculate the Reynolds number on the basis of properties of water at the mean temperature of 30°C. The properties are:

$\quad\quad \mu = 0.001$ Pa–s, $\rho = 1000$ kg/m^3, $k = 0.6$ W/mK, $\quad h_{fg}$ at 50°C = 2375 kJ/kg

$\quad\quad$ Re = $\rho VD/\mu = 10^3 \times 2 \times 0.025/0.001 = 50{,}000$, a turbulent flow. Pr = 7.0.

The heat transfer coefficient at the inside surface can be calculated by:

$\quad\quad$ Nu = 0.023 Re$^{0.8}$ Pr$^{0.3}$ = 0.023 (50000)$^{0.8}$ (7)$^{0.3}$ = 236.828

and $\ h_i = 236.828 \times 0.6/0.025 = 5684$ W/m^{2}K.

The overall heat transfer coefficient based on the outer diameter,

$$U = 1/(0.028/(0.025 \times 5684) + 1/5500 + 0.014 \ln(28/25)/300)$$
$$= 2603.14 \text{ W/m}^2\text{K}$$
$$\Delta T_a = (50 - 25) = 25; \quad \Delta T_b = (50 - 35) = 15;$$
$$\therefore \quad \text{LMTD} = (25 - 15)/\ln(25/15) = 19.576.$$

Assuming one shell pass and one tube pass, Q = UA (LMTD)

or $\qquad A = 2375 \times 10^3/(2603.14 \times 19.576) = 46.6 \text{ m}^2$

Mass of circulating water = $Q/(c_p \Delta T) = 2375/(4.182 \times 10) = 56.79$ kg/s

also, $\quad \dot{m}_w = \rho \times \text{area} \times V \times n,$ where n is the number of tubes
$$n = 56.79 \times 4/(2 \times \pi \times 0.025 \times 0.025 \times 1000) \equiv 58 \text{ tubes}$$

Surface area, $46.6 = n \times \pi \times d \times L$

and $\qquad L = 46.6/(58 \times \pi \times 0.025) = 10.23$ m. Hence more than one pass should be used.

Example 10.8 A heat exchanger is used to heat water from 20°C to 50°C when thin walled water tubes (inner diameter 25 mm, length 15 m) are laid beneath a hot spring water pond, temperature 75°C. Water flows through the tubes with a velocity of 1 m/s. Estimate the required overall heat transfer coefficient and the convective heat transfer coefficient at the outer surface of the tube.

Solution: Water flow rate, $\dot{m} = \rho \times V \times A = 10^3 \times 1 \times (\pi/4)(0.025)^2 = 0.49$ kg/s.

Heat transferred to water, $Q = \dot{m} c (\Delta T) = 0.49 \times 4200 \times 30 = 61740$ W.

Since the temperature of the water in the hot spring is constant,
$$\theta_1 = (75 - 20) = 55; \quad \theta_2 = (75 - 50) = 25;$$
$$\text{LMTD} = (55-25)/\ln(55/25) = 38$$

Overall heat transfer coefficient, U = Q/ (A × LMTD)
$$= 61740/(38 \times \pi \times 0.025 \times 15)$$
$$= 1378.94 \text{ W/m}^2\text{K}.$$

The properties of water at the mean temperature (20 + 50)/2 = 35°C are:
$$\mu = 0.001 \text{ Pa-s}, \quad k = 0.6 \text{ W/mK} \quad \text{and} \quad Pr = 7.0$$

Reynolds number, Re = $\rho Vd/\mu = 1000 \times 1.0 \times 0.25/0.001 = 25000$, turbulent flow
$$Nu = 0.023 \, (Re)^{0.8} \, (Pr)^{0.33}$$
$$= 0.023 \, (25000)^{0.8} \times (7)^{0.33} = 144.2$$

and $\qquad h_i = 144.2 \times k/d = 144.2 \times 0.6/0.025 = 3460.8 \text{ W/m}^2\text{K}.$

Neglecting the resistance of the thin tube wall,
$$1/U = 1/h_i + 1/h_o; \qquad\qquad \therefore 1/h_o = 1/1378.94 - 1/3460.8$$
or, $\qquad h_o = 2292.3 \text{ W/m}^2\text{K}.$

Example 10.9 A hot fluid at 200°C enters a heat exchanger at a mass rate of 10000 kg/h. Its specific heat is 2000 J/kg K. It is to be cooled by another fluid entering at 25°C with a mass flow rate 2500 kg/h and specific heat 400 J/kgK. The overall heat transfer coefficient based on outside area of 20 m² is 250 W/m²K. Find the exit temperature of the hot fluid when the fluids are in parallel flow.

Solution: From Eq(10.3a), $-U\, dA(1/C_h + 1/C_c) = d(\Delta T)/\Delta T$
Upon integration,

$$-U A (1/C_h + 1/C_c) = \ln(\Delta T)\big|_1^2$$

$$= \ln(T_{h_0} - T_{c_0})/(T_{h_i} - T_{c_i})$$

The values are: $U = 250$ W/m²K
$$A = 20 \text{ m}^2$$
$$1/C_h = 3600/(10000 \times 2000) = 1.8 \times 10^{-4}$$
$$1/C_c = 3600/(2500 \times 400) = 3.6 \times 10^{-3}$$
$$-UA(1/C_h + 1/C_c) = -250 \times 20(1.8 \times 10^{-4} + 3.6 \times 10^{-3}) = -18.9$$

$\therefore\quad (T_{h_0} - T_{c_0})/(200 - 25) = \exp(-18.9)°0.0;$ Or, $T_{h_0} = T_{c_0}$

By making an energy balance,

$$10000 \times 2000(200 - T_{h_0}) = 2500 \times 400(T_{c_0} - 25)$$

$$= 2500 \times 400(T_{h_0} - 25) \quad \text{and} \quad 21\, T_{h_0} = 20 \times 200 + 25$$

Or, $\qquad T_{h_0} = 191.67 \,°C.$

Example 10.10 Cold water at the rate of 4 kg/s is heated from 30°C to 50°C in a shell and tube heat exchanger with hot water entering at 95°C at a rate of 2 kg/s. The hot water flows through the shell. The cold water flows through tubes 2 cm inner diameter, velocity of flow 0.38 m/s. Calculate the number of tube passes, the number of tubes per pass if the maximum length of the tube is limited to 2.0 m and the overall heat transfer coefficient is 1420 W/m²K.

Solution: Let T be the temperature of the hot water at exit. By making an energy balance: 4c (50 – 30) = 2c (95 – T);   ∴ T = 55 °C
For a counter-flow arrangement:
$$\Delta T_a = (95 - 50) = 45, \qquad \Delta T_b = (55 - 30) = 25,$$
∴   LMTD = (45 – 25)/1n(45/25) = 34; Q = m C (ΔT) = 4 × 4.182 × 20 = 334.56kW
Since the cold water is flowing through the tubes, the number of tubes, n is given by
$$\dot{m} = n \times \rho \times \text{Area} \times \text{velocity; the cross-sectional area } 3.142 \times 10^{-4} \text{m}^2$$
$$4 = n \times 1000 \times 3.142 \times 10^{-4} \times 0.38; \qquad \therefore\ n = 33.5, \text{ or } 34 \text{ (say)}$$
Assuming one shell and two tube pass, we use Fig. 10.9(a).
$$P = (50 - 30)/(95 - 30) = 0.3; \qquad Z = (95 - 55)/(50 - 30) = 2.0$$
Therefore, the correction factor, F = 0.88
$$Q = U A F \text{ LMTD}; \quad 34560 = 1420 \times A \times 0.88 \times 34; \quad \text{ or } A = 7.875 \text{ m}^2$$
For 2 tube pass, the surface area of 34 tubes per pass = 2 L π d 34
∴   L = 1.843 m.
Thus we will have 1 shell pass, 2 tube pass, 34 tubes of 1.843 m in length.

Example 10.11 A double pipe heat exchanger is used to cool compressed air (pressure 4 bar, volume flow rate 5 m³/min at 1 bar and 15°C) from 160°C to 35°C. Air flows with a velocity of 5 m/s through thin walled tubes, 2 cm inner diameter. Cooling water flows through the annulus and its temperature rises from 25°C to 40°C. The convective heat transfer coefficient at the inside and outside tube surfaces are 125 W/m²K and 2000 W/m²K respectively. Calculate (i) mass of water flowing through the exchanger, and (ii) number of tubes and length of each tube.

Solution: Air is cooled from 160°C to 35°C while water is heated from 25°C to 40°C and therefore this must be a counter flow arrangement.

Temperature difference at section 1: $(T_{h_i} - T_{c_0}) = (160 - 40) = 120$

Temperature difference at section 2: $(T_{h_0} - T_{c_i}) = (35 - 25) = 10$

$$\text{LMTD} = (120 - 10)/\ln(120/10) = 44.27$$

Mass of air flowing, $\dot{m} = \rho \times \text{Volume} = (10^5/287 \times 288)(5/60) = 0.1$ kg/s

Heat given out by air = Heat taken in by water,

∴ $0.1 \times 1.005 \times (160 - 35) = \dot{m}_w \times 4.182 \times (40 - 25)$; Or $\dot{m}_w = 0.20$ kg/s

Density of air flowing through the tube, $\rho = p/RT$. The mean temperature of air flowing through the tube is $(160 + 35)/2 = 97.5°C = 370.5K$

∴ $\rho = 4 \times 10^5/(287 \times 370.5) = 3.76$ kg/m³. If n is the number of tubes, from the conservation of mass, $\dot{m} = \rho AV$; $0.1 = 3.76 \times (\pi/4)(0.02)^2 \times 5 \times n$

∴ $n = 16.9 \equiv 17$ tubes; $\dot{Q} = UA$ (LMTD)

$U = 1/(1/2000 + 1/125) = 117.65$,   Area for heat transfer $A = \pi DLn$

$Q = UA$(LMTD); $0.1 \times 1005 \times 125 = 117.65 \times 3.142 \times 0.02 \times L \times 17 \times 44.27$

and $L = 2.26$ m.

Example 10.12 A refrigerant (mass rate of flow 0.5 kg/s, $C_p = 907$ J/kgK $k = 0.07$ W/mK, $\mu = 3.45 \times 10^{-4}$ Pa-s) at –20°C flows through the annulus (inside diameter 3 cm) of a double pipe counter flow heat exchanger used to cool water (mass flow rate 0.05 kg/s. $k = 0.68$ W/mK, $\mu = 2.83 \times 10^{-4}$ Pa-s) at 98°C flowing through a thin walled copper tube of 2 cm inner diameter. If the length of the tube is 3m, estimate (i) the overall heat transfer coefficient, and (ii) the temperature of the fluid streams at exit.

Solution: Mass rate of flow, $\dot{m} = \rho AV = \rho(\pi/4)D^2V$; $\rho VD = 4\dot{m}/\pi D$
and, Reynolds number, $\text{Re} = \rho VD/\mu = 4\dot{m}/\pi D\mu$
Water is flowing through the tube of diameter 2 cm,

∴ $\text{Re} = 4 \times 0.05/(3.142 \times 0.02 \times 2.83 \times 10^{-4}) = 1.12 \times 10^4$, turbulent flow.

$\text{Nu} = 0.023\,\text{Re}^{0.8}(\text{Pr})^{0.33} = 0.023\,(1.12 \times 10^4)^{0.8}(1.8)^{0.33}$

$= 48.45$;   and $h_i = \text{Nu} \times k/D = 48.45 \times 0.68/0.02 = 1647.3$ W/m²K.

Refrigerant is flowing through the annulus. The hydraulic diameter is $D_o - D_i$, and the Reynolds number would be, $\text{Re} = 4m/\mu\pi\,(D_o + D_i)$

$\text{Re} = 4 \times 0.5/(3.45 \times 10^{-4} \times 3.142 \times (0.02 + 0.03)$

$= 3.69 \times 10^4$, a turbulent flow.

$\text{Nu} = 0.023\,(\text{Re})^{0.8}(\text{Pr})^{0.33}$, where $\text{Pr} = \mu c/k = 3.45 \times 10^{-4} \times 907/0.07 = 4.47$

$= 0.023(3.69 \times 10^4)^{0.8}(4.47)^{0.33} = 169.8$

∴ $h_o = \text{Nu} \times k/(D_o - D_i) = 169.8 \times 0.07/0.01 = 1188.6$ W/m²K

and, the overall heat transfer coefficient, $U = 1/(1/1647.3 + 1/1188.6)$
$$= 690.43 \text{ W/m}^2\text{K}$$

For a counter flow heat exchanger, from Eq. (10.4), we have,
$$(1/C_c - 1/C_h)\, U\,A = \ln(\Delta T_o/\Delta T_i) = \ln[(T_{h_o} - T_{c_i})/(T_{h_i} - T_{c_o})]$$
$$C_c = 0.5 \times 907 = 453.5; \qquad C_h = 0.05 \times 4182 = 209.1$$
$$1/C_c - 1/C_h)\, UA = (1/453.5 - 1/209.1) \times 690.43 \times 3.142 \times 0.02 \times 3$$
$$= -0.335$$
$$\therefore \qquad (T_{h_o} - T_{c_i})/(T_{h_i} - T_{c_o}) = \exp(-0.335) = 0.715$$

or, $\quad (T_{h_o} + 20)/(98 - T_{c_o}) = 0.715$; By making an energy balance,
$$453.5\,(T_{c_o} + 20) = 209.1\,(98 - T_{h_o})$$

which gives $\qquad\qquad T_{c_o} = 3.12\ °\text{C};\ T_{h_o} = 47.8°\text{C}.$

8. Heat Exchangers Effectiveness — Useful Parameters

In the design of heat exchangers, the efficiency of the heat transfer process is very important. The method suggested by Nusselt and developed by Kays and London is now being extensively used. The effectiveness of a heat exchanger is defined as the ratio of the actual heat transferred to the maximum possible heat transfer.

Let $\dot{m}_h$ and $\dot{m}_c$ be the mass flow rates of the hot and cold fluids, c_h and c_c be the respective specific heat capacities and the terminal temperatures be T_{h_i} and T_{h_o} for the hot fluid at inlet and outlet, T_{c_i} and T_{c_o} for the cold fluid at inlet and outlet. By making an energy balance and assuming that there is no loss of energy to the surroundings, we write
$$\dot{Q} = \dot{m}_h\,c_h\,(T_{h_i} - T_{h_o}) = \dot{C}_h\,(T_{h_i} - T_{c_o}),\text{ and}$$
$$= \dot{m}_c\,c_c\,(T_{c_o} - T_{c_i}) = \dot{C}_c\,(T_{c_o} - T_{c_i}) \qquad (10.13)$$

From Eq. (10.13), it can be seen that the fluid with smaller thermal capacity, C, has the greater temperature change. Further, the maximum temperature change of any fluid would be $(T_{h_i} - T_{c_i})$ and this ideal temperature change can be obtained with the fluid which has the minimum heat capacity rate. Thus,
$$\text{Effectiveness, } \epsilon = \dot{Q}/C_{min}\,(T_{h_i} - T_{c_i}) \qquad (10.14)$$

Or, the effectiveness compares the actual heat transfer rate to the maximum heat transfer rate whose only limit is the second law of thermodynamics. An useful parameter which also measures the efficiency of the heat exchanger is the 'Number of Transfer Units', NTU, defined as
$$\text{NTU} = \text{Temperature change of one fluid/LMTD,}$$

Thus, for the hot fluid: $\quad \text{NTU} = (T_{h_i} - T_{h_o})/\text{LMTD, and}$

for the cold fluid: $\quad \text{NTU} = (T_{c_o} - T_{c_i})/\text{LMTD.}$

Since
$$\dot{Q} = U\,A\,(\text{LMTD}) = C_h(T_{h_i} - T_{h_o})$$
$$= \dot{C}_c\,(T_{c_o} - T_{c_i}),$$

we have $\qquad\qquad \text{NTU}_h = U\,A/C_h \quad$ and $\quad \text{NTU}_c = U\,A/C_c.$

The heat exchanger would be more effective when the NTU is greater, and therefore, $\qquad \text{NTU} = AU/C_{min} \qquad (10.15)$

An another useful parameter in the design of heat exchangers is the ratio of the minimum to the maximum thermal capacity, i.e., $R = C_{min}/C_{max}$

where R may vary between 1 (when both fluids have the same thermal capacity) and 0 (one of the fluids has infinite thermal capacity, e.g., a condensing vapour or a boiling liquid).

9. Effectiveness—NTU Relations

For any heat exchanger, we can write: $\epsilon = f(NTU, C_{min}/C_{max})$. In order to determine a specific form of the effectiveness-NTU relation, let us consider a parallel flow heat exchanger for which $C_{min} = C_h$. From the definition of effectiveness (equation 10.14), we get

$$\epsilon = (T_{h_i} - T_{h_0})/(T_{h_i} - T_{c_i})$$

and, $\qquad C_{min}/C_{max} = C_h/C_c = (T_{c_0} - T_{c_i})/(T_{h_i} - T_{h_0})$

for a parallel flow heat exchanger, from Equation 10.4,

$$\ln\,(T_{h_0} - T_{c_0})/(T_{h_i} - T_{c_i}) = -\,UA(1/C_h + 1/C_c) = \frac{-UA}{C_{min}}\,(1 + C_{min}/C_{max})$$

or, $\qquad (T_{h_0} - T_{c_0})/(T_{h_i} - T_{c_i}) = \exp[-\,NTU\,(1 + C_{min}/C_{max})]$

But, $\qquad (T_{h_0} - T_{c_0})/(T_{h_i} - T_{c_i}) = (T_{h_0} - T_{h_i} + T_{h_i} - T_{c_0})/(T_{h_i} - T_{c_i})$

$$= [(T_{h_0} - T_{h_i}) + (T_{h_i} - T_{c_i}) - \{R(T_{h_i} - T_{h_0})\}]/(T_{h_i} - T_{c_i})$$

$$= -\,\epsilon + 1 - R\epsilon = 1 - \epsilon\,(1 + R)$$

Therefore, $\quad \epsilon = [1 - \exp\,\{-\,NTU(1 + R)\}]\,/\,(1 + R)$

$$NTU = -\,\ln[1 - \epsilon\,(1 + R)]/(1 + R)$$

Similarly, for a counterflow exchanger, $\quad \epsilon = \dfrac{[1 - \exp\{-NTU(1 - R)\}]}{[1 - R\,\exp\{-NTU(1 - R)\}]}$;

and, $\qquad NTU = [1/(R - 1)]\,\ln\,[(\epsilon - 1)/(\epsilon R - 1)]$

Heat Exchanger Effectiveness Relation

Flow arrangement	Relation
Concentric tube	
Parallel flow	$\epsilon = \dfrac{1 - \exp\,[-N(1 + R)]}{(1 + R)}$; $R = C_{min}/C_{max}$
Counter flow	$\epsilon = \dfrac{1 - \exp\,[-N(1 - R)]}{1 - R\,\exp\,[-N(1 - R)]}$; $R < 1$.
	$\epsilon = N/(1 + N)$ for $R = 1$.
Cross flow (Single pass)	
Both fluids unmixed	$\epsilon = 1 - \exp\,[(1/R)\,(N)^{0.22}\,\{\exp\,(-\,R(N)^{0.78}) - 1\}]$
C_{max} mixed, C_{min} unmixed	$\epsilon = (1/R)\,[1 - \exp\,\{-\,R(1 - \exp\,(-N)\}]$
C_{min} mixed, C_{max} unmixed	$\epsilon = 1 - \exp\,[-\,R^{-1}\{1 - \exp\,(-\,R\,N)\}]$
All exchangers ($R = 0$)	$\epsilon = 1 - \exp\,(-\,N)$

Kays and London have presented graphs of effectiveness against NTU for various values of R applicable to different heat exchanger arrangements, Fig. 10.11 to Fig. (10.15).

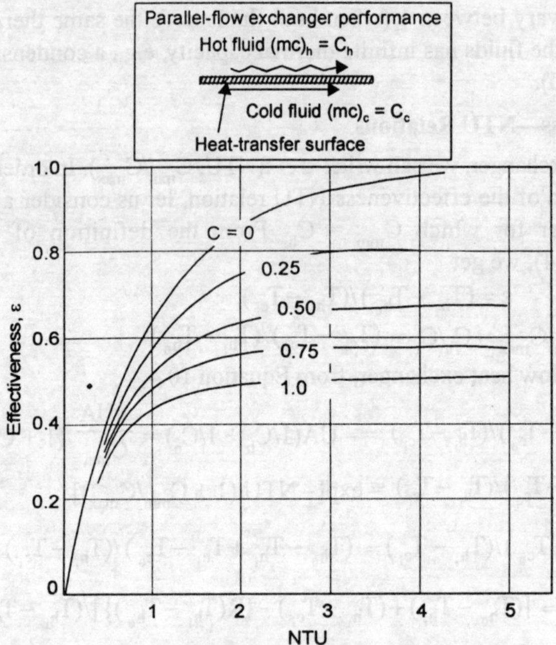

Fig 10.11 Heat exchanger effectiveness for parallel flow

Example 10.13 A single pass shell and tube counter flow heat exchanger uses exhaust gases on the shell side to heat a liquid flowing through the tubes (inside diameter 10 mm, outside diameter 12.5 mm, length of the tube 4 m). Specific heat capacity of gas 1.05 kJ/kgK, specific heat capacity of liquid 1.5 kJ/kgK, density of liquid 600 kg/m^3, heat transfer coefficient on the shell side and on the tube sides are: 260 and 590 W/m^{2}K respectively. The gases enter the exchanger at 675 K at a mass flow rate of 40 kg/s and the liquid enters at 375 K at a mass flow rate of 3 kg/s. If the velocity of liquid is not to exceed 1 m/s, calculate (i) the required number of tubes, (ii) the effectiveness of the heat exchanger, and (iii) the exit temperature of the liquid. Neglect the thermal resistance of the tube wall.

Solution: Volume flow rate of the liquid = 3/600 = 0.005 m^3/s. For a velocity of 1 m/s through the tube, the cross-sectional area of the tubes will be 0.005 m^2. Therefore, the number of tubes would be

$$n = (0.005 \times 4)/(3.142 \times 0.01^2) = 63.65 = 64 \text{ tubes.}$$

The overall heat transfer coefficient based on the outside surface area of the tubes, after neglecting the thermal resistance of the tube wall, is

$$U = 1/(1/h_o + r_o/r_i h_i) = 1/[1/260 + 12.5/(10 \times 590)] = 167.65 \text{ W/m}^2\text{K}$$

$$C_{max} = 40 \times 1.05 = 42; \quad C_{min} = 3 \times 1.5 = 4.5; \quad R = 4.5/42 = 0.107$$

$$NTU = AU/C_{min} = 3.142 \times 0.0125 \times 4 \times 64 \times 167.65/(4.5 \times 1000)$$

$$= 0.374$$

From Fig. 10.12, for R = 0.107, and NTU = 0.374, ϵ = 0.35 approximately

Therefore, $0.35 = (T_{c_0} - 375)/(675 - 375)$ or $T_{c_0} = 207°C$.

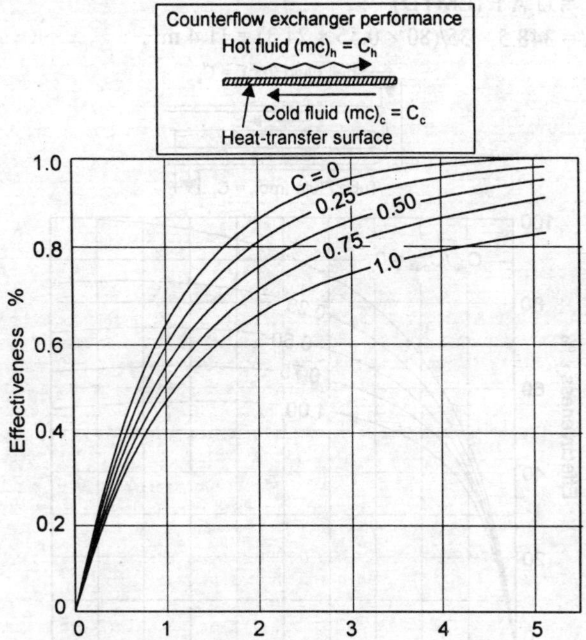

Fig 10.12 Heat exchanger effectiveness for counterflow

Example 10.14 Air at 25°C, mass flow rate 20 kg/min, flows over a cross-flow heat exchanger and cools water from 85°C to 50°C. The water flow rate is 5 kg/min. If the overall heat transfer coefficient is 80 W/m²K and air is the mixed fluid, calculate the exchanger effectiveness and the surface area.

Solution: Let the specific heat capacity of air and water be 1.005 and 4.182 kJ/kgK. By making an energy balance:

$$\dot{m}_c \times c_c \times (T_{c_0} - T_{c_i}) = \dot{m}_h \times c_h \times (T_{h_i} - T_{h_0})$$

or, $5 \times 4182 \times (85 - 50) = 20 \times 1005 \times (T_{c_0} - 25)$,

i.e., the air will come out at 61.4 °C.

Heat capacity rates for water and air are:

$$C_w = 4182 \times 5/60 = 348.5; \qquad C_a = 1005 \times 20/60 = 335$$

$$R = C_{min}/C_{max} = 335/348.5 = 0.96$$

The effectiveness on the basis of minimum heat capacity rate is

$$\epsilon = (61.4 - 25) / (85 - 25) = 0.6$$

From Fig. 10.13, for R = 0.96 and ϵ = 0.6,   NTU = 2.5

Since   NTU = AU/C$_{min}$;   A = 2.5 × 335/80 = 10.47 m².

Since all the four terminal temperatures are easily obtained, we can also use the LMTD approach. Assuming a simple counter flow heat exchanger,

$$LMTD = (25 - 23.6)/\ln(25/23.6) = 24.3$$

The correction factor for using a cross-flow heat exchanger with one fluid mixed and the other unmixed, from Fig. 10.10(d), F = 0.55

$$\dot{Q} = U\,A\,F\,(LMTD)$$

Therefore,   A = 348.5 × 35/(80 × 0.55 × 24.3) = 11.4 m².

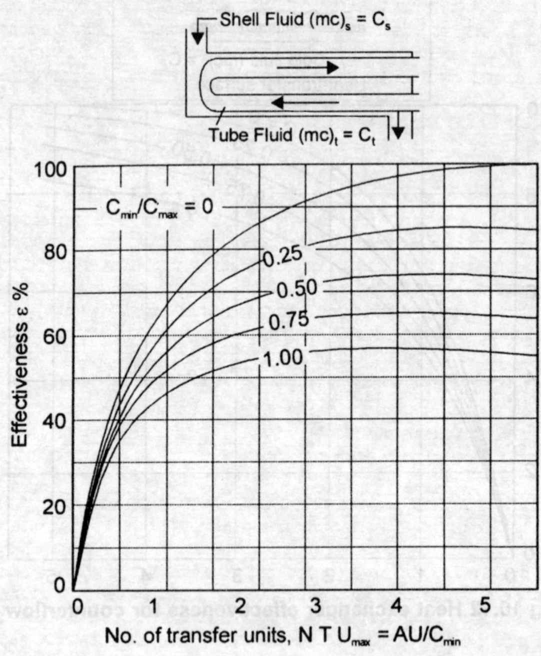

Fig. 10.13 Heat exchanger effectiveness for shell and tube heat exchanger with one shell pass and two, or a multiple of two, tube passes

Example 10.15 Steam at 20 kPa and 70°C enters a counter flow shell and tube exchanger and comes out as subcooled liquid at 40°C. Cooling water enters the condenser at 25°C and the temperature difference at the pinch point is 10°C. Calculate the (i) amount of water to be circulated per kg of steam condensed, and (ii) required surface area if the overall heat transfer coefficient is 5000 W/m²K and is constant.

Solution: The temperature profile of the condensing steam and water is shown in the accompanying sketch.

The saturation temperature corresponding to 20 kPa is 60°C and as such the temperature of the cooling water at the pinch pint is 50°C. The condensing unit may be considered as a combination of three sections:

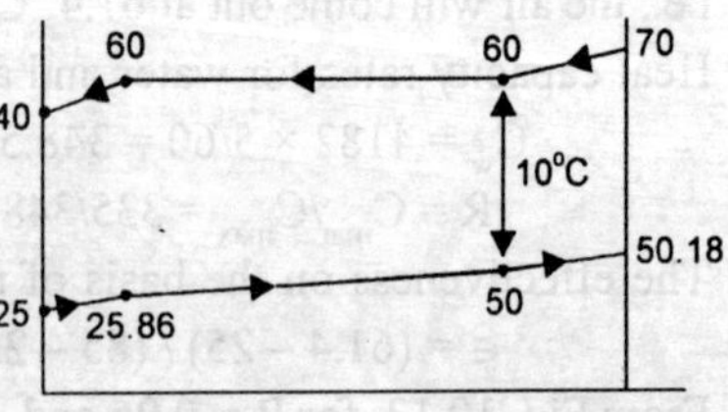

(i) desuperheater – the superheated steam is condensed to saturated steam from 70°C to 60°C.
(ii) the condenser – saturated steam is condensed into saturated liquid.
(iii) subcooler – saturated liquid at 60°C is cooled to 40°C.

Assuming that the specific heat capacity of superheated steam is 1.8 kJ/kgK, heat given out in the desuperheater section is $1.8 \times (70 - 60) = 18000$ J/kg. Heat given out in the condenser section = 2358600 J/kg (= hfg)

Heat given out in the subcooler = $4182 \times (60 - 40) = 83640$ J/kg

By making an energy balance, for subcooler and condenser section, we have
$$\dot{m}_w \times 4182 \times (50 - 25) = (83640 + 2358600);$$
$\therefore$ Mass of water circulated, $\dot{m}_w = 23.36$ kg/kg steam condensed.

The temperature of water at exit $= 25 + (83640 + 2358600 + 18000)/$
$$(23.36 \times 4182) = 50.18\,°C$$

LMTD for desuperheater section $= [(70 - 50.18) - (60 - 50)]/\ln(20.18/10)$
$$= 14.5$$

LMTD for condenser section $= [(60 - 50) - (60 - 25.86)]/\ln(10/34.14)$
$$= 19.66$$

LMTD for subcooler section $= [(34.14 - 15)]/\ln(34.14/15) = 23.27$

Since U is constant through out,

Surface area for subcooler section $= 83640/(5000 \times 23.27) = 0.7188$ m^2

Surface area for condenser section $= 2358600/(5000 \times 19.66) = 23.9939$ m^2

Surface area for desuperheater section $= 18000/(5000 \times 14.5) = 0.2483$ m^2

$\therefore$ Total surface area = 24.96 m^2 and average temperature difference = 19.71°C.

Example 10.16 In an economiser (a cross flow heat exchanger, both fluids unmixed) water, mass flow rate 10 kg/s, enters at 175°C. The flue gas mass flow rate 8 kg/s, specific heat 1.1 kJ/kgK, enters at 350°C. Estimate the temperature of the flue gas and water at exit, if U = 500 W/m^{2}K, and the surface area 20 m^2. What would be the exit temperature if the mass flow rate of flue gas is (i) doubled, and (ii) halved.

Solution: The heat capacity rate of water = $4182 \times 10 = 41820$ W/K

The heat capacity rate of flue gas = $1100 \times 8 = 8800$ W/K
$$C_{min}/C_{max} = 8800/41820 = 0.21$$
$$NTU = AU/C_{min} = 500 \times 20/8800 = 1.136$$

From Fig. 10.14, for NTU = 1.136 and $C_{min}/C_{max} = 0.21$, $\epsilon = 0.62$

Therefore, $0.62 = (350 - T)/(350 - 175)$ and T = 241.5°C

The temperature of water at exit, $T_w = 175 + 8800 \times (350 - 241.5)/41820$
$$= 197.83\,°C$$

When the mass flow rate of the flue gas is doubled. $C_{gas} = 17600$ W/K

$C_{min}/C_{max} = 0.42$, NTU $= AU/C_{min} = 0.568$

$\epsilon = 0.39 = (350 - T)/(350 - 175)$;

T = 281.75°C, an increase of 40°C

and $T_w = 175 + 28.72 = 203.72$°C, an incease of about 6°C.

When the mass flow rate of the flue gas is halved, $C_{min} = 4400$ W/K

$C_{min}/C_{max} = 0.105$, NTU = 2.272, and from the figure, $\epsilon = 0.83$, an increase
and $T_g = 204.75$ and $T_w = 190.3$°C.

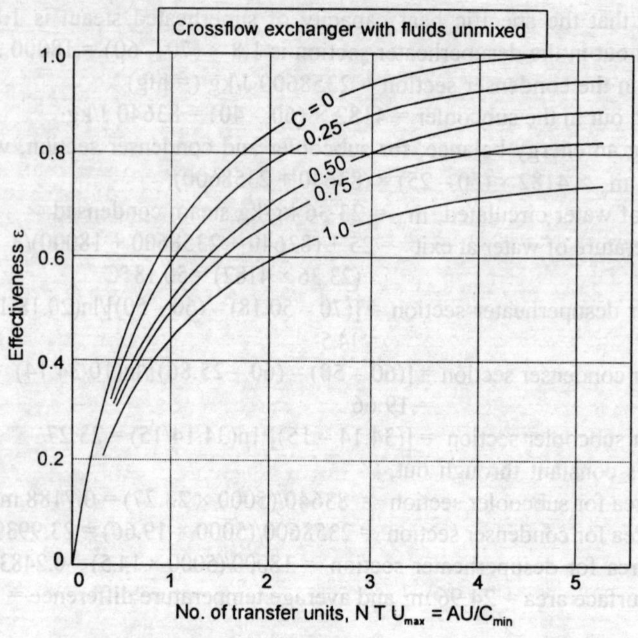

Fig 10.14

Example 10.17 In a tubular condenser, steam at 30 kPa and 0.95 dry condenses on the external surfaces of tubes. Cooling water flowing through the tubes has mass flow rate 5 kg/s, inlet temperature 25°C, exit temperature 4 0°C. Assuming n o s ubcooling o f the c ondensate, estimate the rate of condensation of steam, the effectiveness of the condenser and the NTU.

Solution: Since there is no subcooling of the condensate, the steam will lose its latent heat of condensation = 0.95 × h_{fg} = 0.95 × 2336100 = 2.22 × 10^6 J/kg

At pressure, 30kPa, saturation temperature is 69.124°C.

Steam condensation rate × 2.22 × 10^6 = Heat gained by water

$$= 5 × 4182 × (40 - 25) = 313650 \text{ J}$$

Therefore, $\dot{m}_s$ = 313650/2.22 × 10^6 = 847.7 kg/hour.

When the temperature of the evaporating or condensing fluid remains constant, the value of LMTD is the same whether the system is having a parallel flow or counter flow arrangement, therefore,

$$\text{LMTD} = [(69.124 - 25) - (69.124 - 40)]/\ln(44.124/29.124) = 36.1$$

$$Q = UA(\text{LMTD})$$

Therefore,   UA = 5 × 4182 × (40 − 25)/36.1

$$= 8688.36 \text{ W/K}$$

$$\text{NTU} = UA/C_{min} = 8688.36/(5 × 4182) = 0.4155$$

Effectiveness = Actual temp. difference/Maximum possible temp. difference

$$= (40 - 25)/(69.124 - 25) = 34\%.$$

Example 10.18 A single shell 2 tube pass steam condenser is used to cool steam entering at 50°C and releasing 2 000 MW of heat energy. The cooling water, mass flow rate 3×10^4 kg/s, enters the condenser at 25°C. The condenser has 30,000 thin walled tube of 30 mm diameter. If the overall heat transfer coefficient is 4000 W/m^{2}K, estimate the (i) rise in temperature of the cooling water, and (ii) length of the tube per pass.

Solution: By making an energy balance:

Heat released by steam = heat taken in by cooling water,

or, $\qquad 2000 \times 10^6 = 3 \times 10^4 \times 4182 \times (\Delta T); \qquad\qquad \therefore \Delta T = 15.94$°C.

Since in a condenser, heat capacity rate of condensing steam is usually very large in comparison with the heat capacity rate of cooling water, the effectiveness

$$\epsilon = (T_{c_0} - T_{c_i})/(T_{h_i} - T_{c_i}) = 15.94/(50-25) = 0.6376$$

And, for $C_{min}/C_{max} = 0$, $\epsilon = 1 - \exp(-NTU)$

$\therefore \quad \exp(-NTU) = 1.0 - 0.6376 = 0.3624$

And, $NTU = 1.015 = AU/C_{min} = (2 \times 3.142 \times 0.03 \times L \times 30000) \times 4000/(1.25 \times 108)$

$\qquad\qquad L = 5.546$m.

10. Heat Exchanger Design—Important Factors

A comprehensive design of a heat exchanger involves the consideration of the thermal, mechanical and manufacturing aspect. The choice of a particular design for a given duty depends on either the selection of an existing design or the development of a new design. Before selecting an existing design, the analysis of its performance must be made to see whether the required performance would be obtained within acceptable limits.

In the development of a new design, the following factors are important:

(a) *Fluid Temperature* – the temperature of the two fluid streams are either specified for a given inlet temperature, or the designer has to fix the outlet temperature based on flow rates and heat transfer considerations. Once the terminal temperatures are defined, the effectiveness of the heat exchanger would give an indication of the type of flow path-parallel or counter or cross-flow.

(b) *Flow Rates* – The maximum velocity (without causing excessive pressure drops, erosion, noise and vibration, etc.) in the case of liquids is restricted to 8 m/s and in case of gases below 30 m/s. With this restriction, the flow rates of the two fluid streams lead to the selection of flow passage cross-sectional area required for each of the two fluid steams.

(c) *Tube Sizes and Layout* – Tube sizes, thickness, lengths and pitches have strong influence on heat transfer calculations and therefore, these are chosen with great care. The sizes of tubes vary from 1/4" O.D. to 2" O.D; the more commonly used sizes are: 5/8", 3/4" and 1" O.D. The sizes have to be decided after making a compromise between higher heat transfer from smaller tube sizes and the easy cleanability of larger tubes. The tube thickness will depend pressure, corrosion and cost. Tube pitches are to be decided on the basis of heat transfer

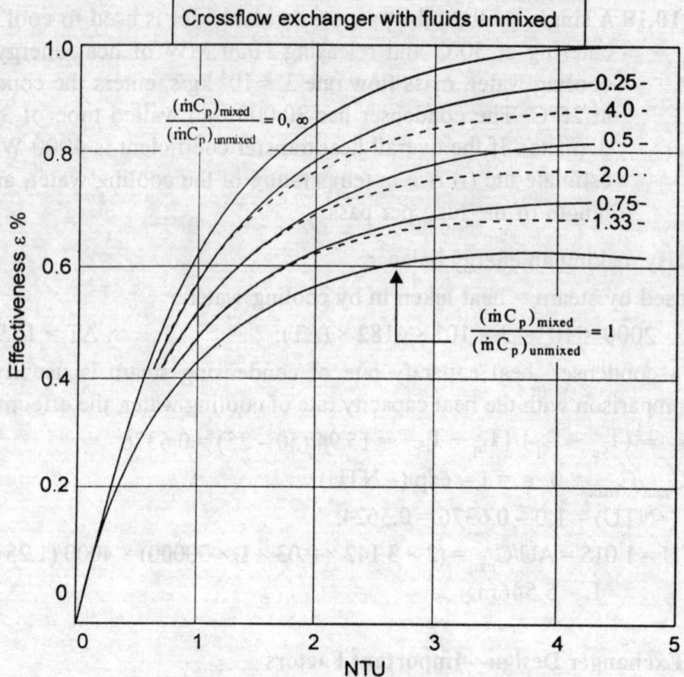

Fig. 10.15 Heat exchanger effectiveness for crossflow with one fluid mixed and the other unmixed

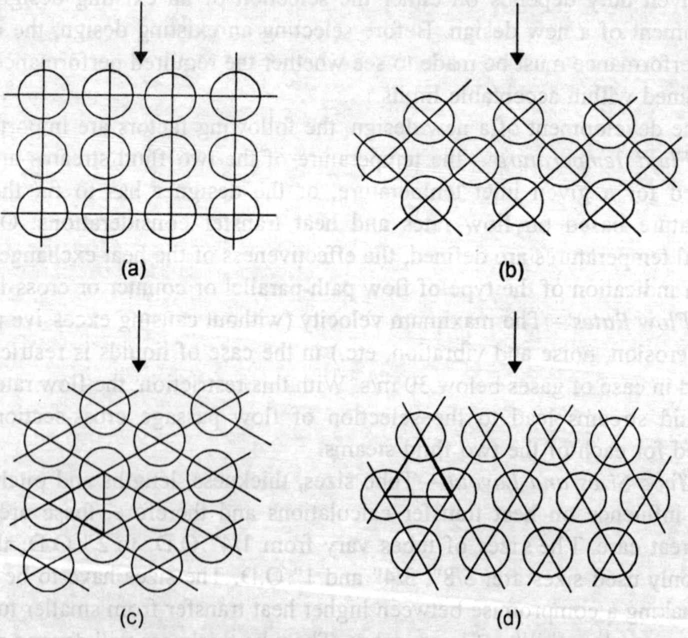

Fig. 10.16 Several arrangements of tubes in bundles: (a) in-line arrangement with square pitch, (b) staggered arrangement with square pitch, (c) and (d) staggered arrangement with triangular pitches

calculations and difficulty in cleaning. Fig. 10.16 shows several arrangements for tubes in bundles. The two standard types of pitches are the square and the triangle. the usual number o f tube passes in a given shell r anges from one to eight. In multipass designs, even numbers of passes are generally used because they are simpler to design.

Fig. 10.17 shows three types of transverse baffles used to increase velocity on the shell side. The choice of baffle spacing and baffle cut is a variable and the optimum ratio of baffle cuts and spacing cannot be specified because of many uncertainties and insufficient data.

(d) *Dirt F actor a nd F ouling* – the accumulation of dirt or deposits affects significantly the r ate of heat transfer a nd the pressure drop. P roper allowance for the fouling factor and dirt factor should receive the greatest attention in

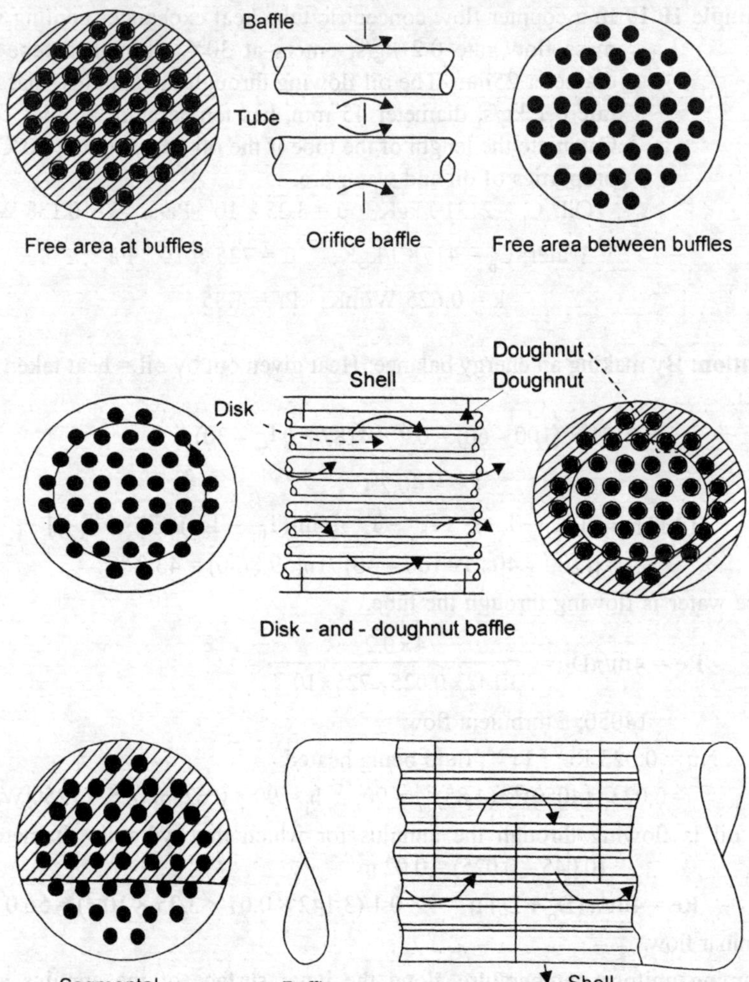

Fig. 10.17 Three types of transverse baffles

design because they cannot be avoided. A heat exchanger requires frequent cleaning. Mechanical cleaning will require removal of the tube bundle for cleaning. Chemical cleaning will require the use of non-corrosive materials for the tubes.

(e) *Size and Installation* – In designing a heat exchanger, it is necessary that the constraints on length, height, width, volume and weight is known at the outset. Safety regulations should also be kept in mind when handling fluids under pressure or toxic and explosive fluids.

(f) *Mechanical Design Consideration* – While designing, operating temperatures, pressures, the differential thermal expansion and the accompanying thermal stresses require attention.

And, above all, the cost of materials, manufacture and maintenance cannot be ignored.

Example 10.19 In a counter flow concentric tube heat exchanger cooling water, mass flow rate 0.2 kg/s, enters at 30°C through a tube inner diameter 25mm. The oil flowing through the annulus, mass flow rate 0.1 kg/s, diameter 45 mm, has temperature at inlet 100°C. Calculate the length of the tube if the oil comes out at 60°C. The properties of oil and water are:

$$\text{Oil: } C_p - 2131 \text{ J/kgK}, \quad \mu = 3.25 \times 10^{-2} \text{ Pa-s}, \quad k = 0.138 \text{ W/mK};$$

$$\text{Water; } C_p = 4178 \text{ J/kgK}, \quad \mu = 725 \times 10^{-6} \text{ Pa-s},$$

$$k = 0.625 \text{ W/mK}, \quad Pr = 4.85$$

Solution: By making an energy balance: Heat given out by oil = heat taken in by water.

$$0.1 \times 2131 \times (100 - 60) = 0.2 \times 4187 \times (T_{c_0} - 30)$$

$$\therefore \qquad T_{c_0} = 40.2°C$$

$$\text{LMTD} = [(T_{h_i} - T_{c_0}) - (T_{h_0} - T_{c_i})] / \ln[(T_{h_i} - T_{c_0})/(T_{h_0} - T_{c_0})]$$

$$= [(100 - 40.2) - (60 - 30)]/\ln(59.8/30) = 43.2 \text{ °C}.$$

Since water is flowing through the tube,

$$\text{Re} = 4\dot{m}/\pi D\mu = \frac{4 \times 0.2}{3.142 \times 0.025 \times 725 \times 10^{-6}}$$

$$= 14050, \text{ a turbulent flow}$$

$$\therefore \qquad \text{Nu} = 0.023 \, \text{Re}^{0.8} \, \text{Pr}^{0.4}, \text{ fluid being heated.}$$

$$= 0.023 \, (14050)^{0.8} \, (4.85)^{0.4} = 90; \therefore h_i = 90 \times 0.625/0.025 = 2250 \text{ W/m}^2\text{K}$$

The oil is flowing through the annulus for which the hydraulic diameter is:

$$(0.045 - 0.025) = 0.02 \text{ m}$$

$$\text{Re} = 4\dot{m}/\pi (D_o + D_i) \mu = 4 \times 0.1/(3.142 \times 0.07 \times 3.25 \times 10^{-2}) = 56.0$$

a laminar flow.

Assuming uniform temperature along the inner surface of the annulus and a perfectly insulated outer surface.

Nu = 5.6, by interpolation (chapter 6)

$h_o = 5.6 \times 0.138/0.02 = 38.6$ W/m²K.

The overall heat transfer coefficient after neglecting the tube wall resistance,

$U = 1/(1/2250 + 1/38.6) = 38$ W/m²K

$\dot{Q} = U\,A\,(LMTD)$,   where $A = \pi D_i \times L$

$L = (0.1 \times 2131 \times 40)/(38 \times 3.142 \times 0.025 \times 43.2) = 66.1$ m

requires more than one pass.

Example 10.20 A double pipe heat exchanger has an effectiveness of 0.5 for the counter flow arrangement and the thermal capacity of one fluid is twice that of the other fluid. Calculate the effectiveness of the heat exchanger if the direction of flow of one of the fluids is reversed with the same mass flow rates as before.

Solution: For a counter flow arrangement and R = 0.5, $\in$ = 0.5

$$NTU = [1/(R - 1)]\ln(\in - 1)/(\in R - 1) = -2.0\ \ln(0.5/0.75) = 0.811$$

For parallel flow, $\in = [1 - \exp\{-NTU(1 + R)\}]/(1 + R)$

$\qquad\qquad\quad = [1 - \exp(-0.811 \times 1.5)]/1.5$

$\qquad\qquad\quad = 0.469.$

Example 10.21 Oil is cooled in a cooler from 65°C to 54°C by circulating water through the cooler. The cooling load is 200 kW and water enters the cooler at 27°C. If the overall heat transfer coefficient, based on the outer surface area of the tube is 740 W/m²K and the temperature rise of cooling water is 11°C, calculate the mass flow rate of water, the effectiveness and the heat transfer area required for a single pass in a parallel flow and in a counter flow arrangement.

Solution: Cooling load = 200 kW = mass of water × sp. heat × temp rise

Mass of water = 200/(4.2 × 11) = 4.329 kg/s

(i) Parallel flow:

From the temperature profile:

LMTD = (38 – 16)/ln(38/16) = 25.434

$Q = U\,A\,(LMTD)$;

Area A = 200 × 10³/(740 × 25.434) = 10.626 m²

Effectiveness, $\in$ = (38 – 27)/(54 – 27) = 0.407.

(ii) Counter flow:

From the temperature profile:

LMTD = mean temperature difference = 27°C

Area A = 200 × 10³/(740 × 27) = 10 m²

Effectiveness, $\in$ = (38 – 27)/(65 – 27) = 0.289.

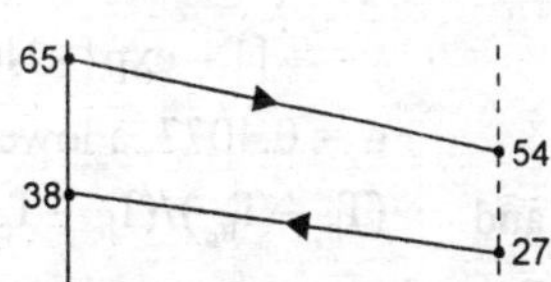

Example 10.22 Oil (mass flow rate 1.5 kg/s, C_p = 2 kJ/kgK) is cooled in a single pass shell and tube heat exchanger from 65 to 42°C. Water (mass flow rate 1 kg/s, C_p = 4.2 kJ/kgK) has an inlet temperature of 28°C. If the overall heat transfer coefficient is 700 W/m²K, calculate heat transfer area for a counter flow arrangement using $\in$ – NTU method.

Solution: Heat capacity rate of oil = $1.5 \times 2.0 = 3$ kW/K

Heat capacity rate of water = $1 \times 4.2 = 4.2$ kW/K

$$C_{min} = 3.0 \text{ kW/K and } R = C_{min}/C_{max} = 3/4.2 = 0.714$$

For a counter flow arrangement, NTU = $[1/(R-1)]\ln[(\epsilon - 1)/(\epsilon R - 1)]$

Effectiveness, $\epsilon = (65 - 42)/(65 - 28) = 0.6216$

and $\qquad$ NTU = $1.346 = A\,U/C_{min}$; $\quad$ A = $1.346 \times 3000/700 = 5.77$ m^2

By making an energy balance, we can compute the water temperature at outlet.

or $\quad 3.0 \times (65 - 42) = 4.2 \times (T - 28)$, $\;$ T = 44.428

LMTD for a counter flow arrangement:

LMTD = $(20.572 - 14)/\ln(20.572/14)$

$\qquad = 17.076$

Area, $\quad$ A = $\dot{Q}/U \times$ (LMTD) = $3 \times 10^3 \times (65 - 42)/(700 \times 17.076) = 5.77$ m^2.

Example 10.23 $\;$ A fluid (mass flow rate 1000 kg/min, sp. heat capacity 3.6 kJ/kgK) enters a heat exchanger at 700 C. Another fluid (mass flow rate 1200 kg/min, sp. heat capacity 4.2 kJ/kgK) enters at 100 C. If the overall heat transfer coefficient is 420 W/m^{2}K and the surface area is 100m^2, calculate the outlet temperatures of both fluids for both counter flow and parallel flow arrangements.

Solution: Heat capacity rate for the hot fluid

$$= 1000 \times 3.6 \times 10^3/60 = 60 \times 10^3 \text{ W/K}$$

Heat capacity rate for the cold fluid = $1200 \times 4.2 \times 10^3/60 = 84 \times 10^3$ W/K

$$R = C_{min}/C_{max} = 60/84 = 0.714, \text{ NTU} = U\,A/C_{min} = 420 \times 100/60000 = 0.7$$

(i) $\;$ For counter flow heat exchanger:

$\qquad \epsilon = [1 - \exp\{-N(1 - R)\}]/[1 - R\exp\{-N(1 - R)\}]$

$\qquad\qquad [1 - \exp\{-0.7(1 - 0.714)\}]/1[1 - 0.714\exp\{-0.7(1 - 0.714)\}] = 0.4367$

Since heat capacity rate of the hot fluid is lower, $\;\epsilon = (700 - T_{h_0})/(700 - 100)$

and $\quad T_{h_0} = 700 - 0.4367 \times 600 = 438°$C

By making an energy balance, $60 \times 10^3 (700 - 438) = 84 \times 10^3 (T_{c_0} - 100)$

or, $\qquad T_{c_0} = 60 \times 262/84 + 100 = 87.14°$C

(ii) $\;$ For parallel flow heat exchanger:

$\qquad \epsilon = [1 - \exp\{-N(1 + R)\}]/(1 + R) = [1 - \exp\{-0.7(1 + 0.714)\}]/(1.714)$

$\qquad \epsilon = 0.4077$, a lower value

and $\quad (T_{h_i} - T_{h_0})/(T_{h_i} - T_{c_0}) = 0.4077 = (700 - T_{h_0})/(700 - T_{c_0})$

By making an energy balance: $60 \times 10^3 \times (700 - T_{h_0}) = 84 \times 10^3 \times (T_{c_0} - 100)$

or, $\qquad (700 - T_{c_0}) = (700 - T_{h_0})/0.4077$

and $\quad 84 \times (T_{c_0} - 100)/60 = (1.4\,T_{c_0} - 140)$

Therefore, $\qquad T_{c_0} = 237.5°$C,

and $\qquad\qquad T_{h_0} = 511.4°$C.

Example 10.24 Steam enters the surface condenser at 100°C and water enters at 25°C with a temperature rise of 25°C. Calculate the effectiveness and the NTU for the condenser. If the water temperature at inlet changes to 35 C, estimate the temperature rise for water.

Solution: Effectiveness, $\epsilon = 25/(100 - 25) = 0.33$

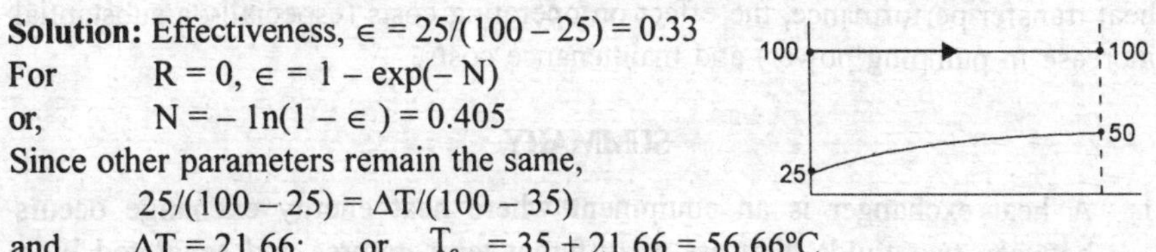

For $R = 0, \epsilon = 1 - \exp(- N)$

or, $N = - \ln(1 - \epsilon) = 0.405$

Since other parameters remain the same,

$$25/(100 - 25) = \Delta T/(100 - 35)$$

and $\Delta T = 21.66;$ or, $T_{c_0} = 35 + 21.66 = 56.66°C.$

11. Increasing the Heat Transfer Coefficient

For a heat exchanger, the heat load is equal to $Q = UA$ (LMTD). The effectiveness of the heat exchanger can be increased either by increasing the surface area for heat transfer or by increasing the heat transfer coefficient. Effectiveness versus NTU(AU/C_{min}) curves, Fig. 10.10 – 15, reveal that by increasing the surface area beyond a certain limit (the knee of the curves), there is no appreciable improvement in the performance of the exchangers. Therefore, different methods have been employed to increase the heat transfer coefficient by increasing turbulence, improved mixing, flow swirl or by the use of extended surfaces. The heat transfer enhancement techniques is gaining industrial importance because it is possible to reduce the heat transfer surface area required for a given application and that leads to a reduction in the size of the exchanger and its cost, to increase the heating load on the exchanger and to reduce temperature differences.

The different techniques used for increasing the overall conductance U are:

(a) *Extended Surfaces* – these are probably the most common heat transfer enhancement methods. The analysis of extended surfaces has been discussed in Chapter 2. Compact heat exchangers use extended surfaces to give the required heat transfer surface area in a small volume. Extended surfaces are very effective when applied in gas side heat transfer. Extended surfaces find their application in single phase natural and forced convection pool boiling and condensation.

(b) *Rough Surfaces* – the inner surfaces of a smooth tube is artificially roughened to promote early transition to turbulent flow or to promote mixing between bulk flow and the various sub-layer in fully developed turbulent flow. This method is primarily used in single phase forced convection and condensation.

(c) *Swirl Flow Devices* – twisted strips are inserted into the flow channel to impart a rotational motion about an axis parallel to the direction of bulk flow. The heat transfer coefficient increases due to increased flow velocity, secondary flows generated by swirl, or increased flow path length in the flow channel. This technique is used in flow boiling and single phase forced flow.

(d) *Treated Surfaces* – these are used mainly in pool boiling and condensation. Treated surfaces promote nucleate boiling by providing bubble nucleation sites. The rate of condensation increases by promoting the formation of droplets, instead of a liquid film on the condensing surface. This can be accomplished by coating the surface with a material that makes the surface non-wetting.

All of these techniques lead to an increase in pumping work (increased frictional losses) and any practical application requires the economic benefit of increased overall conductance. That is, a complete analysis should be made to determine the increased first cost because of these techniques, increased heat exchanger heat transfer performance, the effect on operating costs (especially a substantial increase in pumping power) and maintenance costs.

SUMMARY

1. A heat exchanger is an equipment where heat energy exchange occurs between two fluids, that are at different temperatures and separated by a solid wall.

2. Heat exchangers are typically classified according to the flow arrangement: parallel flow, counter flow, cross flow, etc, and type of construction: shell and tube and compact exchangers which are used to achieve a very large ($\geq$ 700 m^2/m^3) heat transfer area per unit volume.

3. Log Mean Temperature Difference (LMTD) for parallel and counter flow exchangers is defined as

$$\text{LMTD} (\theta_1 - \theta_2)/\ln(\theta_1/\theta_2)$$

    where θ_1 and θ_2 are the temperature difference between the two fluids at the two ends of the heat exchanger. The above relationship is based on the assumption that (i) axial conduction along the tubes are negligible, (ii) potential and kinetic energy changes are negligible, (iii) fluids have constant specific heat, and (iv) the overall heat transfer coefficient is constant.

4. In order to obtain the suitable mean temperature difference for multipass and cross-flow exchangers, the LMTD calculated for a simple counter flow exchanger is to be multiplied by an appropriate correction factor, F.

5. The efficiency of a heat exchanger is measured by its effectiveness obtained on the basis of the fluid with smaller heat capacity rate and is defined as:

$$\epsilon = Q/C_{min} (T_{h_i} - T_{c_i})$$

6. Number of Transfer Units, NTU, also measures the efficiency of the heat exchanger and is defined as:

$$\text{NTU} = AU/C_{min}$$

    where A is the heat transfer surface area and U is the overall heat transfer coefficient.

7. For any heat exchanger, we have $\epsilon = f(\text{NTU}, C_{min}/C_{max})$ and analytical expressions have been developed for a variety of heat exchanger configurations. However, in heat exchanger design calculations, it is more convenient to work with relations of the form

$$\text{NTU} = f(\epsilon, C_{min}/C_{max})$$

    When $C_{min}/C_{max} = 0$, the heat exchanger behaviour is independent of the flow arrangement,

$$[\epsilon = 1 - \exp(-\text{NTU}), \text{ or, NTU} = -\ln(1-\epsilon)]$$

8. The important factors affecting the heat exchanger design are:
    (i) heat transfer requirements – temperature of the fluids and their mass flow rates,
    (ii) physical size – tube sizes and their layouts,

(iii) pressure drop characteristics, differential thermal expansion, etc,

(iv) initial cost and maintenance cost.

9. The different techniques used for increasing the overall heat transfer coefficient are: use of extended surfaces, roughening the surfaces swirl flow devices and coating the surfaces.

MULTIPLE CHOICE QUESTIONS

1. A designer chooses the values of fluid flow rates and specific heats in such a manner that the heat capacities of the two fluids are equal. The hot fluid enters the counter-flow heat exchanger at 100°C and leaves at 60°C. The cold fluid enters at 40°C. The mean temperature difference between the two fluids is

   (a) (100 + 60 + 40)/3   (b) 60   (c) 40   (d) 20   **(d)**

2. For evaporators and condensers, under the given conditions, LMTD for counterflow will be

   (a) greater than parallel flow

   (b) equal to parallel flow

   (c) less than parallel flow

   (d) very much larger than parallel flow   **(b)**

3. Consider the following statements:

   In a shell and tube heat exchanger, baffles are provided on the shell side to

   1. prevent the stagnation of the shell side fluid

   2. improve heat transfer

   3. provide support for tubes

   4. prevent fouling of tubes.

   Of these statements

   (a) 1, 2, 3, and 4 are correct   (b) 1, 2 and 3 are correct

   (c) 1 and 2 are correct   (d) 2 and 3 are correct   **(b)**

4. Match List I with List II and select the correct answer using the code given below

List I	List II
A. number of transfer units	1. recuperators
B. periodic flow heat exchangers	2. regenerators
C. pinch point	3. a measure of heat exchangers size
D. deposition on heat exchanger surface	4. fouling factor
	5. condensers

Code

	A	B	C	D
(a)	1	2	3	4
(b)	4	3	5	2
(c)	3	2	5	4
(d)	3	1	4	5

    (c)

5. A counterflow shell and tube heat exchanger is used to heat water with hot exhaust gases. The water ($C = 4180$ J/kg 'C) flows at a rate of 2 kg/s and the exhaust gases ($C = 1030$ J/kg 'C) flow at the rate of 5.25 kg/s. If the heat transfer surface area is 32 m^2 and the overall heat transfer coefficient is 200 W/m^{2o}C. The NTU for this arrangement is
   (a) 1.2   (b) 2.4   (c) 4.5   (d) 8.6   **(a)**

6. A heat exchanger with heat transfer surface area A and overall heat transfer coefficient U handles two fluids of heat capacities C_1 and C_2 such that $C_1 > C_2$. The NTU of the heat exchanger is given by
   (a) AU/C_1 (b) $\exp(-AU/C_2)$
   (c) $\exp(-AU/C_1)$ (d) AU/C_2 **(d)**

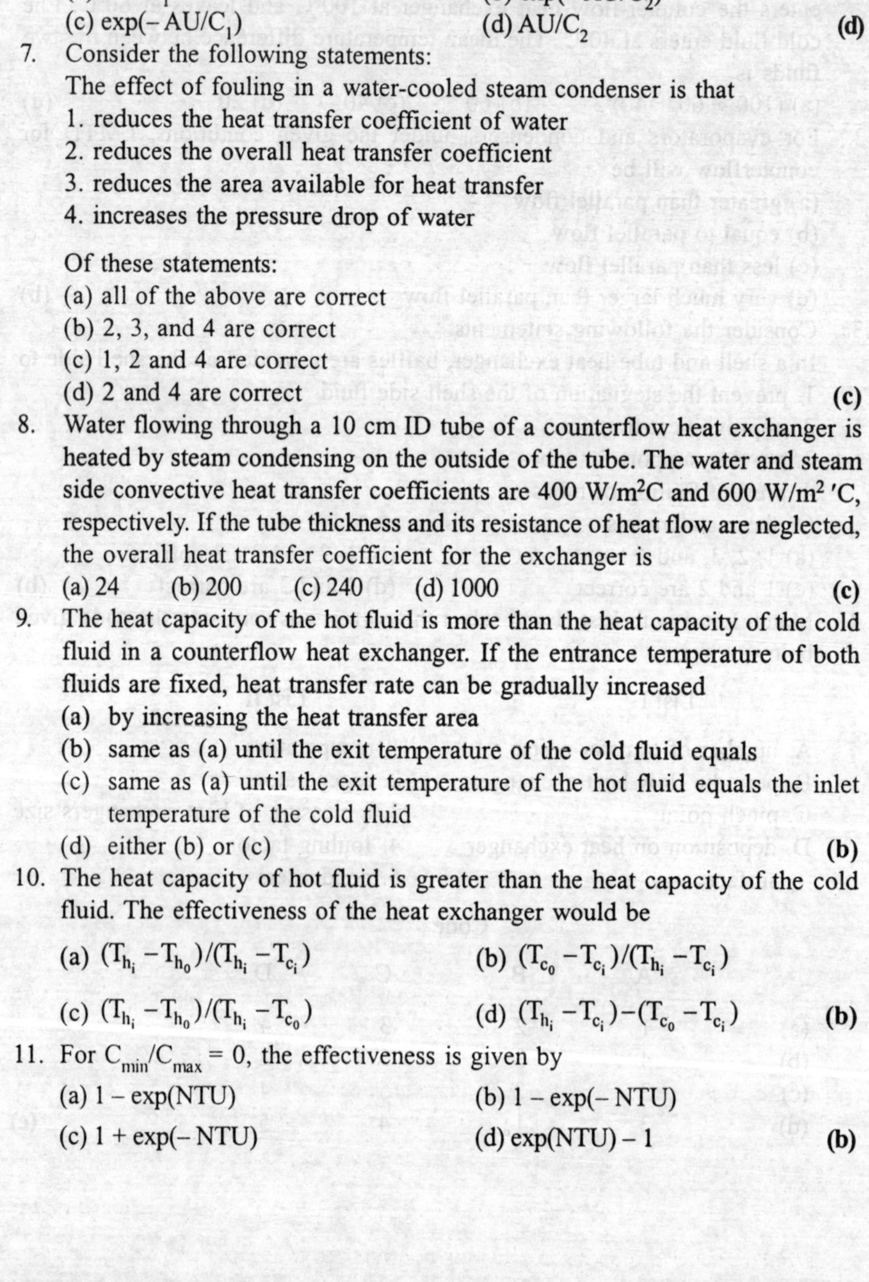

7. Consider the following statements:
   The effect of fouling in a water-cooled steam condenser is that
   1. reduces the heat transfer coefficient of water
   2. reduces the overall heat transfer coefficient
   3. reduces the area available for heat transfer
   4. increases the pressure drop of water

   Of these statements:
   (a) all of the above are correct
   (b) 2, 3, and 4 are correct
   (c) 1, 2 and 4 are correct
   (d) 2 and 4 are correct   **(c)**

8. Water flowing through a 10 cm ID tube of a counterflow heat exchanger is heated by steam condensing on the outside of the tube. The water and steam side convective heat transfer coefficients are 400 W/m^{2}C and 600 W/m^2 'C, respectively. If the tube thickness and its resistance of heat flow are neglected, the overall heat transfer coefficient for the exchanger is
   (a) 24   (b) 200   (c) 240   (d) 1000   **(c)**

9. The heat capacity of the hot fluid is more than the heat capacity of the cold fluid in a counterflow heat exchanger. If the entrance temperature of both fluids are fixed, heat transfer rate can be gradually increased
   (a) by increasing the heat transfer area
   (b) same as (a) until the exit temperature of the cold fluid equals
   (c) same as (a) until the exit temperature of the hot fluid equals the inlet temperature of the cold fluid
   (d) either (b) or (c)   **(b)**

10. The heat capacity of hot fluid is greater than the heat capacity of the cold fluid. The effectiveness of the heat exchanger would be
    (a) $(T_{h_i} - T_{h_o})/(T_{h_i} - T_{c_i})$ (b) $(T_{c_o} - T_{c_i})/(T_{h_i} - T_{c_i})$
    (c) $(T_{h_i} - T_{h_o})/(T_{h_i} - T_{c_o})$ (d) $(T_{h_i} - T_{c_i}) - (T_{c_o} - T_{c_i})$ **(b)**

11. For $C_{min}/C_{max} = 0$, the effectiveness is given by
    (a) $1 - \exp(NTU)$ (b) $1 - \exp(-NTU)$
    (c) $1 + \exp(-NTU)$ (d) $\exp(NTU) - 1$ **(b)**

12. Assertion (A): If $C_{max} = C_c$, it is not possible that T_{c_0} be equal to T_{h_i} .

    Reasoning (R): Because this would require $(T_{h_i} - T_{h_0})$ greater than $(T_{h_i} - T_{c_i})$

    Code: (a) Both A and R are true   (b) Both A and R are false
           (c) A is true, R is false   (d) A is false, R is true   **(a)**

13. Assertion (A): Regenerators are being used successfully as air preheaters in gas turbines.

    Reasoning (R): Because they provide a high heat transfer effectiveness per unit weight and space.

    Code: (a) Both A and R are true   (b) A is true, R is false
           (c) A is false, R is true   (d) Both A and R are false   **(a)**

14. The thermal analysis of a shell and tube heat exchanger assumes :
    1. The overall heat transfer coefficient U is uniform throughout the heat exchanger surface.
    2. The physical properties of the fluids do not vary with temperature.
    3. U can be calculated satisfactorily from the available correlation for predicting individual heat transfer coeffecients:

    of these statements
    (a) only 1 is correct   (b) 1 and 3 are correct
    (c) 2 and 3 are correct   (d) 1, 2 and 3 are correct   **(d)**

15. Which one of the following devices are not used for increasing the overall heat transfer coefficient in designing a condenser:
    (a) treating the surface   (b) swirl flow devices
    (c) extended surfaces   (d) roughening the surface   **(b)**

NUMERICALS

1. In a counter flow heat exchanger, 680 kg/h of oil flows through a thin walled tube, 1 cm in diameter. 550 kg/h of water flows through the annulus. The oil enters at 150°C and leaves at 70°C while water enters at 10°C. Calculate the length of the tube required. The heat transfer coefficient between oil and tube is 2.3 kW/m²K and between water and tube 5.7 kW/m²K, the specific heat capacity of the oil is 2.18 kJ/kgK. Estimate the length of the tube for the parallel flow arrangement.

    (8.73m, 13.67m)

2. Steam condenses on the outside of a thin 2.5 cm diameter copper tube, 5.5m long. The steam pressure being 0.07 bar. Water flows through the tube at the rate of 1.2 m/s, entering at 12°C and leaving at 24°C. Calculate the overall heat transfer coefficient and the heat transfer coefficient between water and tube if the heat transfer coefficient between steam and tube is 17 kW/m²K.

    (3.35 kW/m²K, 4.17 kW/m²K)

3. A heat exchanger consists of ten 2 m long pipes of 2.5 cm bore and 2.9 cm outer diameter, each running inside concrete pipes. Hot oil, specific heat capacity 1.68 kJ/kgK, flow rate 900 kg/h and initially at 250°C flows in the opposite direction to cold oil flow rate 1350 kg/h, initial temperature of 15°C.

Assuming an average overall heat transfer coefficient of 170 W/m²K, calculate the outet temperature of the oil.

(147.6°C, 83.3°C)

4. A heat exchanger cools 55000 kg/h of alcohol (c = 3760 J/kgK) from 60°C to 40°C using 4000 kg/h water entering at 5°C. Calculate the surface area required for (a) parallel flow (b) counter flow when the overall heat transfer coefficient is 580 W/m²K and c for water is 4180 J/kgK.

(136m², 82.5m²)

5. In a double pipe heat exchanger, hot gas enters the pipe (75 mm outside diameter) at 350°C and is cooled to 100°C. The gas flow rate is 200 kg/h and the specific heat capcity is 1.13 kJ/kgK. Water flow rate 1400 kg/h specific heat capacity 4.19 kJ/kgK, enters at 10°C. The heat transfer coefficients for the gases and water may be taken as 0.3 and 1.5 kW/m²K, and the pipe thickness may be taken to be as negligible. Calculate the required pipe length for (i) parallel flow, and (ii) counter flow.

(1.48m, 1.44m)

6. A fluid, mass flow rate 11.8 kg/s, density 1100 kg/m³, specific heat capacity 4.6 kJ/kgK, is to be heated from 65°C to 100°C in a shell and tube heat exchanger. The fluid flows through tubes 25 mm inner diameter at about 1.2 m/s. The fluid is being heated by wet steam at 115°C. If the heat transfer coefficient at the inside and outside surface of the tube is 5 and 10 kW/m²K and the length of the tube does not exceed 3.5 m, estimate the number of tubes and the number of tube passes required. Neglect the thermal resistance of the tube wall.

(18, 4)

7. An oil engine develops 300 kW and the specific fuel consumption is 0.21 kg/kWh. The exhaust from the engine is used in a tubular water heater, flowing through 25 mm diameter tubes, entering with a velocity of 12 m/s at 340°C and leaving at 90°C. The water enters the heater of 10°C and leaves at 90°C, flowing in counter flow to the hot gases. The air fuel ratio of the engine is 20, and the exhaust pressure is 1.01 bar. The overall heat transfer coefficient of the heat exchanger when designed is found to be 56 W/m²K, but after running for some time a fouling factor of 0.5 m²K/kW must be assumed. Taking the specific heat capacity and the gas constant for the gases as 1.11 kJ/kgK and 0.29 kJ/kgK, and the specific heat capacity for the water as 4.19 kJ/kgK, calculate (i) the mass flow rate of water, (ii) the number of tubes required, and (iii) the required tube length.

(1096 kg/h, 110, 1.457 m)

8. Steam at 1 bar enters the shell of a surface condenser (h_{fg} = 2260 kJ/kg). Cooling water (flow rate 216 kg/h) enters the condenser at 20°C and leaves at 75°C. The overall heat transfer coefficient is 250 W/m²K. Calculate the effectiveness, length of the tube and steam condensation rate.

(0.687, 12.4 m, 22.9 kg/h)

9. In a cross flow heat exchanger pressurized water, mass flow rate 1.5 kg/s, enters at 35°C and leaves at 125°C. The hot gases enter at 300°C and come

out at 100°C. The overall heat transfer coefficient is 110 W/mK. Using NTU approach, calculate the effectiveness and the required surface area.

(0.754, 54m²)

10. Dry and saturated steam a t 50°C enters a surface condenser a t the rate of 3000 kg/h. Water enters the tube (inner diameter 25 mm, outer diameter 30 mm, k = 110W/mK) with a velocity of 2 m/s. If the convective heat transfer coefficient for steam side is 5500 W/m²K, calculate the number and length of water tubes for a single pass. Water enters at 25°C and comes out at 35°C. For water, $\rho = 1000$ kg/m³, c = 4.2 kJ/kgK, $\nu = 1 \times 10{-6}$ m²/s, k = 0.6 W/mK.

(48, 2.55 m)

11. A cross-flow heat exchanger with air (c = 1.006 kJ/kgK) at 25°C and mass flow rate o f 14 k g/min is u sed to c ool water, f low rate 4 kg/min a nd c = 4.186 kJ/kgK. Water is cooled from 99°C and 60°C. Water flows through a number o f s eparate p assages w ithin t he h eat e xchanger. The o verall h eat transfer coefficient is 80 W/m²K. Calculate the required heat transfer area and the effectiveness.

(7.33 m², 0.63)

12. A feed water heater having 5.74 m² area is used to heat water, mass flow rate 2.5 kg/s, from 30°C. The temperature of the condensing steam is 120°C. The exit temperature of water is 90°C. Calculate the value of overall heat transfer coefficient. After several years of operation, the outlet temperature of water is 80°C. Estimate the value of the fouling factor assuming that other parameters remain the same.

(0.177×10⁻³ Km²/W)

13. Water mass flow rate 2.5 kg/s flows through the tubes (25 mm diameter) of a shell and tube heat exchanger. Water is heated from 15 to 85°C. Oil at 160°C enters the shell side having convection heat transfer coefficient of 400 W/m²K on the outside of the tubes. Oil comes out at 100°C and the exchanger has 8 tube passes, with ten tubes per pass. Estimate the oil flow rate and the length of the tubes. The properties of oil and water are:

Oil: c = 2350 J/kgK,

Water: c = 4181 J/kgK, $\mu = 548 \times 10^{-6}$ Pa-s,   k = 0.643 W/mK, Pr = 3.56

(5.2 kg/s, 38 m)

14. Steam condenses inside a one-shell two pass condenser. Water flows through tubes (30,000 in number, inner diameter 0.025m, wall thickness negligible) at the rate of 3×10^4 Kg/s. The water enters at 20°C and steam condenses at 50°C. The steam side heat transfer coefficient is 11.0 KW/m². Calculate the temperature of the cooling water at exit and the length of the tube per pass.

μ for water = 855×10^{-6} Pa-s,   Pr = 5.83,   k = 0.613 W/mK,

(36°C, 4.5 m)

CHAPTER 11

Heat Transfer with Change of Phase

1. Condensation and Boiling

Heat energy is being converted into electrical energy with the help of water as a working fluid. Water is first converted into steam when heated in a heat exchanger and then the exhaust steam coming out of the steam turbine/engine is condensed in a condenser so that the condensate (water) is recycled again for power generation. Therefore, the condensation and boiling processes involve heat transfer with change of phase. When a fluid changes its phase, the magnitude of its properties like density, viscosity, thermal conductivity, specific heat capacity, etc., change appreciably and the processes taking place are greatly influenced by them. Thus, the condensation and boiling processes must be well understood for an effective design of different types of heat exchangers being used in thermal and nuclear power plants, and in process cooling and heating systems.

2. Condensation—Filmwise and Dropwise

Condensation is the process of transition from a vapour to the liquid or solid state. The process is accompanied by liberation of heat energy due to the change in phase. When a vapour comes in contact with a surface maintained at a temperature lower than the saturation temperature of the vapour corresponding to the pressure at which it exists, the vapour condenses on the surface and the heat energy thus released has to be removed. The efficiency of the condensing unit is determined by the mode of condensation that takes place:

Filmwise – the condensing vapour forms a continuous film covering the entire surface,

Dropwise – the vapour condenses into small liquid droplets of various sizes.

The dropwise condensation has a much higher rate of heat transfer than filmwise condensation because the condensate in dropwise condensation gets removed at a faster rate leading to better heat transfer between the vapour and the bare surface.

It is therefore desirable to maintain a condition of dropwise condensation in commercial application. Dropwise condensation can only occur either on highly polished surfaces or on surfaces contaminated with certain chemicals. Filmwise condensation is expected to occur in most instances because the formation of dropwise condensation is greatly influenced by the presence of non-condensable gases, the nature and composition of surfaces and the velocity of vapour past the surface.

3. Filmwise Condensation Mechanism on a Vertical Plane Surface—Assumption

Let us consider a plane vertical surface at a constant temperature, T_s on which a pure vapour at saturation temperature, T_g, $(T_g > T_s)$ is condensing. The coordinates are: X-axis along the plane surface with its origin at the top edge and Y-axis is normal to the plane surface as shown in Fig. 11.1. The condensing liquid would wet the solid surface, spread out and form a continuous film over the entire condensing surface. It is further assumed that

(i)   the continuous film of liquid will flow downward (positive X-axis) under the action of gravity and its thickness would increase as more and more vapour condenses at the liquid-vapour interface,

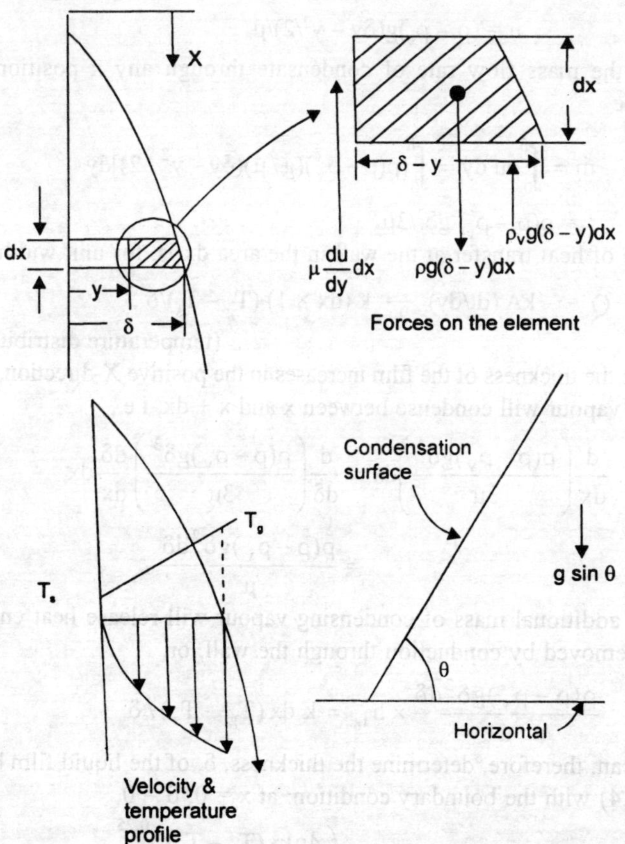

Fig. 11.1 Filmwise condensation on a vertical and inclined surface

(ii) the continuous film so formed would offer a thermal resistance between the vapour and the surface and would reduce the heat transfer rates,

(iii) the flow in the film would be laminar,

(iv) there would be no shear stress exerted at the liquid vapour interface,

(v) the temperature profile would be linear, and

(vi) the weight of the liquid film would be balanced by the viscous shear in the liquid film and the buoyant force due to the displaced vapour.

4. An Expression for the Liquid Film Thickness and the Heat Transfer Coefficient in Laminar Filmwise Condensation on a Vertical Plate

We choose a small element, as shown in Fig. 11.1 and by making a force balance, we write

$$\rho g(\delta - y)dx = \mu(du/dy)dx + \rho_v g (\delta - y)dx \tag{11.1}$$

where ρ is the density of the liquid, ρ_v is the density of vapour, μ is the viscosity of the liquid, δ is the thickness of the liquid film at any x, and du/dy is the velocity gradient at x.

Since the no-slip condition requires u = 0 at y = 0, by integration we get:

$$u = (\rho - \rho_v)g(\delta y - y^2/2)/\mu \tag{11.2}$$

And the mass flow rate of condensate through any x position of the film would be

$$\dot{m} = \int_0^\delta \rho u \, dy = \int_0^\delta [\rho(\rho - \rho_v)(g/\mu)(\delta y - y^2/2)]dy$$

$$= \rho(\rho - \rho_v) \, g\delta^3/3\mu \tag{11.3}$$

The rate of heat transfer at the wall in the area dx is, for unit width,

$$\dot{Q} = -kA \, (dt/dy)_{y=0} = k \, (dx \times 1) \, (T_g - T_s)/\delta \,,$$

$$\text{(temperature distribution is linear)}$$

Since the thickness of the film increases in the positive X-direction, an additional mass of vapour will condense between x and x + dx, i.e.,

$$\frac{d}{dx}\left(\frac{\rho(\rho - \rho_v)g\delta^3}{3\mu}\right)dx = \frac{d}{d\delta}\left(\frac{\rho(\rho - \rho_v)g\delta^3}{3\mu}\right)\frac{d\delta}{dx}dx$$

$$= \frac{\rho(\rho - \rho_v)g\delta^2 d\delta}{\mu}$$

This additional mass of condensing vapour will release heat energy and that has to removed by conduction through the wall, or,

$$\therefore \quad \frac{\rho(\rho - \rho_v)g\delta^2 d\delta}{\mu} \times h_{fg} = k \, dx \, (T_g - T_s)/\delta \tag{11.4}$$

We can, therefore, determine the thickness, δ, of the liquid film by integrating Eq. (11.4) with the boundary condition: at x = 0, δ = 0,

$$\text{or,} \qquad \delta = \left(\frac{4\mu kx(T_g - T_s)}{g \, h_{fg}\rho(\rho - \rho_v)}\right)^{0.25} \tag{11.5}$$

The rate of heat transfer is also related by the relation,

$$h \, dx \, (T_g - T_s) = k \, dx \, (T_g - T_s)/\delta \; ; \; or, \; h = k/\delta$$

which can be expressed in dimensionless form in terms of Nusselt number,

$$Nu = hx/k = \left[\frac{\rho(\rho - \rho_v)gh_{fg}x^3}{4\mu k(T_g - T_s)} \right]^{0.25} \tag{11.6}$$

The average value of the heat transfer coefficient is obtained by integrating over the length of the plate:

$$\bar{h} = (1/L) \int_0^L h_x \cdot dx = (4/3)h_{x = L}$$

$$Nu_L = 0.943 \left[\frac{\rho(\rho - \rho_v)gh_{fg}L^3}{k\mu(T_g - T_s)} \right]^{0.25} \tag{11.7}$$

The properties of the liquid in Eq. (11.7) and Eq. (11.6) should be evaluated at the mean temperature, $T = (T_g + T_s)/2$.

The above analysis is also applicable to a plane surface inclined at angle θ with the horizontal, if g is everywhere replaced by g . $\sin\theta$.
Thus:

Local
$$Nu_x = 0.707 \left[\frac{\rho(\rho - \rho_v)h_{fg}x^3 g \sin\theta}{\mu k(T_g - T_s)} \right]^{0.25}$$

and the average
$$Nu_L = 0.943 \left[\frac{\rho(\rho - \rho_v)h_{fg}L^3 g \sin\theta}{\mu k(T_g - T_s)} \right]^{0.25} \tag{11.8}$$

These relations should be used with caution for small values of θ because some of the assumptions made in deriving these relations become invalid; for example, when θ is equal to zero, (a horizontal surface) we would get an absurd result. But these equations are valid for condensation on the outside surface of vertical tubes as long as the curvature of the tube surface is not too great.

5. An Expression for 'h' in Turbulent Liquid Film

Like all convection phenomena, filmwise condensation may have a turbulent character. If the vertical surface is sufficiently long such that the liquid film becomes thick enough to cause transition to turbulent flow, the assumptions made in deriving equation (11.6) will not be valid. Therefore, in order to determine the onset of turbulence, a condensation film Reynolds number is defined as,

$$Re = D_h \rho V/\mu = 4 A \rho V/P\mu$$

where D_h is the hydraulic diameter and is equal to 4 A/P,

A is area of flow, P is the wetted perimeter and V is the average velocity of flow.

Since it is a little difficult to determine the average velocity of flow, the Reynolds number is defined on the basis of mass rate of flow,

or,
$$Re = 4 \, \dot{m}/P\mu \qquad (\therefore \dot{m} = \rho A \, V) \tag{11.9}$$

By making an energy balance, the heat transfer coefficient is given by,

$$h\,A\,(T_g - T_s) = \dot{m}\,h_{fg}; \text{ and } \dot{m} = h\,A\,(T_g - T_s)/h_{fg}$$

and
$$Re = 4\,h\,A\,(T_g - T_s)/(P\,\mu\,h_{fg})$$

For a vertical plate of height L and width W, the heat flow area, A = LW and the wetted perimeter, P = W.

For a vertical tube of height L and diameter D, A = πDL, and P = πD,

Thus for both cases, A/P = L, and, Re = 4 h L $(T_g - T_s)/ \mu\, h_{fg}$ $\qquad$ (11.10)

and the flow is considered to be turbulent for Re > 1800.

Experimental investigations reveal that ripples start developing in the film when the Reynolds number is around 40. And, the average value of the heat transfer coefficient increases by about 20 percent. McAdams has suggested that the increased heat transfer coefficient can be calculated by the relation:

$$h = 1.13 \left[\frac{\rho(\rho - \rho_v)gh_{fg}k^3}{\mu(T_g - T_s)} \right]^{0.25} \qquad (11.11)$$

When turbulence is encountered in the film (Re > 1800), an empirical relation by Kirkbride may be used:

$$h = \left[\frac{k^3\rho(\rho - \rho_v)g}{\mu^2} \right]^{1/3} \times 0.0077\,Re^{0.4} \qquad (11.12)$$

6. Condensation of Superheated Vapour

When a superheated vapour at temperature T' condenses on a cold surface at temperature T_s, the quantity of heat to be removed per kg of condensing vapour is given by

$$h'_{fg} = h_{fg} + C_p(T' - T_g)$$

where C_p is the specific heat of the superheated vapour. Therefore h_{fg} in Eq. (11.7) should be replaced by h'_{fg}.

Since the contribution made by the quantity $C_p(T' - T_g)$ is much smaller than h_{fg}, it is fairly accurate to use Eq. (11.7) for evaluating the heat transfer coefficient of superheated vapour condensing on isothermal surfaces.

7. The Effect of Non-condensable Gases and the Vapour Velocity

Experiments show that the presence of non-condensable gases significantly decreases the heat transfer coefficient and that varies almost linearly with the mass fraction of the non-condensable gases. Because, when a vapour containing non-condensable gas condenses, the non-condensable gas is left at the surface. Any further condensation at the surface will take place only after the incoming vapour diffuses through the non-condensable gas collected in the vicinity of the surface. The non-condensable gas, thus acts as a thermal resistance to the condensation process.

The vapour velocity changes the velocity distribution of the liquid film and gives rise to shearing stress at the liquid-vapour interface. Further, if the vapour flow is upward, the thickness of the film would increase and the thickness of the liquid film will decrease for downward flow of vapour. And, the transition from laminar to turbulent flow occurs at condensate Reynolds number of the order of 300 when the vapour velocity is high. One of the assumptions made in deriving Eq. (11.6), Nusselt's Film Theory', is that the shearing stress at the liquid vapour interface is negligible, and therefore, Eq. (11.6) ceases to be valid when the velocity of the un-condensed vapour is significant in comparison with the velocity of the liquid at the liquid vapour interface.

8. Improvement of 'h' in Filmwise Condensation

The filmwise condensation heat transfer can be substantially improved by decreasing the liquid film thickness which acts as a thermal resistance and by inducing turbulence. Thus, finned tubes and rough surfaces have been designed to improve the condensate mass flow rate. Further, the non-condensable gases should be extracted as far as possible because their presence reduces the heat transfer coefficient by more than fifty percent.

Example 11.1   A vertical plate 30 cm wide and 1.2 m high is maintained at 70°C and is exposed to saturated steam at 1 atm pressure. Calculate the heat transfer coefficient and the total mass of steam condensed per hour.

Solution: The mean temperature of the film is $(70 + 100)/2 = 85°C$. The properties of water at this temperature are:

$$\rho = 968 \text{ kg/m}^3, \qquad \mu = 3.37 \times 10^{-4} \text{ Pa-s}, \qquad k = 0.674 \text{ W/mK}.$$

Assuming laminar flow, the average heat transfer coefficient is given by Eq. (11.7), which can be written as:

$$h = 0.943 \left[\frac{\rho(\rho - \rho_v)gh_{fg}k^3}{L\mu(T_g - T_s)} \right]^{0.25} ; \qquad \text{Since } \rho_v \ll \rho, \text{ we have}$$

$$h = 0.943 \left[\frac{\rho^2 h_{fg} k^3 g_\lambda}{L\mu(T_g - T_s)} \right]^{0.25}$$

$$h = 0.943 \left[\frac{(968)^2 \times 9.81 \times 2255 \times 10^3 \times (0.674)^3}{1.2 \times 3.37 \times 10^{-4} \times 30} \right]^{0.25} = 4.51 \times 10^3 \text{ W/m}^2\text{K}$$

$$Re = 4hL\frac{(T_g - T_s)}{(h_{fg}\mu)} = \frac{4 \times 4.51 \times 10^3 \times 1.2 \times 30}{2255 \times 10^3 \times 3.37 \times 10^{-4}}$$

$$= 854.6 < 1800, \text{ hence a laminar flow.}$$

Heat transfer through the plate, $Q = hA (\Delta T) = 4.51 \times 10^3 \times (1.2 \times 0.3) \times 30$
$$= 48.7 \text{ kW}.$$

Mass flow rate of condensate, $\dot{m} = \dot{Q}/h_{fg} = (48.7/2255) \times 3600$

$$= 77.76 \text{ kg/hr.}$$

(Since the Reynolds number is greater than 40, ripples are likely to appear, and the average heat transfer coefficient would increase by about 20 percent, but the use of Eq. (11.7), is preferred because it gives a conservative value of the average heat transfer coefficient and it provides the factor of safety in the design problems.)

If the plate of Ex. 11.1 is inclined at $30°$ to the vertical ($\theta = 60°$), the average heat transfer coefficient is given by

$$\bar{h} = 0.943 \left[\frac{\rho^2 h_{fg} k^3 g \sin\theta}{L\mu(T_g - T_s)} \right]^{0.25}$$

$$= 0.943 \left[\frac{(968)^2 \times 9.81 \times \sin 60^0 \times 2255 \times 10^3 \times (0.674)^3}{1.2 \times 3.37 \times 10^{-4} \times 30} \right]^{0.25}$$

$$= 4.35 \times 10^3 \text{ W/m}^2\text{K, a lower value.}$$

Example 11.2 A vertical square plate 40 cm by 40 cm at 70°C is exposed to saturated steam at 1 atm pressure. Calculate the film thickness, the maximum velocity and the local heat transfer coefficient at the bottom edge of the plate.

Solution: Since the mean film temperature is 85°C, we use the properties from Ex. 11.1 The film thickness is given by Eq. (11.5),

$$\delta = \left[\frac{4\mu kL(T_g - T_s)}{g\, h_{fg}\rho^2} \right]^{0.25} = \left[\frac{4 \times 3.37 \times 10^{-4} \times 0.674 \times 0.4(30)}{9.81 \times 2255 \times 10^3 \times (968)^2} \right]$$

$$= 0.151 \text{ mm}$$

The maximum velocity at the bottom edge of the plate is obtained by using Eq. (11.2) which can be written as:

$$u = (\pi g/\mu)(\delta y - y^2/2); \quad \text{and} \quad \text{at } y = \delta, \quad u_{max} = (\rho g \delta^2/2\mu)$$

or, $$u_{max} = \frac{968 \times 9.81 \times (0.151 \times 10^{-3})^2}{2 \times 3.37 \times 10^{-4}} = 0.32 \text{ m/s}$$

The local heat transfer coefficient at $x = L$, is given by

$$h_L = \left[\frac{\rho^2 g h_{fg} k^3}{4\mu L(T_g - T_s)} \right]^{0.25} = \left[\frac{(968)^2 \times 9.81 \times 2255 \times 10^3 \times (0.674)^3}{4 \times 3.37 \times 10^{-4} \times 0.4 \times 30} \right]^{0.25}$$

$$= 4.45 \text{ kW/m}^2\text{K; and the average heat transfer coefficient would be}$$

$$h_{av} = (4/3)h_{x=L} = 5.934 \text{ kW/m}^2\text{K}$$

And, the total heat flow rate would be, $\dot{Q} = h_{av} A (\Delta T)$

$$= 5.934 \times 0.16 \times 30 = 28.48 \text{ kW}$$

The mass of the condensate, $\dot{m} = \dot{Q}/h_{fg} = 28.48 \times 3600/2255 = 45.47 \text{ kg/hr.}$

Example 11.3   Saturated steam at 1 atmosphere pressure condenses on the outer surface of a vertical tube (outer diameter 80 mm, length 1 m). The tube surface is maintained at 50°C by the flow of cold water through the tube. Estimate the rate of heat transfer to the cold water and the rate of condensation of steam.

Solution: Properties of water at mean temperature (75°C) are:

$$\rho = 975 \text{ kg/m}^3; \quad \mu = 375 \times 10^{-6} \text{ P}_{a-s}; \quad k = 0.668 \text{ W/mK}; \quad C_p = 4193 \text{ J/kgK}.$$

Properties of saturated vapour: $\rho_v = 0.596 \text{ kg/m}^3; \quad h_{fg} = 2257 \text{ kJ/kg}$

The average value of heat transfer coefficient over a vertical plate can be determined by using Eq. (11.7). The same equation can also be applied for the evaluation of heat transfer coefficient over a vertical tube provided the thickness of the film at the bottom of the vertical tube is much less than the radius of the vertical cylinder.

The film thickness, δ, is given by Eq. (11.5), or,

$$\delta = [\{4 \times 0.668 \times 375 \times 10^{-6} \times (100 - 50) \times 1.0\}/\{9.8 \times 975(975 - 0.596)2257 \times 10^3\}]^{\frac{1}{4}}$$

$$= 0.22 \text{ mm which is much less than the radius of the tube (40 mm)}$$

From Eq (11.7), we have the average heat transfer coefficient as:

$$h = 0.943 \, [\{g \, \rho \, (\rho - \rho_v) \, k^3 \, h_{fg}\}/\{\mu(T_g - T_s)L\}]^{\frac{1}{4}}$$

$$= 0.943 \, [9.8 \times 975 \, (975 - 0.596) \times 0.668^3 \times 2257 \times 10^3] \, /$$

$$[375 \times 10^{-6} (100 - 50) \times 1]^{\frac{1}{4}}$$

$$= 4031.5 \text{ W/m}^2\text{K}$$

Rate of heat transfer to the cold water, $\dot{Q} = h \, A \, (\Delta T)$

$$= 4031.5 \times 3.142 \times 0.08 \times 1 \times 50 = 50.67 \text{kW}$$

Condensation rate, $\dot{m} = \dot{Q}/h_{fg} = 50.67 \times 3600/2257 = 80.82 \text{ kg/hr}$

Eq. (11.7) assumes laminar flow conditions and therefore we must calculate the Reynolds number.

$$\text{Re} = 4 \times \dot{m}/(\mu \times \text{width of the surface})$$

$$= 4 \times (80.82/3600)/(375 \times 10^{-6} \times 3.142 \times 0.08) = 952.68$$

Since the Reynolds number is greater than 30 and less than 1800, a wave free laminar flow is not possible.

Since the density of the liquid is much more than the density of the vapour Kutateladze recommends a correlation of the form, to be used in the laminar wavy region:

$$\frac{h(v^2/g)^{1/3}}{k} = \text{Re}/(1.08\text{Re}^{1.22} - 5.2), \quad \text{where} \quad \text{Re} = 4g \, \rho^2\delta^3/3\mu^2$$

$$\text{Re} = 4 \times 9.8 \times 975^2 \times (0.22 \times 10^{-3})^3/ \, 3(375 \times 10^{-6})^2 = 940$$

and $\qquad h = [940/(1.08 \times 940^{1.22} - 5.2)] \times k/(v^2/g)^{1/3} = 5548.6 \text{ W/m}^2\text{K}$

Rate of heat transfer, $\dot{Q} = 5548.6 \times 3.142 \times 0.08 \times 1 \times 50 = 69.73 \text{ kW}$

and $\qquad\qquad\qquad \dot{m} = \dot{Q}/h_{fg} = 0.03 \text{ kg/s}.$

Rohsenow has recommended that h_{fg} in Eq (11.7) should be replaced by $h'_{fg} = h_{fg} \, (1 + 0.68 \, \text{Ja})$ where Ja, the Jakob number $= C_p \, (T_g - T_s); T_g$ being the saturation temperature of the condensing vapour, in order to get a better correlation.

Example 11.4   A vertical plate 2.5 m high is maintained at 40°C and is exposed to saturated steam at 1atm pressure. Calculate the heat transfer rate per unit width of the plate.

Solution: The mean temperature of the film = $(100 + 40)/2 = 70°C$, and the properties of water are:

$$\rho = 978 \text{ kg/m}^3, \quad \mu = 0.4 \times 10^{-3} \text{ Pa–s}, \quad k = 0.662 \text{ W/mK}$$

The properties of saturated vapour at 100°C are: $\rho_v = 0.596 \text{ kg/m}^3$

$$h_{fg} = 2255 \text{ kJ/kg}$$

$$Re = 4 \, L \, h \, (T_g - T_s)/h_{fg} \, \mu$$

and for turbulent flow, $h = 0.0077 \, Re^{0.4} \, [k^3\rho(\rho-\rho_v)g/\mu^2]^{1/3}$

and therefore, $$Re = \frac{4L(T_g - T_s)}{\mu \, h_{fg}} \times 0.0077 \, Re^{0.4} \, [k^3 \, \rho(\rho-\rho_v) \, g/\mu^2]^{1/3}$$

or, $$Re^{0.6} = \left(\frac{4 \times 2.5 \times (100-40)}{0.4 \times 10^{-3} \times 2255 \times 10^3}\right) \times 0.0077 \left[\frac{(0.662)^3 \times 978(978-0.596)9.81}{(0.4 \times 10^{-3})^2}\right]^{1/3}$$

or, $Re = 3414.5 > 1800$

From Eq (11.12), $h = 0.0077 \, Re^{0.4} \, [k^3\rho \, (\rho - \rho_v) \, g/\mu^2]^{1/3}$

$$= 5.12 \times 10^3 \text{ W/m}^2\text{K}$$

and $$\dot{Q} = h \, A \, (\Delta T) = 5.12 \times 10^3 \times (2.5 \times 1) \, (60) = 768.5 \text{ kW}.$$

Example 11.5   Calculate the rate of heat transfer for the vertical plate of Ex. 11.1 when the plate is exposed to superheated steam at a pressure of 1 bar and 150°C, and the plate is maintained at 40°C.

Solution: The properties at the mean film temperature $(150 + 40)/2 = 95°C$ are

$$\rho = 962 \text{ kg/m}^3, \quad \mu = 2.96 \times 10^{-4} \text{ Pa–s}, \quad k = 0.68 \text{ W/mK}$$

For steam at 1 bar and 150°C, from steam tables:

$h_g = 2276.3 \text{ kJ/kg}, \quad h_f = 417.51 \text{ and } h'_{fg} = 2358.8 \text{ kJ/kg (superheated steam)}$

Assuming turbulent flow, we have

$$Re^{0.6} = \frac{4L(T_g - T_s)}{\mu \, h_{fg}} \times 0.0077 \, [k^3\rho \, (\rho - \rho_v) \, g/\mu^2]^{1/3}$$

$$= \frac{4 \times 1.2 \times 110 \times 0.0077}{2.96 \times 10^4 \times 2358.8 \times 10^3} \times \left[\frac{(0.68)^3(962)^2 9.81}{(2.96 \times 10^{-4})^2}\right]^{1/3}$$

or, $Re = 6155.31$ (a turbulent flow).

$$h = 0.0077 \, Re^{0.4} \, [k^3\rho \, (\rho - \rho_v) \, g/\mu^2]^{1/3} = 8056 \text{ W/m}^2\text{k}$$

$$\dot{Q} = h \, A \, (\Delta T) = 8056 \times (1.2 \times 0.3) \times 110 = 318.97 \text{ kW}$$

and $\dot{m} = 318.97/h'_{fg} = 486.81 \text{ kg/hr}.$

9. Filmwise Condensation on Horizontal Cylinders—Empirical Relations

When the vapour condenses on the outside surface of the horizontal cylinder, it has been observed that the film starts with zero thickness at the top of the cylinder, as shown in Fig. 11.2 (a). The average heat transfer coefficient for a pure vapour condensing on the outside surface of a horizontal tube can be evaluated by the same method which was used in deriving Eq. (11.7). And, the relevant relation is

$$Nu = 0.725 \, [\rho(\rho - \rho_v)] \, D^3 \, g \, h_{fg} / [\mu k \, (T_g - T_s)]^{0.25} \qquad (11.13)*$$

where D is the outside diameter of the cylinder. All properties are to taken at the mean film temperature and h_{fg} is to be evaluated at the saturation temperature. In most practical applications, the flow over the tubes is always laminar.

If we have a vertical array of horizontal cylinders placed on over another, Fig. 11.2 (b), the condensate from one tube falls on the tube just below and the condensate film on the second tube would start with a finite thickness. This results in a higher film resistance and a lower heat transfer coefficient. Thus each successive tube in a vertical array will yield a lower average heat transfer coefficient. Nusselt has analysed this problem and has suggested that the average heat transfer coefficient for an array of N horizontal tubes placed one above the other is given by

$$Nu = 0.725 \, [\{\rho \, (\rho - \rho_v) \, g \, D^3 \, h_{fg}\} / \{N\mu k \, (T_g - T_s)\}]^{0.25} \quad (11.14)$$

It has been experimentally observed that it is advantageous to stagger the tubes, as shown in Fig. 11.2(c), wherever it is possible.

By comparing Eqs. (11.7) and (11.13), we can obtain the relative values of the heat transfer coefficients for horizontal and vertical tubes:

or, $\qquad h_h/h_v = (0.725/0.943) \, (L/D)^{0.25} = 0.768 \, (L/D)^{0.25} \qquad (11.15)$

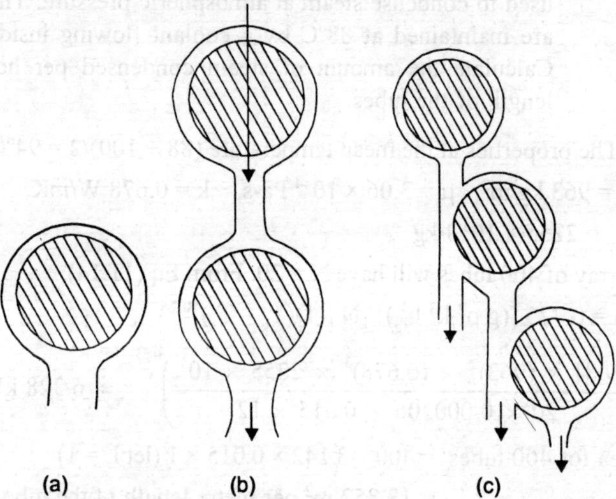

Fig. 11.2 Film condensation on horizontal cylinders

Example 11.6 A tube 1.5 cm outside diameter and 150 cm long, is used for condensing steam at 40 kPa. Calculate the average heat transfer coefficient when the tube is (a) horizontal (b) vertical and its surface is maintained at 50°C.

---

*Eq (11.13) can be used to evaluate the average heat transfer coefficient of a pure saturated vapour condensing on the outside surface of a sphere. The constant will be 0.815 instead of 0.725.

Solution: (a) *Tube Horizontal:* The mean film temperature is $(50 + 76) = 63°C$, and the properties are:

$$\rho = 980 \text{ kg/m}^3, \quad \mu = 0.432 \times 10^{-3} \text{ Pa–s}, \quad k = 0.66 \text{ W/mK},$$

$$h_{fg} = 2320 \text{ kJ/kg}, \quad \rho >> \rho_v$$

$$h = 0.725 \left[(\rho^2 \, h_{fg} \, k^3 \, g)/\mu \, D \, (T_g - T_s) \right]^{0.25}$$

$$= 0.725 \left[(980)^2 \times 2320 \times 10^3 \times (0.66)^3 \times 9.81/(0.432 \times 10^{-3} \times 0.015 \times 26) \right]^{0.25}$$

$$= 10 \text{ kW/m}^2\text{K}.$$

(b) *Tube Vertical:* Eq (11.7) should be used if the film thickness is very small in comparison with the tube diameter.

The film thickness, $\delta = \left[\{ (4\mu k L \, (T_g - T_s)) \} / \{ g \, h_{fg} \, \rho \, (\rho - \rho_v) \} \right] 0.25$

$$= \left[\frac{(980)^2 \times 9.81 \times 2320 \times 10^3 \times (0.66)^3}{0.432 \times 10^{-3} \times 1.5 \times 26} \right]^{0.25}$$

$$= 0.212 \text{ mm} << 15.0 \text{ mm, the tube diameter}$$

Therefore, the average heat transfer coefficient would be

$$h_v = h_h / [0.768 \, (L/D)^{0.25}] = 10/2.429 = 4.11 \text{ kW/m}^2\text{K}.$$

(Thus, the performance of horizontal tubes for filmwise laminar condensation is much better than vertical tubes and as such horizontal tubes are preferred.)

Example 11.7 A square array of four hundred tubes, 1.5 cm outer diameter is used to condense steam at atmospheric pressure. The tube walls are maintained at 88°C by a coolant flowing inside the tubes. Calculate the amount of steam condensed per hour per unit length of the tubes.

Solution: The properties at the mean temperature $(88 + 100)/2 = 94°C$ are:

$$\rho = 963 \text{ kg/m}^3, \quad \mu = 3.06 \times 10^{-4} \text{ Pa–s}, \quad k = 0.678 \text{ W/mK},$$

$$h_{fg} = 2255 \times 10^3 \text{ J/kg}$$

A square array of 400 tubes will have $N = 20$. From Eq (11.14),

$$h = 0.725 \left[(g \, \rho^2 \, k^3 \, h_{fg}) \right] / [N \, \mu \, D(T_g - T_s)]^{0.25}$$

$$= 0.725 \left(\frac{9.81 \times (963)^2 \times (0.678)^3 \times 2255 \times 10^3}{20 \times 0.000306 \times 0.015 \times 12} \right)^{0.25} = 6.328 \text{ kW/m}^2\text{K}$$

Surface area for 400 tubes $= 400 \times 3.142 \times 0.015 \times 1$ (let $L = 1$)

$$= 18.852 \text{ m}^2 \text{ per metre length of the tube}$$

$$\dot{Q} = hA \, (\Delta T) = 6.328 \times 18.852 \times 12 = 1431.56 \text{ kW}$$

$$\dot{m} = \dot{Q}/h_{fg} = 1431.56 \times 3600/2255 = 2285.4 \text{ kg/hr per metre length.}$$

10. Condensation Inside Tubes—Empirical Relation

The condensation of vapours flowing inside a cylindrical tube is of importance in chemical and petro-chemical industries. The average heat transfer coefficient for vapours condensing inside either horizontal or vertical tubes can be determined, within 20 percent accuracy, by the relations:

$$\text{For } Re_g < 5 \times 10^4, \quad Nu_d = 5.03 \, (Re_g)^{1/3} \, (Pr)^{1/3}$$
$$\text{For } Re_g > 5 \times 10^4, \quad Nu_d = 0.0265 \, (Re_g)^{0.8} \, (Pr)^{1/3} \tag{11.16}$$

where Re_g is the Reynolds number defined in terms of the mass velocity, or, $Re_g = DG/\mu$, G being the mass rate of flow per unit cross-sectional area.

11. Dropwise Condensation—Merits and Demerits

In dropwise condensation, the condensation is found to appear in the form of individual drops. These drops increase in size and combine with another drop until their size is great enough that their weight causes them to run off the surface and the condensing surface is exposed for the formation of a new drop. This phenomenon has been observed to occur either on highly polished surfaces or on surface coated/contaminated with certain fatty acids. The heat transfer coefficient in dropwise condensation is five to ten times higher than the filmwise condensation under similar conditions. It is therefore, desirable that conditions should be maintained for dropwise condensation in commercial applications. The presence of non-condensable gases, the nature and composition of the surface, the vapour velocity past the surface have great influence on the formation of drops on coated/contaminated surfaces and it is rather difficult to achieve dropwise condensation.

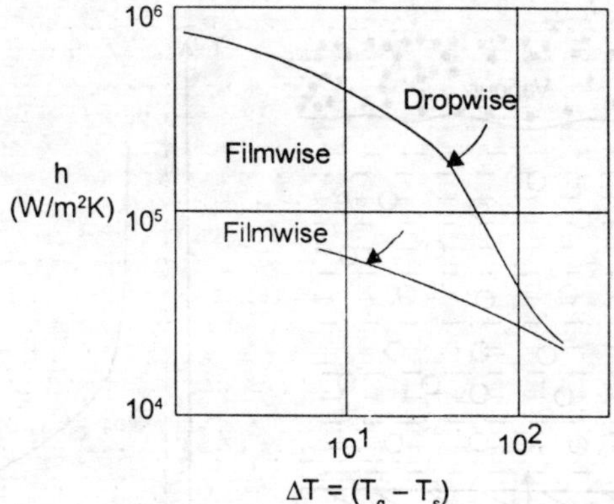

Fig. 11.3 Comparison of h for filmwise and dropwise condensation

Several theories have been proposed for the analysis of dropwise condensation. They do give explanations of the process but do not provide a relation to determine the heat transfer coefficient under various conditions. Fig 11.3 shows the comparison of heat transfer coefficient for filmwise and dropwise condensation.

12. Principles of Boiling Heat Transfer

The conversion of a liquid into its vapour by heating in an evaporator is of great importance in energy conversion systems. When a liquid at its saturation temperatures receives heat energy at its free surface, vapours are produced and the process is known as evaporation. But when a saturated liquid receives heat energy from a submerged solid surface at a temperature higher than the saturation temperature of the liquid, there is an intensive vapourization in the entire volume of the liquid accompanied by the formation of vapour bubbles rising from the heated surface to the free surface of the liquid and the process is called boiling. Heat transfer to the boiling liquid is a convection process and depends upon many factors such as, heating surface characteriscits, the surface tension, the latent heat of vapourization, the pressure, the density, thermal conductivity, specific heat of the liquid, etc.

13. Regimes of Boiling

Let us consider a heating surface (a wire or a flat plate) submerged in a pool of water which is at its saturation temperature. If the temperature of the heated surface exceeds the temperature of the liquid, heat energy will be transferred from the solid surface to the liquid. From Newton's law of cooling, we have:

$$\dot{Q}/A = \dot{q} = h\,(T_w - T_s)$$

where $\dot{Q}/A$ is the heat flux, T_w is the temperature of the heated surface and T_s is the temperature of the liquid, and the boiling process will start.

(i) *Pool Boiling* – Pool boiling occurs only when the temperature of the heated surface exceeds the saturation temperature of the liquid. The liquid above the hot surface is quiescent and its motion near the surface is due to free convection.

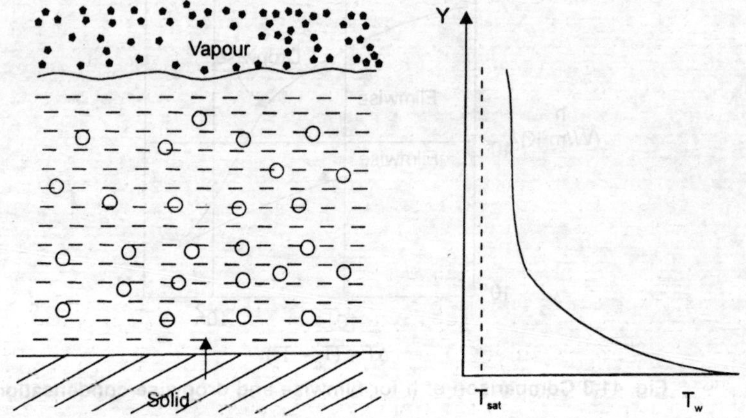

Fig. 11.4 Temperature distribution in pool boiling at liquid-vapour interface

Bubbles grow at the heated surface, get detached and move upward toward the free surface due to buoyancy effect. If the temperature of the liquid is lower than the saturation temperature, the process is called 'subcooled or local boiling'. If the temperature of the liquid is equal to the saturation temperature, the process is known as 'saturated or bulk boiling'. The temperature distribution in saturated pool boiling is shown in Fig. 11.4. When T_w exceeds T_s by a few degrees, the convection currents circulate in the superheated liquid and the evaporation takes place at the free surface of the liquid.

(ii) *Nucleate Boiling* – Fig. 11.5 illustrates the different regimes of boiling where the heat flux ($\dot{Q}/A$) is plotted against the temperature difference ($T_w - T_s$). When the temperature T_w increases a little more, vapour bubbles are formed at a

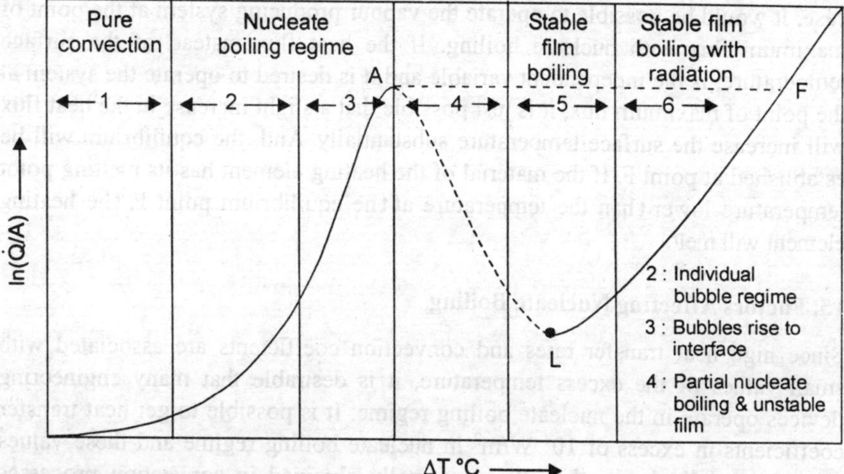

Fig. 11.5 Heat Flux – Temperature difference curve for boiling water heated by a wire (Nukiyama's boiling curve for saturated water at atmospheric pressure) (L is the Laidenfrost Point)

number of favoured spots on the heating surface. The vapour bubbles are initially small and condense before they reach the free surface. When the temperature is raised further, their number increases and they grow bigger and finally rise to the free surface. This phenomenon is called 'nucleate boiling'. It can be seen from the figure (11.5) that in nucleate boiling regime, the heat flux increases rapidly with increasing surface temperature. In the latter part of the nucleate boiling, (regime 3), heat transfer by evaporation is more important and predominating. The point A on the curve represents 'critical heat flux'.

(iii) *Film Boiling* – When the excess temperature, $\Delta T = (T_w - T_s)$ increases beyond the point A, a vapour film forms and covers the entire heating surface. The heat transfer takes place through the vapour which is a poor conductor and this increased thermal resistance causes a drop in the heat flux. This phase is film boiling'. The transition from the nucleate boiling regime to the film boiling regime is not a sharp one and the vapour film under the action of circulating currents collapses and rapidly reforms. In regime 5, the film is stable and the heat flow rate is the lowest.

(iv) *Critical Heat Flux and Burnout Point* – For ΔT beyond 550°C (regime 6) the temperature of the heating metallic surface is very high and the heat transfer occurs predominantly by radiation, thereby, increasing the heat flux. And finally, a point is reached at which the heating surface melts – point F in Fig. 11.5. It can be observed from the boiling curve that the whole boiling process remains in the unstable state between A and F. Any increase in the heat flux beyond point A will cause a departure from the boiling curve and there would be a large increase in surface temperature.

14. Boiling Curve — Operating Constraints

The boiling curve, shown in Fig. 11.5, is based on the assumption that the temperature of the heated surface can be maintained at the desired value. In that case, it would be possible to operate the vapour producing system at the point of maximum flux with nucleate boiling. If the heat flux instead of the surface temperature, is the independent variable and it is desired to operate the system at the point of maximum flux, it is just possible that a slight increase in the heat flux will increase the surface temperature substantially. And, the equilibrium will be established at point F. If the material of the heating element has its melting point temperature lower than the temperature at the equilibrium point F, the heating element will melt.

15. Factors Affecting Nucleate Boiling

Since high heat transfer rates and convection coefficients are associated with small values of the excess temperature, it is desirable that many engineering devices operate in the nucleate boiling regime. It is possible to get heat transfer coefficients in excess of 10^4 W/m^2 in nucleate boiling regime and these values are substantially larger than those normally obtained in convection processes with no phase change. The factors which affect the nucleate boiling are:

(a) *Pressure* – Pressure controls the rate of bubble growth and therefore affects the temperature difference causing the heat energy to flow. The maximum allowable heat flux for a boiling liquid first increases with pressure until critical pressure is reached and then decreases.

(b) *Heating Surface Characteristic*s – The material of the heating element has a significant effect on the boiling heat transfer coefficient. Copper has a higher value than chromium, steel and zinc. Further, a rough surface gives a better heat transfer rate than a smooth or coated surface, because a rough surface gets wet more easily than a smooth one.

(c) *Thermo-mechanical Properties of Liquids* – A higher thermal conductivity of the liquid will cause higher heat transfer rates and the viscosity and surface tension will have a marked effect on the bubble size and their rate of formation which affects the rate of heat transfer.

(d) *Mechanical Agitation* – The rate of heat transfer will increase with the increasing degree of mechanical agitation. Forced convection increases mixing of bubbles and the rate of heat transfer.

16. Empirical Correlations

(i) *Free Convection Boiling* – The rate of heat transfer during free convection (interface evaporation) can be obtained by using the relation

$$Nu = f(Gr, Pr)$$

and the functional relationship will depend upon the geometry of the heating surface. When a liquid is forced through a channel or over a surface at a temperature greater than the saturation temperature of the liquid, forced convection boiling occurs.

McAdams has suggested the following relationship for low pressure boiling water:

$$\dot{Q}/A = 2.253 \, (\Delta T_x)^{3.96} \text{ W/m}^2, \quad \text{for } 0.2 < p < 0.7 \text{ MPa} \qquad (11.17)$$

For high pressure boiling, the recommended relationship is

$$\dot{Q}/A = 283.2\ p^{4/3}\ (\Delta T_x)^3\ W/m^2, \quad \text{for } 0.7 < p < 14\ MPa \tag{11.18}$$

where ΔT_x is in °C and p is in MPa.

(ii) *Nucleate Boiling* – The relationship for nucleate pool boiling as suggested by Roshenow is:

$$\frac{C\Delta T_x}{h_{fg}Pr^s} = C_{sf}\left[\frac{\dot{Q}/A}{\mu h_{fg}}\sqrt{\left(\frac{\sigma}{g(\rho-\rho_v)}\right)}\right]^{0.33} \tag{11.19*}$$

where   C = specific heat of saturated liquid, J/kg °C

ΔT_x = temperature excess, $T_w - T_s$

h_{fg} = enthalpy of vapourization, J/kg

Pr = Prandtl number of the saturated liquids

s = 1.0 for water and 1.7 for other liquids

$\dot{Q}/A$ = heat flux per unit area, W/m²

μ = viscosity of the liquid, pa-s

σ = surface tension of liquid-vapour interface, N/m

g = acceleration due to gravity, m/s²

ρ = density of the saturated liquid, kg/m³

ρ_v = density of saturated vapour

C_{sf} = a constant to be determined experimentally.

Table 11.1 gives the values of vapour-liquid surface tension for water and Table 11.2 gives the values of the coefficient C_{sf} for various liquid-surface combinations.

(iii) *Maximum Heat Flux in Nucleate Boiling* – The maximum heat flux condition tells us that bubbles nearly form an insulating blanket over the heating surface.

Table 11.1 Vapour-liquid surface tension for water

Saturation temperature °C	Surface tension N/m
0	75.6 × 10⁻³
15.56	73.3
37.78	69.8
60	66.0
93.33	60.1
100	58.8
160	46.1
226.67	32.0
293.33	16.2
360	1.46
374.1	0

*Eq. (11.19) requires very accurate values of the property. The heat flux is very sensitive to Prandtl number values.

Table 11.2 Values of the coefficient C_{sf} for various liquid-surface combinations

Fluid-heating surface combination	C_{sf}
Water – copper	0.013
Water – brass	0.0060
Water – platinum	0.013
Water – ground and polished stainless steel	0.0080
Water-mechanically polished stainless steel	0.0130
Benzene – chromium	0.010
Ethylalcohol – chromium	0.027
n-Butyl alcohol – copper	0.027
Isopropyl alcohol – copper	0.00225
Water – Teflon pitted stainless steel	0.0058

Zuber has suggested the following relationship for the peak heat flux in nucleate boiling:

$$\left(\dot{Q}/A\right)_{max} = (\pi/24)h_{fg}\rho_v\left[\frac{\sigma g(\rho-\rho_v)}{\rho_v^2}\right]^{1/4}(1+\rho_v/\rho)^{1/2} \qquad (11.20)*$$

The above relation is in good agreement with the experimental data. In general, the type of surface material does not affect the peak heat flux, but cleanliness of the surface has an influence on the peak heat flux (dirty surfaces cause an increase of about 15% in the peak value).

(iv) *Stable Film Boiling* – Bromley has suggested the following relation for determining the heat transfer coefficients in the stable film boiling region on a horizontal tube:

$$h_b = 0.62\left[\frac{k_g^3(\rho-\rho_v)\rho_v g(h_{fg}+0.4C_{pv}\Delta T_x)}{D\cdot\mu v\cdot\Delta T_x}\right]^{0.25} \qquad (11.21)$$

where D is the tube diameter. This heat transfer coefficient does not include the effects of radiation. The total heat transfer coefficient can be calculated from the empirical relation:

$$h = h_b(h_b/h)^{1/3} + h_r; \quad \text{or} \quad h = h_b + 0.75\,h_r \qquad (11.22)$$

where h_r is the radiation heat transfer coefficient and can be obtained as

$$h_r = \frac{\sigma\varepsilon(T_w^4-T_s^4)}{(T_w-T_s)}; \sigma \text{ is Stefan-Boltzmann Constant} \qquad (11.23)$$

The properties of vapour in Eq. (11.21) should be evaluated at temperature $T = (T_w + T_s)/2$.

---

* Since h_{fg} for water is very large, water will sustain a larger heat flux than any other common liquid. Further the heat flux will decrease in a reduced gravitational field and in a field of zero gravity, vapour will not leave the solid and critical heat flux tends towards zero.

(v) *Simplified Relations for Boiling Heat Transfer with Water:* Since water is being converted into steam for its use as a working fluid in steam prime-movers of conventional thermal power plants, many empirical relations have been developed to determine the boiling heat transfer coefficients for water. Jakob and Hawkins have presented simple relations for the evaluation of heat transfer coefficients for water boiling on the outside surfaces of submerged bodies at atmospheric pressure:

(i) Horizontal surface: $h = 1042\,(\Delta T_x)^{1/3}$, $\dot{Q}/A < 16\ \text{kW/m}^2$.

$\qquad\qquad\qquad h = 5.56\,(\Delta T_x)^3$, $\quad 16 < \dot{Q}/A < 240,\ \text{kW/m}^2$ (11.24)

(ii) Vertical surface: $\quad h = 537\,(\Delta T_x)^{1/7}$, $\dot{Q}/A < 3\ \text{kW/m}^2$

$\qquad\qquad\qquad h = 7.96\,(\Delta T_x)^3$, $\quad 3 < \dot{Q}/A < 63\ \text{kW/m}^2$.

The influence of pressure on heat transfer coefficients can be obtained by using the relation:

$$h_p = h_1\,(p/p_1)^{0.4} \qquad\qquad (11.25a)$$

(heat transfer coefficient increases with pressure)

where h_p is the heat transfer coefficient at the desired pressure p; h_1 is the heat transfer coefficient at the atmospheric pressure p_1.

The flow of a boiling liquid inside a vertical tube is accompanied by a continuous increase of the vapour fraction (bubbles appearing at the surface grow and are carried into the mainstream of the liquid) and decrease in the liquid fraction of a two-phase mixture. The hydrodynamic structure of the flow keeps on changing both along the tube and over the cross-section and therefore, the rate of heat transfer also changes. For forced convection local boiling inside vertical tubes, the following relation has been recommended by Jakob:

$$h = 2.54\,(\Delta T_x)^3\ \exp(p/1.551)\ \text{W/m}^2\,{}^\circ\text{C} \qquad\qquad (11.25b)$$

where ΔT_x is the temperature difference between the heated surface and the saturated liquid in °C and p is the pressure in MPa. The above equation is valid for a pressure range of 5 to 170 atm pressure.

Example 11.8 It is desired to generate 100 kg/hr of saturated steam at 100°C using a heating element of copper, surface area 5 m². Calculate the convective heat transfer coefficient, the temperature of the heating surface and the critical heat flux.

Solution: Mass of steam to be evaporated is 100 kg/hr and the enthalpy of vapourization at 100°C is 2255 kJ/kg.

$\therefore \qquad\qquad \dot{Q} = 2255 \times 1000 \times 100/3600 = 62639\ \text{J/s or W}$

$\qquad\qquad \dot{Q}/A = 62639/5 = 12528\ \text{W/m}^2$.

The relevant properties at 100°C are:

$\quad \rho = 958.4\ \text{kg/m}^3$, $\rho_v = 0.598\ \text{kg/m}^3$, $\mu = 0.283 \times 10^{-3}\ \text{Pa-s}$, $C = 4217\ \text{J/kgK}$

$\quad Pr = 1.75$, $\qquad s = 1.0$ for water, $C_{sf} = 0.013$, $\sigma = 58.8 \times 10^{-3}\ \text{N/m}$

From Eq (11.19)

$$\Delta T_x = \frac{C_{sf} \times h_{fg} \times Pr^s}{C}\left[\frac{\dot{Q}/A}{\mu h_{fg}}\sqrt{\left(\frac{\sigma}{g(\rho-\rho_v)}\right)}\right]^{0.33}$$

$$= \frac{0.013 \times 2255 \times 10^3 \times 1.75}{4217} \left[\frac{\left\{ 12528(58.8 \times 10^{-3} /(9.81 \times 957.8)\right\}^{1/2}}{0.283 \times 10^{-3} \times 2255 \times 10^3} \right]^{0.33}$$

$$= 4.5^\circ C$$

or, $T_w = 104.5^\circ C$

$$\dot{Q}/A = h\,(\Delta T), \qquad\qquad \therefore \quad h = 12528/4.5 = 2784 \text{ W/m}^2{}^\circ C$$

The critical heat flux is obtained by Eq (11.20),

$$(\dot{Q}/A)_{max} = (\frac{\pi}{24} \times 2255 \times 10^3 \times 0.598 \left[\frac{58.8 \times 10^{-3} \times 9.81 \times 957.8}{(0.598)^2} \right]^{0.25} \left[1 + \frac{0.598}{958.4} \right]^{1/2}$$

$$= 1107.15 \text{ kW/m}^2.$$

Example 11.9 Saturated steam is being generated at 50 bar under nucleate boiling conditions. Calculate the critical heat flux and the temperature of the ground a nd polished s tainless steel h eating surface.

Solution: The saturation temperature corresponding to 50 bar is 263.91 °C. The property values at this temperature are: from steam tables

$$\rho_v = 25.36 \text{ kg/m}^3, \quad \rho = 777.7 \text{ kg/m}^3, \quad h_{fg} = 1639.7 \text{ kJ/kg}, \quad \sigma = 24 \times 10^{-3} \text{ N/m},$$
$$C = 4731 \text{ J/kg C}, \quad \mu = 2.19 \times 10^{-4} \text{ Pa-s}, \quad Pr = 0.83, \quad C_{sf} = 0.0080$$

From Eq (11.20), the critical heat flux is given by

$$\left(\dot{Q}/A \right)_{max} = (\pi/24) \times 1639.7 \times 10^3 \times 25.36$$

$$\left[\frac{24 \times 10^{-3} \times 9.81(777.7 - 25.36)}{25.36 \times 25.36} \right]^{1/4} \left(1 + \frac{25.36}{777.7} \right)^{1/2}$$

$$= 4.0 \times 10^6 \text{ W/m}^2$$

From Eq (11.19)

$$\Delta T_x = \frac{0.008 \times 1639.7 \times 10^3 \times 0.83}{4731} \left[\frac{4.0 \times 10^6 \times 10^4}{2.19 \times 1639700} \left(\frac{0.024}{9.81(777.7 - 25.36)} \right)^{1/2} \right]^{1/3}$$

$$= 6.25^\circ C$$

$$\therefore \; T_w = 263.91 + 6.25 = 270.16^\circ C.$$

Example 11.10 Water at atmospheric pressure is boiling on a Teflon-coated stainless steel surface heated electrically. Calculate the heat flux when the surface temperature is 105°C.

Solution: From Table (11.2) $C_{sf} = 0.0058$; From steam tables the properties are:

$$h_{fg} = 2255 \text{ kJ/kg}, \quad \rho = 958.4 \text{ kg/m}^3, \quad \rho_v = 0.598 \text{ kg/m}^3, \quad \mu = 0.283 \times 10^{-3} \text{ Pa-s},$$
$$C = 4217 \text{ J/kgK}, \quad Pr = 1.75, \quad s = 1.0 \text{ for water}, \quad \sigma = 58.8 \times 10^{-3} \text{ N/m}; \quad \Delta T = 5^\circ C$$

Eq (11.19) can be written as:

$$\dot{Q}/A = (C\,\Delta T/C_{sf}\,h_{fg}\,Pr)^3\,\mu\,h_{fg}\,(g(\rho - \rho_v)/\sigma)^{1/2}$$

$$= (4217 \times 5/0.0058 \times 2255 \times 10^3 \times 1.75)^3 \times 0.283 \times 10^{-3}$$

$$\times\, 2255 \times 10^3 \times \left[9.8\left(\frac{958.4 - 0.598}{0.0588} \right) \right]^{1/2}$$

$$= 199.33 \text{ kW/m}^2.$$

We can also calculate the critical heat flux by using Eq (11.20)

$$(\dot{Q}/A)_{max} = (3.142/24) \times 2255 \times 10^3 \times 0.598$$

$$\times \left[\frac{0.0588 \times 9.8 \times (958.4 - 0.598)}{(0.598)^2} \right] \left(1 + \frac{0.598}{958.4} \right)^{1/2}$$

$$= 1107.15 \text{ kW/m}^2.$$

Thus when ΔT is $5°$, the heat flux is much less than the critical value. Or, we are having nucleate pool boiling. Further, Eq. (11.19) tells us that the heat flux is proportional to $(C_{sf})^{1/3}$ and any other material which has a higher C_{sf} value, will give a lower heat flux. However, we can increase the heat flux by increasing the temperature difference. The required temperature difference for critical heat flux condition would be

$$\Delta T = (1107.15/199.33)^{1/3} \times 5 = 8.85°.$$

Example 11.11 A wire 2 mm in diameter and 20 cm long is submerged horizontally in a pool of water. The wire can carry a current of 40 amps. Estimate the voltage applied if the heat flux does not exceed 240 kW/m², and determine the wire surface temperature if the water boils at 8 bar.

Solution: The surface area of the wire $= \pi DL = 3.142 \times 0.002 \times 0.2$

$$= 1.2568 \times 10^{-3} \text{ m}^2$$

$$\dot{Q} = \text{area} \times 240 \times 10^3 \text{ (W)} = 1.2568 \times 10^{-3} \times 240 \times 10^3 = 301.63 \text{ W}$$

Voltage applied $= \dot{Q}/I = 301.63/40 = 7.54$ V

$$\dot{Q}/A = h\,(\Delta T); \quad \text{From Eq (11.24)}, \quad h = 5.56\,(\Delta T)^3$$

or, $\qquad \Delta T = (\dot{Q}/A/5.56)^{1/4}, \quad \therefore \quad \Delta T = (240 \times 10^3/5.56)^{1/4} = 14.41°C$

Therefore, $\quad T_w = 170.41 + 14.41 = 184.82°C.$

Example 11.12 Calculate the heat flux if the temperature of the heating surface of Ex. 11.8 is raised to 400°C.

Solution: In Ex. 11.8, the temperature of the heating surface is 270.16°C under critical flux conditions. Therefore, when the temperature of the surface is 400°C, we should expect film boiling. The property values are to be taken at the mean temperature, $T = (263.91 + 400)/2 = 331.5°C$. The value are: $\rho_v = 20.23$ kg/m³,

$$\rho = 777.7 \text{ kg/m}^3, \quad h_{fg} = 1639.7 \text{ kJ/kg}, \quad \sigma = 6.91 \times 10^{-3} \text{ N/m},$$

$$k_v = 0.0476 \text{ W/mK}, \quad C_{pv} = 9881 \text{ J/kgK}, \quad \mu_v = 20.93 \times 10^{-6} \text{ Pa–s}.$$

Equation (11.21) predicts heat transfer rate in the stable film boiling region on a horizontal tube. For flat surfaces, the relation is:

$$h_b = 0.425 \left[\frac{(\rho - \rho_v)^{1.5} g \rho_v k_v^{\,3} (h_{fg} + 0.68 C_{pv} \Delta T)}{\sigma^{0.5} \mu_v \Delta T} \right]^{0.25} \qquad (11.26)$$

$$= 0.425 \left[\frac{(777.7 - 20.23)^{1.5} 9.81 \times 20.23 \times (0.0476)^3}{(6.91 \times 10^{-3})^{0.5} \times 20.93 \times 10^{-6} \times 136.1} \right]^{0.25}$$

$$(1639.7 \times 10^3 + 0.68 \times 9881 \times 136.1)^{0.25}$$

$$= 268.2 \text{ W/m}^2{}^\circ\text{C}$$

Assuming emissivity of the surface as 0.8, the heat transfer coefficient due to radiation is,

$$h_r = \epsilon \sigma (T_w^4 - T_s^4)/(T_w - T_s) = 0.8 \times 5.67 \, [(6.73)^4 - (5.36)^4]/(673 - 536)$$
$$= 40.59 \text{ W/m}^2{}^\circ\text{C}.$$

From Eq. (11.22), total $h = 268.2 + 0.75 \times 40.59 = 298.65$;
and $Q/A = h \, (\Delta T) = 40.91$ kW/m^2 (a much lower value in comparison with critical heat flux having nucleate boiling).

Example 11.13 Saturated steam is being generated at the rate of 500 kg/hr at atmospheric pressure in a vessel having a surface area of 10 m^2. Estimate the heat transfer coefficient when the pressure is raised to 15 atmosphere and the temperature difference is 25°C.

Solution: Mass of steam to be evaporated is 500 kg/hr and the enthalpy of vapourization at 100°C is 2255 kJ/kg. Therefore,

$$\dot{Q}/A = 2255 \times 10^3 \times 500/(3600 \times 10) = 31319.44 \text{ W/m}^2$$

Assuming that the heating surface is horizontal, from Eq. (11.24)

$$h = 5.56 \, (\Delta T)^3 \text{ for } Q/A > 16 \text{ kW/m}^2$$

Since $\dot{Q}/A = h(\Delta T)$, $\dot{Q}/A = 5.56 \, (\Delta T)^4$ and $\Delta T = (31319/5.56)^{1/4} = 8.66$
∴ Temperature of the heating surface = 108.66°C.

The influence of pressure on heat transfer coefficient is given by Eq. (11.25)
or, $h_p = h_1 \, (p/p_1)^{0.4}$, $h_1 = Q/A(\Delta T) = 31319.44/8.66 = 3.616$ kW/m^2 °C
∴ $h_p = 3.616 \times (15)^{0.4} = 10.684$ kW/m^2 °C
Since Q/A is proportional to $(\Delta T)^3$, (from Eq. 11.19)
h_p, when (ΔT) is raised to 25°C from 8.66°C would be

$$= 10.684 \times (25/8.66)^3 = 257.04 \text{ kW/m}^2 \text{ °C}.$$

Example 11.14 Water at 10 atm flows inside a tube of 5 cm inner diameter under local boiling conditions where the tube wall temperature is 15°C above the saturation temperature. Estimate the heat transfer per metre length of the tube.

Solution: The pressure $p = 10 \times 1.0132 \times 10^5$ N/m$^2 = 1.0132$ MPa
From Eq. (11.25b), $h = (2.54) \, (15)^3 \exp(1.0132/1.551)$
$$= 16.47 \text{ kW/m}^2 \text{ °C}$$
The surface area of the tube per metre length is

$$A = \pi d l = (3.142) \, (0.05) \, (1) = 0.1571 \text{ m}^2$$

And the heat transfer is $Q = hA\,(\Delta T)$
$$= 16.47 \times 0.1571 \times 15$$
$$= 38.81 \text{ kW/m length.}$$

17. Forced Convection Boiling Mechanism

The difference between the pool boiling and the forced convection boiling mechanism lies in their fluid flow motion. In pool boiling, the fluid flow is primarily governed by the buoyancy driven motion of bubbles that form at the heated surface. In forced convection boiling, fluid flow is created by both the bulk motion and the buoyancy effect. The motion of fluid particles depends strongly on the geometry of the surface and we have external forced convection for flow over heated plates or cylinders, and internal forced convection boiling, or two phase flow, for flows through a heated tube or duct, where the bubbles form at the inner surface of the heated tube.

In external flows, the heat flux can be estimated by standard forced convection correlations before the onset of boiling. When the plate temperature increases, nucleate boiling is likely to occur and the heat flux will increase. Rohsenow has suggested that the total heat flux can be estimated in terms of components associated with pure forced convection and pool boiling.

In internal flows, bubble growth and separation are strongly influenced by the flow velocity and the hydrodynamic effects are substantially different from those corresponding to pool boiling. The process gets complicated by the existence of different two-phase flow patterns and it is not possible to predict all the characteristics of this process quantitatively.

Let us consider a vertical heated tube. A fluid at a temperature lower than its saturation temperature corresponding the the system pressure enters the tube. The liquid will get heated and progressive vapourization will occur. Fig. 11.6 shows schematically the characteristics of forced-convection vapourization – types of flow regimes and the variation of heat transfer coefficient with quality.

The liquid enters the tube by forced convection and the heat transfer to the subcooled liquid in the entrance region can be predicted from the correlations given in Chapter 7. Since the liquid is being heated, bubbles appearing at the surface grow in size and are carried into the main stream of the liquid, Fig. 11.6(a), and there is a sharp increase in the convective heat transfer coefficient in the bubbly flow regime. With further heating, the vapour-volume fraction increases and individual bubbles coalesce to form slugs of vapour, Fig. 11.6 (b), and is called 'slug-flow' regime. The mass fraction of vapour is less than 1% whereas the volume fraction of vapour is about 50% and the fluid velocity in the slug-flow regime may increase appreciably.

When the liquid flows farther along in the tube, the quality improves, and a third flow regime, commonly known as the 'annular-flow regime', Fig. 11.6 (c) is obtained. The liquid forms a film and the film moves along the inner surface. The vapour moves at a higher velocity through the core of the tube and the heat transfer coefficient continues to increase. But dry spots eventually appear on the inner surface and this leads to a decrease in the convection coefficient. The transition regime is dominated by the growth of the dry spots and the surface

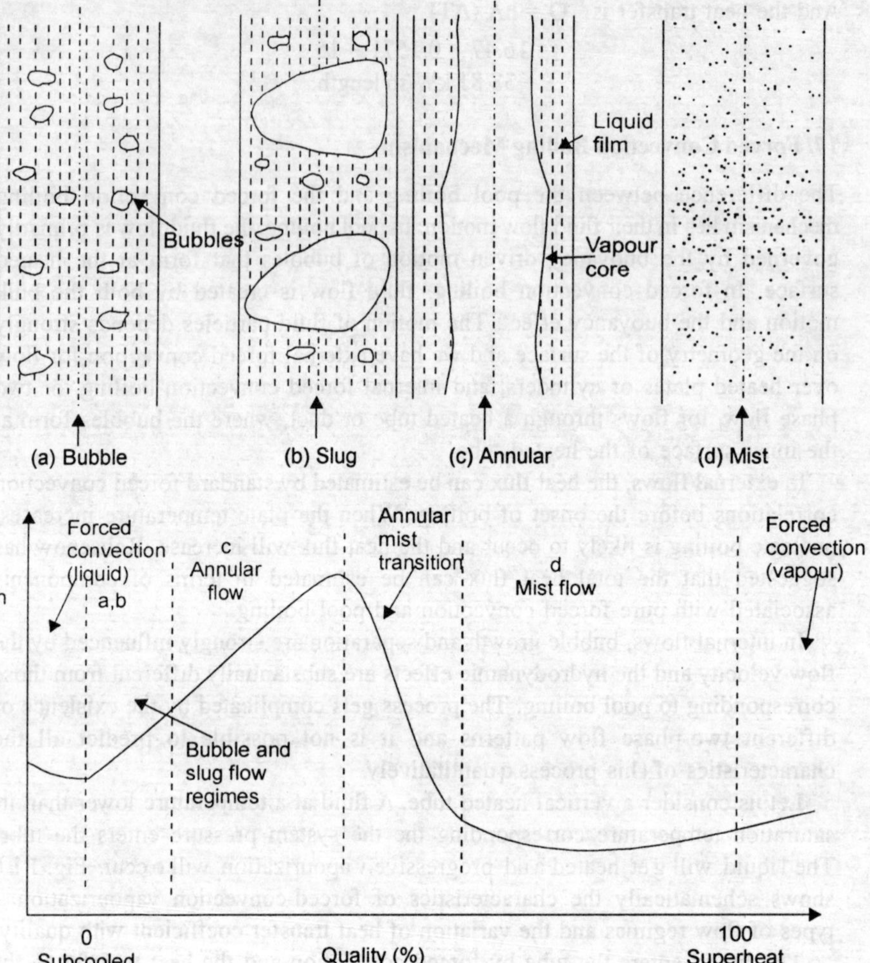

Fig. 11.6 Forced convection vapourization in a vertical tube: types of flow regime and heat transfer coefficient with quality

becomes completely dry and all remaining liquid is converted into the vapour core. The convection coefficient continues to decrease through this regime and the quality range is strongly affected by fluid properties and geometry. It is generally believed that transition to the next flow regime called 'mist flow' regime occurs at qualities of about 25% or higher. Finally the vapour is superheated by forced convection from the surface.

The heat transfer and pressure drop characteristics of forced convection boiling play an important role in the design of boiling nuclear reactors and and many advanced power-production systems.

18. Conduction with Phase Change

We have a change of phase in freezing of foods, solidification of metal castings, welding, ice manufacture, etc., and therefore, it is of practical interest that one should be able to predict the rate of solidification, freezing or melting and to determine the rate of heat transfer necessary to achieve the resulting phase change rate.

Let us consider water as a semi-infinite body (water in a sea) initially at a uniform temperature, T_i. In a cold weather, a strong wind at subfreezing temperature flows over the surface of the sea water freezing the water at the free

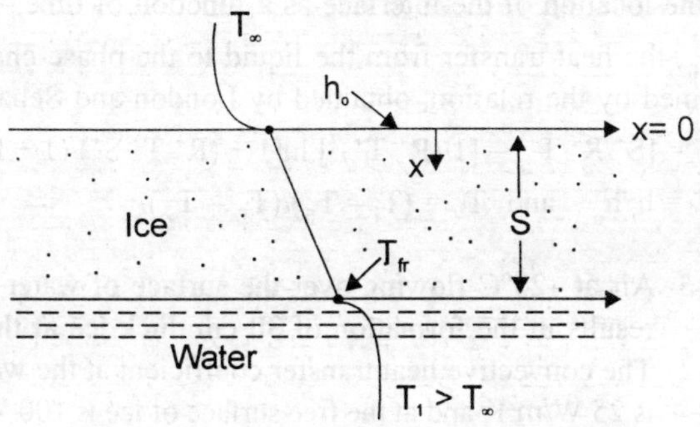

Fig. 11.7 Freezing of water with a convective boundary condition

surface and ice formation takes place progressively at the solid liquid interface because of the heat transfer from the water through the ice to the cold air. The temperature distribution is shown in Fig. 11.7.

Heat energy will first flow by convection from water to the water-ice interface, the temperature at the interface being the freezing temperature of water. Then heat will flow by conduction through the ice slab, the temperature at the free surface of the ice being lower than the freezing temperature, T_{fr}. And, finally, the heat will flow by convection to the cold air.

Since a closed form solution of this problem is very difficult, an approximate solution has been obtained by making the assumption that the heat capacity of ice is negligible (a major approximation) and the heat transfer coefficient at the free surface, h_0 and the temperature at the free surface, T, and the properties of the ice are constant. Thus the heat flux can be written as:

$$Q/A = (T_{fr} - T_\infty) / (1/h_0 + S/k) \tag{11.27}$$

where S is the thickness of the ice slab and k is the thermal conductivity of ice.

Under these conditions, there is no heat removal due to the subcooling of ice (and none conducted through the liquid) so that this heat flux must originate due to freezing at the interface. That is,

$$Q/A = \rho_s L \, dS/dt \tag{11.28}$$

where ρ_s is the density of ice, L, is the latent heat of fusion and dS/dt is the volume rate of solidification per unit area. From Eqs. (11.27) and (11.28) and by separating the variables, we get

$$(1/h_0 + S/k) \, dS = [(T_{fr} - T_\infty)/\rho_s L]dt \tag{11.29}$$

The parameters of the above equation are non-dimensionalised by setting

$$S^+ = h_o\, S/k; \quad \text{and} \quad t^+ = h_o^2\, t\, (T_{fr} - T) / (\rho_s L k)$$

Substituting these in Eq. (11.29), we have

$$(1 + S^+)\, dS^+ = dt^+$$

Integrating and applying the condition that $S = S^+ = 0$ at $t = t^+ = 0$,

$$S^+ = (1 + 2t^+)^{1/2} - 1 \tag{11.30}$$

which gives the location of the interface as a function of time.

When $T_i > T_{fr}$, the heat transfer from the liquid to the phase-change interface at $x = S$ is obtained by the relation, obtained by London and Seban:

$$t^+ = [S^+/R^+\, T^+] - [1/(R^+\, T^+)^2]\, \ln[1 + (R^+\, T^+\, S^+)/(1 + R^+\, T^+)] \tag{11.31}$$

where $\quad R^+ = h_i/h_o \quad$ and $\quad T^+ = (T_i - T_{fr})/(T_{fr} - T_\infty).$

Example 11.15 Air at $-25°C$ flowing over the surface of water in a lake at $5°C$ results in the formation of 30 cm thick ice at the water surface. The convective heat transfer coefficient at the water ice interface is 25 W/m^2K and at the free surface of ice is 100 W/m^2K. Estimate the time required for freezing of water. The properties are: $k = 2.2$ W/mK, $\rho_s = 920$ kg/m^3, $L = 334$ kJ/kg.

Solution: $S^+ = h_o\, S/k = 100 \times 0.3/2.2 = 13.636; \quad R^+ = h_i/h_o = 25/100 = 0.25$

$$T^+ = (T_i - T_{fr})/(T_{fr} - T_\infty) = (5 - 0)/(0 + 25) = 0.2$$

Since $T_i > T_{fr}$, we use Eq (11.31), therefore,

$$t^+ = (13.636/0.25 \times 0.2) - [1/(0.25 \times 0.2)^2]\, \ln\left[1 + \frac{0.25 \times 0.2 \times 13.636}{1 + 0.25 \times 0.2}\right]$$

$$= 72.57, \qquad \text{and } t = 54.5 \text{ hour.}$$

Example 11.16 Ice forms in thin layers on a horizontal rotating drum partly submerged in water at $5°C$. The cylinder is internally refrigerated with a brine spray at $-12°C$. Ice formed on the exterior surface is peeled off as the revolving drum surface emerges from the water. The heat transfer coefficient at the liquid surface is 55 $W/m^2\,°C$ and is 550 $W/m^2\,°C$ at the brine and ice (including the metal wall) interface. Estimate the time required to form an ice layer of 2 mm thick.

Solution: $\qquad R^+ = h_i/h_o = 55/550 = 0.1$

$$T^+ = (T_i - T_{fr})/(T_{fr} - T_\infty)$$

$$= (5 - 0)/[0 - (-12)] = 0.4167$$

$$S^+ = h_o\, S/k = 550 \times 2 \times 10^{-3}/2.2 = 0.5$$

Since the thickness of ice is very small in comparison with the radius of curvature of the drum, ice can be assumed as a sheet and Eq. (11.31) can be applied.

Thus, $\qquad t^+ = -[1/(0.04167)^2]\, \ln[1 + 0.02835/(1 + 0.04167)] + 0.5/0.04167$

$$= 0.5937$$

and $\qquad t = 0.5937 \times 920 \times 334000 \times 2.2/(550)^2 \times 12$

$$= 110.56 \text{ seconds.}$$

Example 11.17 The skin of a person climbing a mountain is exposed to a strong wind at –30°C. If the convective heat transfer coefficient at the surface of the skin is 50 W/m²°C, estimate the skin thickness affected by frost bite after 30 min. The temperature of the body may be taken as 36°C.

Solution: Assuming that the fluid adjacent to the skin freezes when the temperature is 0°C. And, that the thermal conductivity of the fluid is the same a that of ice, we have

$$t^+ = h_o^{\,2} \, (T_{fr} - T_\infty) \, t/(\rho L k)$$
$$= 50 \times 50 \times [0 - (-30)] \times (30 \times 60)/(920 \times 334000 \times 2.2)$$
$$= 0.1997$$

From Eq. (11.30), $S^+ = (1 + 2t^+)^{1/2} - 1.0 = (1 + 0.3994)^{1/2} - 1.0 = 0.18296$

∴ $S = S^+ \, k/h_o = 0.18296 \times 2.2/50 = 8$ mm.

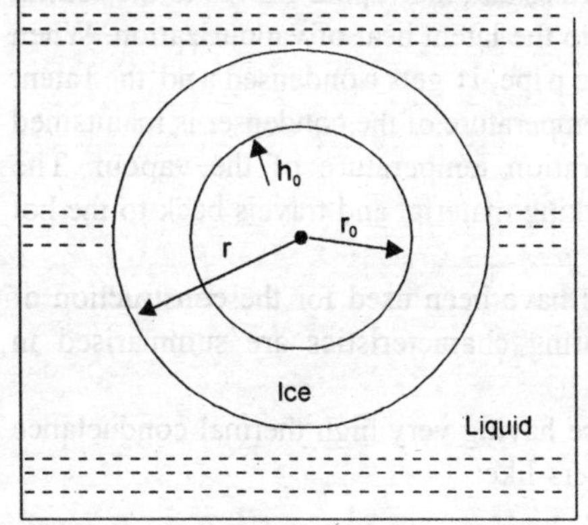

Solidification on the outside of the cylinder

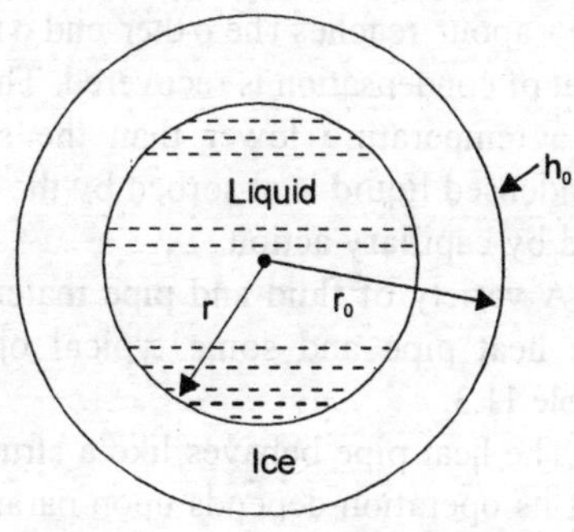

Solidification on the inside of the cylinder

Fig. 11.8 Freezing of liquid on the surface of the cylinder

London and Seban, using the approximation made in deriving Eq. (11.31), have derived equations for rate of ice formation on the inside and outside of cylinders of radius r_o, Fig. 11.8. The equations are:

Freezing inwards: $t^+ = 0.5 \, r^{+2} \ln r^+ + (1 - r^{+2}) (0.25 + 1/2R^+)$ (11.32)

Freezing outwards: $t^+ = 0.5 \, r^{+2} \ln r^+ + (r^{+2} - 1) (0.25 + 1/2R^+)$ (11.33)

where $t^+ = (T_{fr} - T_\infty) \, kt/(\rho L \, r_o^2)$; $r^+ = r/r_o$ and $R^+ = h_o \, r_o/k$.

19. The Working of a Heat Pipe

A heat pipe is a heat exchanger transferring a large amount of heat energy through a relatively small cross-sectional area pipe and with very small temperature difference. Fig. 11.9 illustrates the basic configuration of a heat pipe. It consists of a closed pipe lined with a wicking material that contains a condensible gas. The middle portion of the pipe is insulated and the non-insulated ends respectively serve as evaporators and condensers. When heat energy is added to the

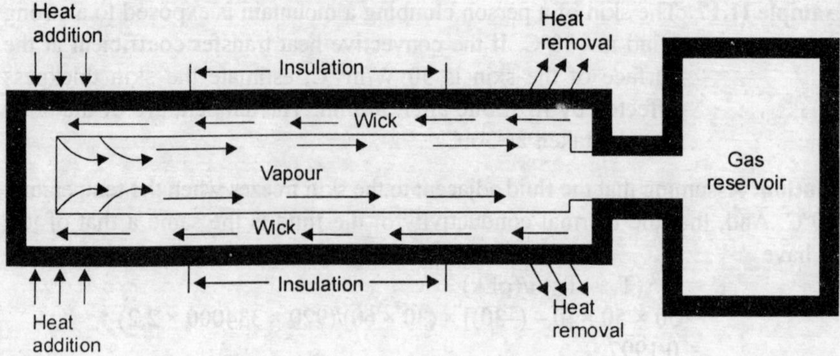

Fig 11.9 A heat pipe

evaporator, the liquid in the wick vapourizes and the vapour moves to the central core. The heat energy supplied is equal to the latent heat of vapourization. When the vapour reaches the other end of the pipe, it gets condensed and the latent heat of condensation is recovered. The temperature of the condenser is maintained at a temperature lower than the saturation temperature of the vapour. The condensed liquid is absorbed by the wicking material and travels back to the hot end by capillary action.

A variety of fluid and pipe materials have been used for the construction of the heat pipe and some typical operating characteristics are summarised in Table 11.3.

The heat pipe behaves like a structure having very high thermal conductance but its operation depends upon parameters like:

Table 11.3 Some typical operating characteristics of heat pipe

Temperature range °C	Working fluid	Pipe material	Axial heat flux, kW/cm²	Surface heat flux, W/cm²
–200 to –80	liquid N$_2$	stainless steel	0.067 @ –163°C	1.01 @ -163°C
–70 to +60	Liquid ammonia	Nickel, Aluminium, Stainless steel	0.295	2.95
– 45 to +120	Methanol	Copper, Nickel, S. Steel	0.45 @ 100°C	75.5 @ 100°C
+5 to +230	water	Copper, Nickel	0.67 @ 200°C	146 @ 170°C
190 to 550	Mercury+ 0.02% Mg	Stainless Steel	25.1 @ 360°C	181 @ 360°C
400 to 800	Potassium	Nickel, S.S	5.6 @ 750°C	181 @ 750°C
500 to 900	Sodium	- do -	9.3 @ 850°C	224 @ 760°C
900 to 1500	Lithium	Niobium + 1% Zirconium	2.0 @ 1250°C	207 @ 1250°C
1500 to 2000	Silver	Tantalum + 5% Tungsten	4.1	413

(i) The pressure differential required to return the liquid from the condenser to the evaporator, to move the vapour from the evaporator to the condenser and due to the difference in elevation between the evaporator and the condenser.

(ii) Area and the height of the wick; and length of the heat pipe—a function of the evaporator and the condenser temperatures and the quantity of heat energy to be transferred.

SUMMARY

1. Both condensation and boiling processes are essential to all closed-loop power and refrigeration cycles. These processes occuring at solid-liquid interfaces, involve fluid motion and are associated with phase change.

2. Through boiling and condensation, large heat transfer rates may be obtained through small temperature differences. Because of the combined latent heat and buoyancy-driven flow effect, the heat transfer coefficients are generally much higher than the typical convection heat transfer without phase change.

3. The heat transfer coefficient, h, can be expressed as a function of the following variables

$$h = f\,[\Delta T,\, g(\rho - \rho_v),\, h_{fg},\, \sigma,\, L,\, \rho,\, C_p,\, k,\, \mu]$$

   which in terms of dimensionless numbers takes the form

$$Nu = f(Gr,\, Pr,\, Ja,\, Bo)$$

   where, Nu = Nusselt number = hL/k, Gr = Grashof number = $\rho g(\rho - \rho_v)L^3/\mu^2$.

   Ja = Jakob number = $C_p\Delta T/h_{fg}$, is the ratio of maximum sensible energy absorbed by the liquid (vapour) to the latent energy absorbed by the liquid (vapour) during condensation (boiling). Its numerical value is very small.

   Bo = Bond number = $g(\rho - \rho_v)L^3/\sigma$, is the ratio of the gravitational body forces to the surface tension forces and represents the effect of buoyancy-induced fluid motion on heat transfer.

4. Surface condensation occurs when a vapour comes in contact with a colder solid surface (temperature of the surface lower than the saturation temperature of the vapour). Filmwise condensation takes place on clean and uncontaminated surfaces. Dropwise condensation takes place on a surface coated with a substance that inhibits wetting.

5. In dropwise condensation, most of the heat transfer takes place through drops of less than $100 - \mu m$ diameter, and the heat transfer rates are much higher than those associated with filmwise condensation. Silicones, Teflone, and an assortment of waxes and fatty acids when coated on surfaces, promote dropwise condensation but such coatings gradually lose their effectiveness after prolonged use.

6. Condenser design calculations are often based on the assumption of filmwise condensation because it is difficult to maintain dropwise condensation. and filmwise condensation may have a turbulent character. If the Reynolds number,

$$Re = 4\,h\,L\,(T_g - T_s)/\mu\,hfg$$

is less than 30, we have a wave free laminar region. For increased Re, ripples or waves form on the condensate film and at Re = 1800, the transition from laminar to turbulent flow is complete.

8. During filmwise condensation, the performance of horizontal tubes is much better than vertical tubes. However, for a tube with length to diameter ratio of 2.86, equal amounts of heat transfer will occur for both horizontal and vertical orientation. With increasing L/D ratio, greater heat transfer is possible with a horizontal tube.

9. Boiling occurs under different conditions. In pool boiling, the liquid is quiescent and its motion near the surface is due to pure convection and the excess temperature $\Delta T = (T_W - T_{sat}) \approx 5°C$. Nucleate boiling exists in the range $5° \leq \Delta T \leq 30°$. Transition boiling (unstable film boiling or partial film boiling) occurs in the range $30° \leq \Delta T \leq 120°$. For $\Delta T > 120°$, the heating surface is completely covered by a vapour blanket and it is stable film boiling.

10. Since nucleate boiling permits high heat transfer rates and convection coefficients are associated with small values of the excess temperature, many engineering devices operate in the range of nucleate boiling.

11. In transition boiling, surface condition oscillates between film and nucleate boiling and the thermal conductance of vapour is much less than that of the liquid, and therefore, the heat flux decreases with increasing ΔT and this regime is of little practical interest.

12. The heat flux is minimum in the stable film boiling regime because the heat is transferred from the heating surface to the liquid by conduction through the vapour. If the heat flux drops below the minimum value, the film will collapse causing the surface to cool and nucleate boiling gets reestablished.

13. A heat pipe is a device to transfer a very large quantity of heat energy through small surface area and utilises both evaporation and condensation processes. Heat pipes are particularly useful in energy-conservation equipments where hot gases coming out of the combustion chamber can pass through one end of a heat pipe while the other end of the heat pipe lies submerged in the water of a boiler. The heat released during condensation of vapour in the heat pipe will convert the boiler water into steam.

14. A simple mathematical model can be used to predict, quite accurately, the phenomenon of frost-bite and the rate of solidification and freezing of ice.

MULTIPLE CHOICE QUESTIONS

1. When a liquid flows through a tube with subcooled or saturated boiling, the process is known as

    (a) pool boiling   (b) bulk boiling

    (c) convection boiling   (d) forced-convection boiling   **(d)**

2. Inspite of large heat transfer coefficients in boiling liquids, cavities are used advantageously when the entire surface is exposed to

    (a) nucleate boiling   (b) film boiling

    (c) transition boiling   (d) all regimes of boiling   **(a)**

3. Consider the following statements regarding nucleate boiling
    (1) The temperature of the surface is greater than the saturation temperature of the liquid.
    (2) Bubbles are created by the expansion of entrapped gases or vapour at small cavities in the surface.
    (3) The temperature is greater than that of film boiling.
    (4) The heat transfer from the surface of the liquid is greater than that in film boiling.
    Of these statements:
    (a) 1, 2 and 4 are correct   (b) 1 and 3 are correct
    (c) 1, 2 and 3 are correct   (d) all the above are correct   **(a)**

4. Assertion (A): The typical boiling curve cannot be obtained with electrical heating.
    Reasoning (R): Because with an electrically heated wire, both nucleate and film boiling exist simultaneously on different portions of the wire.
    Code: (a) Both A and R are false   (b) A is true, R is false
          (c) A is false, R is true   (d) Both A and R are true   **(d)**

5. The slope of the typical boiling curve (Nukiyama boiling curve) in the nucleate boiling regime is
    (a) zero   (b) positive finite
    (c) negative finite   (d) infinity   **(b)**

6. Boiling of milk in an open container when the usual phenomenon of milk spilling over the vessel occurs, is termed as
    (a) saturated nucleate boiling   (b) film boiling
    (c) interface evaporation   (d) pool boiling   **(b)**

7. Consider the following statements regarding condensation heat transfer:
    (1) For a single tube, the horizontal position is preferred over vertical position for better heat transfer.
    (2) Heat transfer coefficient decreases with increasing velocity of vapour stream.
    (3) Condensation of steam on oily surface is dropwise.
    (4) Condensation of pure benzene vapour is always dropwise.

    Of these statements:
    (a) 1 and 2 are correct   (b) 2 and 4 are correct
    (c) 1 and 3 are correct   (d) 3 and 4 are correct   **(c)**

8. Assertion (A) : Thermal conductance of heat pipe is several times more than that of the best available metal conductor under identical conditions.
    Reasoning (R): The value of latent heat is much greater than that of specific heat.
    Code: (a) Both A and R are true   (b) A is true, R is false
          (c) A is false R is true   (d) Both A and R are false   **(a)**

9. Assertion (A): The rate of condensation over a rusty surface is less than that of over a polished surface.
    Reasoning (R): The polished surface promotes dropwise condensation.
    Code: (a) Both A and R are true   (b) Both A and R are false
          (c) A is true, R is false   (d) A is false, R is true   **(a)**

10. Assertion (A): The rate of heat transfer drops significantly when the condensing vapours contain non-condensable gases.
Reasoning (R): The non-condensable gases serve as an obstacle to vapours reaching the surface.
Code: (a) Both A and R are false   (b) Both A and R are true
      (c) A is true, R is false   (d) A is false, R is true   **(b)**

11. Assertion (A): Surface condensers are designed on the basis of filmwise condensation.
Reasoning (R): It is not possible to have dropwise condensation in surface condensers.
Code: (a) Both A and R are true   (b) Both A and R are false
      (c) A is true, R is false   (d) A is false, R is true   **(c)**

12. Assertion (A): The mechanism of heat transfer during solidification of water in lake during really cold weather is conduction only.
Reasoning (R): Cold air (temperature lower than the freezing temperature) works as a heat sink.
Code: (a) Both A and R are true   (b) Both A and R are false
      (c) A is false and R is true   (d) A is true, R is false   **(c)**

13. During filmwise condensation on a vertical place, if δ is the thickness of the liquid film at a distance x from the leading edge, the local heat transfer coefficient is given by
(a) δ/k (b) $\delta \times k$
(c) k/δ (d) none of the above   **(c)**

14. During filmwise condensation, the film thickness δ and the heat transfer coefficient h, varies with the distance x from the leading edge as:
(a) δ decreases, h increases   (b) both δ and h increase
(c) δ increases, h decreases   (d) both δ and h increase   **(c)**

15. Consider the following statements:
1. If a condensing liquid wets a surface, dropwise condensation takes place.
2. Dropwise condensation has a higher heat transfer coefficient than filmwise condensation.
3. Reynolds number of condensing liquid is calculated on its mass flow rate.
4. Suitable coating is used to promote filmwise condensation.

Of these statements:
(a) 1, 2, 3 and 4 are correct   (b) 2, 3 and 4 are correct
(c) 2 and 3 are correct   (d) 1 and 4 are correct   **(c)**

16. When water is boiling on the outside surfaces of submerged bodies
1. The heat transfer coefficient is proportional to $(\Delta T)^3$
2. The heat flux is proportional to $(\Delta T)^4$
3. The heat transfer coefficient increases with increasing pressure
Of the above statements:
(a) 1, 2 and 3 are correct   (b) only 1 and 2 are correct
(c) 1 and 3 are correct   (d) only 3 is correct   **(a)**

17. Assertion (A): In forced convection boiling inside a tube, the heat transfer coefficient increases sharply in the bubble flow regime.

    Because (R): Because the bubbles appearing at the surface grow and are carried away into the mainstream of the liquid.

    Code: (a) A is true, R is false   (b) Both A and R are true

           (c) Both A and R are false   (d) A is false, R is true   **(b)**

18. For filmwise condensation, the heat transfer coefficient will be equal whether the tube is horizontal or vertical, when the ratio of length to diameter is

    (a) 1.3   (b) 2.86

    (c) 0.77   (d) more than 10   **(b)**

NUMERICALS

1. A vertical plate 40 cm long, 40 cm wide is at 40°C and is exposed to saturated steam at 1 atm pressure. Calculate the following:

    (a) the thickness of the film at the bottom edge of the plate

    (b) the maximum velocity at the bottom edge of the plate,

    (c) the heat transfer coefficient, and

    (d) the mass of steam condensed per hour,

    the properties are: $\rho = 978$ kg/m^3, $\mu = 0.4 \times 10^{-3}$ Pa-s,

                   $k = 0.667$ W/mK, $h_{fg} = 2255$ kJ/kg.

    (0.186 mm, 0.414 m/s, 4.77 kW/m^{2}K, 73 kg/hr)

2. A 1.5 cm outside diameter tube, 2 m long is used to condense steam at 76°C. Estimate the heat transfer coefficient for the tube when it is (a) vertical and (b) horizontal. The tube wall surface is maintained at 54°C.

    The properties of condensate at the mean film temperature are:

    $\rho = 980.4$ kg/m^3, $\rho_v = 0.25$ kg/m^3, $\mu = 0.43 \times 10^{-3}$ Pa–s,

    $k = 0.66$ W/mK, $h_{fg} = 2320$ kJ/kg

    (4.03 kW/m^{2}K, 10.52 kW/m^{2}K)

3. Dry saturated steam at 2.5 bar condenses on the surface of a vertical tube, length 1.5 m. The tube surface is maintained at 110°C. Estimate the thickness of the film and the local heat transfer coefficient at a distance of 0.25 m from the upper end of the tube.

    The properties are: $\rho = 942$ kg/m^3, $\rho_v = 1.4$ kg/m^3, $h_{fg} = 2181$ kJ/kg,

                 $\mu = 2.27 \times 10^{-4}$ Pa-s, $k = 0.69$ W/mK.

    (1.09 $\times$ 10^{-4} m, 6.338 kW/m^{2}K)

4. 400 tubes of 6 mm diameter are arranged in a square array and exposed to dry and saturated steam at 0.1 bar. The tube surface temperature is maintained at 24°C. Calculate the heat transfer coefficient and length of each tube for condensing 600 kg/hr of steam. (T_s at 0.1 bar 45.74°C) The properties are:

    $\rho = 994$ kg/m^3, $\rho_v = 0.0676$ kg/m^3, $h_{fg} = 2394$ kJ/kg,

    $k = 0.625$ W/mK, $\mu = 0.73 \times 10^{-3}$ Pa-s.

    (5.376 kW/m^{2}K, 0.453 m)

5. Water at atmospheric pressure is being boiled in an aluminium pan by means of an electric heater. If the bottom surface of the pan is at 110°C and 25 cm in diameter, estimate
   (i) power required to boil the water
   (ii) rate of evaporation of water during boiling process
   (iii) critical heat flux for these conditions.
   The properties of water are:
   $\rho = 960$ kg/m^3, $\quad C_p = 4.182$ kJ/kgK, $\quad \mu = 0.28 \times 10^{-3}$ Pa–s, $\quad$ Pr $= 1.75$
   $h_{fg} = 2255$ kJ/kg, $\quad \sigma = 0.059$ N/m, $\quad \rho_v = 0.596$ kg/m^3, $\quad C_{sf} = 0.01$
   $$\textbf{(14.73 kW, 23.52 kg/h, 1.52 MW/m}^2\textbf{)}$$

6. Water is being boiled at the rate of 30 kg/h in a polished stainless steel pan 250 mm in diameter at atmospheric pressure. Estimate the temperature of the bottom surface of the pan assuming nucleate boiling. Use the properties of water from Ex.5, $C_{sf} = 0.013$.
   $$\textbf{(114°C)}$$

7. A nickel wire 1.25 mm in diameter, 500 mm long carrying a current of 175 amp is submerged in a water bath under atmospheric conditions. Calculate the voltage required for the critical heat flux condition. The properties of water are: $\quad \rho = 960$ kg/m^3, $\quad \rho_v = 0.596$ kg/m^3,
   $\quad h_{fg} = 2255$ kJ/kg, $\quad \sigma = 0.058$ N/m.
   $$\textbf{(12.37 V)}$$

8. Water at 15 atm flows inside a tube of 6 cm inner diameter under local boiling conditions where the tube wall temperature is 100C above the saturation temperature. Estimate the heat transfer per metre length of the tube.
   $$\textbf{(43.055 kW/m)}$$

9. Air at –30°C blowing over the surface of water in a lake at 7°C results in the formation of 25 cm thick ice at the water surface. Estimate the time required for freezing of water. The properties are:
   The convective heat transfer coefficient at the interface = 25 W/m^{2}K, and at the free surface of the ice = 100 W/m^{2}K, $k = 2.2$ W/mK, $\rho = 920$ kg/m^3, $L = 334$ kJ/kg.
   $$\textbf{(9.959 hour)}$$

10. The skin of a person climbing a mountain is exposed to a strong wind at –40°C. If the convective heat transfer coefficient at the surface of the skin is 75 W/m^2°C, estimate the time required for 10 mm thick skin to be affected by the frost bite. The temperature of the body may be taken as 37°C. The properties are: $\rho = 920$ kg/m^3, $\quad L = 334$ kJ/kg, $\quad k = 2.2$ W/mK.
    $$\textbf{(19.8 min)}$$

CHAPTER 12

Mass Transfer

---

1. Mass Transfer and its Applications

Air is a mixture of various gases. Whenever we have a multicomponent system with a concentration gradient, one constituent of the mixture gets transported from the region of higher concentration to the region of lower concentration till the concentration gradient reduces to zero. This phenomenon of the transport of mass as a result of concentration gradient is called 'Mass Transfer'.

The mass transfer phenomenon is analogous to heat transfer phenomenon. In heat transfer – heat energy flows in a direction of decreasing temperature gradient and ceases when the temperature gradient reduces to zero. In mass transfer – the transfer of mass takes place in the direction of decreasing concentration gradient and ceases when the concentration gradient is zero.

The common examples of mass transfer in our everyday life and in many industries are:
- diffusion of smoke discharged by tall chimney into the atmosphere,
- a drop of ink diffusing in a glass of still water,
- evaporation of a drop of perfume in a room,
- humidification of air flowing over a spray pond or cooling tower,
- mixing of diesel or petrol with air inside an internal combustion engine,
- diffusion welding of metals,
- diffusion of neutron in a nuclear reactor.

2. Different Modes of Mass Transfer

There are basically two modes of mass transfer:

(i) *Mass Transfer by Diffusion* – the transport of mass by random molecular motion in quiescent or laminar flowing fluids is known as mass transfer by 'diffusion' and is analogous to heat transfer by conduction. Mass transfer by diffusion occurs due to (a) concentration gradient, (b) temperature gradient, and (c) hydrostatic pressure difference.

(ii) *Convective Mass Transfer* – the rate of molecular diffusion of mass can be accelerated by the bulk motion of the fluid. Mass can be transported between the boundary of a surface and a moving fluid (drying of clothes, molecular diffusion of a sugar cube in a cup of coffee by stirring, moist air flowing over the surface of an ocean and precipitation on a dry land etc.), or between two moving fluids which are relatively immiscible (formation of clouds, vapourisation of water in a tea kettle). This mechanism of mass transfer is called 'convective mass transfer' and is analogous to heat transfer by convection (free or forced).

3. Dalton's Law of Partial Pressure

Each constituent of a multicomponent system contributes to the total pressure by an amount which is known as the 'partial pressure' of the constituent. The relationship between the partial pressures of the constituents is expressed by Dalton's Law:

The pressure of a mixture of gases is equal to the sum of the partial pressure of the constituents. The partial pressure of each constituent is that pressure which the gas would exert if it occupied alone that volume occupied by the mixture at the same temperature.

For a mixture of ideal gases, we have
$$P = P_A + P_B + \ldots + P_K;$$
where P_A is the partial pressure of the species A and so on.

$$= \sum_i P_i \qquad (12.1)$$

Dalton's law was reformulated by Gibbs to include a second statement on the properties of mixtures. The combined statement is Gibbs-Dalton law:

The internal energy, enthalpy and entropy of a gaseous mixture are respectively equal to the sum of the internal energies, enthalpies, and entropies of the constituents. The internal energy, enthalpy and entropy which a constituent would have if it occupied alone that volume occupied by the mixture at the temperature of the mixture.

4. Molar Density, Mass Density, Mass Fraction and Mole Fraction

There are a number of ways by which the concentration for a species in a multi-component mixture can be defined:

(i) Molar Density or Molar Concentration, C_A = number of moles of the species A per unit volume of mixture, $kg-mol/m^3$

(ii) Mass Density or Mass Concentration, ρ_A = mass of the species A per unit volume of the mixture, kg/m^3.

(iii) Mass Fraction, m_A = mass concentration of component A / total mass density of the mixture.

(iv) Mole Fraction, X_A = number of moles of species A / total number of moles of the mixture. $= C_A / C$

Therefore, the following summation rules hold true:

$$C_A + C_B + \ldots + C_K = C$$
$$\rho_A + \rho_B + \ldots + \rho_K = \rho$$
$$X_A + X_B + \ldots + X_K = 1$$
$$m_A + m_B + \ldots + m_K = 1 \qquad (12.2)$$

Since the number of moles = mass of species/molecular weight, we have
$$C_A = \rho_A / M_A$$
For a perfect gas, we have:
$$P_A V = n_A R_0 T, \quad \text{where } R_0 \text{ is the universal gas constant,}$$
and
$$C_A = n_A / V = P_A / R_0 T$$
$$\therefore \qquad X_A = C_A / C = P_A / P \quad \text{and} \quad C = p / R_0 T.$$

Example 12.1 A perfect gas mixture consists of 3 kg of nitrogen and 5 kg of carbon dioxide at a pressure of 2.5 bar and 25°C. Calculate the (a) mole fraction of each constituent, (b) equivalent molecular weight of the mixture, (c) equivalent gas constant of the mixture, (d) partial pressure and partial volume, and (e) volume and the density of the mixture.

Solution: Number of moles of nitrogen = 3/28 = 0.107,
Number of moles of $\qquad CO_2 = 5/44 = 0.1136$

(a) Mole fraction of $N_2 = 0.107 / (0.107 + 0.1136) = 0.485$

Mole fraction of $CO_2 = 0.1136 / (0.107 + 0.1136) = 0.515$

(b) The equivalent molecular weight $M = x_a M_a + x_b M_b$
$$= 0.485 \times 28 + 0.515 \times 44 = 36.24 \text{ kg/kmol}$$

(c) The equivalent gas constant, $R = (3/8)(8314/28) + (5/8)(8314/44)$
$$= 296.92 \text{ J/kgK}$$

(d) Partial pressure = mole fraction x total pressure,
$$\therefore \qquad p_{N_2} = 0.485 \times 2.5 = 1.2125 \text{ bar;}$$
$$p_{CO_2} = 0.515 \times 2.5 = 1.2875 \text{ bar.}$$
$$\text{Volume of } N_2 = 0.107 \times 8314 \times 298 / (2.5 \times 10^5) = 1.06 \text{m}^3$$
$$\text{Volume of } CO_2 = 0.1136 \times 8314 \times 298 / (2.5 \times 10^5) = 1.1258 \text{m}^3$$

(e) Volume of the mixture, V = Volume of N_2 / Mole fraction of N_2
$$= 1.06/0.485$$
$$= 2.1858 \text{ m}^3$$
and the density of the mixture = 8/2.1858 = 3.66 kg/m^3.

Example 12.2 A vessel contains a mixture of 2 kmol of CO_2 and 4.5 kmol of air at 1 bar and 25°C. If air contains 21% oxygen and 79% nitrogen by volume, calculate for the mixture:
(i) the mass of CO_2, O_2 and N_2, and the total mass;
(ii) the percentage carbon content by mass;
(iii) the molar mass and the gas constant for the mixture;
(iv) the specific volume of the mixture.

Solution: (i) Number of moles of $O_2 = 0.21 \times 4.5 = 0.945$ kmol,

Number of moles of $N_2 = 0.79 \times 4.5 = 3.55$ kmol

Mass of $CO_2 = 2 \times 44 = 88$kg;   Mass of $O_2 = 0.945 \times 32 = 30.24$ kg

Mass of $N_2 = 3.55 \times 28 = 99.54$ kg

The total mass = 88 + 30.24 + 99.54 = 217.48 kg

(ii) Percentage of carbon in the mixture = $(24/217.48) \times 100 = 11.035\%$ by mass.

(iii) Total number of moles $= n_{CO_2} + n_{O_2} + n_{N_2} = 2 + 0.945 + 3.555$

$$= 6.5 \text{ kmol}$$

$$\text{Molar mass} = \sum \frac{n_i}{n} m_i$$
$$= (2/6.5) \times 44 + (0.945/6.5) \times 32 + (3.555/6.5) \times 28$$
$$= 33.5 \text{ kg/kmol}$$

and the gas constant of the mixture $= 8314/33.5 = 248.18$ J/kgK

(iv) Specific volume of the mixture, $v = RT/p = 248.18 \times 298 / (1 \times 10^5)$
$$= 0.7395 \text{ m}^3/\text{kg}.$$

Example 12.3 A vessel contains a gaseous mixture of composition by volume 80% H_2 and 20% CO. It is desired that the mixture should be made in the proportion 50% H_2 and 50% CO by removing some of the mixture and adding some CO. Calculate per kilomole of mixture the mass of the mixture to be removed, and the mass of CO to be added. The pressure and temperature of the vessel remain constant during the process.

Solution: Since the pressure and temperature remain constant, the number of kilomole in the vessel would remain the same throughout. Thus, the number of kilomoles of the mixture removed is equal to the number of kilomoles of CO added. Let x kg of the mixture be removed and y kg of CO be added. The molecular weight of the mixture, $M = (0.8 \times 2) + (0.2 \times 28) = 7.2$ kg/kmol

Number of kilomoles of the mixture removed $= x/7.2$ kmol

Number of kmol of CO added $= y/28 = x/7.2$

or, Number of kmol of H_2 in the mixture removed $= 0.8 \times x/7.2 = x/9$ kmol

Initial number of kmol of $H_2 = 0.8 \times 1 = 0.8$ kmol

Therefore, the number of kmol of H_2 remaining in the vessel $= (0.8 - x/9)$

Since 1 kmol of the mixture contains 50% H_2 and 50% CO, we have

$$(0.8 - x/9) = 0.5; \text{ which gives } x = 2.7 \text{ kg.}$$

And, then, $\qquad y = 28 \, x/7.2 = 10.5$ kg

i.e., $\qquad$ mass of CO added $= 10.5$ kg.

5. Mass Average and Molar Average Velocities and Different Types of Fluxes

Velocities: In a multicomponent mixture, the bulk velocity of the mixture can be defined on the basis of mass average or molar average velocity. Let V_A be the velocity of the species A and ρ_A is the mass density of the species A, then the mass average velocity would be:

$$V = \frac{\rho_A V_A + \rho_B V_B + \cdots}{\rho_A + \rho_B} = \frac{\rho_A V_A + \rho_B V_B + \cdots}{\rho}$$
$$= m_A V_A + m_B V_B + \ldots \tag{12.3}$$

Similarly, the molar average velocity would be:

$$U = \frac{C_A V_A + C_B V_B + \dots}{C_A + C_B} = \frac{C_A V_A + C_B V_B + \dots}{C} = X_A V_A + X_B V_B + \dots$$

Since mass transfer requires the diffusion of a species with respect to a plane moving with an average velocity, diffusion will take place when the diffusion velocity is in excess of the average velocity. Thus

Mass diffusion velocity of the species A $= V_A - V$ (12.4)

Molar diffusion velocity of the species A $= V_A - U$ (12.5)

Fluxes: The mass flux of species A can be expressed relative to either a fixed observer or an observer moving with the bulk velocity. For a stationary observer, the absolute flux of any species A will be equal to the sum of the flux due to the molecular diffusion and that due to the bulk motion.

Thus,   Absolute flux $= \rho_A V_A$ and, Diffusion flux $= \dot{m}/A$

           Flux due to bulk motion $= \rho_A V$

$\therefore$ $\rho_A V_A = \dot{m}/A + \rho_A V,$ or $\dot{m}/A = \rho_A (V_A - V)$ (12.6)

Similarly,   molar diffusion flux $= C_A(V_A - V)$ (12.7)

Example 12.4   A vessel contains 4 kmol of hydrogen and 4 kmol of oxygen. Oxygen is moving in the X-direction with a velocity of 1 m/s while hydrogen is stationary. Calculate the average velocity, diffusion fluxes and the flux with respect to a stationary surface.

Solution: Total number of moles = 8 kmol,

Mole fraction of hydrogen = 0.5,   Mole fraction of oxygen = 0.5

Mass of oxygen = 4 × 32 = 128 kg,   Mass of hydrogen = 4 × 2 = 8 kg.

Mass fraction of O_2 = 128/136 = 0.9412

Mass fraction of H_2 = 8/136 = 0.0588

Equivalent molecular weight = 0.5 × 2 + 0.5 × 32 = 17 kg/kmol.

Mass average velocity $= m_A V_A + m_B V_B,$ $V = 0.9412 × 1.0 + 0.0 = 0.9412$ m/s

Molar average velocity $= x_A V_A + x_B V_B,$ $V = 0.5 × 1.0 + 0.0 = 0.5$ m/s

Mass diffusion velocity of O_2 = 1.0 – 0.9412 = 0.0588 m/s

Mass diffusion velocity of H_2 = 0 – V = – 0.9412 m/s

Across a plane moving with mass average velocity, the fluxes are:

    $O_2 : \rho_A(V_A - V) = \rho \times$ mass fraction $\times (V_A - V)$

                    $= \rho \times 0.9412 \times 0.0588 = 0.0553 \, \rho$ kg/s–m^2

    $H_2 : \rho \times 0.0588 \times (- 0.9412) = - 0.0553 \, \rho$ kg/s–m^2

(We observe that the mass diffusion flux of oxygen is exactly equal and opposite to the mass diffusion of hydrogen. This is because of the fact that the net mass crossing the plane moving with mass average velocity should be zero.)

Flux with respect to stationary plane:

For O_2: $= \rho \times$ mass fraction of $O_2 \times$ velocity of O_2

       $= 0.9412 \times 1.0 \times \rho = 0.9412$ kg/s–m^2.

For H_2: $= \rho \times 0.0588 \times 0.0 = 0.0$ kg/s–m^2.

6. Fick's Law of Diffusion*

The fundamental equation (one-dimensional) of molecular diffusion is known as Fick's law. It has been derived from the kinetic theory of gases, and can be written for a binary mixture as

$$J_A = - D_{AB} \, (d \, C_A/dx) \qquad (12.8)$$

where D_{AB} = diffusion coefficient of species A with respect to species B, J_A = molar flux in the X-direction relative to the molar average velocity,

$$dC_A/dx = \text{Concentration gradient in X-direction.}$$

Let us consider a two compartment tank as shown in Fig. 12.1. One compartment contains gas A and the other compartment contains gas B and both the compartments are initially at a uniform pressure and temperature throughout. When the partition between the compartments is removed, the two gases will diffuse through each other until equilibrium is established and the concentration of the gases is uniform throughout the tank. Fig. 12.2 illustrates the dependence of diffusion on the concentration profile.

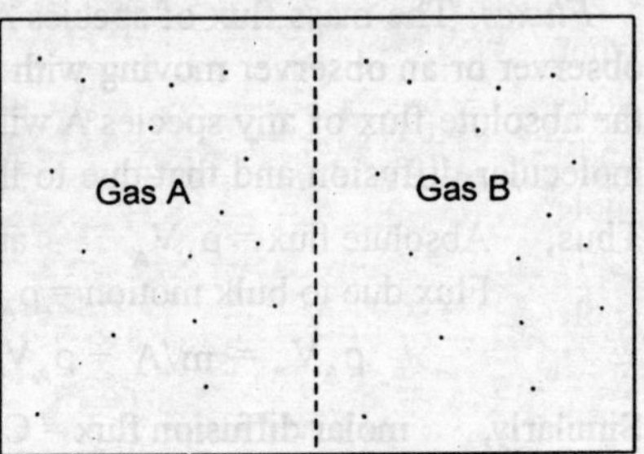

Fig. 12.1 Diffusion of species A into species B

The concentration of the species A on the left side of the imaginary plane is greater than that on the right side. As such, more molecules will cross the plane per unit time from left to right. This would lead to a net transfer of mass from the region of higher concentration to the region of lower concentration.

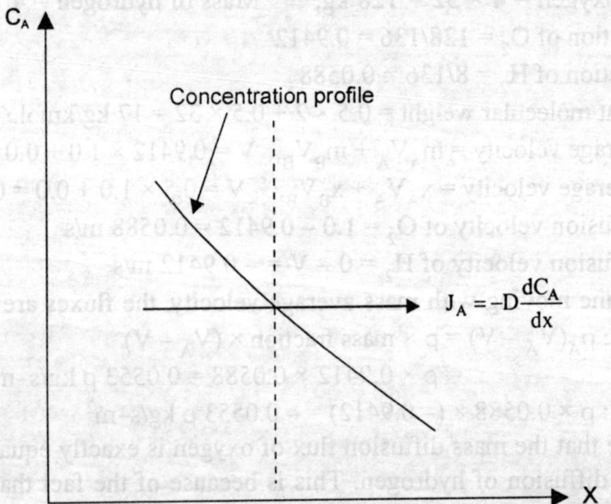

Fig. 12.2 Dependence of diffusion on concentration profile

---

* This law assumes that fluxes are measured relative to the coordinates that move with the average velocity of the mixture.

7. Diffusion in Gases, Liquids and Solids

(i) *Diffusion in Gases* – the diffusion rates in gases are dependent on the molecular speed which is a function of temperature and therefore, the diffusion coefficient depends upon the temperature of gases.

Gilliland has proposed a semi-empirical equation for diffusion coefficient in a binary gas mixture –

$$D = 435.7 \frac{T^{3/2}}{p(V_A^{1/3} + V_B^{1/3})^2} \left[\frac{1}{M_A} + \frac{1}{M_B} \right]^{1/2} \tag{12.9}$$

where D is in square centimeters per second, T is in Kelvin, p is the total pressure of the system in pascals, V_A and V_B are the molecular volumes of the species A and B as calculated from the atomic volumes in Table 12.1, M_A, and M_B are the molecular weights of species A and B.

Diffusion coefficients for gases depend upon pressure, temperature and other molecular properties of diffusing gases. At two different pressure and temperature, we have

$$D_2/D_1 = (p_1/p_2) \cdot (T_2/T_1)^{3/2} \tag{12.10a}$$

Table 12.1 Atomic volumes*

Air	29.9	In secondary amines	1.2
Bromine	27.0	Oxygen, molecule (O_2)	7.4
Carbon	14.8	Coupled to two other elements:	
Carbon dioxide	34.0	In aldehydes and ketones	7.4
Chlorine		In methyl esters	9.1
Terminal as in R-Cl	21.6	In ethyl esters	9.9
Medial as in R-CHCl-R	24.6	In higher esters & ethers	11.0
Flourine	8.7	In acids	12.0
Hydrogen, molecule (H_2)	14.3	In union with S, P, N	8.3
in compounds	3.7	Phosphorous	27.0
Iodine	37.0	Sulphur	25.6
Nitrogen, molecule (N_2)	15.6	Water	18.8
in primary amines	10.5		

*(For three numbered ring like ethylene oxide, deduct 6.0, for four numbered ring like cyclobutane, deduct 8.5, for six numbered ring like benzene, deduct 15.6, for napthelene ring, deduct 30.0.)

Example 12.5   Calculate the diffusion coefficient of CO_2 in air at 30°C and at 1 atm pressure.

Solution: From Table 12.1,   the atomic volumes are:

CO_2: Volume 34;   molecular weight 44

Air: Volume 29.9; molecular weight 28.9

From Eq. (12.9), we get

$$D = 435.7 \frac{(273+30)^{1.5}}{(1.0135 \times 10^5)(34^{1/3} + 29.9^{1/3})^2} (1/44 + 1/28.9)^{1/2}$$

$$= 0.1 \ cm^2/s.$$

Example 12.6   Estimate the diffusivity of ethyl alcohol ($C_2 H_5 OH$) in air at 25°C and 1 atm pressure.

Solution: Kopp's law of additive atomic volumes applies in gases where compounds are involved.

For ethyl alcohol, the volume would be: $2(14.8) + 6(3.7) + 7.4 = 59.2$

From Eq. (12.9), we have for ethyl alcohol,

$$D = 435.7 \frac{(273+30)^{1.5}}{(1.0135 \times 10^5)(59.2^{1/3} + 29.9^{1/3})^2} (1/46 + 1/28.9)^{1/2}$$

$$= 0.753 \text{ cm}^2/\text{s}$$

Similarly, for methane CH_4, the volume would be: $14.8 + 4(3.7) = 29.6$

Using Eq (12.9), for methane, $D = 1.114 \text{ cm}^2/\text{s}$.

(ii) *Diffusion in Liquids and Solids* – Diffusion in liquids occurs at much slower rate than in gases. Since kinetic theory of liquids is not as much developed as that of gases, it is usually assumed as a first approximation that equations of the same general form are applicable to the diffusion of a solute in a solvent as to the diffusion in gases, i.e., Fick's law is assumed valid for liquids.

Diffusion coefficient for most of the common organic and inorganic materials in the usual solvents such as water, alcohol and benzene at room temperature lie in the range of 1.79×10^{-3} to 1.075×10^{-7} cm^2/s.

Diffusion in solids is much slower than in liquids. Diffusion of solids in solid has limited engineering applications but diffusion of fluids in solids have extensive applications. Fick's law is sometimes used, with an empirically determined effective diffusivity which takes care of the structure of solid. A typical problem of liquid transfer in a solid, of interest, is drying of solids.

8. The Equivalence of Diffusion Coefficient

Fick's law (Eq. 12.8) can also be expressed in terms of mass flux per unit area or mass concentration or in terms of molal concentrations and fluxes. For gases, the law may be expressed in terms of partial pressures by making use of the perfect gas e quation of s tate:

$$p = \rho RT$$

Since the characteristic gas constant of a gas is: $R_A = R_O/M_A$,
we have $\rho_A = p_A M_A / R_O T$

and $m_A/A = - D_{AB}(M_A/R_O T)dp_A/dx$ for isothermal diffusion. (12.10 b)

Similarly, the diffusion of the component B, for the system shown in Fig. 12.1, we can write

$$\frac{\dot{m}_B}{A} = D_{BA} \frac{M_B}{R_O T} \frac{dp_B}{dx} \tag{12.11}$$

When we have equimolal counter diffusion, shown in Fig. 12.3 (a, b), the steady state molal diffusion rates of the species A and B, represented by N_A and N_B will be given by

$$N_A = \frac{\dot{m}_A}{M_A} = -D_{AB}\left(\frac{A}{R_0 T}\right)\left(\frac{dp_A}{dx}\right) \qquad (12.12)$$

and
$$N_B = \frac{\dot{m}_B}{M_B} = +D_{BA}\left(\frac{A}{R_0 T}\right)\left(\frac{dp_B}{dx}\right) \qquad (12.13)$$

The total pressure of the system remains constant at steady state,

or, $\qquad p = p_A + p_B;$ and $dp_A/dx + dp_B/dx = 0$

as $\qquad dp_A/dx = -dp_B/dx$

Since each molecule of A is replaced by a molecule of B, the molal diffusion rates must be equal. Thus: $N_A = -N_B$, and

$$-D_{AB}\left(\frac{A}{R_0 T}\right)\left(\frac{dp_A}{dx}\right) = -D_{BA}\left(\frac{A}{R_0 T}\right)\left(\frac{dp_A}{dx}\right) \qquad (12.14)$$

or $\qquad D_{AB} = D_{BA} = D$

This fact is known as the equivalence of diffusion coefficients or diffusivities in binary mixtures, and is a property of the binary mixture.

By integrating Eq. (12.10), we can obtain the mass flux of the species A as:

$$\frac{\dot{m}_A}{A} = -\frac{DM_A}{R_0 T} \times \left(p_{A_2} - p_{A_1}\right)\Big/\Delta x \qquad (12.15)$$

corresponding to the nomenclature used in Fig. 12.3 (a, b). Table 12.2 gives the values of the binary diffusion coefficients.

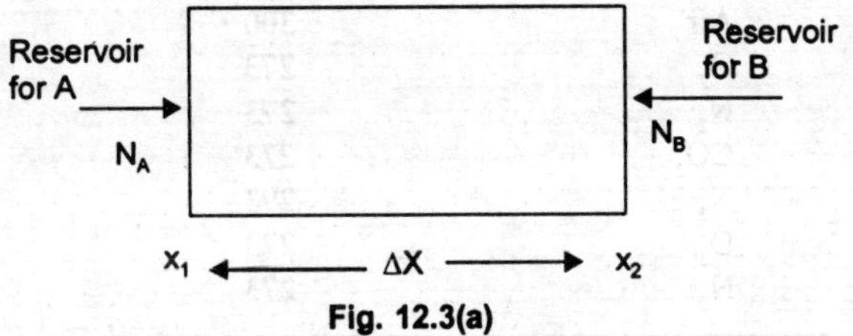

Fig. 12.3(a)

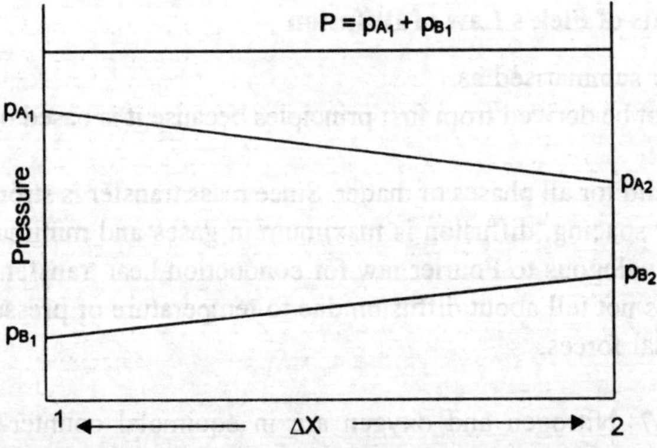

Fig. 12.3(b) Equimolal counter-diffusion (partial pressure profile)

Table 12.2 Values of binary diffusion coefficient

Component A	Component B	$T(K)$	$D_{AB}(m^2/s)$
Solids			
O_2	Rubber	298	0.21×10^{-9}
CO_2	Rubber	298	0.11×10^{-9}
He	S_iO_2	293	0.4×10^{-18}
Cd	Cu	293	0.27×10^{-18}
Al	Cu	293	0.13×10^{-33}
Dilute Solutions			
Caffeine	H_2O	298	0.63×10^{-9}
Ethanol	H_2O	298	0.12×10^{-8}
Glucose	H_2O	298	0.69×10^{-9}
Acetone	H_2O	298	0.20×10^{-8}
O_2	H_2O	298	0.24×10^{-8}
H_2	H_2O	298	0.63×10^{-8}
N_2	H_2O	298	0.26×10^{-8}
Gases			
NH_3	Air	298	0.28×10^{-4}
H_2O	Air	298	0.26×10^{-4}
CO_2	Air	298	0.16×10^{-4}
H_2	Air	298	0.41×10^{-4}
O_2	Air	298	0.21×10^{-4}
Acetone	Air	273	0.11×10^{-4}
Naphthalene	Air	300	0.62×10^{-5}
H_2	O_2	273	0.70×10^{-4}
H_2	N_2	273	0.68×10^{-4}
H_2	CO_2	273	0.55×10^{-4}
CO_2	N_2	293	0.16×10^{-4}
CO_2	O_2	273	0.14×10^{-4}
O_2	N_2	273	0.18×10^{-4}

9. Main Points of Fick's Law of Diffusion

These can be summarised as:

(i) It cannot be derived from first principles because it is based on experimental evidence.

(ii) It is valid for all phases of matter. Since mass transfer is strongly influenced by molecular spacing, diffusion is maximum in gases and minimum in solids.

(iii) It is analogous to Fourier law for conduction heat transfer.

(iv) It does not tell about diffusion due to temperature or pressure gradient or due to external forces.

Example 12.7 Nitrogen and oxygen are in equimolal counter-diffusion. The total pressure is 1 bar and temperature 25°C. The diffusion coefficient is 0.2 cm²/s. If the partial pressures at two planes perpendicular to the direction of diffusion are 25 kPa and 5 kPa and the planes are separated by a distance of 2.5 cm, calculate the rate of diffusion of the mixture.

Solution: From Eq. (12.15), we have $(\dot{m}_A/A) = -D(M_A/R_oT)\,(p_{A_2} - p_{A_1})/\Delta x$

For N_2: $\quad p_{A_1} = 25$ kPa, $\ p_{A_2} = 5$ kPa, $\ M_A = 28$

$$\therefore \quad \frac{\dot{m}_{N_2}}{A} = -0.2 \times 10^{-4} \times 28/(8314 \times 298) \times (5 - 25) \times 10^3/(2.5 \times 10^{-2})$$

$$= 0.00018 \ \text{kg/s} - \text{m}^2 \equiv 6.43 \times 10^{-6} \ \text{kg mol/m}^2\text{s}$$

For O_2 : $\quad p_{A_1} = 95$ kPa, $\ p_{A_2} = 75$ kPa, $\ $ Mol. wt. $= 32$

$$\therefore \quad \frac{\dot{m}_{O_2}}{A} = -0.2 \times 10^{-4} \times 32/(8314 \times 298) \times (75 - 95) \times 10^3/(2.5 \times 10^{-2})$$

$$= -0.000207 \ \text{kg/s} - \text{m}^2 \equiv 6.43 \times 10^{-6} \ \text{kg.mol/m}^2\text{s}.$$

Example 12.8 Estimate the rate of burning of a pulverized carbon particle in a furnace if the diameter of the particle is 4 mm, pressure 1 bar. The oxygen is available at 1100 K. Assume that a fairly large layer of CO_2 surrounds the carbon particle. Take $D = 1 \ \text{cm}^2/\text{s}$.

Solution: The combustion equation is $C + O_2 \rightarrow CO_2$, i.e., there will be an equimolal counter-diffusion between O_2 and CO_2.
Since a fairly large blanket of carbon dioxide surrounds the carbon particle, the partial pressure of carbon dioxide at the surface of the carbon particle will be 1 bar and the partial pressure of oxygen will be zero. Similarly, the partial pressure of carbon dioxide far outside will be zero and the partial pressure of oxygen will be 1 bar.

From Eq. (12.12), we have: $\quad \dfrac{N_A}{A} = -D\dfrac{1}{R_oT}\dfrac{dp_A}{dx}$

or, $\quad \dfrac{N_A}{4\pi r^2} = -\dfrac{D}{R_oT}\dfrac{dp_A}{dr}$

Separating the variables and integrating, we get

$$\frac{N_A R_o T}{4\pi D}\int_{r=r_1}^{\infty} dr/r^2 = -\int_1^0 dp_A$$

or, $\quad p_{A_1} = \dfrac{N_A R_o T}{4\pi D}\cdot\dfrac{1}{r_1} \quad$ and $\quad N_{CO_2} = \dfrac{4\pi \times 1 \times 10^{-4} \times 10^5 \times 2 \times 10^{-3}}{8314 \times 1100}$

$$= 2.748 \times 10^{-8} \ \text{kgmol/s}.$$

Since 1 mol of carbon will produce 1 mol of CO_2,
the rate of burning of carbon will be
$$= 2.748 \times 10^{-8} \times 12 = 3.298 \times 10^{-7} \ \text{kg/s}.$$

10. An Expression for Isothermal Evaporation of Water Vapour into Stagnant Air from a Surface

Let us consider a tank containing water which is exposed to air in the tank as shown in Fig. 12.4. We assume that:
   (i)   the system is isothermal,
   (ii) the total pressure remains constant,
   (iii) the system is in steady state. Since there has to be a little movement of air over the top of the tank to remove the water vapour that diffuses to that point, the air movement does not create any turbulence to alter the concentration profile in the tank, and
   (iv) air and water vapour both behave like ideal gases.

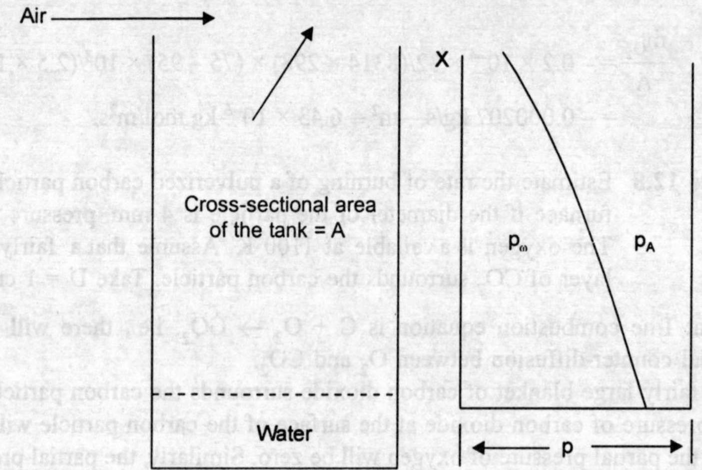

Fig. 12.4 Diffusion of water vapour in air

From Eq. (12.10), the downward diffusion of air can be written as

$$\dot{m}_A = -(DAM_A/R_0T)(dp_A/dx) \tag{12.16}$$

and this has to be balanced by the bulk mass transfer upward. Therefore,

$$-\rho_A AV = -\frac{p_A M_{air}}{R_0 T} AV \; ; \text{ where V is the upward bulk mass velocity}$$

$$\therefore \qquad V = \frac{D}{p_A} \frac{dp_A}{dx} \tag{12.17}$$

The mass diffusion of water vapour upward is

$$\dot{m}_w = -DA \frac{M_w}{R_0 T} \frac{dp_w}{dx} \tag{12.18}$$

and the bulk transport of water vapour would be

$$\rho_w AV = \frac{p_w M_w}{R_0 T} AV \tag{12.19}$$

And, the total mass transport is then,

$$\dot{m}_{w_{total}} = -\frac{DAM_w}{R_0T}\frac{dp_w}{dx} + \frac{p_wM_w}{R_0T}A\frac{D}{p_A}\frac{dp_A}{dx}$$

Since the total pressure remains constant, by Dalton's law we get

$$p = p_A + p_\omega \quad \text{or}, \quad dp_A/dx = -dp_\omega/dx$$

$$\dot{m}_{w_{total}} = -\frac{DAM_w}{R_0T}\frac{dp_w}{dx}\left[1+\frac{\dot{p}_w}{p_a}\right]$$

$$\therefore \quad \dot{m}_{total} = -\frac{DM_wA}{R_0T}\frac{p}{p-p_w}\frac{dp_w}{dx} \tag{12.20}$$

This relation is called the Stefan's law. Upon integration,

$$\dot{m}_{w_{total}} = +\frac{DpM_wA}{R_0T(x_2-x_1)}\log_e\frac{p-p_{w_2}}{p-p_{w_1}} = \frac{DpM_wA}{R_0T(x_2-x_1)}\log_e\frac{p_{A_2}}{p_{A_1}} \tag{12.21}$$

Example 12.9   Estimate the diffusion rate of water from the bottom of a well 10m deep and 1.5m in diameter into atmospheric air at 25°C when (i) the air is completely dry, and (ii) the relative humidity of air is 0.5.

Solution: The partial pressure of water vapour at the water surface is equal to the saturation pressure corresponding to 25°C = 0.03169 bar
(i)   When the air is dry, the partial pressure of water vapour is zero.
From Eq (12.21)

$$\dot{m}_w = \frac{D(1\times10^5)\times18\times\pi\times(0.75)^2}{8314\times298\times(10)}\ln\left[(1-0.0)/(1-0.03169)\right]$$

if D = 0.256 × 10⁻⁴ m²/s, $\dot{m}_w$ = 1.236 × 10⁻⁷ kg/s.
(ii) When the humidity is 0.5, partial pressure of water vapour would be
    0.5 × 0.03169 = 0.0158 bar.
and

$$\dot{m}_w = \frac{0.256\times10^{-4}(1\times10^5)\times18\times\pi\times(0.75)^2}{8314\times298\times10}\ln\left[(1-0.0158)/(1-0.03169)\right]$$

$$= 5.33\times10^{-8}\text{ kg/s}.$$

Example 12.10 A 50 mm deep pan contains water to a level of 20 mm and is exposed to dry air at 1 bar and 40°C. If the diffusion coefficient of water is 0.25 cm²/s, estimate the time required for all the water to evaporate.

Solution: $x_2 - x_1 = 50 - 20 = 30$ mm = 0.03 m
                T = 273 + 40 = 313K,   D = 0.25 × 10⁻⁴ = m²/s
At the water surface, the partial pressure of water vapour will be equal to the saturation pressure corresponding to 40°C = 0.07384 bar. And, at the top of the pan, the partial pressure of water vapour is zero. From Eq. (12.21),

$$\dot{m}_w = \frac{0.25 \times 10^{-4} \times 1 \times 10^5 \times 18 \times 1}{8314 \times 313 \times 0.03} \cdot \ln\left[(1-0.0)/(1-0.07384)\right]$$

$$= 4.42 \times 10^{-5} \text{ kg/s per m}^2 \text{ area of the pan}$$

The amount of water to be evaporated per m² area of the pan is

$$\text{Volume} \times \text{Density} = 0.02 \times 1 \times 1000 = 20 \text{ kg/m}^2 \text{ area}$$

Therefore, the time required $= 20/4.42 \times 10^{-5} = 4.52 \times 10^5$ s

$$= 125.65 \text{ hours.}$$

Example 12.11 A petrol station attendant accidentally spills 10 litres of gasoline over a level concrete floor of 3 sq. metre. Estimate the time required for the gasoline to evaporate into still dry air; D = 0.65 m²/hr. The temperature is 25°C and it may be assumed that the evaporation takes place through a film 15 cm thick. The vapour pressure of gasoline is 0.14 bar.

Solution: Film theory concept has been often useful in solving mass transfer processes. This concept visualizes an imaginary film of stagnant gas adjacent to the liquid surface, as shown in Fig 12.5. The thickness of the stagnant film should be so chosen that the gas film offers the same resistance to diffusion as encountered in the combined process of molecular diffusion and diffusion by mixing of the moving fluid.

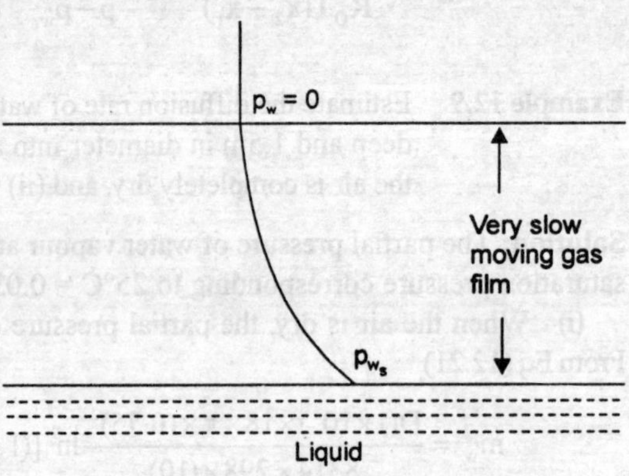

Fig. 12.5 Imaginary film model for mass transfer (Ex 12.11)

Thus by assuming that the film is atleast 15 cm thick, with the help of Eq. (12.21) we write,

$$\dot{m}_g = [(0.65/3600) \times (10^5) \times 118 \times 3]/(8314 \times 298 \times 0.15) \ln(1.0/0.86)$$

Because the vapour pressure at the gasoline surface is 0.14 bar and at the edge of the imaginary stagnant film the vapour pressure is zero.

$$\therefore \qquad \dot{m}_g = 2.594 \times 10^{-3} \text{ kg/s}$$

The total amount of gasoline to be evaporated = density × volume of gasoline

$$= 760 \times 10 \times 10^{-3} = 7.6 \text{ kg}$$

Therefore, the time required, $t = 7.6/2.594 \times 10^{-3} = 0.8138$ hours.

11. An Expression for Steady State Diffusion through a Plane Membrane

Let us consider a thin plane membrane of thickness L. The mass concentrations on the two sides of the membrane is as shown in Fig. 12.6. Since the mechanism of mass transfer by molecular diffusion is analogous to the heat transfer mechanism by conduction, the one-dimensional mass transfer molecular diffusion under steady state can be written as

$$\frac{d^2 C_A}{dx^2} = 0 \qquad (12.22)$$

where C_A is the mass concentration of the species A. The boundary conditions are:

at $x = 0$, $C_A = C_{A_1}$; and at $x = L$, $C_A = C_{A_2}$

The concentration profile and the mass transfer rate would be given respectively by

$$C_A = C_{A_1} + (C_{A_2} - C_{A_1}) x/L \; ; \text{ and}$$

$$\dot{m}_A/A = -D(dC_A/dx) = D(C_{A_1} - C_{A_2})/L \qquad (12.23)$$

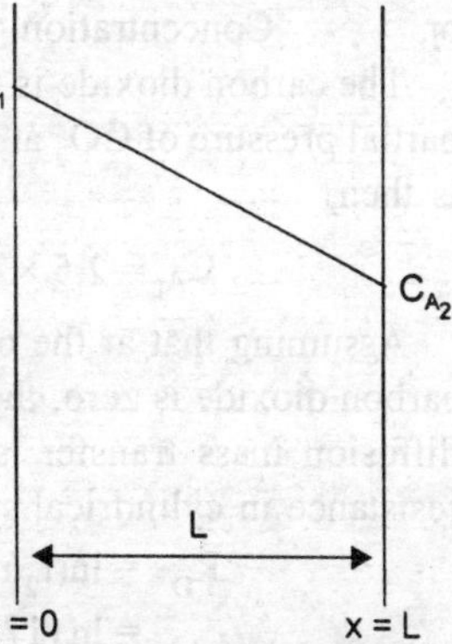

Fig. 12.6 Steady-state concentration profile in a thin plane membrane

Since the surface concentrations cannot by easily measured, it is more convenient to express the rate of mass transfer in terms of vapour pressure on the two sides of the membrane instead of concentrations at the surface. And, a new parameter called permeability $\wp$ is sometimes used and is defined as:

$$\wp = (\dot{m}_A/A)[(p_{A_1} - p_{A_2})/L] \text{ (seconds)} \qquad (12.24)$$

On the basis of analogy between conduction heat transfer and diffusion mass transfer, we can define a diffusional resistance as:

$$R_D = (C_{A_1} - C_{A_2})/\dot{m}_A = L/AD \qquad (12.25)$$

And, diffusion through composite membranes placed in series and/or in parallel can be treated similar to thermal or electrical resistances in series and/or parallel.

Example 12.12 Helium diffuses through Pyrex glass 2.5 mm thick. The mass concentration of helium at the inside and outside surface of the glass is 0.2 kg/m³ and 0.06 kg/m³ respectively. Calculate the diffusion flux of helium through the glass if the diffusion coefficient for helium-glass combination is 0.4×10^{-13} m²/s.

Solution: Assuming steady state conditions, the mass transfer rate is given by Eq (12.23).

Or, $\dot{m}_A / A = D(C_{A_1} - C_{A_2})/L$

$$= 0.4 \times 10^{-13} (0.2 - 0.06) / 2.5 \times 10^{-3} = 2.24 \times 10^{-12} \text{ kg/m}^2\text{s}.$$

Example 12.13 Carbon dioxide at 30°C and at a pressure of 2.5 bar is flowing through a rubber pipe, inside diameter 25 mm and thickness 5 mm. The coefficient of diffusion of carbon dioxide-rubber combination is 0.11×10^{-9} m²/s and the solubility of carbon dioxide in rubber is 4×10^{-2} kmol/m³ bar. Calculate the loss of CO_2 by diffusion per unit length of the pipe.

Solution: The species concentration at the gas-solid interface is obtained in terms of the partial pressure of the gas adjacent to the solid surface and a solubility factor, S.

or,   Concentration = Partial pressure × S

The carbon dioxide is flowing through the rubber tube at 2.5 bar. As such, the partial pressure of CO_2 at the rubber-gas interface is 2.5 bar and the concentration is then,

$$C_{A_1} = 2.5 \times 4 \times 10^{-2} \text{ kmol/m}^3 = 4.4 \text{ kg/m}^3.$$

Assuming that at the outer surface of the rubber pipe, the partial pressure of carbon dioxide is zero, the concentration at the outer surface is zero. Further, the diffusion mass transfer is analogous to conduction heat transfer, the diffusion resistance in cylindrical system can be written as,

$$R_D = \ln(r_2/r_1) / (2\pi \, L \, D)$$
$$= \ln (17.5 / 12.5) / (2 \times 3.142 \times 1 \times 0.11 \times 10^{-9})$$
$$= 4.867 \times 10^8$$

and $\dot{m}_{CO_2} = (C_{A_1} - C_{A_2}) / R_D = (4.4 - 0.0) / 4.867 \times 10^8$
$$= 9.04 \times 10^{-9} \text{ kg/s, or } 3.25 \times 10^{-5} \text{ kg/hr.}$$

Example 12.14 Hydrogen gas is maintained at 4 bar and 1 bar on the opposite sides of a plastic membrane 0.4 mm thick. The temperature is 25°C and the binary diffusion coefficient of hydrogen in the plastic is 8.7×10^{-8} m²/s. The solubility of hydrogen in the membrane is 1.5×10^{-3} kmol/m³ bar. Calculate the mass diffusive flux of hydrogen through the membrane.

Solution: Assuming steady-state, one-dimensional conditions, we use Eq. (12.23):
$$\dot{m}_A / A = D(C_{A_1} - C_{A_2})/L$$
where $C_{A_1} = 1.5 \times 10^{-3} \times 4 \text{ (bar)} \times 2 \text{ (kg/kmol)} = 1.2 \times 10^{-2}$ kg/m³
and $C_{A_2} = 1.5 \times 10^{-3} \times 1 \text{ (bar)} \times 2 \text{ (kg/kmol)} = 0.3 \times 10^{-2}$ kg/m³
∴ $\dot{m}_A / A = 8.7 \times 10^{-8} (1.2 \times 10^{-2} - 0.3 \times 10^{-2}) / 0.4 \times 10^{-3}$
$$= 1.957 \times 10^{-6} \text{ kg/m}^2\text{s.}$$

Example 12.15 In order to maintain a pressure close to 1 bar, a pipeline carrying ammonia gas is vented to atmosphere. Venting is achieved by tapping the pipe and inserting a 4 mm diameter tube, which extends for 25 m into the atmosphere. Calculate the mass rate of ammonia lost to the atmosphere and the mole and mass fraction of air in the pipe when the ammonia flow rate through the pipeline is 6 kg/hr. The entire system is operating at 25°C.

Solution: Assuming steady state, one-dimensional diffusion in tube and ideal gas behaviour, we use Eq (12.15) and take $D = 0.28 \times 10^{-4}$ m²/s

$$\dot{m}_A/A = D(M_A/R_0 T)(p_{A_1} - p_{A_2})/L$$

where $p_{A_1} = 1$ bar,   and $p_{A_2} = 0$

$$\frac{\dot{m}_A}{A} = \frac{0.28 \times 10^{-4} \times 17}{8314 \times 298} \frac{(1-0.0) \times 10^5}{25}$$

$$= 7.685 \times 10^{-7} \text{ kg/m}^2\text{s}$$

$$\dot{m}_A = 7.685 \times 10^{-7} \times \frac{\pi \times (0.004)^2}{4} \times 3600$$

$$= 1.1 \times 10^{-8} \text{ kg/hr}$$

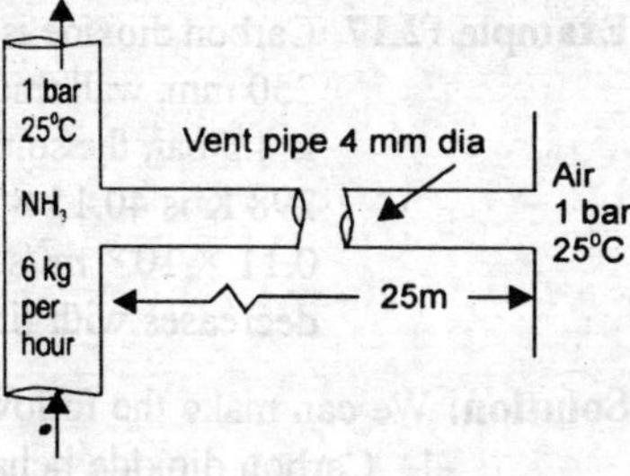

Diffusion rate expressed in terms of number
 of moles would be $= 6.47 \times 10^{-10}$ kmol/hr

For equimolal diffusion, $N_B = -N_A$, and therefore the number of mole of air diffusing
through the tube will also be $= 6.47 \times 10^{-10}$ kmol/hr

Mass of air diffusing through $= 6.46 \times 10^{-10} \times 28.97$

$$= 1.87 \times 10^{-8} \text{ kg/hr}$$

Since ammonia is flowing at the rate of 6 kg/hr, mass fraction of air would be

$$= 1.87 \times 10^{-8}/6 = 0.3117 \times 10^{-8}$$

and the mole fraction of air would be

$$= 6.47 \times 10^{-10}/(6/17) = 1.833 \times 10^{-9}.$$

Example 12.16 The air pressure inside a synthetic rubber ball (400 mm inside
diameter and 15 mm thick) decreases from 3.5 bar to 3.45 bar in
seven days. Estimate the coefficient of diffusion of air in synthetic
rubber if the temperature is 25°C and the solubility of air in the
rubber is 1.8×10^{-3} kmol/m³.bar.

Solution: Since the pressure change is very small during a period of seven days,
the problem can be treated as quasi-steady. The initial mass of air inside the ball

$$m_1 = p_1 V / RT = \frac{3.5 \times 10^5 \times (4/3)\pi(0.2)^3}{287 \times 298} = 0.137 \text{ kg}$$

The final mass, $m_2 = \dfrac{3.45 \times 10^5 \times (4/3)\pi(0.2)^3}{287 \times 298} = 0.1352 \text{ kg}$

The rate of leakage $= \dfrac{0.137 - 0.1352}{7 \times 24 \times 3600} = 2.976 \times 10^{-9} \text{ kg/s}$

The average pressure inside the ball $= (3.45 + 3.5)/2 = 3.475$ bar

Concentration inside the ball $\quad = p_1 \times S = 3.475 \times 1.8 \times 10^{-3} \times 29, \text{kg/m}^3$

$$= 0.1814 \text{ kg/m}^3$$

Concentration at the outside surface $= p_2 \times S = 1 \times 1.8 \times 10^{-3} \times 29$

$$= 0.0522 \text{ kg/m}^3.$$

Since conduction heat transfer is analogous to diffusion mass transfer, the
diffusive resistance for the spherical shell can be written as

$$R_D = (r_2 - r_1) / (4\pi D r_1 r_2), \text{ and}$$

$$\dot{m}_A = (C_{A_1} - C_{A_2}) / R_D = 4\pi D r_1 r_2 (C_{A_1} - C_{A_2}) / (r_2 - r_1)$$

or, $\qquad D = \dot{m}_A (r_2 - r_1)/[4\pi r_1 r_2 (C_{A_1} - C_{A_2})]$

$$= \frac{2.976 \times 10^{-9} \times 0.015}{4 \times 3.142 \times 0.2 \times 0.215 \times (0.1814 - 0.0522)} = 6.4 \times 10^{-10} \text{ m}^2/\text{s}.$$

Example 12.17 Carbon dioxide is stored in a spherical rubber ball (inside diameter 250 mm, wall thickness 2 mm). The initial pressure inside the ball is 4.5 bar, the solubility of carbon dioxide in rubber at temperature 298 K is 40.15×10^{-3} kmol/m^3 bar and coefficient of diffusion is 0.11×10^{-9} m^2/s. Estimate the initial rate at which the pressure decreases with time.

Solution: We can make the following assumptions:

1. Carbon dioxide behaves like a perfect gas.
2. The variation of pressure inside the vessel is sufficiently low such that steady state condition for diffusion is valid.
3. The diameter of the vessel is very large in comparison with the wall thickness and as such the diffusion can be approximated as one-dimensional through a plane wall.
4. The partial pressure of carbon dioxide outside the vessel is z

The initial mass of the gas = density of the gas x volume

The rate of change of mass, $\dot{m} = dm/dt = - d(\rho V)/dt$
$$= - dP/dt \cdot (M \cdot V)/R_o T; \quad (M = \text{mol.wt})$$

The rate of diffusion of mass, $\dot{m} = A \times D \times (C_{A_1} - C_{A_2})/L$

The concentration at the inside surface of the vessel

$$C_{A_1} = \text{Partial pressure} \times \text{solubility} \times \text{molecular weight}$$

$$= p \times S \times M; \quad \text{and} \quad C_{A_2} = 0.0$$

Therefore, $dp/dt = \dfrac{-A \times D \times p \times S \times M \times R_o \times T}{L \times M \times V}$

Since the surface area, $A = 4\pi r^2$ and volume $= (4/3)\pi r^3$; $A/V = 3/r$

or, $\qquad dp/dt = \dfrac{-0.11 \times 10^{-9} \times 4.5 \times 40.15 \times 10^{-3} \times 831 \times 4 \times 298 \times 3}{0.002 \times 0.125}$

$$= 0.59 \times 10^{-5} \text{ bar/s.}$$

12. Expression for Transient Diffusion in a Semi-infinite Medium

Transient diffusion occurs in processes in which the concentration at a given point varies with time.

By making an analogy to one-dimensional transient heat conduction problem, we write:

$$\partial C_A/\partial t = D \, \partial^2 C_A/\partial x^2 \qquad (12.26)$$

With the boundary conditions:

(i) $C_A (0,t) = C_{A_s}$, $\qquad$ (ii) $C_A(\infty,t) = C_{A_i}$, $\qquad$ (iii) $C_A(x, 0) = C_{A_i}$

The first boundary condition requires that the mass concentration C_A be held constant at the surface; the second requires that the core of the body, at a large distance from the surface, remains at its initial mass concentration. The initial condition states that at any location in the medium at the time, $t = 0$, the mass concentration is constant C_{A_i}. The solution for the mass concentration and mass transfer would be

$$(C_A - C_{A_s})/(C_{A_i} - C_{A_s}) = \text{erf}\,(x/\sqrt{4Dt}) = \text{erf}\,(x/2\sqrt{Dt}) \qquad (12.27)$$

and $\qquad \dot{m}_A/A = -D\partial C_A/\partial x\big|_{x=0} = -D(C_{A_s} - C_{A_i})/\sqrt{\pi Dt} \qquad (12.28)$

The solution is a function of a single dimensionless parameter $\eta_D = \dfrac{x}{2\sqrt{Dt}}$

which is a combination of two independent parameters x and t. Thus, the rate of penetration of any concentration depends on $t^{-1/2}$. The concentration profiles as a function of time are shown in Fig. 12.7.

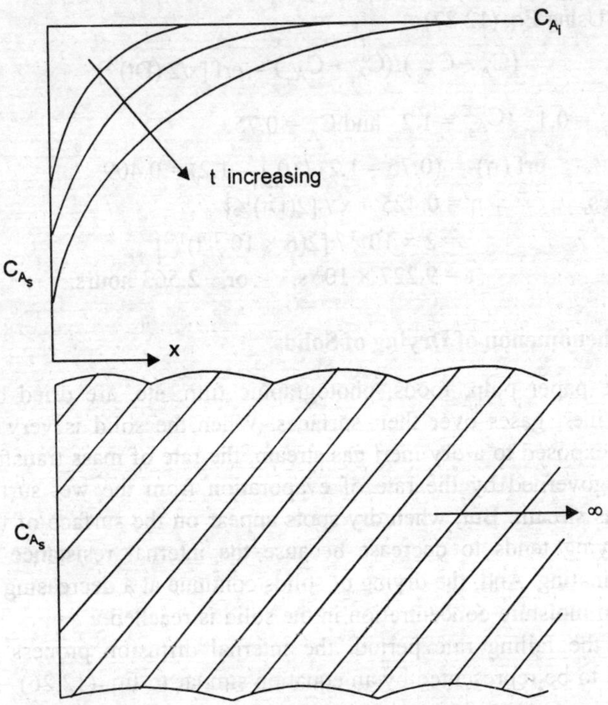

Fig 12.7 Transient diffusion in a semi-infinite medium and concentration distribution

Example 12.18 Estimate the concentration of the surface of the earth at a depth of 50 cm after a 24 hour period when the dry surface of the earth experiences a sudden flood. Take the coefficient of diffusion for the soil as 0.0012 m²/hr.

Solution: Initially the surface of the earth is dry. The concentration $C_A = 0.0$.

After the sudden flood, $C_{A_s} = 1.0$; $\quad t = 24$ hour

The variable, $\qquad \eta = x/2(Dt)^{1/2} = 0.5/2(0.0012 \times 24)^{1/2} = 1.47$

From tables $\qquad \text{erf}\,(\eta) = 0.95 = (C_A - C_{A_s})/(C_{A_i} - C_{A_s})$

or, $(C_A - 1.0) / (0.0 - 1.0) = 0.95$ and $C_A = 0.05$

Since the density of the soil is not known, the mass concentration can be expressed as 0.05 kg/kg soil.

Example 12.19 A low carbon steel material is subjected to case hardening. Initially it contains 0.1 percent carbon. It is preheated to 1223 K and is packed in a carburizing mixture at the same temperature. If the concentration of carbon at the surface of the material is maintained as 1.2 per cent, estimate the time required for the concentration at a depth of 2 mm to be 0.75 percent. The coefficient of diffusion of carbon in steel at that temperature is 6.0×10^{-10} m^2/s.

Solution: Using Eq. (12.27)

$$(C_A - C_{A_s}) / (C_{A_i} - C_{A_s}) = \text{erf}\,[x/2\,(Dt)^{1/2}]$$

where $C_{A_i} = 0.1$, $C_{A_s} = 1.2$ and $C_A = 0.75$

$\therefore$ $\text{erf}\,(\eta) = (0.75 - 1.2) / (0.1 - 1.2) = 0.409$

From tables, $\eta = 0.425 = x / [2(Dt)^{1/2}]$

$= 2 \times 10^{-3} / [2(6 \times 10^{-10}t)^{1/2}]$

Therefore, $t = 9.227 \times 10^3$ s,   or,   2.563 hours.

13. The Phenomenon of Drying of Solids

Solids like paper pulp, foods, photographic film, etc, are dried by passing a stream of inert gases over their surfaces. When the solid is very wet and the surface is exposed to a dry inert gas stream, the rate of mass transfer (drying of solids) is governed by the rate of evaporation from the wet surface into the flowing gas stream. But, when dry spots appear on the surface of the solid, the rate of drying tends to decrease because the internal resistance to diffusion starts dominating. And, the drying of solids continue at a decreasing rate until an equilibrium moisture concentration in the solid is reached.

During the falling rate period, the internal diffusion process of liquid is considered to be represented by an equation similar to Eq. (12.26)

$$\partial C/\partial t = D\,\partial^2 C/\partial x^2$$

where C is the concentration of the liquid in the solid provided the effects of capillarity play a minor role and are neglected.

Since there is no mass transfer from the surface when the liquid concentration at the surface $C_s = C_e$, a new coefficient at the surface is defined such that

$$\dot{m}_A/A = h_D(C_s - C_e)$$

where C_s and C_e are respectively the liquid concentration in the solid at the surface and at the equilibrium condition.

These equations are analogous to transient heat conduction equations and therefore, the solutions of these mass transfer equations are obtained with the help of Heisler Charts, provided in Chapter 3.

Example 12.20 A slab of wood 5 cm thick has a moisture content $C_i = 30$ per cent (based on dry wood) when the falling rate period begins. Estimate the coefficient of diffusion for water if the mositure content to be 10 percent at a depth of 2.5 cm when the equilibrium moisture content C_e is 5 percent (based on dry wood). Assume that the surface resistance is negligible and the edges and ends are covered with a moisture-resistance coating. Take the drying time as 2 hours.

Solution: Assuming that the wood does not shrink during drying, the values are obtained from the Fig. 3.8 (a)

The ratio of surface diffusion resistance to the internal diffusion resistance, $h_D x/L$, replaces the Biot modulus and for negligible surface resistance $1/Bi$ is equal to zero. The dimensionless concentrations are:

$$(C_A - C_e) / (C_i - C_e) = (10 - 5) / (30 - 5) = 0.2;$$

which gives D $t/L^2 = 0.6$ [from Fig. 3.8(a)]

Since $L = 2.5 \times 10^{-2}$ m, $t = 2$ hours

$$D = 0.6 \times (2.5 \times 10^{-2})^2/2 = 1.875 \times 10^{-4} \text{ m}^2/\text{hr}.$$

14. Convective Mass Transfer Coefficient

When the mass is transported between the boundary of a surface and a moving fluid or between two moving fluids which are relatively immiscible, we have mass transfer by convection. The convective process can be either natural or forced, depending upon the existence of density or pressure gradient respectively, in the medium. The convective mass transfer coefficient is defined in a manner similar to that for convective heat transfer.

Or, $$\dot{m}_A = h_{D_A} A(C_{A_1} - C_{A_2}) \tag{12.29}$$

where $\dot{m}_A$ = diffusive mass flux of species A, h_{D_A} = mass transfer coefficient, and C_{A_1}, C_{A_2} are the concentrations through which diffusion takes place. The convective mass transfer coefficient depends upon the fluid properties, the mechanism of fluid flow, and the geometry of the flow system and is analogous to the convective heat transfer coefficient.

For a steady-state diffusion across a layer of thickness L,

$$\dot{m}_A = DA(C_{A_1} - C_{A_2})/L = h_{D_A} A(C_{A_1} - C_{A_2})$$

therefore, $$h_{D_A} = D/L, \text{ m/s} \tag{12.30}$$

15. Dimensionless Numbers Used in Mass Transfer

The following dimensionless numbers are of significance in convective mass transfer:

Schmidt Number, Sc = momentum diffusivity/mass diffusivity = $v/D = \mu/\rho D$

This number is analogous to Prandtl number and when Sc = 1, there would be complete similarity between momentum and concentration equations and the hydrodynamic results may be applied directly to convective mass transfer problems.

Sherwood number, $Sh = h_D x/D$, and is analogous to Nusselt number.

Lewis Number, $Le = \alpha/D$, the temperature and concentration profiles will be similar when the Lewis number is equal to unity.

Mass Grashof number, $\qquad Gr_m = \dfrac{gL^3(C_s - C_\infty)}{v^2 C_\infty}$

C_s = mass concentration at surface

C_∞ = free stream mass concentration

16. An Expression for the Convective Mass Transfer Coefficient for Laminar Flow Over a Flat Plate

Let us consider a flat plate where the concentration at the surface of the species A is different than its concentration in the free stream. The species A will diffuse into the fluid and a concentration boundary layer will develop as shown in Fig. 12.8. The thickness of the concentration boundary layer is defined in the same manner as that of the hydrodynamic boundary layer or thermal boundary layer.

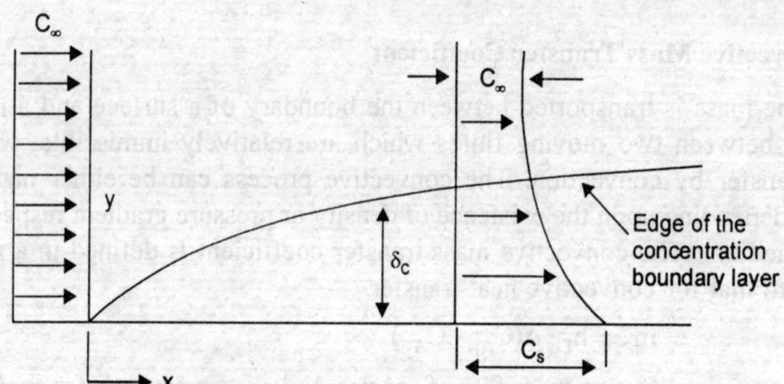

Fig. 12.8 Concentration boundary layer on a flat plate

Since there is a remarkable similarity between the laws governing the boundary layer growth of the three processes: momentum, heat and mass, the governing equation for the concentration boundary layer can be written as:

$$u \cdot \partial C_A/\partial x + v \cdot \partial C_A/\partial y = D \cdot \partial^2 C_A/\partial y^2 \qquad (12.31)$$

and for laminar flow over the flat plate, the average values for the convective mass transfer coefficient can be obtained from the relation:

$$Sh_L = h_D L/D = 0.664\, Re_L^{0.5}\, Sc^{1/3} \text{ (for } Sc \geq 0.6) \qquad (12.32)$$

When the boundary layer is partly laminar and partly turbulent, transition occuring at $Re = 5 \times 10^5$, the correlation for mass transfer would be, similar to heat transfer, and is given by

$$Sh_L = (0.037\, Re_L^{0.8} - 870)\, Sc^{1/3} \qquad (12.33)$$

For flow over smooth flat plates, the Colburn analogy predicts:

(a) Laminar flow: $\quad C_f/2 = 0.664\, Re_L^{-0.5} = h_D/U_\infty\, Sc^{2/3}$

(b) Turbulent flow: $C_f/2 = (h_D/U_\infty)\, Sc^{2/3} = 0.0296\, Re_L^{-0.2} \qquad (12.34)$

Example 12.21 Air, relative humidity 40 percent and temperature 25°C, flows over a 2 m long wet plate with a velocity of 5 m/s. Estimate the rate of mass transfer for water vapour in air. The physical properties of air are:

$$D = 0.256 \times 10^{-4} \text{ m}^2/\text{s}, \quad \mu = 2 \times 10^{-5} \text{ Pa--s}, \quad Pr = 0.7,$$
$$C_p = 1.005 \text{ kJ/kgK}, \quad \rho = 1.17 \text{ kg/m}^3.$$

Solution: Schmidt number $Sc = \mu/\rho D = 2 \times 10^{-5}/(1.17 \times 0.256 \times 10^{-4}) = 0.668$

Reynolds number $Re = \rho Vx/\mu \quad \dfrac{1.17 \times 5 \times 2}{2 \times 10^{-5}}$

$$= 5.85 \times 10^5 > 5 \times 10^5, \text{ a turbulent flow.}$$

Using Colburn analogy, Eq. (12.34)

$$(h_D/U_\infty) \cdot Sc^{2/3} = 0.0296 \, Re^{-0.2}$$

$$h_D = 0.0296 \, (5.85 \times 10^5)^{-0.2} \times 5/(0.668)^{2/3}$$

$$= 0.0136 \text{ m/s}$$

Using   Eq (12.33), we get

$$h_D = (D/L) \, (0.037 \times (5.85 \times 10^5)^{-0.2} - 870) / (0.668)^{1/3}$$
$$= 0.0145 \text{ m/s; and the average of the two values} = 0.01405 \text{ m/s}$$

The difference in the two values of the convective mass transfer coefficient can be attributed to the results obtained by empirical relations.

saturation pressure of water at 25°C, from steam tables = 0.03166 bar

Therefore, the concentration at the wet surface $= p/RT = \dfrac{0.03166 \times 18 \times 10^5}{8314 \times 298}$

$$= 0.023 \text{ kg/m}^3$$

Since the relative humidity is 0.4, the partial pressure of water at 25°C = 0.4 × 0.03166 bar, and the concentration is then, 0.4 × 0.023 = 0.0092 kg/m³.

Rate of mass transfer, $\dot{m}/A = h_D \, (C_s - C_\infty)$

$$= 0.01405 \, (0.023 - 0.0092) \times 3600 \text{ kg/m}^2.\text{hr}$$

$$= 0.698 \text{ kg/m}^2 \cdot \text{hr.}$$

Example 12.22 Dry air at 25°C flows over the surface of a swimming pool 4.5 m by 25 m. Estimate the convective mass transfer coefficient for the evaporation of water vapour in air when the velocity of air is 1.5 m/s (i) along the length of the pool, and (ii) along the width of the pool.

Take, $v = 15 \times 10^{-6} \text{ m}^2/\text{s}$, $D = 0.256 \times 10^{-4} \text{ m}^2/\text{s}$

Solution: (i) Flow along the length of the pool:

$$Sc = v/D = 15 \times 10^{-6}/0.256 \times 10^{-4} = 0.586$$
$$Re = UL/v = 1.5 \times 25 / 15 \times 10^{-6} = 2.5 \times 10^6, \text{ a turbulent flow.}$$

From Eq (12.33) $h_D = \dfrac{D}{L} [0.037(2.5 \times 10^6)^{0.8} - 870)(0.586)^{1/3}]$

$$= 0.0034 \text{ m/s}$$

(ii) Flow along the width of the pool:

$$Re = UL/v = 1.5 \times 4.5 / 15 \times 10^{-6}$$
$$= 4.5 \times 10^5, \text{ a laminar flow.}$$

Using Eq (12.32), $h_D = \dfrac{D}{L} [0.664 (4.5 \times 10^5)^{0.5} (0.586)^{1/3}]$
$$= 0.0032 \text{ m/s.}$$

17. Expressions for Convective Mass Transfer Coefficient for Flow through Tubes, Flow over Spheres and Cylinders

We have forced convection mass transfer when a liquid evaporates from the wetted walls of a tube into the gas flowing through that tube. A concentration boundary layer develops inside the tube, similar to the hydrodynamic boundary

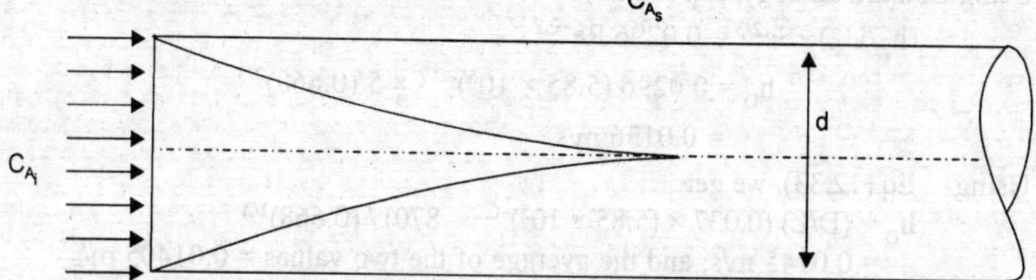

Fig. 12.9 Concentration profile for mass transfer in a tube

layer, as shown in Fig. 12.9.

For fully developed velocity and concentration profiles in a laminar flow through a tube, the following equations have been suggested for the evaluation of mass transfer coefficients:

$$h_D d/D = Sh = 3.66 \text{ for uniform wall mass concentration} \tag{12.35}$$
$$Sh = 4.34 \text{ for uniform wall mass flux} \tag{12.36}$$

In turbulent flow, the concentration of the diffusing component varies with time and space and as such mass transfer coefficients are evaluated either experimentally or with the help of empirical equations. These equations are based on analogy with heat transfer.

Gilliland has proposed the following equation for the vapourization of liquids from the walls of smooth circular tubes when air is forced to flow through the tube: $\quad Sh = 0.023(Re)^{0.83} (v/D)^{0.44}; \quad 2000 < Re < 35,000;$
and for gases $\qquad\qquad 0.6 < Sc < 2.5 \tag{12.37}$

The Reynolds and Colburn analogy can also be used to calculate the mass transfer coefficient from the friction factor. Or,

$$\frac{h_D}{U_\infty} \cdot Sc^{2/3} = f/8 = \frac{C_f}{2} \tag{12.38}$$

When the coefficient of friction C_f is eliminated from Colburn analogy for heat and mass transfer, we get

$$St (Pr)^{2/3} = St_m (Sc)^{2/3}$$

or, $\qquad \dfrac{h}{\rho C_p U_\infty} (Pr)^{2/3} = St_m (Sc)^{2/3}$

$$\therefore \qquad \frac{h}{h_D} = \rho C_p \left(\frac{Sc}{Pr}\right)^{2/3} = \rho C_p \left(\frac{\alpha}{D}\right)^{2/3} = \rho C_p (Le)^{2/3} \qquad (12.39)$$

Eq. (12.39) gives a relation between the mass transfer and heat transfer coefficients. That is, the coefficient of mass transfer can be evaluated from the data available for heat transfer coefficients. Eq. (12.39) is known as Lewis Equation. $Le \approx 1$ in gas-vapour mixture.

Example 12.23 Air at 27°C and at atmospheric pressure containing small quantities of iodine flows with a velocity of 6 m/s through a 2.5 cm inner diameter tube, Determine the mass transfer coefficient from the air stream to the tube surface. The properties of air are: $D = 0.82 \times 10^{-5}$ m^2/s, $\nu = 15.4 \times 10^{-6}$ m^2/s.

Solution: Since a very small quantity of iodine is present in air, it is a dilute solution and we can use the properties of air.

$$Re = Ud/\nu = 6 \times 2.5 \times 10^{-2}/15.4 \times 10^{-6}$$
$$= 0.974 \times 10^4, \text{ a turbulent flow.}$$
$$Sc = \nu/D = 15.4 \times 10^{-6}/0.82 \times 10^{-5}$$
$$= 1.878$$

From Eq (12.37), $Sh = h_D\, d/D = 0.023\,(9740)^{0.83}\,(1.878)^{0.44}$

or, $h_D = 0.0203$ m/s.

Flow over Spheres and Cylinders

The convective mass transfer coefficient for flows over spheres and cylinders are evaluated by using the relation:

$$Sh = C\,(Re)^n\,(Sc)^{1/3} \qquad (12.40)$$

Eq. (12.40) is analogous to the empirical relations for evaluation of convective heat transfer coefficient for flows over cylinders and spheres and the two constants C and n are evaluated experimentally. For evaluating convective mass transfer coefficient for flows over a sphere, Froessling has suggested the following relation:

$$Sh = 2(1 + 0.276\,Re^{1/2}\,Sc^{1/3}) \qquad (12.41)$$

Example 12.24 Air at 27°C and at 1 bar, RH = 0.30, flows with a velocity of 10 m/s over a 50 mm diameter cork ball completely soaked in water and at 25°C. Estimate the mass transfer coefficient and the rate of water vapour transfer into air. The properties are: $\nu = 16 \times 10^{-6}$ m^2/s, $D = 0.256 \times 10^{-4}$ m^2/s.

Solution: Density of air, $\rho = p/RT = 10^5/287 \times 300 = 1.16$ kg/m^3

Schmidt number, $Sc = \nu/D = 16 \times 10^{-6}/0.256 \times 10^{-4} = 0.625$

Reynolds number, $Re = Ud/\nu = 10 \times 50 \times 10^{-3}/16 \times 10^{-6} = 31250$

From Eq. (12.41), $Sh = 2[1 + 0.276\,(31250)^{1/2}\,(0.625)^{1/3}] = 85.43$

Therefore $h_D = 85.43 \times D/d = 85.43 \times 0.256 \times 10^{-4}/50 \times 10^{-3}$
$$= 0.0437 \text{ m/s}$$

For $Re = 31250$,   the values of $C = 0.193$, $n = 0.618$ (from Table 8.1).

From Eq. (12.40), $Sh = 0.193 \, (31250)^{0.618} \, (0.625)^{1/3} = 98.95$

and, $\qquad h_D = 98.95 \times 0.256 \times 10^{-4}/50 \times 10^{-3} = 0.0506$ m/s, a higher value.

At the surface of the cork, saturation pressure corresponding to 25°C is 0.03166 bar, therefore,

$$C_s = P/RT = 0.03166 \times 10^5 \times 18/(8314 \times 298) = 0.023 \text{ kg/m}^3$$

Partial pressure of water vapour in air at 27°C $= 0.3 \times 0.03564$

$$C_\infty = 0.3 \times 0.03564 \times 10^5 \times 18/(8314 \times 300) = 0.0077 \text{ kg/m}^3$$

Taking the mass transfer coefficient as $(0.0437 + 0.0506)/2 = 0.04715$

The rate of transfer of water vapour $= h_D \, A \, (C_s - C_\infty)$

$$= 0.4715 \times 4\pi(0.025)^2(0.023 - 0.0077)$$

$$= 0.0204 \text{ kg/hr.}$$

18. Simultaneous Heat and Mass Transfer Process

Simultaneous heat and mass transfer processes are of importance in vaporization and condensation operations, i.e., processes having a change of phase. Common examples are: cooling towers, dryers, dehumidifying equipments and gas absorption equipments. In some cases, the heat transfer may be insignificant whereas in others, it may dominate the design considerations.

As an example, let us consider the operation of an ordinary wet-bulb thermometer, shown in Fig. 12.10. Atmospheric air at temperature T_1 having some moisture (partial pressure of water p_w) flows over the wick (wet cover) of the thermometer bulb. Water vapour from the surface of the wick will diffuse into air as a result of concentration gradient. Consequently, the temperature of the water will decrease, because the latent heat of vapourization has to be

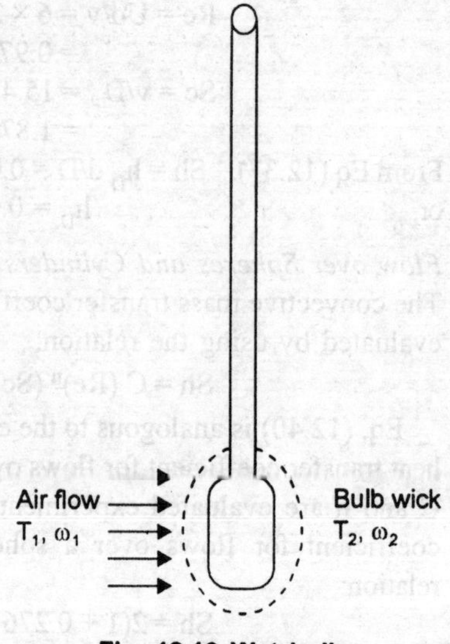

Fig. 12.10 Wet bulb thermometer

supplied by the water, and the heat transfer by convection will take place from the atmospheric air to water. At the equilibrium, the thermometer will record the temperature of water in the wick. By making an energy balance we can write

$$\dot{Q}/A = h \, (T_1 - T_2), \text{ where } T_2 \text{ is the temperature of the wick}$$

$$= \dot{m}/A \, (h_{fg}), \qquad (12.42)$$

where h is the convective heat transfer coefficient, neglecting the effects of radiation, h_{fg} is the latent heat of vapourization of water at T_2, and $\dot{m}$ is the mass of water evaporated per unit time. Also,

$$\dot{m}/A = h_D \, (C_s - C_\infty) \qquad (12.43)$$

and the two coefficients h and h_D are related by Eq. (12.39).

Example 12.25   A wet bulb thermometer reads 20°C when dry air is flowing over the wick of the thermometer. Estimate the temperature of the dry air.

Solution: Saturation pressure corresponding to 20°C = 0.02337 bar

$$\text{Mass concentration,} \quad C_s = \frac{0.02337 \times 10^5 \times 18}{8314 \times 293}$$

$$= 0.01727 \text{ kg/m}^3$$

h_{fg} at 20°C = 2454.3 kJ/kg;   for dry air $C_\infty = 0.0$

Since the properties of air are to be evaluated at the mean temperature, let us assume that the temperature of dry air is 60°C.

Therefore, mean temperature = 40°C and the properties are:

$$C_p = 1.006 \text{ kJ/kgK}, \ \nu = 16 \times 10^{-6} \text{ m}^2\text{/s}, \quad Pr = 0.70 \text{ and } \rho = 1.113 \text{ kg/m}^3.$$

From Eq. (12.39), $h/h_D = \rho C_p (Le)^{2/3}$, $Le = Sc/Pr = 16 \times 10^{-6}/(0.256 \times 10^{-4} \times 0.70)$

$$= 0.893$$

$$\therefore \qquad h/h_D = 1.113 \times 1.006(0.893)^{2/3}$$

$$= 1.0382$$

From Eq. (12.42) and Eq. (12.43), $h/h_D = (C_s - C_\infty) h_{fg}/(T_1 - T_2)$

$$T_1 = 20 + (0.01727 - 0.0) \times 2454.3/1.0382$$

$$= 60.8°C, \text{ a good initial guess.}$$

Example 12.26 The temperature of air of Ex. 12.25 is 50°C after it flows over the wick. Estimate the relative and specific humidity of the air stream.

Solution: The mean film temperature is (50 + 20)/2 = 35°C.
The properties of air are:

$$\rho = 1.13 \text{ kg/m}^3, \ Pr = 0.7, \ \nu = 17 \times 10^{-6} \text{ m}^2\text{/s}, \ C_p = 1.006 \text{ kJ/kgK}$$

and for air-water vapour, D = 0.256 × 10⁻⁴ m²/s; Sc = 0.664
From Eq (12.39), Eq (12.42) and Eq (12.43), we have

$$\rho C_p (Le)^{2/3} (T_1 - T_2) = h_{fg} (C_s - C_\infty)$$

or, $1.13 \times 1.006(0.664/0.7)^{2/3}(50 - 25) = 2454.3 \times (0.01727 - C_\infty)$

$$C_\infty = 0.003855 \text{ kg/m}^3$$

At 50°C, the specific volume of saturated vapour 12.05 m³/kg

$$\therefore \qquad C_s = 1/12.05 = 0.08;$$

The relative humidity of the air stream = 0.003855/0.08

$$= 0.048 \equiv 4.8\%$$

The specific humidity and the relative humidity are related by the relation:

$$w = 0.622 \ \phi \ (p_s/p_a); \quad \text{where } \phi = \text{relative humidity,}$$

$$w = \text{specific humidity,}$$

$$p_s = \text{saturation pressure of vapour}$$

$$P_a = \text{partial pressure of air at 50°C,}$$

$$p_s = 0.12355 \text{ bar}$$

$$p_a = 1.0 - 0.12335 = 0.87665 \text{ bar}$$

Therefore, w = 0.622 × 0.048 × 0.12335/0.87665

$$= 4.2 \text{ gm/kg dry air.}$$

SUMMARY

1. Mass transfer is mass in transit as the result of a species concentration difference in a mixture.
2. Diffusion at microscopic level (Fick's law) is analogous to conduction heat transfer (Fourier' law).
3. For equimolal counter diffusion in a binary mixture, the molal diffusion rates are equal.
4. The isothermal evaporation of water vapour in a stagnant air is a function of logarithmic mean partial pressure.
5. Transient diffusion in a semi-infinite medium can be solved with the help of transient charts of Heisler because the governing equations for heat and mass transfer are analogous.
6. Colburn analogy can be applied to solve convective mass transfer problems.
7. Schmidt number, and Sherwood number are respectively analogous to Prandtl number and Nusselt number.
8. Convective mass transfer coefficient can be evaluated from heat transfer coefficient data (Lewis equation).
9. Simultaneous heat and mass transfer occurs in desert coolers, spray ponds, cooling towers, etc.

MULTIPLE CHOICE QUESTIONS

1. Fick's law of diffusion in terms of mass fraction can be written as:

   (a) Mass flux $= D \dfrac{\partial C_a}{\partial x}$ (b) Mass flux $= -\rho D \dfrac{\partial C_a}{\partial x}$

   (c) Mass flux $= \rho D \dfrac{\partial C_a}{\partial x}$ (d) Mass flux $= -(D/\rho) \dfrac{\partial C_a}{\partial x}$ **(b)**

   where C_a is the mass fraction, ρ is the mass density and D is the mass diffusivity

2. The bulk velocity for a binary mixture of components a and b can be written as:

   (a) $V_x = (C_a V_{ax} + C_b V_{bx})$ (b) $V_x = \rho(C_a V_{ax} + C_b V_{bx})$

   (c) $V_x = (1/\rho)(C_a V_{ax} + C_b V_{bx})$ (d) $V_x = D(C_a V_{ax} + C_b V_{bx})$ **(c)**

   where V_{ix} is the statistical mean velocity of species i and C_i is the mass concentration.

3. Assertion (A): Fick's law of diffusion cannot be applied to diffusion of liquids in solids

   Reasoning (R): Because mass flux is not proportional to concentration gradient

   Code   (a) Both A and R are true   (b) Both A and R are false

           (c) A is false, R is true   (d) A is true, R is false   **(c)**

4. Choose the wrong statement:

   During the isothermal evaporation of water vapour into still air, it is assumed that

   (a) the total pressure remains constant

(b) air and water vapour behave like perfect gases
(c) the air movement creates a little turbulence
(d) the system is in steady state **(c)**

5. During transient diffusion in a semi-infinite medium, the rate of penetration of any concentration depends on

(a) t (b) $\sqrt{t}$ (c) 1/t (d) $1/\sqrt{t}$ **(d)**

6. Match List I with List II and choose the answer from the code given below

List I	List II
A. Schmidt number	1. α/D
B. Sherwood number	2. v/D
C. Lewis number	3. $h_D x/D$
D. Mass Grashof number	4. $gL^3(C_s-C_\infty)/v^2 C_\infty$

CODE

	A	B	C	D
(a)	1	2	3	4
(b)	1	3	2	4
(c)	3	4	2	1
(d)	2	3	1	4	**(d)**

7. The diffusion coefficient is a property of the system. Its value depends upon the system's
(a) pressure (b) temperature
(c) composition (d) all of the above **(d)**

8. If heat and mass transfer take place simultaneously, the ratio of heat transfer coefficient to the mass transfer coefficient is a function of
(a) Schmidt and Reynolds number
(b) Schmidt and Prandtl number
(c) Nusselt and Lewis number
(d) Reynolds and Lewis number **(b)**

9. When there is forced convection with mass transfer over a flat plate, there is a similarity in momentum and concentration boundary layer if
(a) Pr = 1 (b) Sc = 1
(c) Sh = 1 (d) Sc × Re = 1 **(b)**

10. For complete similarity in momentum, heat and mass transfer, we should have
(a) Le = Re = 1 (b) Pr = Sc = 1
(c) Sh = Sc = 1 (d) Pr = Sh = 1 **(b)**

11. The convective mass transfer coefficient for flows over spheres and cylinders are evaluated from the functional relationship between the following dimensionless number
(a) Sh, Re, and Le (b) Sc, Re and Le
(c) Sh, Sc and Pr (d) Sh, Re and Pr. **(a)**

12. For a turbulent flow over a flat plate, the average convective mass transfer coefficient, with increasing Reynolds number,
    (a) increases   (b) remains the same
    (c) decreases   (d) is unpredictable   **(c)**
13. The diffusion coefficient defined by the Fick's law of diffusion in terms of mass concentration is not identical
    (a) in a solid   (b) in a liquid solution
    (c) in a dilute gaseous mixture (d) when the density is not uniform. **(d)**
14. In equimolar counter diffusion in a binary, isothermal ideal gas mixture, we have
    (a) $dp_A/dx = dp_B/dx$ (b) $dp_A/dx = - dp_B/dx$,
    (c) $dp_A/dx > dp_B/dx$ (d) $dp_A/dx < dp_B/dx$ **(b)**
15. Assertion (A): The mass diffusivity of liquids is much smaller than those of gases.
    Reasoning (R): Because the liquids have higher molecular density.
    Code: (a) Both A and R are false   (b) Both A and R are true
          (c) A is true and R is false   (d) A is false and R is true   **(b)**
16. The mass transfer Biot modulus (Bi_m) is defined as $h_m L/D_{AB}$. If R_d is the resistance to species transfer by diffusion in the medium and R_c is the resistance to species transfer by convection at the surface, then Bi_m is
    (a) R_d/R_c (b) R_c/R_d (c) $R_c \times R_d$ (d) $R_c + R_d$ **(a)**

NUMERICALS

1. A mixture of CO and oxygen is to be prepared in the proportion of 7 kg to 4 kg in a vessel of 0.3 m^3 capacity. If the temperature of the mixture is 15°C, calculate the pressure inside the vessel. If the temperature is raised to 40°C, what would be pressure in the vessel? Also calculate the molar mass and the characteristic gas constant.

   (29.94 bar, 32.54 bar, 29.3 kg/kmol; 0.283 kJ/kgK)

2. A vessel of 3 m^3 capacity contains a mixture of nitrogen and carbon dioxide in equal volumes of each. The temperature is 15°C and the total pressure is 3.5 bar. Determine the mass of each constituent. If the mixture is changed so that it contains 70% CO_2 and 30% N_2 by volume, calculate the mass of the mixture to be removed and the mass of CO_2 to be added to give the required mixture at the same temperature and pressure as before.

   (6.14 kg N_2; 9.65 kg CO_2; 6.31 kg, 7.72 kg CO_2)

3. Ammonia diffuses in air at 25°C and 1 atm pressure. Estimate the diffusion coefficient.

   (1880cm^2/s)

4. Calculate the rate of burning of a pulverized carbon particle of 0.25 cm diameter in an atmosphere of pure oxygen at 727°C and 1 atm pressure. Assume that CO_2 forms a very large blanketing layer around the particle.

   (2.3 $\times$ 10^{-7} kg/s)

5. A pan 20 cm in diameter and 10 cm deep contains water at 25°C and is exposed to dry air. If the rate of diffusion of water vapour is 3×10^{-7} kg/s, calculate the diffusion coefficient of water in air.

   $(4.1 \times 10^{-4}$ m^2/s)

6. The air pressure in a rubber tyre (solubility of air in rubber 1.8×10^{-3} kmol/m^3 bar) reduces from 3 bar to 2.99 bar in five days. If the volume of air in the tube is 0.025 m^3, the surface area of the tube permitting diffusion is 0.5 m^2, and the rubber tube is 2mm thick, estimate the diffusion coefficient, temperature being 300 K.

   $(2.6 \times 10^{-11}$ m^2/s)

7. A low carbon steel rod contains 0.2 percent carbon. It is preheated to 900°C and is packed in a carburizing mixture at the same temperature. The concentration of carbon at the surface of the bar is maintained at 1.4 percent. Calculate the time required for the carbon percentage to reach 0.8 percent at a depth of 1 mm. The diffusivity of carbon in steel at that temperature is 5.8×10^{-10} m^2/s.

   (30 min)

8. Dry air at 25°C flows over the surface of a swimming pool 50 m long. Estimate the rate of evaporation of water per unit width of the pool for an air velocity of 2 m/s. Take $D = 0.256 \times 10^{-4}$ m^2/s and $v = 15 \times 10^{-6}$ m^2/s

   (16.56 kg/hour)

9. Air at 25°C and 30% relative humidity flows through a 25 mm inner diameter pipe with a velocity of 5 m/s. The inside surface has a thin water film maintained throughout. Estimate the rate of water evaporated per metre length of the pipe.
   $D = 0.256 \times 10^{-4}$ m^2/s; $v = 15 \times 10^{-6}$ m^2/s.

   (0.152 kg/hr/m)

10. Gaseous hydrogen has been stored, under pressure, in a steel container (walls 10mm thick). The molar concentration of hydrogen in the steel at the inner surface is 1 kmol/m^3 and that at the outer surface is negligible. The binary diffusion coefficient for hydrogen in steel is 26×10^{-14} m^{2}s. Calculate the molar diffusive flux and mass flux for hydrogen

    $(2.6 \times 10^{-11}$ kmol/s-m^2, 5.2×10^{-11} s-m^2)

11. There is an equimolar counter diffusion of CO_2 and air in a circular tube, diameter 50mm and length 1m. The system is at a total pressure of 1 bar and at 25°C temperature. The partial pressure of CO_2 at one end is 0.25 bar and at the other end is 0.125 bar. Estimate the mass transfer rate of air and CO_2.
    Take $D = 0.16 \times 10^{-14}$ m^{2}s.
    Mol. wt. of air = 29.0

    $(-1.67 \times 10^{-3}$ kg/hr, 2.54×10^{-5} kg/hr)

Index